Mathematical Modeling with Multidisciplinary Applications

Mathematical Modeling with Multidisciplinary Applications

Edited by

Xin-She Yang

School of Science and Technology
Middlesex University
United Kingdom

Mathematics and Scientific Computing
National Physical Laboratory
United Kingdom

WILEY

A JOHN WILEY & SONS, INC., PUBLICATION

Published by John Wiley & Sons, Inc., Hoboken, New Jersey.
Published simultaneously in Canada.

For general information on our other products and services please contact our Customer Care Department within the United States at (800) 762-2974, outside the United States at (317) 572-3993 or fax (317) 572-4002.

Wiley also publishes its books in a variety of electronic formats. Some content that appears in print, however, may not be available in electronic formats. For more information about Wiley products, visit our web site at www.wiley.com.

Library of Congress Cataloging-in-Publication Data:

Yang, Xin-She.
Mathematical modeling with multidisciplinary applications / Xin-She Yang.
pages cm
Includes bibliographical references and index.
ISBN 978-1-118-29441-3
1. Differential equations. 2. Mathematical models. I. Title.
QA371.Y28 2013
510.1'1—dc23 2012020899

10 9 8 7 6 5 4 3 2 1

CONTENTS

LIST OF FIGURES

PREFACE

Mathematical modeling is a multidisciplinary endeavor that applies mathematical techniques to study real-world phenomena such as physical, chemical, biological, and economical processes. The quantities of a process of interest are often expressed as variables, while their interactions are often expressed as mathematical relationships or model equations, based on fundamental physical laws such as mass and energy conservation. Such mathematical models can be partial differential equations (PDEs), statistical relationships, or rule-based descriptions, though PDEs are mostly widely used.

Once of the main objectives of mathematical modeling is to model the process and mechanism of interest accurately so as to gain insight and make reasonably accurate predictions. This is a challenging, multidisciplinary task. Often, modeling is an interactive, iterative, time-consuming process. It is rarely the case that a first simple mathematical model will work well; more often, a modeler has to construct a series of mathematical models based on further assumptions, simplifications, adjustments, and improvement so that the revised/improved model can provide better predictions than initial crude models.

Even with the right mathematical models or a right set of differential equations, the task could be even more challenging. First, most mathematical models are highly nonlinear, and their mathematical analysis is often intractable.

Even with some simplified models, analysis is possible, but the mathematical techniques involved are still not straightforward. In most cases, no analytical solution or solutions of any closed form is possible. Secondly, some approximation techniques have to be employed to get some estimates to the true solutions. Approximation methods can be very diverse, though some common techniques such as asymptotic analysis, perturbation methods, and model reduction are often used. In most cases, mathematical analysis and approximations still do not provide sufficient information to construct the exact solutions of the mathematical model. Numerical methods are usually used to provide a fuller picture of the solution characteristics. Numerical methods are also a diverse subject. Solutions of PDEs can be achieved by finite difference methods, finite element methods, finite volume methods, boundary element methods, and spectral methods among others. These topics can fill several books in computational methods if they are described in detail. However, numerical methods are not the main focus of this book, though we will introduce the basics of the numerical methods in relevant chapters.

The aims of this book are twofold: model formulation and analysis, and multidisciplinary applications. We will mainly focus on how to formulate mathematical models for a given process or phenomenon. For a given problem of interest, we will show to build a workable mathematical model, then we will show to do mathematical analysis to obtain solutions (analytical, approximation, or numerical). Another emphasis will be on the diverse mathematical models arisen from multidisciplinary applications such as physics, chemistry, climate, environment, finance, and economics. Though applications are multidisciplinary, key mathematical equations can be the same for different processes. For example, a parabolic PDE can be used to model mass diffusion, heat transfer, reaction-diffusion, pattern formation, and many other phenomena. Similarly, a random walk model can also be used to describe diffusion, search optimization, option pricing, and random samplings in Monte Carlo methods. Therefore, we will demonstrate the above key features throughout this book.

The basic requirement for this book is the good knowledge of basic calculus and mathematical foundations at university level. However, we will briefly review the key concepts of calculus and partial differential equations as well as the fundamental nature of mathematical modelling in the first few chapters. This makes it possible for readers to read all the relevant chapters without much difficulty.

This book strives to provide diverse coverage of multidisciplinary applications with a major focus on mathematical modeling. It divides into three parts. Part I reviews the fundamental of mathematics required for this book. Part II provides the basics of mathematical modeling and numerical methods. Part III covers a diverse range of multidisciplinary applications. Due to the multidisciplinary nature of this book, contributed by multiple authors who are leading experts in their fields, we have strived to make all chapters self-contained with enough background information and further reading materials, and consequently some chapters are more suitable for advanced graduates. A major advantage of this diverse coverage is that readers can choose topics and chapters of their own interest, while skipping some chapters without interrupting the flow and main scheme of modeling and applications. We hope to provide a solid foundation for readers to pursue further studies and research in their chosen area. The other advantage of this book is that all chapters are provided with exercises and answers so that readers can consolidate what they have learned. Thus, this book serves well as a textbook or reference for mathematical modeling courses as well as for self-study.

XIN-SHE YANG

Cambridge and London, UK
December, 2012

ACKNOWLEDGMENTS

I would like to thank all contributing authors for their enthusiastic support for this book. Without their professional contributions, this book would not be possible.

I also would like to thank my Editor, Susanne Steitz-Filler, Associate Editor, Jacqueline Palmieri, Production Editor, Melissa Yanuzzi, Copyeditor, Liz Belmont, and staff at Wiley for their help and professionalism. I also thank my students, Aman Atak, Osaseri O. I. Guobadia and Qichen Xu, at Cambridge University for their help in proofreading some chapters of this book.

Last but not least, I thank my wife and son for their support and help.

X. S. Y.

Editor and Contributors

Editor

Xin-She Yang
School of Science and Technology, Middlesex University, United Kingdom. (x.yang@mdx.ac.uk)

Contributors

Belaid Aouni
School of Commerce and Administration, Faculty of Management, Laurentian University, Sudbury, Ontario, Canada. (baouni@laurentian.ca)

Fred Espen Benth
Center of Mathematics for Applications, University of Oslo, Blindern, Oslo, Norway. (fredb@math.uio.no)

Alfredo Bermúdez
Departamento de Matemática Aplicada, Universidade de Santiago de Compostela, Spain. (alfredo.bermudez@usc.es)

L. A. Boukas
Department of Information and Communication Systems Engineering, University of the Aegean, Greece.

Cinzia Colapinto
Department of Management, Ca' Foscari University of Venice, San Giobbe Cannaregio, Italy. (cinzia.colapinto@unive.it)

Luz M. García García
Instituto Español de Oceanografía, Spain.

Ivan Jeliazkov
Department of Economics, University of California, Irvine, 3175 Social Science Plaza A, Irvine, CA, USA. (ivan@uci.edu)

Boualem Khouider
Mathematics and Statistics University of Victoria, Victoria, B.C., Canada. (khouider@math.uvic.ca)

Herb Kunze
Department of Mathematics and Statistics, University of Guelph, Guelph, Ontario, Canada. (hkunze@uoguelph.ca)

Davide La Torre
Department of Economics, Business and Statistics, University of Milan, via Conservatorio, Milan, Italy. (davide.latorre@unimi.it)

Thomas Lingefjärd
Department of pedagogical, curricular and professional studies,
University of Gothenburg, Gothenburg, Sweden. (Thomas.Lingefjard@gu.se)

S. Natesan
Department of Mathematics, Indian Institute of Technology Guwahati, Guwahati, India. (natesan@iitg.ernet.in)

E. V. Petracou
Department of Geography, University of the Aegean, Greece.

Ivo Petráš
Technical University of Kosice, Faculty of BERG, URaIVP, Kosice, Slovak Republic. (ivo.petras@tuke.sk)

Robert Piché
Department of Mathematics, Tampere University of Technology, Tampere, Finland. (robert.piche@tut.fi)

Mohammad Arshad Rahman
Department of Economics, University of California, Irvine, CA, USA.

Huajun Tang
Faculty of Management and Administration, Macau University of Science and Technology, Macau. (hjtangcyw@gmail.com)

K. I. Vasileiadis
Laboratory for Financial and Actuarial Mathematics, Department of Statistics and Actuarial - Financial Mathematics, University of the Aegean, Greece.

S. Z. Xanthopoulos
Laboratory for Financial and Actuarial Mathematics, Department of Statistics and Actuarial - Financial Mathematics, University of the Aegean, Greece.

Xin-She Yang
School of Science and Technology, Middlesex University, London, UK.
(x.yang@mdx.ac.uk)

A. N. Yannacopoulos
Department of Statistics, Athens University of Economics and Business, Greece.
(ayannaco@aueb.gr)

Chuang Zheng
School of Mathematical Science, Beijing Normal University, Beijing, China.
(chuang.zheng@bnu.edu.cn)

PART I

INTRODUCTION AND FOUNDATIONS

CHAPTER 1

DIFFERENTIAL EQUATIONS

XIN-SHE YANG

School of Science and Technology, Middlesex University, London, UK
Also Mathematics and Scientific Computing, National Physical Laboratory, UK

The main requirement for this book is the basic knowledge of calculus and statistics as covered by most undergraduate courses in engineering and science subjects. However, we will provide a brief review of mathematical foundations in the first few chapters so as to help readers to refresh some of the most important concepts.

Most mathematical models in physics, chemistry, biology and many other applications are formulated in terms of differential equations. If the variables or quantities (such as velocity, temperature, pressure) change with other independent variables such as spatial coordinates and time, their relationship can in general be written as a differential equation or even a set of differential equations.

Mathematical Modeling with Multidisciplinary Applications.
Edited by Xin-She Yang

1.1 ORDINARY DIFFERENTIAL EQUATIONS

An ordinary differential equation (ODE) is a relationship between a function $y(x)$ of an independent variable x and its derivatives y', y'', ..., $y^{(n)}$. It can be written in a generic form

$$\Psi(x, y, y', y'', ..., y^{(n)}) = 0, \tag{1.1}$$

where Ψ is a function of $x, y, ...,$ and $y^{(n)}$. The solution of the equation is a function $y = f(x)$, satisfying the equation for all x in a given domain Ω. The order of the differential equation is equal to the order n of the highest derivative in the equation. Thus, the so-called Riccati equation

$$y' + a(x)y^2 + b(x)y = c(x), \tag{1.2}$$

is a first-order ODE, and the following equation of Euler-type

$$x^2y'' + a_1xy' + a_0y = 0, \tag{1.3}$$

is a second order. The degree of an equation is defined as the power to which the highest derivative occurs. Therefore, both the Riccati equation and the Euler equation are of the first degree.

An equation is called linear if it can be arranged into the form

$$a_n(x)y^{(n)} + ... + a_1(x)y' + a_0(x)y = \phi(x), \tag{1.4}$$

where all the coefficients depend on x only, not on y or any of its derivatives. If any of the coefficients is a function of y or any of its derivatives, then the equation is nonlinear. If the right-hand side is zero or $\phi(x) = 0$, the equation is homogeneous. It is called nonhomogeneous if $\phi(x) \neq 0$.

To find a solution of an ordinary differential equation is not always easy, and it is usually very complicated for nonlinear equations. Even for linear equations, solutions can be found in a straightforward way for only a few simple cases. The solution of a differential equation generally falls into three types: closed form, series form and integral form. A closed form solution is the type of solution that can be expressed in terms of elementary functions and some arbitrary constants. Series solutions are the ones that can be expressed in terms of a series when a closed form is not possible for certain types of equations. The integral form of solutions or quadrature is sometimes the only form of solution that is possible. If all these forms are not possible, the alternatives are to use approximate and numerical solutions.

1.1.1 First-Order ODEs

1.1.1.1 Linear ODEs A first-order linear differential equation can generally be written as

$$y' + a(x)y = b(x), \tag{1.5}$$

where $a(x)$ and $b(x)$ are the known functions of x. Multiplying both sides of the equation by $\exp[\int a(x)dx]$, called the integrating factor, we have

$$y'e^{\int a(x)dx} + a(x)ye^{\int a(x)dx} = b(x)e^{\int a(x)dx}, \tag{1.6}$$

which can be written as

$$[ye^{\int a(x)dx}]' = b(x)e^{\int a(x)dx}. \tag{1.7}$$

By simple integration, we have

$$ye^{\int a(x)dx} = \int b(x)e^{\int a(x)dx}dx + C. \tag{1.8}$$

So its solution becomes

$$y(x) = e^{-\int a(x)dx} \int b(x)e^{\int a(x)dx}dx + Ce^{-\int a(x)dx}, \tag{1.9}$$

where C is an integration constant.

■ **EXAMPLE 1.1**

For example, from $y'(x) - y(x) = e^{-x}$, we have $a(x) = -1$ and $b = e^{-x}$, so the solution is

$$y(x) = e^{-\int(-1)dx} \int e^{-x}e^{\int(-1)dx} + Ce^{-\int(-1)dx}$$

$$= e^{x} \int e^{-2x}dx + Ce^{x} = -\frac{1}{2}e^{-x} + Ce^{x}. \tag{1.10}$$

1.1.1.2 Nonlinear ODEs For some nonlinear first-order ordinary differential equations, sometimes a transform or change of variables can convert it into the standard first-order linear equation (1.5). This is better demonstrated by an example.

The Bernoulli's equation can be written in the generic form

$$y' + p(x)y = q(x)y^n, \qquad n \neq 1. \tag{1.11}$$

In the case of $n = 1$, it reduces to a standard first-order linear ordinary differential equation. By dividing both sides by y^n and using the change of variables

$$u(x) = \frac{1}{y^{n-1}}, \qquad u' = \frac{(1-n)y'}{y^n}, \tag{1.12}$$

we have

$$u' + (1-n)p(x)u = (1-n)q(x), \tag{1.13}$$

which is a standard first-order linear differential equation whose general solution is given earlier in (1.9).

■ **EXAMPLE 1.2**

In the simpler case when $p(x) = 2x$, $q(x) = -1$ and $n = 2$, we have

$$u' - 2xu = 1, \qquad u(x) = \frac{1}{y(x)}.$$

For the initial condition $y(0) = 1$, we have $u(0) = 1$. Using solution (1.9), we have

$$u(x) = \frac{\sqrt{\pi}}{2} e^{x^2} \operatorname{erf}(x) + Ae^{x^2},$$

where A is the integration constant to be determined.

If we further set $u(0) = 1$ as an initial condition, we have $A = 1$. Thus, the solution for $y(x)$ becomes

$$y(x) = \frac{2e^{-x^2}}{(\sqrt{\pi} \operatorname{erf}(x) + 2)}.$$

In general, such transformations are not always possible.

1.1.2 Higher-Order ODEs

Higher-order ODEs are more complicated to solve even for the linear equations. For the special case of higher-order ODEs where all the coefficients $a_n, ..., a_1, a_0$ are constants,

$$a_n y^{(n)} + \cdots + a_1 y' + a_0 y = f(x), \tag{1.14}$$

its general solution $y(x)$ consists of two parts: a complementary function $y_c(x)$ and a particular integral or particular solution $y_p^*(x)$. We have

$$y(x) = y_c(x) + y_p^*(x). \tag{1.15}$$

The complementary function which is the solution of the linear homogeneous equation with constant coefficients can be written in a generic form

$$a_n y_c^{(n)} + a_{n-1} y_c^{(n-1)} + \cdots + a_1 y_c' + a_0 = 0. \tag{1.16}$$

Assuming $y = Ae^{\lambda x}$ where A is a constant, we get the characteristic equation as a polynomial

$$a_n\lambda^n + a_{n-1}\lambda^{(n-1)} + \cdots + a_1\lambda + a_0 = 0, \tag{1.17}$$

which has n roots in the general case. Then, the solution can be expressed as the summation of various terms $y_c(x) = \sum_{k=1}^{n} c_k e^{\lambda_k x}$ if the polynomial has n distinct zeros $\lambda_1, ...\lambda_n$. For complex roots, and complex roots always occur in pairs $\lambda = r \pm i\omega$, the corresponding linearly independent terms can then be replaced by $e^{rx}[A\cos(\omega x) + B\sin(\omega x)]$.

The particular solution $y_p^*(x)$ is any $y(x)$ that satisfies the original inhomogeneous equation (1.14). Depending on the form of the function $f(x)$, the particular solutions can take various forms. For most of the combinations of basic functions such as $\sin x, \cos x$, e^{kx}, and x^n, the method of the undetermined coefficients is widely used. For $f(x) = \sin(\alpha x)$ or $\cos(\alpha x)$, then we can try $y_p^* = A\sin\alpha x + B\sin\alpha x$. We then substitute it into the original equation (1.14) so that the coefficients A and B can be determined. For a polynomial $f(x) = x^n$ where $n = 0, 1, 2, ..., N$, we then try $y_p^* = A + Bx + ... + Qx^n$ (polynomial). For $f(x) = e^{kx}x^n$, we can try $y_p^* = (A + Bx + ...Qx^n)e^{kx}$. Similarly, for $f(x) = e^{kx}\sin\alpha x$ or $f(x) = e^{kx}\cos\alpha x$, we can use $y_p^* = e^{kx}(A\sin\alpha x + B\cos\alpha x)$. More general cases and their particular solutions can be found in various textbooks.

A very useful technique is to use the method of differential operator D. A differential operator D is defined as

$$D \equiv \frac{d}{dx}. \tag{1.18}$$

Since we know that $De^{\lambda x} = \lambda e^{\lambda x}$ and $D^n e^{\lambda x} = \lambda^n e^{\lambda x}$, so they are equivalent to $D \mapsto \lambda$, and $D^n \mapsto \lambda^n$. Thus, any polynomial $P(D)$ will map to a corresponding $P(\lambda)$. On the other hand, integral operator $D^{-1} = \int dx$ is just the inverse of differentiation. The beauty of the differential operator form is that one can factorize it in the same as for a polynomial, then solve each factor separately. The differential operator is very useful in finding out both the complementary functions and particular integral. This method also works for $\sin x, \cos x, \sinh x$ and others, and this is because they are related to $e^{\lambda x}$ via $\sin\theta = \frac{1}{2i}(e^{i\theta} - e^{-i\theta})$ and $\cosh x = (e^x + e^{-x})/2$.

Higher-order differential equations can conveniently be written as a system of differential equations. In fact, an nth-order linear equation can always be written as a linear system of n first-order differential equations. A linear system of ODEs is more suitable for mathematical analysis and numerical integration.

1.1.3 Linear System

For an nth order linear equation (1.16), it can always be written as a linear system

$$\frac{dy}{dx} = y_1, \quad \frac{dy_1}{dx} = y_2, \; ..., \quad \frac{dy_{n-1}}{dx} = y_{n-1},$$

$$-a_n(x)y_{n-1}' = a_{n-1}(x)y_{n-1} + ... + a_1(x)y_1 + a_0(x)y + \phi(x), \tag{1.19}$$

which is a system for $u = [y \; y_1 \; y_2 \; ... \; y_{n-1}]^T$. If the independent variable x does not appear explicitly in y_i, then the system is said to be autonomous with important properties. For simplicity and in keeping with the convention, we use $t = x$ and $\dot{u} = du/dt$ in our following discussion. A general linear system of nth order can be written as

$$\begin{pmatrix} \dot{u}_1 \\ \dot{u}_2 \\ \vdots \\ \dot{u}_n \end{pmatrix} = \begin{pmatrix} a_{11} & a_{12} & \dots & a_{1n} \\ a_{21} & a_{22} & \dots & a_{2n} \\ \vdots & \vdots & \ddots & \vdots \\ a_{n1} & a_{n2} & \dots & a_{nn} \end{pmatrix} \begin{pmatrix} u_1 \\ u_2 \\ \vdots \\ u_n \end{pmatrix}, \tag{1.20}$$

or

$$\dot{\mathbf{u}} = \mathbf{A}\mathbf{u}. \tag{1.21}$$

If we $\mathbf{u} = \mathbf{v}\exp(\lambda t)$, then this becomes an eigenvalue problem,

$$(\mathbf{A} - \lambda\mathbf{I})\mathbf{v} = \mathbf{0}, \tag{1.22}$$

which will have non-null solution only if

$$\det(\mathbf{A} - \lambda\mathbf{I}) = 0. \tag{1.23}$$

1.1.4 Sturm-Liouville Equation

One of the commonly used second-order ordinary differential equations is the Sturm-Liouville equation in the interval $x \in [a, b]$

$$\frac{d}{dx}\left[p(x)\frac{dy}{dx}\right] + q(x)y + \lambda r(x)y = 0, \tag{1.24}$$

with the boundary conditions

$$y(a) + \alpha y'(a) = 0, \qquad y(b) + \beta y'(b) = 0, \tag{1.25}$$

where the known function $p(x)$ is differentiable, and the known functions $q(x), r(x)$ are continuous. The parameter λ to be determined can only take certain values λ_n, called the eigenvalues, if the problem has solutions. For the obvious reason, this problem is called Sturm-Liouville eigenvalue problem.

Sometimes, it is possible to transform a nonlinear equation into a standard Sturm-Liouville equation, and this is better demonstrated by an example.

■ EXAMPLE 1.3

The Riccati equation can be written in the generic form

$$y' = p(x) + q(x)y + r(x)y^2, \qquad r(x) \neq 0.$$

If $r(x) = 0$, then it reduces to a first-order linear ODE. By using the transform

$$y(x) = -\frac{u'(x)}{r(x)u(x)},$$

or

$$u(x) = e^{-\int r(x)y(x)dx},$$

we have

$$u'' - P(x)u' + Q(x)u = 0,$$

where $P(x) = -r'(x)/r(x) + q(x)$ and $Q(x) = r(x)p(x)$.

For each eigenvalue λ_n, there is a corresponding solution ψ_{λ_n}, called an eigenfunction. The Sturm-Liouville theory states that for two different eigenvalues $\lambda_m \neq \lambda_n$, their eigenfunctions are orthogonal. That is

$$\int_a^b \psi_{\lambda_m}(x)\psi_{\lambda_n}(x)r(x)dx = 0, \quad \text{or} \quad \int_a^b \psi_{\lambda_m}(x)\psi_{\lambda_n}(x)r(x)dx = \delta_{mn},$$

where $\delta_{mn} = 1$ if $m = n$, otherwise $\delta_{mn} = 0$ if $m \neq n$. It is possible to arrange the eigenvalues in an increasing order

$$\lambda_1 < \lambda_2 < ... < \lambda_n < ... \to \infty.$$

Now let us study a real-world problem using differential equations. Many fluid flow problems are related to flow through a pipe, including the water flow through a pipe, oil in an oil pipeline. Let us look at the Poiseuille flow in a cylindrical pipe.

■ EXAMPLE 1.4

The laminar flow of a viscous fluid through a pipe with a radius $r = a$ is under a constant pressure gradient (see Fig. 1.1)

$$\nabla p = \Delta P/L = (P_o - P_i)/L,$$

where P_i and P_o ($< P_i$) are the pressures at inlet and outlet, respectively. L is the length of the pipe. The drag force is balanced by pressure change, and this leads to the following second-order ordinary differential

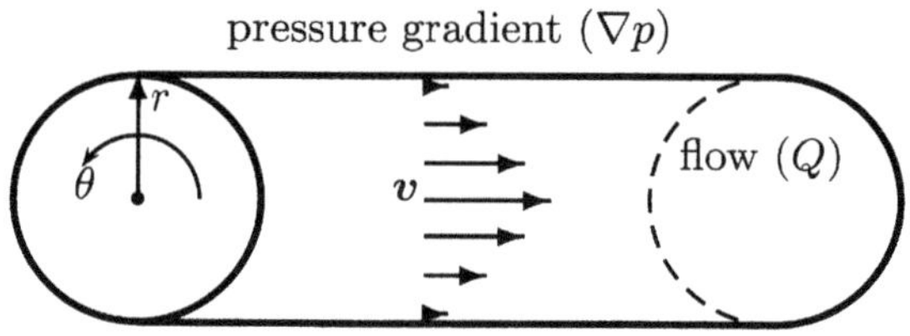

Figure 1.1 Flow through a pipe under pressure gradient.

equation

$$\frac{\Delta P}{L} = \eta \frac{1}{r} \frac{d}{dr} \Big[r \frac{dv(r)}{dr} \Big],$$

where η is the viscosity of the fluid. This equation implies that the flow velocity v is not uniform, it varies with r. Integrating the above equation twice, we have

$$v(r) = \frac{\Delta P}{4\eta L} r^2 + A \ln r + B,$$

where A and B are integrating constants. The velocity must be finite at $r = 0$, which means that $A = 0$. The no-slip boundary $v = 0$ at $r = a$ requires that

$$\frac{\Delta P}{4\eta L} a^2 + B = 0.$$

Thus, the velocity profile is

$$v(r) = -\frac{\Delta P}{4\eta L}(a^2 - r^2).$$

Now the total flow rate Q down the pipe is given by integrating the flow over the whole cross section. We have

$$Q = \int_0^a 2\pi r v(r) dr = -\frac{\pi \Delta P}{2\eta L} \int_0^a (a^2 r - r^3) dr = -\frac{\pi \Delta P}{8\eta L} a^4. \tag{1.26}$$

Here the negative sign means the flow down the pressure gradient.

We can see that the flow rate is proportional to the pressure gradient, inversely proportional to the viscosity. Double the radius of the pipe, and the flow rate will increase to 16 times.

1.2 PARTIAL DIFFERENTIAL EQUATIONS

Partial differential equations are much more complicated compared with ordinary differential equations. There is no universal solution technique for nonlinear equations, even numerical simulations are usually not straightforward. Thus, we will mainly focus on the linear partial differential equations and equations of special interest.

A partial differential equation (PDE) is a relationship containing at least one partial derivative. Similar to the ordinary differential equation, the highest nth partial derivative is referred to as the order n of the partial differential equation. The general form of a partial differential equation can be written as

$$\psi\Big(u, x, y, ..., \frac{\partial u}{\partial x}, \frac{\partial u}{\partial y}, \frac{\partial^2 u}{\partial x^2}, \frac{\partial^2 u}{\partial y^2}, \frac{\partial^2 u}{\partial x \partial y}, ...\Big) = 0. \tag{1.27}$$

where u is the dependent variable, and $x, y, ...$ are the independent variables.

A simple example of partial differential equations is the linear first-order partial differential equation, which can be written as

$$a(x, y)\frac{\partial u}{\partial x} + b(x, y)\frac{\partial u}{\partial y} = f(x, y). \tag{1.28}$$

for two independent variables and one dependent variable u. If the right-hand side is zero or simply $f(x, y) = 0$, then the equation is said to be homogeneous. The equation is said to be linear if a, b and f are functions of x, y only, not u itself.

For simplicity in notation in the studies of PDEs, compact subscript forms are often used in the literature. They are

$$u_x \equiv \partial_x u \equiv \frac{\partial u}{\partial x}, \quad u_y \equiv \partial_y u \equiv \frac{\partial u}{\partial y}, \quad u_{xx} \equiv \frac{\partial^2 u}{\partial x^2}, \quad u_{xy} \equiv \frac{\partial^2 u}{\partial x \partial y}, \quad ... \tag{1.29}$$

and thus we can write (1.28) as

$$au_x + bu_y = f. \tag{1.30}$$

1.2.1 First-Order PDEs

A first-order linear partial differential equation can be written as

$$a(x, y)u_x + b(x, y)u_y = f(x, y), \tag{1.31}$$

which can be solved using the method of characteristics in terms of a parameter s

$$\frac{dx}{ds} = a, \quad \frac{dy}{ds} = b, \quad \frac{du}{ds} = f, \tag{1.32}$$

which essentially forms a system of first-order ordinary differential equations. The simplest example of first-order linear partial differential equations is the first-order hyperbolic equation

$$u_t + cu_x = 0, \tag{1.33}$$

where c is a constant. It has a general solution

$$u = \psi(x - ct), \tag{1.34}$$

which is a travelling wave along the x-axis with a constant speed c. If the initial shape is $u(x, 0) = \psi(x)$, then $u(x, t) = \psi(x - ct)$ at time t, therefore the shape of the wave does not change with time though its position is constantly changing.

1.2.2 Classification of Second-Order PDEs

A linear second-order partial differential equation can be written in the generic form in terms of two independent variables x and y,

$$au_{xx} + bu_{xy} + cu_{yy} + gu_x + hu_y + ku = f, \tag{1.35}$$

where a, b, c, g, h, k and f are functions of x and y only. If $f(x, y, u)$ is also a function of u, then we say that this equation is quasi-linear.

If $\Delta = b^2 - 4ac < 0$, the equation is elliptic. One famous example is the Laplace equation $u_{xx} + u_{yy} = 0$.

If $\Delta > 0$, it is hyperbolic. A good example is the wave equation $u_{tt} = c^2 u_{xx}$.

If $\Delta = 0$, it is parabolic. Diffusion and heat conduction are of the parabolic type $u_t = \kappa u_{xx}$.

1.3 CLASSIC MATHEMATICAL MODELS

Three types of classic partial differential equations are widely used and they occur in a vast range of applications. In fact, almost all books or studies on partial differential equations will have to deal with these three types of basic partial differential equations.

Laplace's and Poisson's Equation: In heat transfer problems, the steady state of heat conduction with a source is governed by the Poisson equation

$$k\nabla^2 u = f(x, y, t), \qquad (x, y) \in \Omega, \tag{1.36}$$

or

$$u_{xx} + u_{yy} = q(x, y, t), \tag{1.37}$$

for two independent variables x and y. Here k is thermal diffusivity and $f(x, y, t)$ is the heat source. Ω is the domain of interest, usually a physical region. If there is no heat source ($q = f/\kappa = 0$), it becomes the Laplace equation. The solution of a function is said to be harmonic if it satisfies Laplace's equation.

In order to determine the temperature u completely, the appropriate boundary conditions are needed. A simple boundary condition is to specify the temperature $u = u_0$ on the boundary $\partial\Omega$. This type of problem is the Dirichlet problem.

On the other hand, if the temperature is not known, but the gradient $\partial u/\partial \boldsymbol{n}$ is known on the boundary where $\boldsymbol{n}$ is the outward-pointing unit normal, this forms the Neumann problem. Furthermore, some problems may have a mixed type of boundary conditions in the combination of

$$\alpha u + \beta \frac{\partial u}{\partial \boldsymbol{n}} = \gamma,$$

which naturally occurs as a radiation or cooling boundary condition.

Parabolic Equation: Time-dependent problems, such as diffusion and transient heat conduction, are governed by the parabolic equation

$$u_t = k u_{xx}. \tag{1.38}$$

Written in the n-dimensional case $x_1 = x, x_2 = y, x_3 = z, ...$, it can be extended to the reaction-diffusion equation

$$u_t = k\nabla^2 u + f(u, x_1, .., x_n, t). \tag{1.39}$$

Wave Equation: The vibration of strings and travelling seismic waves are governed by the hyperbolic wave equation.

The 1D wave equation in its simplest form is

$$u_{tt} = c^2 u_{xx}, \tag{1.40}$$

where c is the velocity of the wave. Using a transformation of the pair of independent variables

$$\xi = x + ct, \tag{1.41}$$

and

$$\eta = x - ct, \tag{1.42}$$

for $t > 0$ and $-\infty < x < \infty$, the wave equation can be written as

$$u_{\xi\eta} = 0. \tag{1.43}$$

Integrating twice and substituting back in terms of x and t, we have

$$u(x, t) = f(x + ct) + g(x - ct), \tag{1.44}$$

where f and g are functions of $x + ct$ and $x - ct$, respectively. We can see that the solution is composed of two independent waves. One wave moves to the right and one travels to the left at the same constant speed c.

1.4 OTHER MATHEMATICAL MODELS

We have shown examples of the three major equations of second-order linear partial differential equations. There are other equations that occur frequently in engineering and science. We will give a brief description of some of these equations.

Elastic Wave Equation: A wave in an elastic isotropic homogeneous solid is governed by the following equation in terms of displacement **u**,

$$\rho \frac{\partial^2 \mathbf{u}}{\partial t^2} = \mu \nabla^2 \mathbf{u} + (\lambda + \mu)\nabla(\nabla \cdot \mathbf{u}) + \mathbf{f}, \tag{1.45}$$

where ρ is density, λ and μ are Lamé constants, and $\mathbf{f}$ is body force. Such an equation can describe two types of wave: transverse wave (S wave) and longitudinal or dilatational wave (P wave). The speed of the longitudinal wave is

$$v_p = \sqrt{(\lambda + 2\mu)/\rho}, \tag{1.46}$$

and the transverse wave has the speed

$$v_s = \sqrt{\mu/\rho}. \tag{1.47}$$

Reaction-Diffusion Equation: The reaction-diffusion equation is an extension of heat conduction with a source f

$$u_t = D\nabla^2 u + f(x, y, z, u), \tag{1.48}$$

where D is the diffusion coefficient and f is the reaction rate. One example is the combustion equation

$$u_t = D u_{xx} + Q u e^{-\lambda/u}, \tag{1.49}$$

where Q and λ are constants.

Navier-Stokes Equations: The Navier-Stokes equations for incompressible flow in the absence of body forces can be written, in terms of the velocity $\boldsymbol{u}$ and the pressure p, as

$$\nabla \cdot \mathbf{u} = 0, \qquad \rho[\mathbf{u}_t + (\mathbf{u} \cdot \nabla)\mathbf{u}] = \mu \nabla^2 \mathbf{u} - \nabla p, \tag{1.50}$$

where ρ and μ are the density of the fluid and its viscosity, respectively. In computational fluid dynamics, most simulations are mainly related to these equations. We can define the Reynolds number as $\mathrm{Re} = \rho U L/\mu$ where U is the typical velocity and L is the length scale.

In the limit of $\mathrm{Re} \ll 1$, we have the Stokes flow governed by

$$\mu \nabla^2 \mathbf{u} = \nabla p. \tag{1.51}$$

In the other limit of Re $\gg 1$, we have the inviscous flow

$$\nabla \cdot \mathbf{u} = 0, \qquad \rho[\mathbf{u}_t + (\mathbf{u} \cdot \nabla)\mathbf{u}] = -\nabla p, \tag{1.52}$$

where there is still a nonlinear term $(\mathbf{u} \cdot \nabla)\mathbf{u}$.

Groundwater Flow: The general equation for three-dimensional groundwater flow is

$$S_\sigma \frac{\partial p}{\partial t} = \frac{k}{\mu} \nabla^2 p - S_\sigma B \frac{\partial \sigma}{\partial t} + Q, \tag{1.53}$$

where $\sigma = \sigma_{kk}/3$ is the mean stress, p is the pore water pressure, and Q is source or sink term. S_σ is the specific storage coefficient and B is the Skempton constant. k is the permeability of the porous medium and μ is the viscosity of water. This can be considered as the inhomogeneous diffusion equation for pore pressure.

1.5 SOLUTION TECHNIQUES

Each type of equation usually requires different solution techniques. However, there are some methods that work for most of the linearly partial differential equations with appropriate boundary conditions on a regular domain. These methods include separation of variables, method of series expansion and transform methods such as the Laplace and Fourier transforms.

1.5.1 Separation of Variables

The separation of variables attempts a solution of the form

$$u = X(x)Y(y)Z(z)T(t), \tag{1.54}$$

where $X(x), Y(y), Z(z), T(t)$ are functions of x, y, z, t, respectively. By determining these functions that satisfy the partial differential equation and the required boundary conditions in terms of eigenvalue problems, the solution of the original problem is then obtained.

As a classic example, we now try to solve the 1D heat conduction equation in the domain $x \in [0, L]$ and $t \geq 0$

$$u_t = k u_{xx}, \tag{1.55}$$

with the initial value and boundary conditions

$$u(0,t) = 0, \qquad \frac{\partial u(x,t)}{\partial x}\Big|_{x=L} = 0, \qquad u(x,0) = \psi(x). \tag{1.56}$$

Letting $u(x,t) = X(x)T(t)$, we have

$$\frac{X''(x)}{X} = \frac{T'(t)}{kT}. \tag{1.57}$$

As the left-hand side depends only on x and the right-hand side only depends on t, therefore, both sides must be equal to the same constant, and the constant can be assumed to be $-\lambda^2$. The negative sign is just for convenience because we will see below that the finiteness of the solution $T(t)$ requires that eigenvalues $\lambda^2 > 0$ or λ are real. Hence, we now get two ordinary differential equations

$$X''(x) + \lambda^2 X(x) = 0, \qquad T'(t) + k\lambda^2 T(t) = 0, \tag{1.58}$$

where λ is the eigenvalue. The solution for $T(t)$ is

$$T = A_n e^{-\lambda^2 kt}. \tag{1.59}$$

The basic solution for $X(x)$ is simply

$$X(x) = \alpha \cos \lambda x + \beta \sin \lambda x. \tag{1.60}$$

So the fundamental solution for u is

$$u(x,t) = (\alpha \cos \lambda x + \beta \sin \lambda x)e^{-\lambda^2 kt}, \tag{1.61}$$

where we have absorbed the coefficient A_n into α and β because they are the undetermined coefficients anyway. As the value of λ varies with the boundary conditions, it forms an eigenvalue problem. The general solution for u should be derived by superposing solutions of (1.61), and we now have

$$u = \sum_{n=1}^{\infty} X_n T_n = \sum_{n=1}^{\infty} (\alpha_n \cos \lambda_n x + \beta_n \sin \lambda_n x)e^{-\lambda_n^2 kt}. \tag{1.62}$$

From the boundary condition $u(0,t) = 0$ at $x = 0$, we have

$$0 = \sum_{n=1}^{\infty} \alpha_n e^{-\lambda_n^2 kt}, \tag{1.63}$$

which leads to $\alpha_n = 0$ since $\exp(-\lambda^2 kt) > 0$.

From $\frac{\partial u}{\partial x}\Big|_{x=L} = 0$, we have

$$\lambda_n \cos \lambda_n L = 0, \tag{1.64}$$

which requires

$$\lambda_n L = \frac{(2n-1)\pi}{2}, \qquad (n = 1, 2, ...). \tag{1.65}$$

Therefore, λ cannot be continuous, and it only takes an infinite number of discrete values, called eigenvalues.

Each eigenvalue $\lambda = \lambda_n = \frac{(2n-1)\pi}{2L}$, $(n = 1, 2, ...)$ has a corresponding eigenfunction $X_n = \sin(\lambda_n x)$. Substituting into the solution for $T(t)$, we have

$$T_n(t) = A_n e^{-\frac{[(2n-1)\pi]^2}{4L^2} kt}. \tag{1.66}$$

By expanding the initial condition into a Fourier series so as to determine the coefficients, we have

$$u(x,t) = \sum_{n=1}^{\infty} \beta_n \sin\left(\frac{(2n-1)\pi x}{2L}\right) e^{-[\frac{(2n-1)\pi}{2L}]^2 kt},$$

$$\beta_n = \frac{2}{L} \int_0^L \psi(x) \sin\left[\frac{(2n-1)\pi x}{2L}\right] dx. \tag{1.67}$$

■ EXAMPLE 1.5

In the special case when initial condition $u(x, t = 0) = \psi = u_0$ is constant, the requirement for $u = u_0$ at $t = 0$ becomes

$$u_0 = \sum_{n=1}^{\infty} \beta_n \sin \frac{(2n-1)\pi x}{2L}. \tag{1.68}$$

Using the orthogonal relationships

$$\int_0^L \sin \frac{m\pi x}{L} \sin \frac{n\pi x}{L} dx = 0, \qquad m \neq n,$$

and

$$\int_0^L (\sin \frac{n\pi x}{L})^2 dx = \frac{L}{2}, \qquad (n = 1, 2, ...),$$

and multiplying both sides of Eq.(1.68) by $\sin[(2n-1)\pi x/2L]$, we have the integration

$$\beta_n \frac{L}{2} = \int_0^L \sin \frac{(2n-1)\pi x}{2L} u_0 dx = \frac{2u_0 L}{(2n-1)\pi}, \quad (n = 1, 2, ...),$$

which leads to

$$\beta_n = \frac{4u_0}{(2n-1)\pi}, \qquad n = 1, 2, ...,$$

and thus the solution becomes

$$u = \frac{4u_0}{\pi} \sum_{n=1}^{\infty} \frac{1}{(2n-1)} e^{-\frac{(2n-1)^2 \pi^2 kt}{4L^2}} \sin \frac{(2n-1)\pi x}{2L}. \tag{1.69}$$

This solution is essentially the same as the classical heat conduction problem discussed by Carslaw and Jaeger in 1959. This same solution can also be obtained using the Fourier series of u_0 in $0 < x < L$.

1.5.2 Laplace Transform

The integral transform can reduce the number of the independent variables. For the 1D time-dependent case, it transforms a partial differential equation into an ordinary differential equation. By solving the ordinary differential equation and inverting it back, we can obtain the solution for the original partial differential equation. As an example, we now solve the heat conduction problem over a semi-infinite interval $[0, \infty)$,

$$u_t = k u_{xx}, \qquad u(x,0) = 0, \;\; u(0,t) = T_0. \tag{1.70}$$

■ **EXAMPLE 1.6**

Let $\bar{u}(x,s) = \int_0^{\infty} u(x,t) e^{-st} dt$ be the Laplace transform of $u(x,t)$, then Eq.(1.70) becomes

$$s\bar{u} = k \frac{d^2 \bar{u}}{dx^2}, \qquad \bar{u}_{x=0} = \frac{T_0}{s},$$

which is an ordinary differential equation whose general solution can be written as

$$\bar{u} = A e^{-\sqrt{\frac{s}{k}} x} + B e^{\sqrt{\frac{s}{k}} x}.$$

The finiteness of the solution as $x \to \infty$ requires that $B = 0$, and the boundary condition at $x = 0$ leads to

$$\bar{u} = \frac{T_0}{s} e^{-\sqrt{\frac{s}{k}} x}.$$

By using the inverse Laplace transform, we have

$$u = T_0 \operatorname{erfc}\Big(\frac{x}{2\sqrt{kt}}\Big),$$

where $\operatorname{erfc}(x)$ is the complementary error function.

The Fourier transform works in a similar manner to the Laplace transform.

1.5.3 Similarity Solution

Sometimes, the diffusion equation

$$u_t = \kappa u_{xx}, \tag{1.71}$$

can be solved by using the so-called similarity method by defining a similar variable

$$\eta = \frac{x}{\sqrt{\kappa t}}, \quad \text{or} \quad \zeta = \frac{x^2}{\kappa t}. \tag{1.72}$$

One can assume that the solution to the equation has the form

$$u = (\kappa t)^{\alpha} f\Big[\frac{x^2}{(\kappa t)^{\beta}}\Big]. \tag{1.73}$$

By substituting it into the diffusion equation, the coefficients α and β can be determined. For most applications, one can assume $\alpha = 0$ so that $u = f(\zeta)$. In this case, we have

$$4\zeta u'' + 2u' + \zeta\beta(\kappa t)^{\beta-1} u' = 0, \tag{1.74}$$

where $u' = du/d\zeta$. In deriving this equation, we have used the chain rules of differentiations

$$\frac{\partial}{\partial x} = \frac{\partial}{\partial \zeta}\frac{\partial \zeta}{\partial x}, \quad \frac{\partial}{\partial t} = \frac{\partial}{\partial \zeta}\frac{\partial \zeta}{\partial t}. \tag{1.75}$$

Since the original equation does not have time-dependent terms explicitly, this means that all the exponents for any t-terms must be zero. Therefore, we have

$$\beta = 1. \tag{1.76}$$

Now, the diffusion equation becomes

$$\zeta f''(\zeta) = -\Big(\frac{1}{2} + \frac{\zeta}{4}\Big) f'. \tag{1.77}$$

Using $(\ln f')' = f''/f'$ and integrating the above equation once, we get

$$f' = \frac{K e^{-\zeta/4}}{\sqrt{\zeta}}. \tag{1.78}$$

Integrating it again and using the substitution $\zeta = 4\xi^2$, we obtain

$$u = A\int_0^{\xi} e^{-\xi^2} d\xi = C \text{erf}\Big(\frac{x}{\sqrt{4\kappa t}}\Big) + D, \tag{1.79}$$

where C and D are constants that can be determined from appropriate boundary conditions.

For the same problem as (1.70), the boundary condition as $x \to \infty$ implies that $C + D = 0$, while $u(0, t) = T_0$ means that $D = -C = T_0$. Therefore, we finally have

$$u = T_0\Big[1 - \text{erf}\Big(\frac{x}{\sqrt{4\kappa t}}\Big)\Big] = T_0 \text{erfc}\Big(\frac{x}{\sqrt{4\kappa t}}\Big).$$

1.5.4 Change of Variables

In some cases, the partial differential equation cannot be written in any standard form; however, it can be converted into a known standard equation by a change of variables. For example, the following simple reaction-diffusion equation

$$\frac{\partial u}{\partial t} = k\frac{\partial^2 u}{\partial x^2} - \alpha u, \tag{1.80}$$

describes the heat conduction along a wire with a heat loss term $-\alpha u$. Carslaw and Jaeger show that it can be transformed into a standard equation of heat conduction using the following change of variables

$$u = ve^{-\alpha t}, \tag{1.81}$$

where v is the new variable. By simple differentiations, we have

$$\frac{\partial u}{\partial t} = \frac{\partial v}{\partial t}e^{-\alpha t} - \alpha v e^{-\alpha t} = \frac{\partial v}{\partial t}e^{-\alpha t} - \alpha u, \qquad \frac{\partial^2 u}{\partial x^2} = \frac{\partial^2 v}{\partial x^2}e^{-\alpha t}, \tag{1.82}$$

we have

$$\frac{\partial u}{\partial t} = \underbrace{\frac{\partial v}{\partial t}e^{-\alpha t}} - \alpha u = k\frac{\partial^2 u}{\partial x^2} - \alpha u = k\underbrace{\frac{\partial^2 v}{\partial x^2}e^{-\alpha t}} - \alpha u, \tag{1.83}$$

which becomes

$$\frac{\partial v}{\partial t}e^{-\alpha t} = k\frac{\partial^2 v}{\partial x^2}e^{-\alpha t}. \tag{1.84}$$

After dividing both sides by $e^{-\alpha t} > 0$, we have

$$\frac{\partial v}{\partial t} = k\frac{\partial^2 v}{\partial x^2}, \tag{1.85}$$

which is the standard heat conduction equation for v.

For given initial (usually constant) and boundary conditions (usually zero), we can use all the techniques for solving the standard equation to get solutions. However, for some boundary conditions such as $u = u_0$, a more elaborate form of change of variables is needed. Crank introduced Danckwerts's method by using the following transform

$$u = \alpha \int_0^t v e^{-\alpha \tau} d\tau + v e^{-\alpha t}. \tag{1.86}$$

Noting that $\frac{\partial u}{\partial t} = \alpha v e^{-\alpha t} - \alpha v e^{-\alpha t} + \frac{\partial v}{\partial t} e^{-\alpha t}$, it is straightforward to show

$$\frac{\partial u}{\partial t} + \alpha u = k \frac{\partial^2 u}{\partial x^2}. \tag{1.87}$$

For the boundary condition $u = u_0$, we have $v = v_0 = u_0$, and this is because

$$u = u_0 = \alpha v_0 \int_0^t e^{-\alpha \tau} d\tau + v_0 e^{-\alpha t} = v_0 - v_0 e^{-\alpha t} + v_0 e^{-\alpha t} = v_0, \tag{1.88}$$

which is the same boundary condition as that for u.

There are other important methods for solving partial differential equations. These include Green's function, series methods, asymptotic methods, approximate methods, perturbation methods and naturally the numerical methods.

EXERCISES

1.1 The so-called Coriolis force or effect exists in a rotational system, which makes the falling object lands slightly to the east (without considering air resistance). Assume the falling height is h, estimate the distance deviation to the east due to this Coriolis acceleration $a = 2\omega v$ where ω is the angular velocity of the Earth's rotation and v is its falling velocity.

1.2 Find the general solution $x^2 y'' - y = 0$ for $x > 0$.

1.3 The governing equation for the damped simple harmonic motion can be written as a general second-order ordinary differential equation

$$\ddot{u} + 2\eta\omega_0 \dot{u} + \omega_0^2 u = 0,$$

where ω_0 is the so-called undamped frequency, and η is called damping coefficient. Show that $\eta > 1$ and $\eta < 1$ will lead to different characteristics in the system.

1.4 The Laplace equation is often written as $\Delta u = 0$ or $\frac{\partial^2 u}{\partial x^2} + \frac{\partial^2 u}{\partial y^2} = 0$ in 2D case. Define a polar coordinate system (r, θ) so that $x = r\cos\theta$ and $y = r\sin\theta$, and then write the Laplace equation in the polar coordinates.

1.5 The FitzHugh-Nagumo equation occurs in many applications such as biology, genetics and heat transfers. In the 1D case, it can be written as

$$\frac{\partial u}{\partial t} = \frac{\partial^2 u}{\partial x^2} + u(u-1)(\lambda - u),$$

where λ is a constant. Show that this equation supports a traveling wave solution

$$u(x,t) = \frac{A\exp(\eta_1) + \lambda B \exp(\eta_2)}{A\exp(\eta_1) + B\exp(\eta_2) + K},$$

where

$$\eta_1 = \left(\frac{1}{2} - \lambda\right)t \pm \frac{x}{\sqrt{2}}, \quad \eta_2 = \lambda\left(\frac{\lambda}{2} - 1\right)t \pm \frac{\lambda x}{\sqrt{2}},$$

and A, B and K are arbitrary constants.

1.6 The Klein-Gordon equation $\frac{\partial^2 u}{\partial t^2} = a^2 \frac{\partial^2 u}{\partial x^2} - bu$ occurs in quantum field theory and other applications. Verify that $u(x,t) = \sin(\lambda x)[A\cos(\omega t) + B\sin(\omega t)]$ is a solution if $b = -a^2\lambda^2 + \omega^2$. If $u(x,t) = \exp(\pm\lambda x)[A\cos(\omega t) + B\sin(\omega t)]$ is also a solution, what is the relationship between a, b, λ and ω.

1.7 In many applications, partial differential equations can be rewritten in other forms so that they can be linked with other well-known equations. For example, the so-called telegraph equation

$$\frac{\partial^2 u}{\partial t^2} + v\frac{\partial u}{\partial t} = a^2\frac{\partial^2 u}{\partial x^2} + bu, \quad v > 0, b < 0,$$

can be transformed into the Klein-Gordon equation by a transform $u(x,t) = \exp(-\frac{1}{2}vt)w(x,t)$. Show that this is true.

1.8 The Burgers equation in one-dimensional case is often written as

$$\frac{\partial u}{\partial t} = \frac{\partial^2 u}{\partial x^2} + u\frac{\partial u}{\partial x}.$$

Show that it can be transformed into the standard linear diffusion equation by the so-called Hopf-Cole transformation $u(x,t) = \frac{2}{\phi}\frac{\partial \phi}{\partial x}$.

REFERENCES

1. Berger, A. L., Long term variations of the Earth's orbital elements, *Celestial Mechanics*, **15**, 53–74 (1977).
2. Carrrier, G. F. and Pearson, C. E., *Partial Differential Equations: Theory and Technique*, 2nd Edition, Academic Press (1988).
3. Carslaw, H. S. and Jaeger, J. C., *Conduction of Heat in Solids*, 2nd Edition, Oxford University Press, Oxford (1986).
4. Crank, J., *Mathematics of Diffusion*, Clarendon Press, Oxford (1970).
5. Fowler, A. C., *Mathematical Models in the Applied Sciences*, Cambridge University Press, Cambridge (1997).
6. Jeffrey, A., *Advanced Engineering Mathematics*, Academic Press, Waltham, MA (2002).
7. Kreyszig, E., *Advanced Engineering Mathematics*, 6th Edition, Wiley & Sons, New York (1988).
8. Riley, K. F., Hobson, M. P., and Bence, S. J., *Mathematical Methods for Physics and Engineering*, Cambridge University Press, Cambridge (2006).
9. Selby, S. M., *Standard Mathematical Tables*, CRC Press, Cleveland, Ohio (1975).

CHAPTER 2

MATHEMATICAL MODELING

XIN-SHE YANG

School of Science and Technology, Middlesex University, London, UK
Also Mathematics and Scientific Computing, National Physical Laboratory, UK

2.1 MATHEMATICAL MODELING

Mathematical modeling is the process of formulating an abstract model in terms of mathematical language to describe the complex behavior of a real system. Mathematical models are quantitative models and often expressed in terms of ordinary differential equations and partial differential equations. Mathematical models can also be statistical models, fuzzy logic models and empirical relationships. In fact, any model description using mathematical language can be called a mathematical model. Mathematical modeling is widely used in natural sciences, computing, engineering, meteorology, and industrial applications. For example, theoretical physics is essentially all about the modeling of real world processes using several basic principles (such as the

Mathematical Modeling with Multidisciplinary Applications.
Edited by Xin-She Yang

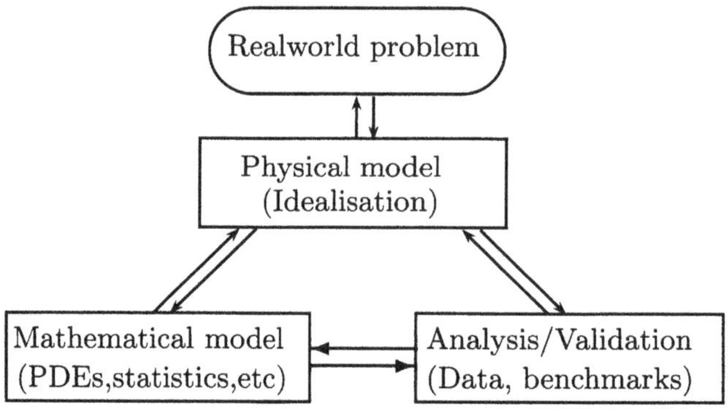

Figure 2.1 Mathematical modeling.

conservation of energy, momentum) and a dozen important equations (such as the wave equation, the Schrödinger equation, the Einstein equation). Almost all these equations are partial differential equations (PDEs).

An important feature of mathematical modeling and numerical algorithms is its interdisciplinary nature. It involves applied mathematics, computer sciences, physical and biological sciences, and others. Mathematical modeling in combination with scientific computing is an emerging interdisciplinary technology. Many international companies use it to model physical processes, to design new products, to find solutions to challenging problems, and increase their competitiveness in international markets.

The basic steps of mathematical modeling can be summarized as meta-steps shown in Figure 2.1. The process typically starts with the analysis of a real world problem so as to extract the fundamental physical processes by idealization and various assumptions. Once an idealized physical model is formulated, it can then be translated into the corresponding mathematical model in terms of partial differential equations (PDEs), integral equations, and statistical models. Then, the mathematical model should be investigated in great detail by mathematical analysis (if possible), numerical simulations and other tools so as to make predictions under appropriate conditions. Then, these simulation results and predictions will be validated against the existing models, well-established benchmarks, and experimental data. If the results are satisfactory (which they rarely are at first), then the mathematical model can be accepted. If not, both the physical model and mathematical model will be modified based on the feedback, and then the new simulations and prediction will be validated again. After a certain number of iterations of the whole process (often many), a good mathematical model can properly be formulated, which will provide great insight into the real-world problem and may also predict the behavior of the process under study.

For any physical problem in physics, chemistry and biology, for example, there are traditionally two ways to deal with it by either theoretical approaches or field observations and experiments. The theoretical approach in terms of mathematical modeling is an idealization and simplification of the real problem and the theoretical models often extract the essential or major characteristics of the problem. The mathematical equations obtained even for such oversimplified systems are usually very difficult for mathematical analysis. On the other hand, the field studies and experimental approach are usually expensive if not impractical. Apart from financial and practical limitations, other constraining factors include the inaccessibility of the locations, the range of physical parameters, and time for carrying out various experiments. As computing speed and power have increased dramatically in the last few decades, a practical third way or approach is emerging, which is computational modeling and numerical experimentation based on the mathematical models. It is now widely acknowledged that computational modeling and computer simulations serve as a cost-effective alternative, bridging the gap or complementing the traditional theoretical and experimental approaches to problem solving.

Mathematical modeling is essentially an abstract art of formulating the mathematical models from the corresponding real-world problems. The mastery of this art requires practice and experience, and it is not easy to teach such skills as the style of mathematical modeling largely depends on each person's own insight, abstraction, type of problems, and experience of dealing with similar problems. Even for the same physical process, different models could be obtained, depending on the emphasis of some part of the process, say, based on your interest in certain quantities in a particular problem, while the same quantities could be viewed as unimportant in other processes and other problems.

2.2 MODEL FORMULATION

Mathematical modeling often starts with the analysis of the physical process and attempts to make an abstract physical model by idealization and approximations. From this idealized physical model, we can use the various first principles such as the conservation of mass, momentum, energy and Newton's law to translate into mathematical equations. Let us look at the example of the diffusion process of sugar in a glass of water. We know that the diffusion of sugar will occur if there is any spatial difference in the sugar concentration. The physical process is complicated and many factors could affect the distribution of sugar concentration in water, including the temperature, stirring, mass of sugar, type of sugar, how you add the sugar, even geometry of the container and others. We can idealize the process by assuming that the temperature is constant (so as to neglect the effect of heat transfer), and that there is no stirring because stirring will affect the effective diffusion coefficient and introduce the advection of water or even vertices in the (turbulent) water

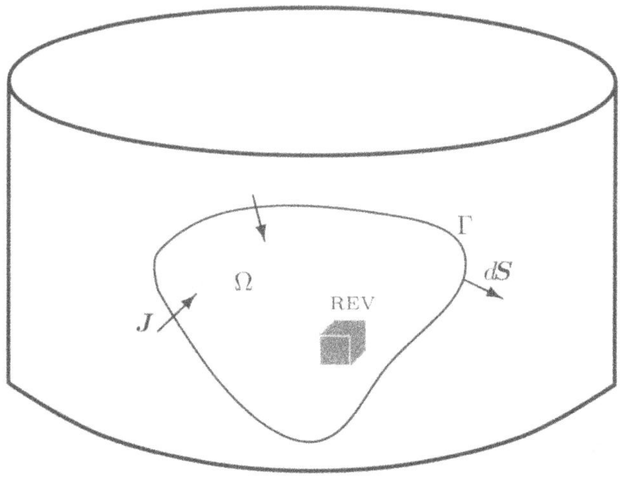

Figure 2.2 Representative element volume (REV).

flow. We then choose a representative element volume (REV) whose size is very small compared with the size of the cup so that we can use a single value of concentration to represent the sugar content inside this REV (if this REV is too large, there is considerable variation in sugar concentration inside this REV). We also assume that there is no chemical reaction between sugar and water (otherwise, we are dealing with something else). If you drop the sugar into the cup from a considerable height, the water inside the glass will splash and thus the fluid volume will change, and this becomes a fluid dynamics problem. So we are only interested in the process after the sugar is added and we are not interested in the initial impurity of the water (or only to a certain degree). With these assumptions, the whole process is now idealized as the physical model of the diffusion of sugar in still water at a constant temperature. Now we have to translate this idealized model into a mathematical model, and in the present case, a parabolic partial differential equation or diffusion equation. Let us look at an example.

■ EXAMPLE 2.1

Let c be the averaged concentration in a representative element volume with a volume dV inside the cup, and let Ω be an arbitrary, imaginary closed volume Ω (much larger than our REV but smaller than the container, see Figure 2.2). We know that the rate of change of the mass of sugar per unit time inside Ω is

$$\delta_1 = \frac{\partial}{\partial t} \iiint_\Omega c dV,$$

where t is time. As the mass is conserved, this change of sugar content in Ω must be supplied in or flow out over the surface $\Gamma = \partial\Omega$, enclosing the region Ω. Let $\boldsymbol{J}$ be the flux through the surface, thus the total mass flux through the whole surface Γ is

$$\delta_2 = \iint_\Gamma \boldsymbol{J} \cdot d\boldsymbol{S}.$$

Thus the conservation of total mass in Ω requires that

$$\delta_1 + \delta_2 = 0,$$

or

$$\frac{\partial}{\partial t} \iiint_\Omega c dV + \iint_\Gamma \boldsymbol{J} \cdot d\boldsymbol{S} = 0.$$

This is essentially the integral form of the mathematical model. Using the Gauss's theorem (discussed later in this book)

$$\iint_\Gamma \boldsymbol{J} \cdot d\boldsymbol{S} = \iiint_\Omega \nabla \cdot \boldsymbol{J} dV,$$

we can convert the surface integral into a volume integral. We thus have

$$\frac{\partial}{\partial t} \iiint_\Omega c dV + \iiint_\Omega \nabla \cdot \boldsymbol{J} dV = 0.$$

Since the domain Ω is fixed (independent of t), we can interchange the differentiation and integration in the first term, we now get

$$\iiint_\Omega \frac{\partial c}{\partial t} dV + \iiint_\Omega \nabla \cdot \boldsymbol{J} dV = \iiint_\Omega \Big[\frac{\partial c}{\partial t} + \nabla \cdot J\Big] dV = 0.$$

Since the enclosed domain Ω is arbitrary, the above equation should be valid for any shape or size of Ω, therefore, the integrand must be zero. We finally have

$$\frac{\partial c}{\partial t} + \nabla \cdot \boldsymbol{J} = 0.$$

This is the differential form of the mass conservation. It is a partial differential equation. As we know that diffusion occurs from the higher concentration to lower concentration, and the rate of diffusion is proportional to the gradient ∇c of the concentration. The flux $\boldsymbol{J}$ over a unit surface area is given by Fick's law

$$\boldsymbol{J} = -D\nabla c,$$

where D is the diffusion coefficient which depends on the temperature and the type of materials. The negative sign means the diffusion is opposite to the gradient. Substituting this into the mass conservation, we have

$$\frac{\partial c}{\partial t} - \nabla \cdot (D \nabla c) = 0,$$

or

$$\frac{\partial c}{\partial t} = \nabla \cdot (D \nabla c).$$

In the simplified case when D is constant, we have

$$\frac{\partial c}{\partial t} = D \nabla^2 c, \tag{2.1}$$

which is the well-known diffusion equation. This equation can be applied to study many phenomena such as heat conduction, pore pressure dissipation, groundwater flow and consolidation if we replace D by the corresponding physical parameters. This will be discussed in greater detail in the related chapters this book.

2.3 PARAMETER ESTIMATION

Another important topic in mathematical modeling is the ability to estimate the orders (not the exact numbers) of certain quantities. If we know the order of a quantity and its range of variations, we can choose the right scales to write the mathematical model in the nondimensional form so that the right mathematical methods can be used to tackle the problem. It also helps us to choose more suitable numerical methods to find the solution over the correct scales. The estimations will often give us greater insight into the physical process, resulting in more appropriate mathematical models.

■ **EXAMPLE 2.2**

We now try to carry out an estimation of the Earth's surface temperature assuming that the Earth is a spherical black body. The incoming energy from the Sun on the Earth's surface is

$$E_{\text{in}} = (1 - \alpha) \pi r_E^2 S, \tag{2.2}$$

where α is the albedo or the planetary reflectivity to the incoming solar radiation, and $\alpha \approx 0.3$. In addition, the total solar irradiance on the Earth's surface (S) is about $S = 1367$ W/m^2. r_E is the radius of the Earth. Here the effective area of receiving sunlight is equivalent to the area of a disk πr_E^2 as only one side of the Earth is constantly facing the Sun. A body at an absolute temperature T will have black-body radiation and the total energy E_b emitted by the object per unit area per unit time obeys the Stefan-Boltzmann law

$$E_b = \sigma T^4, \tag{2.3}$$

where $\sigma = 5.67 \times 10^{-8}$ J/K^4 s m^2 is the Stefan-Boltzmann constant. For example, we know a human body has a typical body temperature of $T_h = 36.8^\circ$C or $273+36.8 = 309.8$ K. An adult in an environment with a constant room temperature $T_0 = 20^\circ$C or $273+20 = 297$ K will typically have a skin temperature $T_s \approx (T_h + T_0)/2 = (36.8 + 20)/2 = 28.4^\circ$C or $273 + 28.4 = 301.4$ K. In addition, an adult can have a total skin surface area of about $A = 1.8$ m^2. Therefore, the total energy per unit time radiated by an average adult is

$$E = A(\sigma T_s^4 - \sigma T_0^2) = A\sigma(T_s^4 - T_0^4)$$
$$= 1.8 \times 5.67 \times 10^{-8} \times (301.4^4 - 297^4) \approx 90 \text{ J/s}, \tag{2.4}$$

which is about 90 watts. This is very close to the power of a 100-watt light bulb.

For the Earth system, the incoming energy must be balanced by the Earth's black-body radiation

$$E_{\text{out}} = A\sigma T_E^4 = 4\pi r_E^2 \sigma T_E^4, \tag{2.5}$$

where T_E is the surface temperature of the Earth, and $A = 4\pi r_E^2$ is the total area of the Earth's surface. Here we have assumed that outer space has a temperature $T_0 \approx 0$ K, though we know from the cosmological background radiation that it has a temperature of about 4 K. However, this has little effect on our estimations.

From $E_{\text{in}} = E_{\text{out}}$, we have

$$(1 - \alpha)\pi r_E^2 S = 4\pi r_E^2 \sigma T_E^4, \tag{2.6}$$

or

$$T_E = \sqrt[4]{\frac{(1 - \alpha)S}{4\sigma}}. \tag{2.7}$$

Plugging in the typical values, we have

$$T_E = \sqrt[4]{\frac{(1 - 0.3) \times 1367}{4 \times 5.67 \times 10^{-8}}} \approx 255 \text{ K}, \tag{2.8}$$

which is about -18°C. This is too low compared with the average temperature 9°C or 282 K on the Earth's surface. The difference implies that the greenhouse effect of the CO_2 is in the atmosphere. The greenhouse gas warms the surface by about 27°C.

You may argue that the difference may also come from the heat flux from the lithosphere to the Earth surface, and the heat generation in the crust. That is partly true, but the detailed calculations for the greenhouse effect are far more complicated, and still form an important topic of active research.

Let us look at an example of Stokes' law which is very important for modeling physical processes such as sedimentation and viscous flow.

■ **EXAMPLE 2.3**

For a sphere of radius r and density ρ_s falling in a fluid of density ρ_f (see Fig. 2.3), the frictional/viscous resistance or drag is given by Stokes' law

$$F_{\rm up} = 6\pi\mu v r, \tag{2.9}$$

where μ is the dynamic viscosity of the fluid. v is the velocity of the spherical particle. The driving force $F_{\rm down}$ of falling is the difference between the gravitational force and the buoyant force or buoyancy. That is the difference between the weight of the sphere and the weight of the displaced fluid by the sphere (with the same volume). We have

$$F_{\rm down} = \frac{4\pi\rho_s g r^3}{3} - \frac{4\pi\rho_f g r^3}{3} = \frac{4\pi(\rho_s - \rho_f) g r^3}{3}, \tag{2.10}$$

where g is the acceleration due to gravity.

The falling particle will reach a uniform velocity v_s, called the terminal velocity or settling velocity, when the drag $F_{\rm up}$ is balanced by $F_{\rm down}$, or $F_{\rm up} = F_{\rm down}$. We have

$$6\pi\mu v_s r = \frac{4\pi(\rho_s - \rho_f) g r^3}{3}, \tag{2.11}$$

which leads to

$$v_s = \frac{2(\rho_s - \rho_f) g r^2}{9\mu} = \frac{(\rho_s - \rho_f) g d^2}{18\mu}, \tag{2.12}$$

where $d = 2r$ is the diameter of the particle.

We know that the typical size of sand particles is about 0.1 mm $=10^{-4}$ m. Using the typical values of $\rho_s = 2000$ kg/m^3, $\rho_f = 1000$ kg/m^3, $g = 9.8$ m/s^2, and $\mu = 10^{-3}$ Pa s, we have $v_s \approx 0.5 \times 10^{-2}$ m/s $= 0.5$ cm/s. Any flow velocity higher than v_s will result in sand suspension in water and long-distance transport.

Stokes' law is valid for laminar steady flows with very low Reynolds number Re, which is a dimensionless number, and is usually defined as $Re = \rho_f v d/\mu = vd/\nu$, where μ is the viscosity or dynamic viscosity, and $\nu = \mu/\rho_f$ is called the kinematic viscosity. Stokes' law is typically for a flow with $Re < 1$, and such flow is often called the Stokes flow.

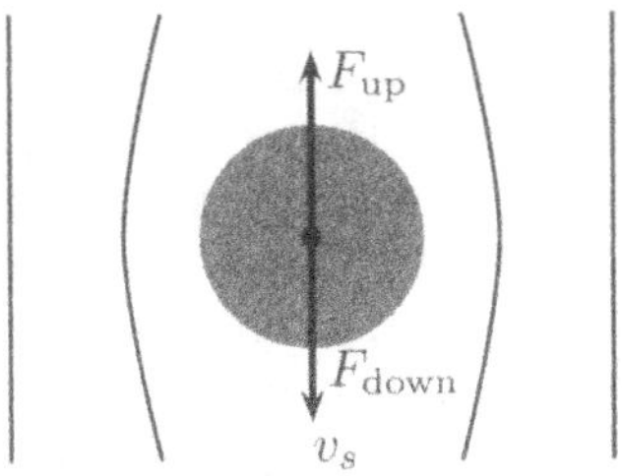

Figure 2.3 Settling velocity of a spherical particle.

From (2.12), we can see that if $\rho_s < \rho_f$, then the particle will move up. When you pour some champagne or sparkling water in a clean glass, you will notice a lot of bubbles of different sizes moving up quickly. The size of a bubble will also increase as it moves up; this is due to the pressure decrease and the nucleation process. Large bubbles move faster than smaller bubbles. If we consider a small bubble with negligible change in size, we can estimate the velocity of the bubbles. The dynamic viscosity and density of champagne are about 1.5×10^{-3} Pa s and 1000 kg/m^3, respectively. For simplicity, we can practically assume the density of the bubbles is zero. For a bubble with a radius of $r = 0.1$ mm or diameter $d = 0.2$ mm $=2 \times 10^{-4}$ m, its uprising velocity can be estimated by

$$v_s = \frac{(1000 - 0) \times 9.8(2 \times 10^{-4})^2}{18 \times 1.5 \times 10^{-3}} \approx 0.015 \text{ m/s} = 1.5 \text{ cm/s}. \qquad (2.13)$$

Of course the choice of typical values is important in order to get a valid estimation. Such a choice will depend on the physical process and the scales of interest. The right choice will be perfected by expertise and practice. We will give many worked examples like this in this book.

2.4 MATHEMATICAL MODELS

2.4.1 Differential Equations

The first step of the mathematical modeling process produces some mathematical equations, often partial differential equations. The next step is to identify the detailed constraints such as the proper boundary conditions and initial conditions so that we can obtain a unique set of solutions. For the sugar diffusion problem discussed earlier, we cannot obtain the exact solution in the actual domain inside the water-filled glass, because we need to know where the sugar cube or grains were initially added. The geometry of the glass also needs to be specified. In fact, this problem needs numerical methods such as finite element methods or finite volume methods. The only possible solution is the long-time behavior: when $t \to \infty$, we know that the concentration should be uniform $c(z, t \to \infty) \to c_\infty$ (=mass of sugar added/volume of water).

You may say that we know this final state even without mathematical equations, so what is the use of the diffusion equation? The main advantage is that you can calculate the concentration at any time using the mathematical equation with appropriate boundary and initial conditions, either by numerical methods in most cases or by mathematical analysis in some very simple cases. Once you know the initial and boundary conditions, the whole system history will be determined to a certain degree. The beauty of mathematical models is that many seemingly diverse problems can be reduced to the same mathematical equation. For example, we know that the diffusion problem is governed by the diffusion equation $\frac{\partial c}{\partial t} = D\nabla^2 c$. The heat conduction is governed by the heat conduction equation

$$\frac{\partial T}{\partial t} = \kappa \nabla^2 T, \qquad \kappa = \frac{K}{\rho c_p}, \tag{2.14}$$

where T is temperature and κ is the thermal diffusivity. K is thermal conductivity, ρ is the density and c_p is the specific heat capacity. Similarly, the dissipation of the pore pressure p in poroelastic media is governed by

$$\frac{\partial p}{\partial t} = c_v \nabla^2 p, \tag{2.15}$$

where $c_v = k/(S\mu)$ is the consolidation coefficient, k is the permeability of the media, μ is the viscosity of fluid (water), and S is the specific storage coefficient.

Mathematically speaking, whether it is concentration, temperature or pore pressure, it is the same dependent variable u. Similarly, it is just a constant κ whether it is the diffusion coefficient D, the thermal diffusivity α or the consolidation coefficient c_v. In this sense, the above three equations are identical to the following parabolic partial differential equation

$$\frac{\partial u}{\partial t} = \kappa \nabla^2 u. \tag{2.16}$$

Suppose we want to solve the following problem. For a semi-infinite domain shown in Figure 2.4, the initial condition (whether temperature or concentration or pore pressure) is $u(x, t = 0) = 0$. The boundary condition at $x = 0$ is that $u(x = 0, t) = u_0$ =const at any time t. Now the question what the distribution of u versus x at t is?

Let us summarize the problem. As this problem is one-dimensional, only the x-axis is involved, and it is time-dependent. So we have

$$\frac{\partial u}{\partial t} = \kappa \frac{\partial^2 u}{\partial x^2}, \tag{2.17}$$

with an initial condition

$$u(x, t = 0) = 0, \tag{2.18}$$

and the boundary condition

$$u(x = 0, t) = u_0. \tag{2.19}$$

Let us start to solve this mathematical problem. How should we start and where to start? Well, there are many techniques to solve these problems, including the similarity solution technique, Laplace's transform, Fourier's transform, separation of variables and others.

Similarity variable is an interesting and powerful method because it neatly transforms a partial differential equation (PDE) into an ordinary differential equation (ODE) by introducing a similarity variable ζ, then you can use the standard techniques for solving ODEs to obtain the desired solution. We first define a similar variable

$$\zeta = \frac{x^2}{4\kappa t}, \tag{2.20}$$

so that $u(x,t) = u(\zeta) = f(\zeta)$. Using the chain rules of differentiations

$$\frac{\partial}{\partial x} = \frac{\partial}{\partial \zeta}\frac{\partial \zeta}{\partial x} = \frac{x}{2\kappa t}\frac{\partial}{\partial \zeta},$$

$$\frac{\partial^2}{\partial x^2} = \left(\frac{x}{2\kappa t}\right)^2 \frac{\partial^2}{\partial \zeta^2} + \frac{1}{2\kappa t}\frac{\partial}{\partial \zeta} = \frac{\zeta}{\kappa t}\frac{\partial^2}{\partial \zeta^2} + \frac{1}{2\kappa t}\frac{\partial}{\partial \zeta},$$

$$\frac{\partial}{\partial t} = \frac{\partial}{\partial \zeta}\frac{\partial \zeta}{\partial t} = -\frac{x^2}{4\kappa t^2}\frac{\partial}{\partial \zeta} = -\frac{\zeta}{t}\frac{\partial}{\partial \zeta}, \tag{2.21}$$

we can write the PDE (2.17) for u as

$$-\frac{\zeta}{t} f' = \kappa \cdot \left[\frac{\zeta}{\kappa t} f'' + \frac{1}{2\kappa t} f'\right], \tag{2.22}$$

where $f' = df/d\zeta$. Multiplying both sides by t/ζ,

$$-f' = f''(\zeta) + \frac{1}{2\zeta} f', \quad \text{or} \quad \frac{f''}{f'} = -\left(1 + \frac{1}{2\zeta}\right). \tag{2.23}$$

Using $(\ln f')' = f''/f'$ and integrating the above equation once, we get

$$\ln f' = -\zeta - \frac{1}{2}\ln \zeta + C, \tag{2.24}$$

where C is an integration constant. This can be written as

$$f' = \frac{K e^{-\zeta}}{\sqrt{\zeta}}, \tag{2.25}$$

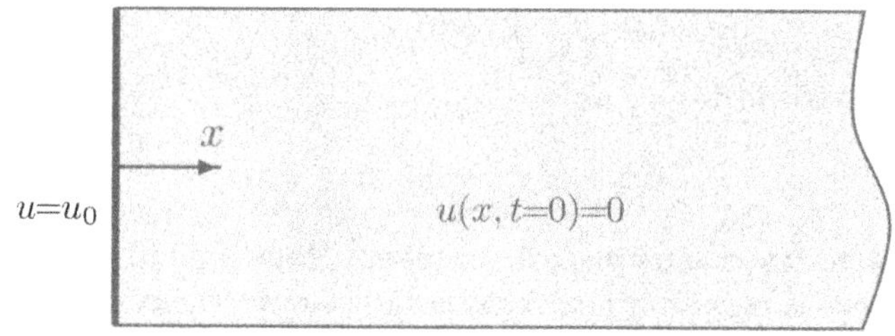

Figure 2.4 Heat transfer through a semi-infinite medium near a dyke in geology.

where $K = e^C$. Integrating it again, we obtain

$$u = f(\zeta) = A\mathrm{erf}\Big(\sqrt{\zeta}\Big) + B = A\mathrm{erf}(\frac{x}{\sqrt{4\kappa t}}) + B, \tag{2.26}$$

where

$$\mathrm{erf}(x) = \frac{2}{\sqrt{\pi}} \int_0^x e^{-\xi^2} d\xi, \tag{2.27}$$

is the error function and ξ is a dummy variable. $A = K\sqrt{\pi}$ and B are constants that can be determined from appropriate boundary conditions. This is the basic solution in the infinite or semi-infinite domain. The solution is generic because we have not used any of the boundary conditions or initial conditions.

■ EXAMPLE 2.4

For the heat conduction problem near a magma dyke in a semi-infinite domain, we can determine the constants A and B. Let $x = 0$ be the center of the rising magma dyke so that its temperature is constant at the temperature u_0 of the molten magma, while the temperature at the far field is $u = 0$ (as we are only interested in the temperature change in this case).

The boundary condition at $x = 0$ requires that

$$A\mathrm{erf}(0) + B = u_0.$$

We know that $\mathrm{erf}(0) = 0$, this means that $B = u_0$. From the initial condition $u(x, t = 0) = 0$, we have

$$A \lim_{t \to 0} \mathrm{erf}\Big(\frac{x}{\sqrt{4\kappa t}}\Big) + u_0 = 0.$$

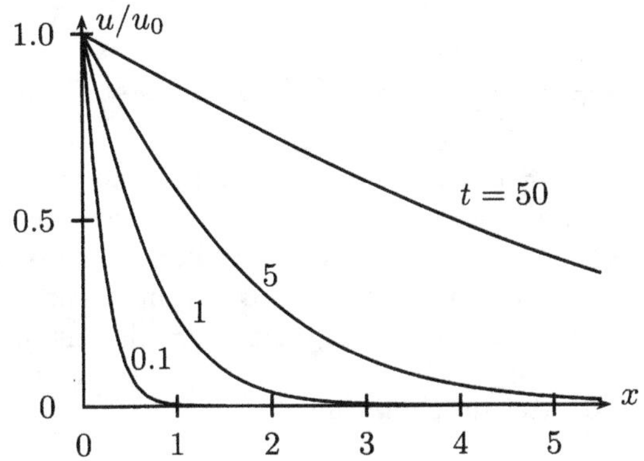

Figure 2.5 Distribution of $u(x,t)/u_0$ with $\kappa = 0.25$.

Since $x/\sqrt{4\kappa t} \to \infty$ as $t \to 0$ and $\text{erf}(\infty) = 1$, we get $A + u_0 = 0$, or $A = -u_0$. Thus the solution becomes

$$u = u_0\Big[1 - \text{erf}\Big(\frac{x}{\sqrt{4\kappa t}}\Big)\Big] = u_0 \text{erfc}\Big(\frac{x}{\sqrt{4\kappa t}}\Big),$$

where $\text{erfc}(x) = 1 - \text{erf}(x)$ is the complementary error function. The distribution of u/u_0 is shown in Fig. 2.5.

From the above solution, we know that the temperature variation becomes significant in the region of $x = d$ such that $d/\sqrt{\kappa t} \approx 1$ at a given time t. That is

$$d = \sqrt{\kappa t}, \tag{2.28}$$

which defines a typical length scale. Alternatively, for a given length scale d of interest, we can estimate the time scale $t = \tau$ at which the temperature becomes significant. That is

$$\tau = \frac{d^2}{\kappa}. \tag{2.29}$$

This means that it will take four times longer if the size of the hot body d is doubled.

Now let us see what it means in terms of typical cooling time of a geological object. We know that the thermal conductivity is $K \approx 3$ W/m K for rock, its density is $\rho \approx 2700$ kg/m^3 and its specific heat capacity $c_p \approx 1000$ J/kg K. Thus, the thermal diffusivity of solid rock is

$$\kappa = \frac{K}{\rho c_p} \approx \frac{3}{2700 \times 1000} \approx 1.1 \times 10^{-6} \text{ m}^2/\text{s}. \tag{2.30}$$

For $d \approx 1$ m, the time scale of cooling is

$$\tau = \frac{d^2}{\kappa} \approx \frac{1}{1.1 \times 10^{-6}} \approx 8.8 \times 10^5 \text{ seconds} \approx 10 \text{ days}. \tag{2.31}$$

For a larger hot body $d = 100$ m, then that time scale is $\tau = 10^5$ days or 270 years. This estimate of the cooling time scale is based on the assumption that no more heat is supplied. However, in reality, there is usually a vast magma reservoir below to supply hot magma constantly, and this means that the cooling time is at the geological time scale over millions of years.

2.4.2 Functional and Integral Equations

Though most mathematical models are written as partial different equations, however, sometimes it might be convenient to write them in terms of integral equations, and these integral forms can be discretized to obtained various numerical methods. For example, the Fredholm integral equation can be generally written as

$$u(x) + \lambda \int_a^b K(x, \eta) y(\eta) d\eta = v(x) y(x), \tag{2.32}$$

where $u(x)$ and $v(x)$ are known functions of x, and λ is constant. The kernel $K(x, \eta)$ is also given. The aim is to find the solution $y(x)$. This type of problem can be extremely difficult to solve and analytical solutions exist in only a few very simple cases.

Sometimes the problem you are trying to solve does not give a mathematical model in terms of dependent variance such as u which is a function of spatial coordinates (x, y, z) and time t, rather they lead to a functional (or a function of the function u); this kind of problem is often linked to the calculus of variations.

For example, finding the shortest path between any given points on the Earth's surface is a complicated geodesic problem. If we idealize the Earth's surface as a perfect sphere, then the shortest path joining any two different points is a great circle through both points. How can we prove this is true? Well, the proof is based on the Euler-Lagrange equation of a functional $\psi(u)$

$$\frac{\partial \psi}{\partial u} = \frac{d}{dx}(\frac{\partial \psi}{\partial u'}), \tag{2.33}$$

where u a function of x, $u' = du/dx$, and ψ a function of $u(x)$. Interested readers can refer to more advanced literature.

2.4.3 Statistical Models

Both differential equations and integral equations are the mathematical models for continuum systems. Other systems are discrete and different mathematical models are needed, though they could reduce to certain forms of differential equations if some averaging is carried out. On the other hand, many systems have intrinsic randomness, thus the description and proper modeling require statistical models.

For example, in order to describe a drop of ink in a water tank or sugar in a tea cup, we can use a diffusion equation discussed earlier. Alternatively, we can also use Brownian motion which is a random walk.

A random walk is a random process which consists of taking a series of consecutive random steps. Mathematically speaking, let S_N denotes the sum of each consecutive random step X_i, then S_N forms a random walk

$$S_N = \sum_{i=1}^{N} X_i = X_1 + ... + X_N, \tag{2.34}$$

where X_i is a random step drawn from a random distribution. This relationship can also be written as a recursive formula

$$S_N = \sum_{i=1}^{N-1} + X_N = S_{N-1} + X_N, \tag{2.35}$$

which means the next state S_N will only depend the current existing state S_{N-1} and how the motion or transition X_N from the existing state to the next state. This is typically the main property of a Markov chain.

Here the step size or length in a random walk can be fixed or varying. Random walks have many applications in physics, economics, statistics, computer sciences, environmental science and engineering.

Consider a scenario: A drunkard walks on a street. At each step, he can randomly go forward or backward. This forms a one-dimensional random walk. If this drunkard walks on a football pitch, he can walk in any direction randomly. This becomes a 2D random walk. Mathematically speaking, a random walk is given by the following equation

$$S_{t+1} = S_t + w_t, \tag{2.36}$$

where S_t is the current location or state at t, and w_t is a step or random variable with a known distribution.

A particle can jump to the right or the left with equal probability 1/2, and each jump can only take one step only. This jump probability, often called transition probability, can be written as

$$w_t = \begin{cases} 1/2 & \text{if } S = +1 \\ 1/2 & \text{if } S = -1 \\ 0 & \text{otherwise} \end{cases}. \tag{2.37}$$

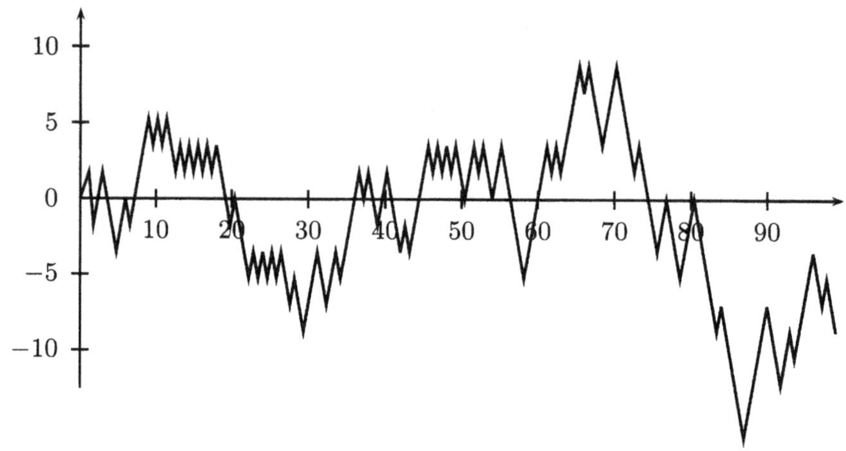

Figure 2.6 Random walk and the path of 100 consecutive steps staring at position 0.

A particle starting at $S_0 = 0$ jumps along a straight line; let us follow its first few steps. Suppose if we flip a coin, the particle moves to the right (or up) if it is a head; otherwise, the particle moves to the left (or down) when the coin is a tail. First, we flop the coin, and we get, say, a head. So the particle moves to the right by a fixed unit step. So $S_1 = S_0 + 1 = 1$. Then, a tail leads a move to the left, that is, $S_2 = S_1 - 1 = 0$. By flipping the coin again, we get a tail. So $S_3 = S_2 - 1 = -1$. We continue the process in the similar manner, and the path of 100 steps or jumps is shown in Figure 2.6. It has been proved theoretically that the probability of returning to the origin or reaching any point approaches 1 when the number N of steps approaches infinity.

Suppose the probability of moving to the right is p, and thus the probability of moving to the left is $q = 1 - p$. The probability of taking k steps to the right among N steps obeys the binomial distribution

$$p(k; N, p) = \binom{N}{k} p^k (1-p)^{N-k}. \tag{2.38}$$

It leaves as exercises to show that the mean number of steps to the right is the mean

$$\langle k \rangle = pN, \tag{2.39}$$

which means the mean number of steps to the left is simply $N - pN = (1-p)N = qN$. The variance associated with k is

$$\sigma_k^2 = p(1-p)N = pqN. \tag{2.40}$$

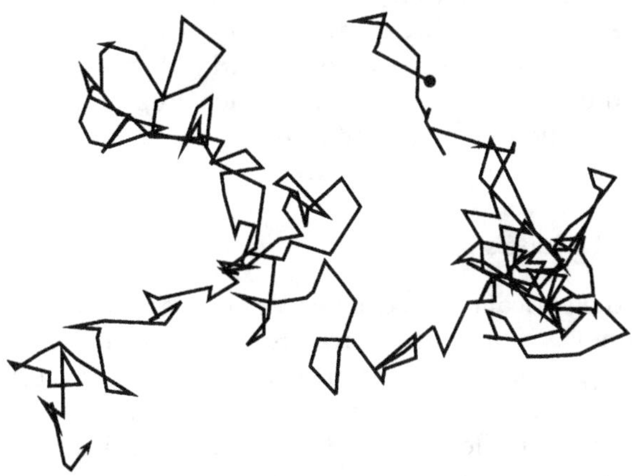

Figure 2.7 Brownian motion in 2D: random walk with a Gaussian step-size distribution and the path of 100 steps starting at the origin $(0, 0)$ (marked with $\bullet$).

As time increases, the number of steps also increases. That is $N = t$ if each step or jump takes a unit time. This means that the variance increases linearly with t, or

$$\sigma^2 \propto t. \tag{2.41}$$

If each step or jump is carried out in the n-dimensional space, the random walk discussed earlier

$$S_N = \sum_{i=1}^{N} X_i, \tag{2.42}$$

becomes a random walk in higher dimensions. In addition, there is no reason why each step length should be fixed. In fact, the step size can also vary according to a known distribution. If the step length obeys the Gaussian distribution, the random walk becomes the Brownian motion (see Figure 2.7). Similar to the one-dimensional case, the variance σ^2 also increases linearly with time t or the total number of steps N.

In theory, as the number of steps N increases, the central limit theorem implies that the random walk (2.42) should approaches a Gaussian distribution. As the mean of particle locations shown in Figure 2.7 is obviously zero, their variance will increase linearly with t. Therefore, the Brownian motion $B(t)$ essentially obeys a Gaussian distribution with zero mean and time-dependent variance. That is,

$$B(t) \sim N(0, \sigma^2 = t), \tag{2.43}$$

where $\sim$ means the random variance obeys the distribution on the right-hand side, or samples should be drawn from the distribution. The diffusion process can be viewed as a series of Brownian motion, and the motion obeys the Gaussian distribution. For this reason, standard diffusion is often referred to as the Gaussian diffusion. If the motion at each step is not Gaussian, then the diffusion is called non-Gaussian diffusion.

2.4.4 Rule-based Models

Sometimes, a mathematical model can be more conveniently expressed as a set of rules. For example, cellular automata are a class of rule-based models. In a very simple case, a cellular automaton can be represented on a 2D grid, and each cell has 8 neighbor cells. The state (say, 0 or 1) of each cell will depend on a set of rules, and thus depends on the states of its neighbor cells and its current state. The well-known "game of life" often used in many computer screensavers, is a cellular automaton. In fact, it has been proved that a cellular automaton can act as a universal computing machine, capable of carrying out very complex computations and simulating real-world phenomena.

In physics and fluid dynamics, the so-called Lattice-Boltzmann method is an extension of the basic cellular automaton ideas. In modern metaheuristics, rule-based algorithms can solve complex optimization problems using inspiration from nature.[1]

2.5 NUMERICAL METHODS

2.5.1 Numerical Integration

In the solution (2.26) of problem (2.17), there is a minor problem in the evaluation of the solution u. That is the error function $\text{erf}(x)$ because it is a special function whose integral cannot be expressed as a simple explicit combination of basic functions, and it can only be expressed in terms of a quadrature. In order to get its values, we have to either use approximations or numerical integration. You can see that even with a seemingly precise solution of a differential equation, it is quite likely that it may involve some special functions.

Let us try to evaluate $\text{erf}(1)$. From advanced mathematics, we know its exact value is $\text{erf}(1) = 0.8427007929...$, but how do we calculate it numerically?

■ **EXAMPLE 2.5**

[1] Xin-She Yang, *Nature-Inspired Metaheuristic Algorithms*, Luniver Press, UK. First Edition (2008), 2nd Edition (2010).

In order to estimate erf(1), we first try to use a naive approach by estimating the area under the curve $f(x) = \frac{2}{\sqrt{\pi}} e^{-x^2}$ in the interval $[0, 1]$ shown in Figure 2.8. We then divide the interval into 5 equally spaced thin strips with $h = \Delta x = x_{i+1} - x_i = 1/5 = 0.2$. We have six values of $f_i = f(x_i)$ at $x_i = hi (i = 0, 1, ..., 5)$, and they are

$$f_0 = 1.1284, \; f_1 = 1.084, \; f_2 = 0.9615,$$

$$f_3 = 0.7872, \; f_4 = 0.5950, \; f_5 = 0.4151.$$

Now we can either use the rectangular area under the curve (which underestimates the area) or the area around the curve plus the area under curve (which overestimates the area). Their difference is the tiny area about the curve which could still make some difference. If we use the area under the curve, we have the estimation of the total area as

$$A_1 \approx 0.2(f_1 + f_2 + f_3 + f_4 + f_5) \approx 0.7686.$$

The other approach gives

$$A_2 \approx 0.2(f_0 + f_1 + f_2 + f_3 + f_4) \approx 0.91125.$$

Both are about 8% from the true value $\text{erf}(1) \approx 0.8247$. If we take the average of these two estimates, we get

$$A_3 \approx \frac{A_1 + A_2}{2} \approx 0.8399,$$

which is much better, but still 0.3% from the true value. This average method is essentially equivalent to using $f_i = (f_{i-1} + f_i)/2$ to approximate the value of $f(x)$ in each interval.

As you can see from this example, the way you discretize the integrand to estimate the integral numerically can have many variants, subsequently affecting the results significantly. There are much better ways to carry out the numerical integration, notably the Gaussian integration which requires only seven points to get the accuracy of about 9th decimal place or 0.0000001%. As this book focuses mainly on the model formulation, interested readers can refer to an advanced book on numerical methods for more details.

2.5.2 Numerical Solutions of PDEs

The diffusion equation (2.1) is a relatively simple parabolic equation. If we add a reaction term (source or sink) to this equation, we get the classical reaction-diffusion equation

$$\frac{\partial u}{\partial t} = D\nabla^2 u + \gamma u(1 - u), \tag{2.44}$$

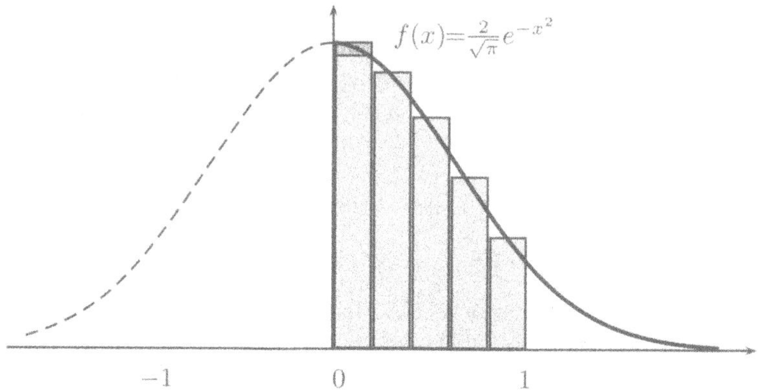

Figure 2.8 Naive numerical integration.

where u can be concentration and any other quantities. $\gamma u(1-u)$ is the reaction term and γ is a constant. This seemingly simple partial differential equation is in fact rather complicated for mathematical analysis because the equation is nonlinear due to the term $-\gamma u^2$. However, a numerical technique can be used and it is relatively straightforward to obtain solutions. This mathematical model can produce intriguing patterns due to its intrinsic instability under appropriate conditions.

In the two-dimensional case, we have

$$\frac{\partial u}{\partial t} = D\Big(\frac{\partial^2 u}{\partial x^2} + \frac{\partial^2 u}{\partial y^2}\Big) + \gamma u(1-u). \tag{2.45}$$

Using the finite difference method to be introduced in the next chapter, we can solve this equation on a 2D domain. Figure 2.9 shows the stable pattern generated by Eq.(2.45) with $D = 0.2$ and $\gamma = 0.5$. The initial condition is completely random, say, $u(x, y, t{=}0) = \mathsf{rand}(n, n) \in [0, 1]$ where $n \times n$ is the size of the grid used in the simulations. The function $\mathsf{rand}()$ is a random number generator and all the random numbers are in the range of 0 to 1.

We can see that a beautiful and stable pattern forms automatically from an initially random configuration. This pattern formation mechanism has been used to explain many pattern formation phenomena in nature, including patterns on zebra skin, tiger skin and sea shell, zebra leaf (green and yellow), and zebra stones. For example, the zebra rocks have reddish-brown and white bands first discovered in Australia. It is believed that the pattern is generated by dissolution and precipitation of mineral bands such as iron oxide as mineral in the fluid percolating through the porous rock.

In the second part of this book, we will introduce in detail various mathematical model formulations and analysis of many processes and phenomena in real-world applications.

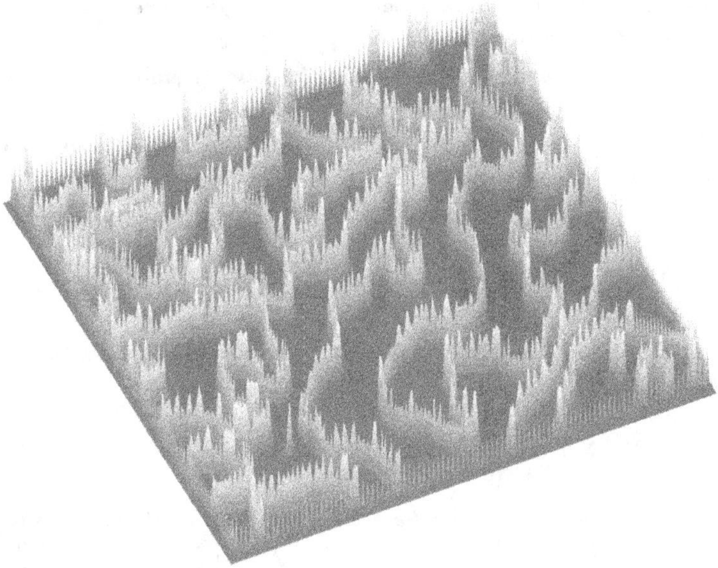

Figure 2.9 Pattern formation of reaction-diffusion equation (2.45) with $D = 0.2$ and $\gamma = 0.5$.

EXERCISES

2.1 From the first principle and theory of gravitation, estimate the escape velocity of a satellite.

2.2 Estimate the velocity of a rain drop, assuming the size of a drop is between 0.1 mm to 0.55 cm.

2.3 For water waves in a lake or ocean, the phase speed or velocity v in most cases is governed by

$$v = \sqrt{\frac{g\lambda}{2\pi} \tanh\left(\frac{2\pi h}{\lambda}\right)},$$

where h is the water depth, λ is the wavelength of the waves, and g is the acceleration due to gravity. Estimate the velocity of a typical tsunami.

2.4 Assume the air pressure is hydrostatic and air is an ideal gas, estimate the pressure variations versus altitude.

REFERENCES

1. Fowler, A. C., *Mathematical Models in the Applied Sciences*, Cambridge University Press, Cambridge (1997).
2. Gershenfeld, N., *Nature of Mathematical Modeling*, Cambridge University Press, Cambridge (1998).
3. Kardestruncer, H. and Norrie, D. H., *Finite Element Handbook*, McGraw-Hill, New York (1987).
4. Kreyszig, E., *Advanced Engineering Mathematics*, 6th Edition, Wiley & Sons, New York (1988).
5. Murch, B. W. and Skinner, B. J., *Geology Today - Understanding Our Planet*, John Wiley & Sons, New York (2001).
6. Press, W. H., Teukolsky, S. A., Vetterling, W. T., and Flannery, B. P., *Numerical Recipes in C++: The Art of Scientific Computing*, 2nd Edition, Cambridge University Press, Cambridge (2002).
7. Smith, G. D., *Numerical Solution of Partial Differential Equations*, Oxford University Press, Oxford (1974).
8. Wang H. F., *Theory of Linear Poroelasticity: With Applications to Geomechanics and Hydrogeology*, Princeton University Press, New Jersey (2000).

CHAPTER 3

NUMERICAL METHODS: AN INTRODUCTION

XIN-SHE YANG

School of Science and Technology, Middlesex University, London, UK

Because it is not always possible to solve differential equations analytically, numerical methods have become an important tool in modeling and simulation. In fact, computational modeling has become the so-called third paradigm, complementing the tradition theoretical and experimental approaches to problem solving. Among many powerful numerical methods, the finite difference method is one of the most popular methods that are used commonly in computer simulations. It has the advantage of simplicity and clarity, especially in 1D configurations and other cases with regular geometry. The finite difference method essentially transforms an ordinary differential equation into a coupled set of algebraic equations by replacing the continuous derivatives with finite difference approximations on a grid of mesh or node points that span the domain of interest based on the Taylor series expansions. In general, the boundary conditions and boundary nodes need special treatment.

Mathematical Modeling with Multidisciplinary Applications.
Edited by Xin-She Yang

3.1 DIRECT INTEGRATION

The second-order or higher-order ordinary differential equations can be written as a first-order system of ODEs. Since the technique for solving a system is essentially the same as that for solving a single equation

$$\frac{dy}{dx} = f(x, y), \tag{3.1}$$

we shall focus on the first-order equation in the rest of this section. In principle, the solution can be obtained by direct integration,

$$y(x) = y_0 + \int_{x_0}^{x} f(x, y(x))dx, \tag{3.2}$$

but in practice it is usually impossible to do the integration analytically as it requires the solution of $y(x)$ to evaluate the right-hand side. Thus, some approximations shall be utilized. Numerical integration is the most common technique for obtaining approximate solutions. There are various integration schemes with different orders of accuracy and convergent rates. These schemes include the simple Euler scheme, Runge-Kutta method, relaxation method, and many others.

3.1.1 Euler Scheme

Using the notation $h = \Delta x = x_{n+1} - x_n$, $y_n = y(x_n)$, $x_n = x_0 + n\Delta x$ ($n = 0, 1, 2, ..., N$), and $' = d/dx$ for convenience, then the explicit Euler scheme can simply be written as

$$y_{n+1} = y_n + \int_{x_n}^{x_{n+1}} f(x, y)dx \approx y_n + hf(x_n, y_n). \tag{3.3}$$

This is a forward difference method as it is equivalent to the approximation of the first derivative

$$y'_n = \frac{y_{n+1} - y_n}{\Delta x}. \tag{3.4}$$

The order of accuracy can be estimated using the Taylor expansion

$$y_{n+1} = y_n + hy'|_n + \frac{h^2}{2}y''|_n + ... \approx y_n + hf(x_n, y_n) + O(h^2). \tag{3.5}$$

Thus, the Euler method is first-order accurate.

For any numerical algorithms, the algorithm must be stable in order to reach convergent solutions. Thus, stability is an important issue in numerical analysis. Defining δy as the discrepancy between the actual numerical solution and the true solution of the Euler finite difference equation, we have

$$\delta y_{n+1} = [1 + hf'(y)] = \xi \delta y_n. \tag{3.6}$$

In order to avoid the discrepancy to grow, it requires the following stability condition $|\xi| \leq 1$. The stability restricts the size of interval h, which is usually small.

One alternative that can use larger h is the implicit Euler scheme, and this scheme approximates the derivative by a backward difference $y'_n = (y_n - y_{n-1})/h$ and the right-hand side of (3.2) is evaluated at the new y_{n+1} location. Now the scheme can be written as

$$y_{n+1} = y_n + hf(x_{n+1}, y_{n+1}). \tag{3.7}$$

The stability condition becomes

$$\delta y_{n+1} = \xi \delta y_n = \frac{\delta y_n}{1 - hf'(y)}, \tag{3.8}$$

which is always stable if $f'(y) = \frac{\partial f}{\partial y} \leq 0$. This means that any step size is acceptable. However, the step size cannot be too large as the accuracy reduces as the step size increases.

Another practical issue is that, for most problems such as nonlinear ODEs, the evaluation of y' and $f'(y)$ requires the value of y_{n+1} which is unknown. Thus, an iteration procedure is needed to march to a new value y_{n+1}, and the iteration starts with a guess value which is usually taken to be zero for most cases. The implicit scheme generally gives better stability.

3.1.2 Leap-Frog Method

The leap-frog scheme is the central difference

$$y'_n = \frac{y_{n+1} - y_{n-1}}{2\Delta x}, \tag{3.9}$$

which leads to

$$y_{n+1} = y_{n-1} + 2hf(x_n, y_n). \tag{3.10}$$

The central difference method is second-order accurate. In a similar way as Eq. (3.6), the leap-frog method becomes

$$\delta y_{n+1} = \delta y_{n-1} + 2hf'(y)\delta y_n, \tag{3.11}$$

or

$$\delta y_{n+1} = \xi^2 \delta y_{n-1}, \tag{3.12}$$

where $\xi^2 = 1 + 2hf'(y)\xi$. This scheme is stable only if $|\xi| \leq 1$, and a special case is $|\xi| = 1$ when $f'(y)$ is purely imaginary. Therefore, the central scheme is not necessarily a better scheme than the forward scheme.

3.1.3 Runge-Kutta Method

We have so far seen that stability of the Euler method and the central difference method is limited. The Runge-Kutta method uses a trial step to the midpoint of the interval by central difference and combines with the forward difference at two steps

$$\hat{y}_{n+1/2} = y_n + \frac{h}{2} f(x_n, y_n), \tag{3.13}$$

$$y_{n+1} = y_n + hf(x_{n+1/2}, \hat{y}_{n+1/2}). \tag{3.14}$$

This scheme is second-order accurate with higher stability compared with previous simple schemes. One can view this scheme as a predictor-corrector method. In fact, we can use multisteps to devise higher-order methods if the right combinations are used to eliminate the error terms order by order. The popular classical Runge-Kutta method can be written as

$$a = hf(x_n, y_n),$$

$$b = hf(x_n + h/2, y_n + a/2),$$

$$c = hf(x_n + h/2, y_n + b/2),$$

$$d = hf(x_n + h, y_n + c),$$

$$y_{n+1} = y_n + \frac{a + 2(b + c) + d}{6}, \tag{3.15}$$

which is fourth-order accurate.

■ **EXAMPLE 3.1**

Let us solve the following nonlinear equation numerically

$$\frac{dy}{dx} + y^2 = -1, \qquad x \in [0, 2]$$

with the initial condition

$$y(0) = 1.$$

We know that it has an analytical solution

$$y(x) = -\tan\left(x - \frac{\pi}{4}\right).$$

On the interval $[0, 2]$, let us first solve the equation using the Euler scheme for $h = 0.5$. There are five points $x_i = ih(i = 0, 1, 2, 3, 4)$. As $dy/dx = f(y) = -1 - y^2$, we have the Euler scheme

$$y_{n+1} = y_n + hf(y) = y_n - h - hy_n^2.$$

From the initial condition $y_0 = 1$, we now have

$$y_1 = y_0 - h - hy_0^2 = 1 - 0.5 - 0.5 \times 1^2 = 0,$$

$$y_2 \approx -0.5, \qquad y_3 \approx -1.125, \qquad y_4 \approx -2.2578.$$

These are significantly different (about 30%) from the exact solutions

$$y_0^* = 1, \;\; y_1^* \approx 0.2934079, \;\; y_2^* = -0.21795809,$$

$$y_3^* = -0.86756212, \;\; y_4^* = -2.68770693.$$

Now let us use the Runge-Kutta method to solve the same equation to see if it is better. Since $f(x_n, y_n) = -1 - y_n^2$, we have

$$a = hf(x_n, y_n) = -h(1 + y_n^2), \qquad b = -h\Big[1 + \big(y_n + \frac{a}{2}\big)^2\Big],$$

$$c = -h\Big[1 + \big(y_n + \frac{b}{2}\big)^2\Big], \qquad d = -h[1 + (y_n + c)^2],$$

and

$$y_{n+1} = y_n + \frac{a + 2(b + c) + d}{6}.$$

From $y_0 = 1$ and $h = 0.5$, we have

$$y_1 \approx 0.29043, \qquad y_2 \approx -0.22062,$$

$$y_3 = -0.87185, \qquad y_4 \approx -2.67667.$$

These values are within about 1% of the analytical solutions y_n^*. We can see that even with the same step size, the Runge-Kutta method is much more efficient and accurate than the Euler scheme.

Generally speaking, higher-order schemes are better than lower-order schemes, but not always.

3.2 FINITE DIFFERENCE METHODS

Numerical solutions of partial differential equations are more complicated than that of ODEs because they involve time and space variables and the geometry of the domain of interest. Usually, boundary conditions are more complex. In addition, nonlinear problems are very common in real-world processes.

We start with the simplest first-order equations and then move onto more complicated cases.

3.2.1 Hyperbolic Equations

For simplicity, we first look at the one-dimensional scalar equation of hyperbolic type,

$$\frac{\partial u}{\partial t} + c\frac{\partial u}{\partial x} = 0, \tag{3.16}$$

where c is a constant or the velocity of advection. By using the forward Euler scheme for time and central scheme for space, we have

$$\frac{u_j^{n+1} - u_j^n}{\Delta t} + c\Big[\frac{u_{j+1}^n - u_{j-1}^n}{2h}\Big] = 0, \tag{3.17}$$

where $t = n\Delta t, n = 0, 1, 2, ..., x = x_0 + jh, j = 0, 1, 2, ...,$ and $h = \Delta x$. In order to see how this method behaves numerically, we use the von Neumann stability analysis.

Assuming the independent solutions or eigenmodes, also called Fourier modes, in spatial coordinate x in the form of $u_j^n = \xi^n e^{ikhj}$ where k is the equivalent wavenumber, and substituting into Eq. (3.17), we have

$$\xi = 1 - i\frac{c\Delta t}{h}\sin(kh). \tag{3.18}$$

The stability criteria $|\xi| \leq 1$ require

$$\Big(\frac{c\Delta t}{h}\Big)^2 \sin^2 kh \leq 0. \tag{3.19}$$

However, this inequality is impossible to satisfy and this scheme is thus unconditionally unstable.

To avoid the difficulty of instability, we can use other schemes such as the upwind scheme and Lax scheme. For the upwind scheme, the equation becomes

$$\frac{u_j^{n+1} - u_j^n}{\Delta t} + c\Big[\frac{u_j^n - u_{j-1}^n}{h}\Big] = 0, \tag{3.20}$$

whose stability condition is

$$|\xi| = \Big|1 - \frac{c\Delta t}{h}[1 - \cos(kh) + i\sin(kh)]\Big| \leq 1, \tag{3.21}$$

which is equivalent to

$$0 < \frac{c\Delta t}{h} \leq 1. \tag{3.22}$$

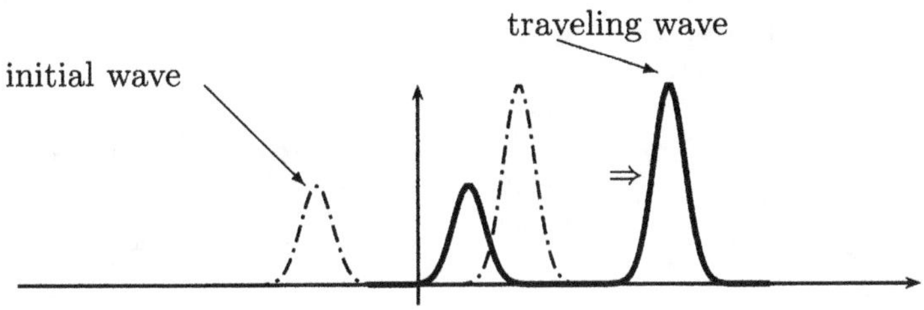

Figure 3.1 First-order hyperbolic equation and its traveling wave solution $u_t + u_x = 0$.

This the well-known Courant-Friedrichs-Lewy stability condition, often referred to as the Courant stability condition. Thus, the upwind scheme is conditionally stable.

The wave propagation of the first-order hyperbolic equation can be demonstrated by the following simple case,

$$u_t + u_x = 0, \qquad 0 \leq x \leq L, \tag{3.23}$$

with an initial condition

$$u(x, 0) = \frac{1}{2} e^{-[(x-\frac{L}{4})/L]^2} + e^{-[(x-\frac{L}{2})/L]^2}, \tag{3.24}$$

and boundary conditions $u(0, t) = u(L, t) = 0$.

Figure 3.1 shows the wave propagation where the dashed curve corresponds to the initial wave profile while the solid curve corresponds to the traveling wave. We can see that the wave profile does not change with time but moves with a constant velocity.

3.2.2 Second-Order Wave Equation

Higher-order equations such as the second-order wave equation can be written as a system of hyperbolic equations and then be solved using numerical integration. They can also be solved by direct discretization using the finite difference scheme. The wave equation

$$\frac{\partial^2 u}{\partial t^2} = c^2 \frac{\partial^2 u}{\partial x^2}, \tag{3.25}$$

consists of second derivatives. If we approximate the first derivatives at each time step n using

$$u'_i = \frac{u^n_{i+1} - u^n_i}{\Delta x}, \qquad u'_{i-1} = \frac{u^n_i - u^n_{i-1}}{\Delta x}, \tag{3.26}$$

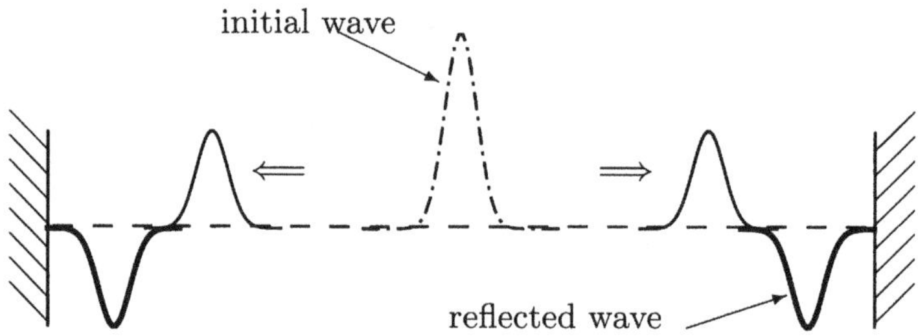

Figure 3.2 Traveling wave solution of the wave equation: $u_{tt} - c^2 u_{xx} = 0$.

then we can use the following approximation for the second derivative

$$u_i'' = \frac{u_i' - u_{i-1}'}{\Delta x} = \frac{u_{i+1}^n - 2u_i^n + u_{i-1}^n}{(\Delta x)^2}. \tag{3.27}$$

This is in fact a central difference scheme of second-order accuracy. If we use the similar scheme for time-stepping, then we get a central difference scheme in both time and space.

Thus, the numerical scheme for this equation becomes

$$\frac{u_i^{n+1} - 2u_i^n + u_i^{n-1}}{(\Delta t)^2} = c^2 \frac{u_{i+1}^n - 2u_i^n + u_{i-1}^n}{(\Delta x)^2}. \tag{3.28}$$

This is a two-level scheme with a second-order accuracy. The idea of solving this difference equation is to express (or to solve) u_i^{n+1} at time step $t = n+1$ in terms of the known values or data u_i^n and u_i^{n-1} at two previous time steps $t = n$ and $t = n - 1$.

Solving the wave equation (3.25) with the initial condition

$$u(x, 0) = e^{-x^2}, \tag{3.29}$$

and wave reflection boundary conditions at both ends $u(-L, t) = u(L, t) = 0$, we have the solution shown in Figure 3.2. We can see that the initial profile is split into two traveling waves: one travels to the left and one to the right.

3.2.3 Parabolic Equation

For the parabolic equation such as the diffusion or heat conduction equation

$$\frac{\partial u}{\partial t} = \frac{\partial}{\partial x}\Big(D\frac{\partial u}{\partial x}\Big), \tag{3.30}$$

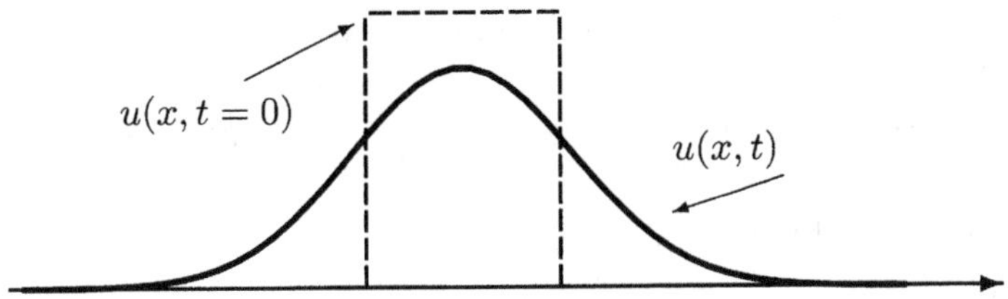

Figure 3.3 The 1D time-dependent diffusion equation: $u_t - \kappa u_{xx} = 0$.

a simple Euler method for the time derivative and centered second-order approximations for space derivatives lead to

$$u_j^{n+1} = u_j^n + \frac{D\Delta t}{h^2}(u_{j+1}^n - 2u_j^n + u_{j-1}^n). \tag{3.31}$$

From the application of von Neumann stability analysis by assuming $u_j^n = \xi^n e^{ikhj}$, the above equation becomes

$$\xi = 1 - \frac{4D\Delta t}{h^2} \sin^2\left(\frac{kh}{2}\right). \tag{3.32}$$

The stability requirement $|\xi| \leq 1$ leads to the constraint on the time step,

$$\Delta t \leq \frac{h^2}{2D}. \tag{3.33}$$

This scheme is thus conditionally stable.

For simplicity, we consider a 1D heat conduction equation $u_t = \kappa u_{xx}$ with an initial condition

$$u(x, 0) = [H(x - L/5) - H(x)]$$

where $H(x)$ is a Heaviside function.

$$H(x) = 1, \quad x \geq 0, \qquad H(x) = 0 \quad x < 0.$$

The evolution of the temperature profile is shown in Figure 3.3 where the initial profile is plotted as a dashed curve. We can see that the profile is gradually smoothed out as time increases and this is the typical behavior of the diffusive system. The time-stepping scheme we used limits the step size of time as larger time steps will make the scheme unstable. There are many ways to improve this, and one of the most widely used schemes is the implicit scheme.

To avoid the limitation due to very small time steps, we now use an implicit scheme for time derivative differencing, and thus we have

$$u_j^{n+1} - u_j^n = \frac{D\Delta t}{h^2}(u_{j+1}^{n+1} + 2u_j^{n+1} + u_{j-1}^{n+1}). \tag{3.34}$$

Applying the stability analysis, we have

$$\xi = \frac{1}{1 + \frac{4D\Delta t}{h^2}\sin^2\frac{kh}{2}}, \tag{3.35}$$

whose norm is always less than unity ($|\xi| \leq 1$). This means the implicit scheme is unconditionally stable for any size of time steps. That is why implicit methods are more desirable in simulations. However, there is one disadvantage of this method, which requires more programming skills because the inverse of a large matrix is usually needed in implicit schemes.

3.2.4 Elliptical Equation

In the parabolic equation, if the time derivative is zero or u does not change with time $u_t = 0$, then we reach a steady-state problem that is governed by the elliptic equation. For the steady-state heat conduction problem, we generally have the Poisson problem,

$$\nabla \cdot [\kappa(u, x, y, t)\nabla u] = f, \tag{3.36}$$

If κ is a constant, this becomes

$$\nabla^2 u = q, \qquad q = \frac{f}{\kappa}. \tag{3.37}$$

There are many methods available to solve this problem, such as the boundary integral method, the relaxation method, and the multigrid method. Two relevant methods are the long-time approximation of the transient parabolic diffusion equations, and the other includes the iteration method.

The long-time approximation method is essentially based on the fact that the parabolic equation

$$\frac{\partial u}{\partial t} + \kappa\nabla^2 u = f, \tag{3.38}$$

evolves with a typical scale of $\sqrt{\kappa t}$. If $\sqrt{\kappa t} \gg 1$, the system is approaching its steady state. Assuming $t \to \infty$ and $\kappa \gg 1$, we then have

$$\nabla^2 u = \frac{f}{\kappa} - \frac{1}{\kappa}u_t \to 0. \tag{3.39}$$

In the case of $\kappa = \text{const}$, it degenerates into the above steady-state equation (3.36) because $u_t \to 0$ as $t \to \infty$. This approximation becomes better if

$\kappa \gg 1$. Thus, the usual numerical methods for solving parabolic equations are valid. However, other methods may obtain the results more quickly.

The iteration method uses the second-order scheme for space derivatives, and Eq. (3.37) in the 2D case becomes

$$\frac{u_{i+1,j} - 2u_{i,j} + u_{i-1,j}}{(\Delta x)^2} + \frac{u_{i,j+1} - 2u_{i,j} + u_{i,j-1}}{(\Delta y)^2} = q. \tag{3.40}$$

If we use $\Delta x = \Delta y = h$, then the above equation simply becomes

$$(u_{i,j+1} + u_{i,j-1} + u_{i+1,j} + u_{i-1,j}) - 4u_{i,j} = h^2 q, \tag{3.41}$$

which can be written as

$$\mathbf{Au} = \mathbf{b}. \tag{3.42}$$

This equation can be solved using the various methods such as the Gauss-Seidel iteration.

EXERCISES

3.1 Use a four-step Runge-Kutta method to solve the following equation:

$$\frac{dy}{dx} = (x^4 + e^x)e^{-y(x)}, \quad y(0) = 0.$$

Then, compare your solution with the analytical solution

$$y(x) = \ln\left(e^x + \frac{x^5}{5}\right),$$

at $x = 1$.

3.2 Sometimes, seemingly simple equations may lead to complex behavior. For example, the so-called logistic differential equation

$$dw/dt = aw - bw^2, \quad a, b > 0,$$

can lead to chaotic behavor. Using an explicit Euler scheme to solve this equation and try to vary a and b and see what happens.

3.3 The Lorentz equations

$$\dot{u} = 10(v - u), \quad \dot{v} = u(20 - w) - v, \quad \dot{w} = uv - 8w/3,$$

can produce a butterfly-shaped attractor if you plot u versus w. Write a simple program to solve this ODE system and plot the trajectory of u vs w evolving with time.

3.4 Implicit schemes work better than explicit schemes. One of the well-known schemes is the Crank-Nicolson method, which uses the central differences. For the simple heat conduction equation $u_t = \lambda u_{xx}$, this method can be written as

$$\frac{u_j^{n+1} - u_j^n}{\Delta t} = \frac{\lambda}{2}\Big(\frac{u_{j+1}^{n+1} - 2u_j^{n+1} + u_{j-1}^{n+1}}{(\Delta x)^2} + \frac{u_{j+1}^n - 2u_j^n + u_{j-1}^n}{(\Delta x)^2}\Big).$$

Show that this scheme is numerical stable and the resulting matrix system is tridiagonal.

3.5 Using an explicit finite difference method to solve the following 2D partial differential equation

$$\frac{\partial u}{\partial t} = D\Big(\frac{\partial^2 u}{\partial x^2} + \frac{\partial^2 u}{\partial y^2}\Big) + \gamma u(1-u).$$

Visualize your solutions as a 3D landscape for the case of $D = 0.2$ and $\gamma = 0.5$ and observe what happens.

REFERENCES

1. Bathe, K. J., *Finite Element Procedures in Engineering Analysis*, Prentice Hall, New Jersey (1982).
2. Cook, R. D., *Finite Element Modeling For Stress Analysis*, Wiley & Sons, New York (1995).
3. Langtangen, H. P., *Computational Partial Differential Equations: Numerical Methods and Diffpack Programming*, Springer, Heidelberg (1999).
4. Press, W. H., Teukolsky, S. A., Vetterling, W. T., and Flannery, B. P., *Numerical Recipes in C++: The Art of Scientific Computing*, 2nd Edition, Cambridge University Press, Cambridge (2002).
5. Strang, G. and Fix, G. J., *An Analysis of the Finite Element Method*, Prentice-Hall, Englewood Cliffs, New Jersey (1973).
6. Yang, X. S., *Introduction to Computational Mathematics*, World Scientific Publishing, Singapore (2008).
7. Zienkiewicz, O. C. and Taylor, R. L., *The Finite Element Method*, Vol. I/II, 4th Edition, McGraw-Hill, New York (1991).

CHAPTER 4

TEACHING MATHEMATICAL MODELING IN TEACHER EDUCATION: EFFORTS AND RESULTS

THOMAS LINGEFJÄRD

University of Gothenburg, Sweden

Can mathematical modeling serve as a vehicle for the learning of mathematics? Mathematical modeling is mentioned in most countries' curricula, yet it is not as valued either in the teaching of school mathematics or in most of the teacher training programs the author knows of. Nevertheless, in mathematical modeling processes, the mathematics really comes into play and many mathematical modeling exercises show the importance of mathematics in many different ways. The purpose of this chapter is to show that also in mathematics education there are many different modeling activities that could be used.

4.1 INTRODUCTION

There are many reasons to give a course on mathematical modeling to lower secondary and upper secondary prospective teachers. To start with, the teach-

Edited by Xin-She Yang

ing of mathematical modeling might allow students to take more responsibility over their own learning and the teaching might be changed into a more undirected teaching manner.

> The heart of applied mathematics is the injunction "Here is a situation; think about it." The heart of our usual mathematics teaching, on the other hand, is: "Here is a problem; solve it" or "Here is a theorem; prove it." We have very rarely, in mathematics, allowed the student to explore a situation for himself and find out what the right theorem to prove or the right problem to solve might be. [26, p. 328]

Second, the concept of models exists in our daily language. *Model* and *modeling* are common expressions with many seemingly different meanings. We are introduced to new car models that we are supposed to feel attracted to, to picture ourselves in possession of the new car. Architects use models of a landscape or a house to illustrate a product they want to sell. In the fashion industry, a model is a person who wears clothes that other people watching can imagine themselves wearing. Fashion models are selected because they possess certain idealized human characteristics, which change from time to time but always refer to ideals such as thinness, height, skin color, and attitude. Children use many models of reality in their toy cars, dolls, trains, and so forth. All modeling activities have at least two aspects in common: They use a model in order to think about or introduce the related reality, and the model is something more or less idealized or simplified.

Third, the concept of mathematical modeling is mentioned in most countries' curricula as something which should be taught at least in secondary schools.

> The Swedish school should, in its teaching of mathematics, strive for that students in projects and in group discussion develop their conceptual capacity and that they learn how to formulate and motivate different methods for solving mathematical problems.
> They should also develop their aptitude to give shape to, refine, and use mathematical models together with a critical estimation of the model's conditions, possibilities and limitations. [27, p. 2, my translation]

A national statement like this addresses many questions, including how to prepare prospective secondary school mathematics teachers to function in an environment where teaching and learning are characterized by processes and activities. The statement further gives a reflection in how to look upon teaching and learning in mathematics. More and more policy documents around the globe talk about quality in mathematical knowledge and of mathematical competencies. For example, Ref. [28] emphasizes that our thinking on learning maybe rooted simultaneously in two different conceptual domains:

> "On the one hand, there is the view of learning as an *acquisition* of some *property*: and on the other hand there is the idea of learning as *becoming a participant* in a certain *practice* or *discourse*". [28, p. 120, italics in original].

For many teachers mathematical modeling is probably easy to confuse with problem solving. Problem solving in mathematics can be seen as applying to either specific problems within mathematics, so-called pure mathematics, or to problems outside mathematics where the field often is called applied mathematics. The process of mathematical modeling also has a variety of definitions. As used in secondary mathematics, it ordinarily entails taking a situation, usually one from the real world, and using variables and one or more elementary functions that fit the phenomena under consideration to arrive at a conclusion that can then be interpreted in light of the original situation. Ref. [28] argued that we seldom challenge students to study a situation and try to make a model of it for analyzing the situation.

A carefully organized course in mathematics is sometimes too much like a hiking trip in the mountains that never leaves the well-worn trails. The tour manages to visit a steady sequence of the "high spots" of the natural scenery. It carefully avoids all false starts, dead-ends, and impossible barriers, and arrives by five o'clock every afternoon at a well-stocked cabin. The order of difficulty is carefully controlled, and it is obviously a most pleasant way to proceed. However, the hiker misses the excitement of risking an enforced camping out, of helping locate a trail, and of making his way cross-country with only intuition and a compass as a guide. "Cross-country" mathematics is a necessary ingredient of a good education. (26, p. 329)

Prospective teachers need to understand a great variety of topics and approaches in mathematics. Today these topics include concepts, principles, methods, and procedures that were not traditionally part of school or college mathematics but that many secondary school students may now address very well through the use of computers and graphing calculators. Applied mathematics as a field, and the process of mathematical modeling in particular, is one part of the mathematical curriculum that may be broadened and enhanced through the use of technology. Henry Pollak's vision of cross-country mathematics may very well be helped by the presence of technology, since so much of the tedious routine calculations today can be done by the technology and much more open-ended problems can be modelled.

For at least three decades, many authors with different perspectives have discussed the role of applications and modeling in the curriculum. The 1979 yearbook of the National Council of Teachers of Mathematics (NCTM), *Applications in School Mathematics*, contains articles illustrating the variety of those perspectives [22]. In recent years, interest in mathematical modeling has increased among mathematics educators. The *Curriculum and Evaluation Standards for School Mathematics* (NCTM, 1989), for example, stressed its importance [23]. Given the potential value of technology for enhancing learning, students can undertake some realistic modeling problems and thereby develop their ideas about and their understanding of mathematics. The intent of the NCTM's recommendations regarding the curriculum is that through a consideration of real-world problems—problems that capture students' interest and that might readily arise in daily life—students will gain both an appreciation of the power of mathematics and some essential mathematical skills. In the report *Heeding the Call for Change* (Steen, 1992), published by the Mathematical Association of America, a group of collegiate mathematics educators suggested that "the key [in selecting such problems] is to have the contexts relate to students' interest, daily life, and likely work settings" [30, p. 100].

There are obviously various arguments as to why applications and modeling belong in the curriculum. Blum and Niss ([1], p. 5) define five arguments that I have termed as follows: formative, critical, practical, cultural, and instrumental.

Applications and modeling should be part of the mathematics curriculum in order to:

1. Foster among students general *creative* and *problem solving* attitudes, activities and competences.

2. Generate, develop and qualify a *critical potential* in students towards the use (and misuse) of mathematics in extra-mathematical contexts.

3. Prepare students to being able to *practice applications and modeling*—in other teaching subjects; as private individuals or as citizens, at present or in the future; or in their professions.

4. Establish a representative and balanced *picture of mathematics*, its character and role in the world. Such a picture must encompass all essential aspects of mathematics, and the application of mathematics and mathematical modeling in other areas *do* form one such aspect.

5. Assist students' *acquisition* and *understanding of mathematical concepts*, notions, methods, results and topics, either to give a fuller body to them, or to provide motivation for the study of certain mathematical disciplines. [1, pp. 23–24]

4.2 THEORETICAL FRAMEWORKS CONNECTED TO MATHEMATICAL MODELING

There are of course almost unlimited numbers of different theoretical frameworks which could be used to analyze and evaluate learning when students are involved in a mathematical modeling process. Since nearly every mathematical modeling process is accompanied or supported by technology, there are good reasons why we could start with sociocultural theory. One of the fundamental concepts of sociocultural theory is the claim that the human mind is mediated. Vygotsky [32] claimed that tools play a significant role for humans' understanding of the world and advocated that humans do not act directly on the physical world without the intermediary of tools. We use symbolic tools or signs, and tools are artifacts created by humans under specific cultural (culture specific) and historical conditions, and as such they carry with them the characteristics of that culture. Tools and artifacts are used as aids in solving problems that cannot be solved in the same way in their absence. Furthermore, they also exert an influence on the individuals who use them in that they give rise to previously unknown activities and previously unknown ways of conceptualizing phenomena in the world.

4.2.1 Instrumental Competence

The rapid development of new technologies manufactures sophisticated, complex instruments that offer quite new challenges and possibilities to their users. Consequently, we think of technology in new and more theoretical ways. When technological tools become more and more sophisticated, the relation between a user and the tool (the instrument) also becomes more and more complicated and sophisticated. That relation might be seen as a manifestation of what the tool offers (affordance) but also of the tool's limits (constraints). Imagine a normal hammer with a hammerhead. The most common use for a hammer is for driving nails, fitting parts, forging metal and breaking up objects. No one would think about a hammer for making a telephone call. So when we see the hammer, we think about it in the frame of affordances and constraints. Gibson [5] defined affordances as all the "action possibilities" latent in the environment, objectively measurable, and independent of the individual's ability to recognize those possibilities.

Obviously, action possibilities related to a tool are dependent on the capabilities of the user and should always be measured in relation to the relevant users. Norman [24] talked about perceived affordance, as distinct from actual affordance. This distinction makes the concept dependent not only on the physical (body related) capabilities of the users but also on their goals, plans, values, beliefs, and past experiences. Effectively, Norman's conception of affordance "suggests" how an object can be interacted with. But when we start to interact with a tool, we also change the way we look at the tool, we adjust the way we understand the tool, and we accordingly adjust the way we use the tool. The concept of instrumental genesis is based on the distinction between artefact and instrument (tool) with the latter having a psychological component (sometimes called scheme), indicating a dialectic relationship between activity and implicit mathematical, or any other type of, knowledge. The activity that employs and is shaped by the use of instruments (instrumented activity) is directed towards the artefact, eventually transforming it for specific aims (instrumentalization):

The subject has to develop the instrumental genesis and efficient procedures in order to manipulate the artefact. During this interaction process, he or she acquires knowledge, which may lead to a different use of it. Similarly, the specific features of instrumented activity are specified: firstly, the constraints inherent to artifacts; secondly, the resources artifacts afford for action; and finally, the procedures linked to the use of artifacts. The subject is faced with constraints imposed by the artefact to identify, understand and manage in the course of this action: some constraints are relative to the transformations this action allows and to the way they are produced. Others imply, more or less explicitly, a prestructuration of the user's action. [6, p. 201]

Let us take an example. A modern technological tool is GeoGebra; see www.geogebra.org. GeoGebra is free and multi-platform dynamic mathematics software for all levels of education that joins geometry, algebra, tables, graphing, statistics and calculus in one easy-to-use package. It has received several educational software awards in Europe and the USA.

When working with a highly complex and sophisticated tool as GeoGebra, the tool itself is a significant part of a complex learning process. We therefore need to bridge the theory of affordances with the perspective of the complex construction of instrumental genesis [31, 3, 6]. In this framework, it is important to distinguish between the utilization of instrumentation—how the tool influences and shapes the thinking of the user—and the mental instrumentalization—where the tool is shaped by the user. A user working with GeoGebra will be affected and behave accordingly in both these ways.

Instrumentation is an evolutionary theoretical construct, and can be characterized as the concept of mental schemes that emerge when users execute a task such as constructing a circle and dragging it around in a dynamical geometry program such as GeoGebra. During that process the user creates some mental schemata. But GeoGebra is also an instrument created with specific utilities that allow the user to engage in activities within the constraints of the artifact. Without much reflective effort, the user would drag the circle around since he or she already knows that GeoGebra allows such flexible manipulations.

Instrumentalization is a psychological process that leads to an internalization of the uses and roles of an artifact; it can be viewed as an organization of the mental schemes, but also includes a personalization and perhaps transformation of the tool, and a differentiation between the complex processes that constitute instrumental genesis and those that are critical for teachers to master [6]. An example of this is when a teacher uses the option to create a new tool in GeoGebra or to save a construction in GeoGebra for later use in her or his teaching of geometry.

The competence Instrumental Genesis occurs when a user with specific knowledge and methods acts on a tool with specific affordances and constraints. The tool brings instrumentation to the user, while the user brings instrumentalization to the instrument. Instrumental genesis can be viewed as occurring in the combination of these two processes. It seems to the author that the presence of dynamic geometry systems has opened up for almost unlimited instrumental genesis to be developed. Consequently, new knowledge is to be developed when students use the tool in their own way.

> One of the main characteristics of the instrumental approach, as we see it, is that it stresses the effort and time that the nontrivial process of instrumental genesis requires. A second important aspect of this approach is the importance of the bilateral relationship between the artifact and the user: while the student's knowledge guides the way the tool is used and in a sense shapes the tool (this is called instrumentalization), the affordances and the constraints of the tool influence the student's problem solving strategies and the corresponding emergent conceptions (this is called instrumentation). In short, the student's thinking is shaped by the artifact, but also shapes the artifact. [9, pp. 205, 206]

4.2.2 The Importance of Variation

Where instrumental genesis or competence covers technology as a whole, another theory is more suitable for dynamic geometry systems where the user can vary different parts of an object. Discernment, variation and simultaneity are the central concepts in the phenomenographic research approach in which learning and awareness are interpreted under a theoretical framework of variation (see for example Ref. [18]). In short, phenomenography is about categorizing the limited number of qualitatively different ways of seeing or experiencing, a phenomenon in a hierarchical fashion. In particular,

> To discern an aspect is to differentiate among the various aspects and focus on the one most relevant to the situation. Without variation there is no discernment.... Learning in terms of changes in or widening in our ways of seeing the world can be understood in terms of discernment, simultaneity and variation [18, 19].

But variation is of course nothing new, and teachers of all times have most likely varied aspects of concepts when teaching different subjects. Kaput [8] phrased it as:

> One very important aspect of mathematical thinking is the abstraction of invariance. But of course, to recognize invariance—to see what stays the same—one must have variation. *Dynamic media inherently make variation easier to achieve.* [8, p. 525, *italics* in the original]

Many technological systems but perhaps most dynamical geometry systems (DGS) are built around the possibility of variation. It is a system where mathematical concepts can be given visual dynamic forms subject to our actions, powerful or not. DGS are a natural experimental ground for one to experience the theory of variation since it has the built-in mechanism that enables us qualitatively different ways of literally seeing a geometrical phenomenon in action. DGS also enables a mathematics teacher to encourage the students to explore mathematical problems in new ways.

Inspired by the old Roman expression *"Repetitio Est Mater Studiorum"* - repetition is the mother of learning—one may be able to summarize what has been said so far about phenomenography as Marton & Trigwell [18] titled their paper: *"Variatio Est Mater Studiorum"*—variation is the mother of learning. Another way to see this is to think about the possibility to learn something as a function of the possibility of variation. All aspects of variation are important when we look at what students experience and learn when working with technology and experimental mathematical investigations.

4.3 MATHEMATICAL MODELING TASKS

Not all mathematical modeling topics need to be correlated to all the criteria in Blum & Niss [1], but it could be nice to have these criteria or other criteria at hand when one select topics or particular problems. Let us look at a simple problem in which we enable technology to help with the calculations. I do not consider this a mathematical modeling problem, but it is a good start for conjecturing, hypothesis, pattern seeking and investigations.

■ EXAMPLE 4.1

The Four Number Challenge Problem: Place any four natural numbers you choose in the first row of an array. You might prefer to use a spreadsheet for this exercise. In the second row, in the first three columns you should write the difference of the two numbers just above and to the right in the first row (the larger minus the smaller). In the fourth column you write the difference of the number above and the one in the first column of that row (again the larger minus the smaller). In other words, each entry is the absolute value of the difference of the two terms in the previous row.

Repeat this procedure for each row, in terms of the numbers in the row just above. Implement it on a spreadsheet by using the command ABS. See Table 4.1. Will every choice of the four numbers you begin with eventually lead to rows of zeros?

Investigation: Can you find four beginning numbers that let you generate more than 10 rows with nonzero values? Can you find four numbers that generates more than 20 rows with nonzero values? What is the underlying structure here? (Note: I am grateful to Jim Wilson, University of Georgia, Athens, Georgia, who showed me this activity many years ago.)

Discussion: Where is this problem present in the list above [1]? I believe that 1) is related. Furthermore, the problem is easy to introduce, explain,

Table 4.1 The four number challenge in Excel, starting with 2, 7, 15, 35.

2	7	15	25
5	8	20	33
3	12	13	28
9	1	15	25
8	14	10	16
6	4	6	8
2	2	2	2
0	0	0	0

and implement. It most likely leads the investigator to making conjectures and testing them (e.g., "Let's try four prime numbers as a start"). I assume that you, who read this, have tried out some values already? A proof that the sequence always goes to a row of zeroes is quite elusive. The same methodology does work for decimal values (subject to some minor issues with rounding).

Expansion could be done by trying with 2 columns (trivial), 3 columns, 5 columns, and/or 6 columns. It appears that the sequence will go to a row of zeros for an even number of columns and oscillates between a row of zeros and ones for an odd number of columns. If the four numbers are related by a function $f(x)$, for some functions the number of nonzero rows can be very limited for a very narrow range of x values.

In terms of theory, I would say that the spreadsheet implementation of this problem enables us to vary and therefore we might see results we otherwise would have missed.

■ EXAMPLE 4.2

If we want students to explore and relate to a larger amount of Blum & Niss' criteria, we need to broaden the concept of mathematical modeling and perhaps collect data or problems from the real world. One issue then is that we will have several more concepts to deal with. Real-life problems easily get very complicated.

Problem: Due to many different factors, it is more and more interesting to use natural gas to heat homes. It is not hard to find figures for weekly gas consumption (m^3) and average outside temperature (°C) for a house before the installation of cavity wall insulation.

The students are provided with data for gas consumption related to outdoor temperature. A rather straightforward problem but one experience from trying this problem with teacher program students is that students easy get lost in a complicated modeling process. We can see this as if the instrumental genesis is examined and that the students have not yet developed efficient procedures in order to manipulate the artifact. Most students I have met do not have any specific problems with finding the first two linear models that are required to solve the problem, but then many students find it difficult to finish the problem because they need a periodic model over one year to illustrate the temperature changes. How do you do that?

Attempt 1: One student: *If I enter the values in CurveExpert and use curve fitting of a model as* $y = a + b\cos(cx + d)$, *and define* $c = 2\pi/12$ *which will force a periodicity of 12 months, then I get* $y = 8 + 4.9\cos(0.62 + \pi/6)$.

Attempt 2: Other students selected a model based on the calendar year, meaning that they ordered the monthly averages of the outside temperature from January to December, resulting in a different output.

In the process of mathematical modeling it is crucial that the modeler relates to how a modeling problem might be encountered within the real world. It seems as if some students learn to do this by instinct which helps them to become careful mathematical modeler.

Another student: *When I see the graph and that the curve is "empty" during the warmest months of the summer, then I realize that it would be very stupid to use gas for heating when it is warmer outside than inside. And I just exclude this interval from my calculation.*

Of course the problem can be analyzed through simple arithmetic. From Table 4.1 we get that the mean temperature 8 months is approximately 7°C, and we also get that the amount of saved gas for that temperature is about 35 m^3 a week. We could guess that the amount of gas saved over a year probably does not exceed 1200 m^3. Further, the information in Table 4.1 shows that 5 months have an average temperature under 7°C so the amount of saved gas is probably not below 700 m^3.

Some students stay out of the modeling process a while, which is a good strategy sometimes.

A third student: *Since I know that the amount of gas saved should be somewhere around 1200* m^3, *I can check my models in order to "fit" the model to the real-world solution, instead of the other way around.*

In my experience, prospective teachers do not always appreciate complex problems as perhaps engineering students do. It seems as if the students' views of open-ended tasks in general are intertwined with their views of the responsibility they have for their learning; similar connections might very well exist between these views and their views about sources of authority. You probably need a critical mind in order to force yourself to sit back and mainly do mental arithmetic before you employ some advance technology in a modeling process. In general, most students seem to very unaware of the concept of authority or where its sources might be. Sometimes they just are swept off by all figures which can be delivered so easily by all sorts of advanced technology. Read more about theories for authority and students believes in computer generated results in Ref. [11], and in Ref. [12]. For more details about this problem, readers can refer to Ref. [13].

■ EXAMPLE 4.3

During the many years I have taught mathematical modeling to prospective teachers, I have also learned that different students appreciate different modeling situations differently. This is of course very logical; we are all different from each other. It is, at least for me, hard to say that there is any winner in that contest, but mathematical modeling within the field of medicine seems rather popular among the students I have met over the years. Somewhat all humans can relate to medical treatment either personally or through some relative.

There are different ways to select mathematical modeling problems in medicine; it can for instance be related to measurement inside medical standard procedures or related to measuring of medical treatment for diseases. I will illustrate with one example, although there are quite many more which I have tried in teaching mathematical modeling over the years.

A Problem From Anaesthesiology: To put it simply, we can say that the function of the heart is to pump blood throughout the body. The blood, in turn, transports oxygen (O_2) from the lungs to various tissues of the body and transport carbon dioxide (CO_2) from the tissues back to the lungs. When constructing a mathematical model of the circulatory system, we also consider it to be a *closed loop*, and assume that the blood is *incompressible*. Consequently the total volume V of blood (measured in liters) in the system is constant.

Naturally, it is important at what rate the blood flows around the circulatory loop. The flow rate is (in principle) measurable (in liters/minute) past any given point in the system. Ordinarily most attention is focused on the heart's condition and the concept *cardiac output CO* is the rate at which blood is pumped out of the heart. The cardiac output of the heart is the product of the **stroke volume** SV—the volume of blood pumped per beat – and the **heart rate** HR—number of beats per minute. Typical values for a 70-kg man are: SV = 70 to 80 cm^3/beat = 0.070 to 0.080 liters/beat, HR = 70 to 80 beats/minute, CO = 5 to 6.5 liters/minute.

To enable a comparison of patients with different body sizes, cardiac output is often considered relative to the body surface area BSA (in square meters). The cardiac index is the ratio CI = CO/BSA measured in litres per minute per square meter. A typical value of CI for a 70 kg man with a body surface area of about 2 m^2 is 2.5 to 3.5 liters/minute/m^2, which is equal to about 5.6 liters/minute for a human male. However, it should be noted that a normal person has a large "reserve capacity" that allows the cardiac output to increase to as much as 25 to 30 liters/minute during strenuous exercise over a short time.

Cardiac output is often monitored during and after surgery (especially in the case of heart surgery). Serial measurements are used to assess the general status of the circulation and to determine the appropriate hemodynamic therapy and estimate its efficacy. Several other useful variables—such as the stroke volume, the left ventricular stroke work index, the systemic vascular resistance, and the stroke index—can be determined once the cardiac output is known.

Cardiac output can be measured by several techniques, all based on the same idea for measuring the flow rate in a fluid loop. A measurable *indicator* is injected into the fluid, and its subsequent *concentrations* at various points in the flow loop are measured. Such a method was first proposed in 1870 by the German physiologist Adolph Fick, who described a means of determining blood flow by measuring overall oxygen intake and content in the blood.

> One determines how much oxygen an animal takes out of the air in a given time.... During the experiment one obtains a sample of arterial and a sample of venous blood. In both the content of oxygen is to be determined. The difference in oxygen content tells us how much oxygen each cubic centimeter of blood takes up in its course through the lungs, and since one knows the total quantity of oxygen absorbed in a given time, one can calculate how many cubic centimeters of blood passed through the lungs in this time [20].

The *indicator dilution method* is a variant of Fick's technique in which a known amount I of an indicator substance is injected into the blood stream and its concentration $C(t)$ (in liters per cubic meter) is measured as a function of time t at a single downstream location. This dilution method was first introduced by the British physician Stewart who, together with a colleague Hamilton, developed the dye solution method and the ***Stewart-Hamilton formula***

$$CO = \frac{I}{\int_0^\infty C(t)dt}, \tag{4.1}$$

which gives the corresponding cardiac output CO (See Ref. [20], p. 1059).

For a much more extended and throughout derivation of the Stewart-Hamilton formula, please see Ref. [14]. Let me just briefly point out the mathematical complexity in the standard procedures of cardiac output. The Stewart-Hamilton formula says simply that what goes in (at the injection site) must eventually be measured at the downstream sensor site.

Because a definite integral of a positive-valued function gives the area under its graph, the Stewart-Hamilton formula says that the equation is that the cardiac output equals the quantity of the indicator injected divided by the area under the concentration-versus-time curve $C = C(t)$ in the tC-plane. At this point in the analysis, the (constant) parameters CO and I, the variables t and C, and the Stewart-Hamilton formula relating them constitute a simple mathematical model for the process of circulation and dilution that ensues upon the injection of the indicator into the circulatory system.

Complications in the Model: It should be noted that today the dye solution method has been almost entirely replaced by variants of a *thermodilution method.* Originally (around 1954) the thermodilution method used an iced or room temperature solution of salt or dextrose in water. Today the method uses a small heating thermistor on the Swan-Ganz catheter (a lung artery catheter). The temperature $T_B(t)$ of the blood at the sensor is measured (rather than the injectate concentration), and the simple Stewart-Hamilton formula discussed above is replaced with the formula [20]

$$CO = \frac{KI(T_B - T_I)}{\int_0^\infty T_B(t)dt}, \tag{4.2}$$

where $T_B - T_I$ is the initial blood-injectate temperature difference ($T_B = T_B(0)$, and K is an empirical constant depending on the catheter size, specific heat and volume of the injectate, and the rate of injection.

This mathematical modeling assignment concentrates the attention towards the original Stewart-Hamilton formula and method. Here we see that a significant complication results from the fact that the circulatory system is a closed loop. Before the indicator concentration curve returns to zero (i.e., when the entire indicator has passed the censor), the indicator concentration exhibits a secondary peak due to recirculation. There are two ways to evaluate cardiac output in the presence of recirculation. One could develop theoretical equations to account explicitly for recirculation, but this approach would require a detailed analysis of the indicator washout curve, rather than simply finding the area under the curve. A simpler and more effective method to account for recirculation is to remove its effect from the observed indicator "washout curve."

The way the circulatory system washes out drugs and anaesthetics from the tissues by means of the blood flow is quite similar to the way a muddy bathtub can clear itself. The mud can be cleaned out by running water in from the tap and draining water out of the tub simultaneously at the same rate of flow. For simplicity, let us assume that the bath water in the tub is constantly stirred so that the concentration of the mud is always uniform. If $Q(t)$ denotes the amount (kg) of mud in the tank at time t, then the concentration at time t is given by $c(t) = Q(t)/V$ (kg/liter) where V is the (constant) volume of bath water in the tub. Hence the change dQ in Q during the short time interval

dt is given by

$$dQ = -rc(t)dt = -r \cdot Q/V \cdot dt = -kQdt. \tag{4.3}$$

Thus $Q(t)$ satisfies the simple differential equation

$$\frac{dQ}{dt} = -kQ, \tag{4.4}$$

with the familiar exponential decay solution

$$Q(t) = Q_0 e^{-kt}, \tag{4.5}$$

where $k = Q/V$ and Q_0 is the initial amount of mud in the tank.

For the purpose of a preliminary analysis of the indicator dilution curve, let us assume that the initial decreasing part of the indicator dilution curve–before recirculation sets in—is similarly exponential in character. Then suppose this dilution curve is plotted on semi-log paper—paper on which the vertical (Q) scale is logarithmic and the horizontal (t) scale is linear. This exponentially decreasing part of the curve then looks like

$$\ln Q = \ln Q_0 - kt. \tag{4.6}$$

Thus the initial "down stroke" is a straight line on this semi-log plot. With the help of suitable software or modern graphing calculators, we may even use measured values of the concentration to fit an exponential curve to the washout part.

With this background, a possible examination task in a mathematical modeling course could be the following: The cardiac output as monitoring devices present it is normally traced out on a paper slip where the paper shows the change in dye concentration as a *deflection* from zero. Normally the measured CO is also printed on the paper.

Student task: Calculate the cardiac output for the patient whose measured data are present in Table 4.2. The dye injection was 5.68 mg. Observe that 55 mm of deflection equals a change in dye concentration of 5 mg/liter.

This is a mathematical modeling situation with a richness of mathematical objects and methods, with mathematical depth and exploring possibilities. For a more extensive and detailed discussion of student's attitudes, attempts, reactions, and accomplishments when handling the cardiac output assignment: please see Ref. [11]. Please also see Ref. [16] for a study of how students handle the treatment of asthma in a mathematical modeling process. Please read more about this problem in Ref. [11].

Besides medicine, the environmental issue is often of high importance for students. Therefore I have decided to select a problem from that area.

EXAMPLE 4.4

Imagine a small lake. Although any lake is receiving and losing water in different ways, we simplify the situation and say that water flows in through a stream A and out through a stream B. Imagine that A and B are at opposite sides of the lake.

At a certain time of the day, as a result of a road accident, a petrol truck overturns and spills a toxic chemical into the stream A at a position X. Thirty minutes later the police and emergency services have brought the situation under control, and an unknown amount z (m^3) of the toxic chemical has leaked into the lake. Develop a mathematical model that you can use to predict the concentration of the pollutant in the lake at any time and use it to estimate (for a range of possible initial pollution amounts z):

1. The maximum pollution level in the lake and the time at which the maximum is reached.

2. The time it will take for the pollution to fall below the safe level of 0.05%.

3. How will your results be affected if a constant rain starts at the same time as the accident? The rain covers the whole geographic area.

Table 4.2 A patient's cardiac output, measured over time.

Time (sec)	Deflection (mm)	Time (sec)	Deflection (mm)
0	0	1	5
2	20	3	50
4	88	5	115
6	122	7	118
8	100	9	80
10	66	11	53
12	41	13	35
14	29	15	24
16	20	17	17
18	15	19	13
20	12	21	13
22	14	23	15
24	16	25	18

Here we have an open and therefore demanding mathematical modeling situation, with not much data given. Instead it has to be invented or gathered by the students. With an open mind and curiosity, they must investigate what kind of accident there was and what amount of toxic substance could possibly leak out in 30 minutes. Students must set up conjectures and a hypothesis and combine their efforts into a mathematical model that will be quite complicated.

The resulting discussions took place with students in groups or in pairs since the examinations were integrated in the process of learning. The following quotations illustrate different results from this exam.

Student 1: *My data is: The flow in and out =0.1 m^3/s, equal to 6 m^3/min. My assumption is that the flow out of the lake will be unchanged, but the flow into the lake increases ($z/30$ m^3/min) during the time the toxic pollution pours into the lake. A small enlargement of the lake when the pollution flows in, but no increased outflow.*

Student 2: *When the toxic pollution has poured into the lake through A, no polluted water has yet managed to flow out of the lake through creek B. After 30 minutes fresh water begins to flow into the lake again through A while polluted water starts to flow out of the lake through creek B. Inflow = 6 + $z/30$ m^3/min = Outflow... then there is 19 m^3 toxic chemical pollution in the lake after 30 minutes. Reasonable, since I thought that 20 m^3 had leaked out of the petrol truck. The missing m^3 probably have disappeared between X and A, or in the lake.*

Student 3: *Inflow = 6 + $z/30$ = Outflow. I have added the original flow of water to and from the lake to the toxic leak from the petrol truck evenly spread out over the thirty minutes. I have chosen to keep the volume of the lake constant, so the Outflow also increases with $z/30$.*

The first two student responses are based on an interpretation of how flow and water level are affected by the discharge. Many of the students described the alteration as a "tidal wave," an additional amount of fluid like a pulse through the system. That led to difficulties when the mathematical model and its behavior are evaluated. Since this line of reasoning was common in the group, we regarded it as a result of the discussions among students in the group during the modeling process. The third student is one of the few students who expressed a more reliable reasoning depending on assumptions made in the modeling process. The instructors' concluded that this student was one of the strong students in the group.

When students respond to open mathematical modeling problems, several components of that student's view of mathematics become visible. Fischbein [14] mentions three different components of mathematics as a human activity: the formal aspect, the algorithmic component, and intuition. In a mathematical modeling process the formal aspect could be the student's knowledge of required mathematical theories, the algorithmic component could be the skills needed in the solving procedures, and the intuition component could be the possibility for the student to validate a complicated mathematical model. There are obviously both interactions and conflicts between these aspects and components in a students' mathematical modeling activity.

There is also a risk that some students might work too hard in an ambition to make the mathematical model "complete," thereby aiming that it should describe and control every single detail of the pollution accident:

Student 4: *The pollution in the lake is affected the most by the size of the flow of toxic chemicals from the petrol truck, and that in turn is related to the cross-sectional area of the aperture.*

This statement was written by a mathematical student, who constructed her mathematical model at a fairly sophisticated level. Her model was detailed over of how the geometrical properties of the aperture in the petrol truck direct the pollution of the lake, which in turn prevented her from making a free validation of the model. This student also divided the lake into 18 discrete zones with a certain cross-sectional area. That detailed solution resulted in the following statement concerning the concentration of pollution in the lake:

Student 4: *The concentration is directly proportional to the cross-sectional area, which yields the highest concentration at the inflow and outflow positions in the lake.*

The impossible aim for such a detailed mathematical model that it has a full correlation with reality, and the search for a "general" algorithmic solution seems to have annihilated this student's intuitive interpretation. This phenomenon is balanced by the following summary by another student, expressing a more balanced view when looking back on how the results of the mathematical model would be affected if a constant rain starts:

Student 5: *When the inflow becomes larger than the outflow, the result will be a rise in the surface level in the lake. The meaning of this is hard to interpret. Will the surface of the water attract more chemicals and then how much? What is a lake's potential to store more water? If the surface level rises, it might result in new streams of outflow. When validating my mathematical model against all assumptions, I realize that my mathematical results may in fact be doubtful.*

Discussion: One conclusion from the coursework is that also when students solve a mathematical modeling problem using data points generated from their "own model," they can yet become confused. Sometimes led by a desire to deliver a sophisticated solution, they become somewhat blind, almost unable to see simple, elegant solutions.

According to Ref. [4], there are situations in which the intuitive understanding prevents or otherwise disturbs formal understanding. However, the instructors argued that they saw that the formal component of the students' mathematical ability sometimes creates an obstacle for the intuitive interpretation of the process that is modeled. Obviously it is essential to find a good balance between solid formal knowledge and insight in how to use this formal knowledge and how to develop a good intuitive capacity. Perhaps this is extra important for prospective teachers, who should be trained to guide their own students towards the goals proposed in the mathematics curriculum.

One result of an open-ended problem as this is that students will come up with different solutions and by sharing these solutions with each other they will become more experienced in the mathematical modeling process. Readers can refer to Ref. [7] for more details.

■ EXAMPLE 4.5

Let me show you one last problem where the possibility to vary dynamical objects led to a surprising result. This is, once again, a mathematical modeling exercise entirely inside mathematics. The possibility of visualizing mathematical concepts and experimenting with numerical and graphical tools allow us to go beyond traditional deductive reasoning methods and to use geometrical modeling for solving geometrical problems. The problem, *Walters Theorem,* is a rather famous problem, since it, in fact, encouraged a ninth grader to discover new mathematical knowledge.

Walter's Theorem: Walter's Theorem is named after Marion Walter, professor of mathematics at the University of Oregon. It seems as if it first was presented in the *Mathematics Teacher* (November 1993) in *Reader Reflections* [2]. The theorem says:

> Take any triangle and trisect the sides. Connect the trisection points to the opposite vertices. The resulting hexagon has an area equal to one-tenth the area of the original triangle.

Two and a half year later, in the May 1996 issue of the *Mathematics Teacher*, Walter's Theorem was mentioned again. This time it was in an article about *Morgan's* theorem [33]. In the article, the 9th grader Ryan Morgan's mathematical investigations were described. The article told that Ryan Morgan's mathematics teacher (Frank Nowosielski) presented Walter's Theorem to his class in the fall of 1993. The school bestows the students with computers and the *GSP* for the investigation.

Ryan Morgan was interested in investigating what would happen if the sides of the triangles were partitioned into more than three congruent segments (see Figure 4.5). Ryan and his teacher called the process "n-secting", and Ryan experimented with different n-sections using *GSP* [33, p. 420]. Inspired by Ryan Morgan's results, we challenged a group of prospective mathematics teachers at the University of Gothenburg with Walter's Theorem in 2001. The students were asked to investigate what happens with the ratio between the inner constructed hexagon and the outer triangle if the outer triangle's sides were n-sectioned, where n is odd. See [15, pp. 123–124] for a description of the problem and of student's accomplishments.

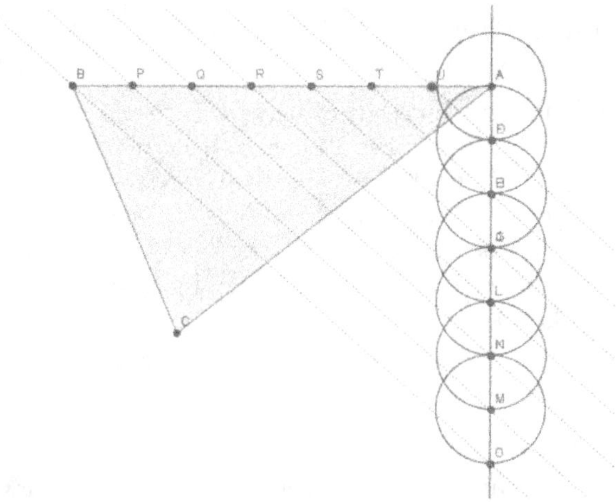

Figure 4.1 A method for seven sectioning a triangle side with GeoGebra.

Results and conclusions: Not all students managed to reach the solution of Ryan Morgan. But most students got fascinated with the problem and in the connections they discovered:

When I divided every side in the triangle in n sections, the relation between the two areas becomes surprisingly $(9n^2 - 1)/8$. *This is amazing, but I cannot prove it.*

DGE in a way invites the user to prove things, yet a strict formal concept of proof seems not to be possible in dynamic geometry software. As King and Schattschneider (1997) put it, "While dynamic geometry software cannot actually produce proofs, the experimental evidence it provides users with produces strong conviction, which can motivate the desire for proof" (Ref. [10], p. xiii). In a complicated modeling process, it is also quite easy to almost get lost and perhaps unable to see what is expected:

I have really investigated a large variety of triangles, but I still don't see any relation. Yes, the larger triangle is larger and larger compared to the hexagon when n-secting more and more. But what's the big deal with that?

An important part of this inquiry based teaching experiment was that there was a:

1. possibility to learn—an open investigation.

2. possibility to use the technological tool, GPS, for the investigation.

3. possibility to seek patterns and making conjectures.

The surprising result illustrating the connection between algebra and geometry can be illustrated by any curve fitting tool or by calculation by hand once you have the ratios. I challenge you, dear reader, to find them by for instance using the free software GeoGebra. The work of Ryan Morgan did actually lead to that he was invited to present the results at a mathematics colloquium at Towson State University in 1994. Interested readers can refer to [15] for more details.

■ EXAMPLE 4.6

My last example has just recently been tried out and I have yet no students' reports to quote from. It is an interesting example of how students can generate the data themselves and thereby perhaps become even more involved in the modeling process, especially since the data comes from inside their own bodies. The responses from the students who tried this a week ago was overwhelming positive.

The target group is once again prospective teachers for secondary school. In my experience, a suitable group size is 3 or 4 students in each group and they should work on the following phenomena under the conditions that one in the group is responsible for filming (by smart phone or a small digital camera) the first part of the exercise which the other students do:

Measure your pulse rate at rest. Then exercise (such as run stairs up and down, run outside, etc.) until your pulse rate has risen to approximately 190 bpm. Stop exercising, rest and measure your pulse rate every minute until it is back to normal. This should be done by more than one person in each group, preferable by all who not are filming.

1. Describe how your pulse rate varies over time with a diagram.

2. Determine average values for every individual in the group for every moment in time, and visualize these in a diagram.

3. Try to determine a mathematical model which describes how the pulse varies with time. Do one or several screen cast movies of your work at the computer.

4. Evaluate the validity of the model

5. Prepare a presentation for the whole student group, in which you present both a film from the gathering of the data and a screen cast generated movie from the mathematical modeling process.

Presentation: The students will be asked to present their activities with a short film (filmed by their cell phones) together with their solution strategies (captured by screen cast technology) as a presentation to the whole class. Part of the activity is an assessment procedure in which the students try to grade their peer's presentations and thereby learn how to grade their future students if involved in similar inquiry based exercises.

Assessment: Both the mathematical correctness in terms of measuring techniques and modeling performance, together with presentation techniques, will be measured. The student's communicative skills will be evaluated. The students will, assisted by the professors, create an evaluation matrix for the assessment of the student's performance on each task and, as well, the students will be asked to report an assessment on their peer's performance.

4.4 CONCLUSIONS

It is evident that several of the examples in this chapter are hard to investigate the way I have described them without technology. The work in technological tools goes very well in hand with learning theories as the variation theory. The possibility to vary and see visually how objects change is an important feature with technology. Noss, Healy and Hoyles [25] explored the relationship between learners' actions, visualizations and the means by which these are articulated in a computer dynamic environment. According to them,

> A central challenge for the design of mathematical learning environments is to make visible that which is normally visible only to the mathematical cognoscenti. (p. 231)

Technology not only has the power to make visible, but even to amplify our dynamic imagination that often contributes significantly to the development of mathematical knowledge. Presmeg [27] remarked that:

> Software facilitates visualization processes ... which may clarify and further the solution to a mathematical problem by providing insight, thus suggesting productive paths for reason and logic." (p. 220)

The possibility to allow students to explore manipulable dynamic objects has a vast potential for the learning of mathematics at several levels in educational systems. The object of learning in these five examples is not entirely to find the correct answer, but more to be part of an investigation. That in turn is related to the fact that mathematics can be thought of and learned in several different ways.

EXERCISES

4.1 Following Example 1, you should try by yourself, which is the purpose of this activity. This one is a nice attempt:

$$1 \quad 60 \quad 168 \quad 367$$

How many nonzero (not all zero) rows are generated for each of these sets of start values?

0 941 2672 5856

0 155 440 964

1000 2550 5400 10642

1000 7000 18037 38338

0 10000000 28394732 62229538

4.2 In Example 2, change the values for heating to more modern figures. Leave the values for heating for the students to find out.

4.3 In Example 3, the only possibility to alter anything in this mathematical modeling activity is to change the values for the patient's Cardiac Output over time measured in mm of deflection or to change the dye injection.

4.4 For details about Example 4, please refer to Ref. [7]. Since this is such an open problem, there are as many paths to walk as there are students. You could of course give details or construct constraints for the initial values which the students have to find.

4.5 In Example 5, what happens if you do not do the n-secting for odd numbers for n? Show that once you have the areas, you can do the curve fitting part by paper or pencil or by Excel.

4.6 Following Example 6, compare the variables of size, weigh, gender, fitness, and age with the models.

REFERENCES

1. Blum, W., & Niss, M. (1989). Mathematical problem solving, modeling, applications, and links to other subjects — state, trends and issues in mathematics instruction. In W. Blum, M. Niss, & I. Huntley (Eds.), *Modeling, Applications and Applied Problem Solving: Teaching Mathematics in a Real Context* (pp. 1–21). London: Ellis Horwood.

2. Cuoco, A., Goldenberg, P., & Mark, J. (1993). Reader reflections: Marion's theorem. *Mathematics Teacher 86*(8), 619.

3. Drijvers, P. (2003). *Learning Algebra in a Computer Algebra Environment.* Doctoral dissertation. Utrecht: Freudenthal Institute.

4. Fischbein, E. (1994). The interaction between the formal, the algorithmic, and the intuitive components in a mathematical activity. In R. Bielhler, R. W. Scholtz, R. Strässer, & B. Winkelmann (Eds)., *Didactics of Mathematics as a Scientific Discipline* (pp. 231–245). Dordrecht: Kluwer.

5. Gibson, J. J. (1977). The theory of affordances. In R. E. Shaw & J. Bransford (Eds.), *Perceiving, Acting, and Knowing: Toward an Ecological Psychology.* Hillsdale, NJ: Lawrence Erlbaum.
6. Guin, D., & Trouche, L. (1999). The complex process of converting tools into mathematical instruments: The case of calculators. *International Journal of Computers for Mathematical Learning,* 3(3), 195–227.
7. Holmquist, M., & Lingefjrd, T. (2003). Mathematical modeling in teacher education. In Q. Ye, W. Blum, S. K. Houston, & Q. Jiang (Eds.), *Mathematical Modeling in Education and Culture ICTMA 10: Applications in Science and Technology* (pp. 197–208). Horwood: Chichester.
8. Kaput, J. (1992). Technology and mathematics education. In D. Grouws (Ed.), *Handbook of Research on Mathematics Teaching and Learning.* (pp. 515–556). New York.: Macmillian.
9. Kieran, C., & Drijvers, P. (2006). The co-emergence of machine techniques, paper-and-pencil techniques, and theoretical reflection: A study of CAS use in secondary school algebra. *International Journal of Computers for Mathematical Learning,* 11(2), 205–263.
10. King, J., & Schattschneider, D. (1997). Preface: Making geometry dynamic. In J. R. King & D. Schattschneider (Eds.), *Geometry Turned On!: Dynamic Software in Learning, Teaching, and Research* (pp. ix–xiv). Washington, D.C.: The Mathematical Association of America.
11. Lingefjrd, T., & Kilpatrick, J. (1998). Authority and responsibility when learning mathematics in a technology-enhanced environment. In D. Johnson & D. Tinsley (Eds.), *Information and Communications Technologies in Mathematics* (pp. 233–236). London: Chapman & Hall.
12. Lingefjärd, T (2000). *Mathematical Modeling by Prospective Teachers Using Technology.* Ph.D. Thesis at University of Georgia.
13. Lingefjärd, T. & Holmquist, M. (2001). Mathematical modeling and technology in teacher education—Visions and reality. In J. Matos, W. Blum, K. Houston, S. Carreira (Eds.), *Modeling and Mathematics Education ICTMA 9: Applications in Science and Technology* (pp. 205–215). Horwood: Chichester.
14. Lingefjärd, T. (2002). Mathematical modeling for preservice teachers: A problem from anesthesiology. *The International Journal of Computers for Mathematical Learning* 7(2), pp. 117–143.
15. Lingefjärd, T. & Holmquist, M. (2003). Learning mathematics using dynamic geometry tools. In S. J. Lamon, W. A. Parker, & S. K. Houston (Eds.), *Mathematical Modeling: A Way of Life. ICTMA 11* (pp. 119-126). Horwood: Chichester.
16. Lingefjärd, T. (2009). Challenges with international collaboration regarding teaching of mathematical modeling. In Blomhøj, M. & S. Carreira, (Eds.) (2009). *Different Perspectives on Research in Teaching and Learning Mathematical Modeling.* Proceeding from Topic Study Group 21 at ICME-11 in Monterrey, Mexico. IMFUFA-text no. 461, Department of science, systems and models, Roskilde University.
17. Marton, F., & Booth, S. (1997). *Learning and Awareness.* Mahwah, N.J.: Law Earlbaum.

18. Marton, F., & Trigwell, K. (2000). Variatio Est Mater Studiorum. *Higher Education Research & Development*, 19, pp. 381–395.

19. Marton, F., Runesson, U., & Tsui, A. B. (2004). The space of learning. In F. Marton, & A. B. Tsui, *Classroom Discourse and the Space of Learning* (pp. 3–40). Mahwah, N.J.: Lawrence Erlbaum Associates.

20. Miller, R. D. (Ed.) (1982). *Anesthesia.* New York, N.Y.: Churchill Livingstone.

21. Microsoft Co. *Microsoft Excel.* Stockholm, Sweden. Microsoft Corporation.

22. NCTM (1979). *Applications in School Mathematics (Yearbook – NCTM: 1979).* Reston: NCTM.

23. NCTM (1989). The *Curriculum and Evaluation Standards for School Mathematics.* Reston: NCTM.

24. Norman, D. (1988). *The Design of Everyday Things.* New York: Basic Books.

25. Noss, R., Healy, L. & Hoyles, C. (1997). The construction of mathematical meanings: connecting the visual with the symbolic. *Educational Studies in Mathematics 33*(2), 203–233.

26. Pollak, H. O. (1970). Applications of mathematics. In E. Begle (Ed.), *The Sixty-Ninth Yearbook of the National Society for the Study of Education* (pp. 311–334). Chicago: University of Chicago Press.

27. Presmeg, N. (2006) Research on visualization in learning and teaching mathematics. In A. Gutirrez & P. Boero (Eds.), *Handbook of Research on the Psychology of Mathematics Education*: Past, Present and Future,(pp.205–235), Sense Publishers: Rotterdam/Taipei.

28. Sfard, A. (1997). From acquisitionist to participationist framework: putting discourse at the heart on learning mathematics. In T. Lingefjrd & G. Dahland (Eds.), *Research in Mathematics Education* (pp. 109–136). Report 1998:02. Gothenburg: University of Gothenburg.

29. Skolverket (2000). [Swedish Board of Education].

30. Steen, Lynn A., (Ed.), *Heeding the Call for Change: Suggestions for Curricular Action*, MAA Notes No. 22, 1992.

31. Verillon, P., & Rabardel P. (1995). Cognition and artifacts: A contribution to the study of thought in relation to instrument activity. *European Journal of Psychology of Education*, 10(1), 77–101.

32. Vygotsky, L. S. (1934/1962). *Thought and Language.* Cambridge, MA: MIT Press.

33. Watanabe, T., Hanson, R., & Nowosielski, F. (1996). Morgan's theorem. *Mathematics Teacher, 89*(5), 420–423.

PART II

MATHEMATICAL MODELING WITH MULTIDISCIPLINARY APPLICATIONS

CHAPTER 5

INDUSTRIAL MATHEMATICS WITH APPLICATIONS

ALFREDO BERMÚDEZ[1] AND LUZ M. GARCÍA GARCÍA[2]

[1]Departamento de Matemática Aplicada, Universidade de Santiago de Compostela, Spain
[2]Instituto Español de Oceanografía, Spain

The introduction of computers in the middle of the last century and the continuous and spectacular increasing in their power is a key issue in the scientific and technological revolution of the last decades. Computers have created needs of mathematical expertise in almost all fields. Close interaction between mathematicians, computer scientists and application field experts is necessary to correctly and fully exploit the enormous potential that the computer has brought. This is true not only for mathematical applications to other sciences but for industrial applications as well.

Mathematical Modeling with Multidisciplinary Applications.
Edited by Xin-She Yang

5.1 INDUSTRIAL MATHEMATICS

Mathematical modeling is nowadays a key tool for innovation in almost every industrial sector (see [10] for a wide variety of examples). By using computer simulation-based on mathematical models, industry is able to create new products and processes and introduce them into the market in much shorter times than by using the conventional trial and error experiment-based methodology. Building a prototype is done by making virtual essays in a computer before it is effectively manufactured and is subjected to the tests required by its specifications sheet. Thus, not only is the time to market considerably reduced but also are the development costs, allowing the company to increase its competitiveness.

Having a mathematical model is also a first step to optimize a product or a process. While some degree of improvement can be achieved by manually changing the design parameters and performing a new simulation, optimization theory and algorithms allow us to do that in an automatic way. Defining new design parameters is done in a rational mathematical way by introducing an objective or cost function of them and then computing its gradient to get a descent direction.

In this chapter two industrial applications of mathematical modeling are described. The first one concerns the metallurgical industry, more specifically, obtaining silicon in electric arc furnaces is considered. The second part deals with the numerical simulation of environmental engineering problems arising from studies of water quality in artificial lakes occupying the space left after exploiting a coal open pit mine. Both parts are structured as follows: firstly, a detailed description of the industrial problem is given. Then, appropriate mathematical models are introduced. Next numerical methods are proposed and, finally, numerical results from real situations are shown.

5.2 NUMERICAL SIMULATION OF METALLURGICAL ELECTRODES

In this section we introduce a thermoelectrical model for numerical simulation of electrodes used in metallurgical electric arc furnaces. First we describe the industrial problem motivating the study. Then we introduce the models and the numerical methods for solving them in a computer. Numerical simulations of real industrial electrodes are presented.

5.2.1 The Industrial Problem: Metallurgy of Silicon

Silicon (Si) is the second most abundant element in the Earth's crust after oxygen. In natural form, it can be found mainly as silicon dioxide (silica, SiO_2) and silicates (silicon combined with other elements). In particular, quartz and sand are two of the most common forms.

Silicon is produced industrially by the reduction of silicon dioxide with carbon by a reaction which can be written in a simple way as follows:

$$\mathrm{Si\ O_2} + 2\mathrm{C} \longrightarrow \mathrm{Si} + 2\mathrm{CO}. \tag{5.1}$$

Depending on its purity, silicon has a wide variety of applications:

- Ferrosilicon (silicon steels, it can contain more than 2% of other materials)
- Metallurgical silicon (e.g. silicon-aluminum alloys, it contains about 1% of other elements)
- Chemical silicon (silicones)
- Solar silicon (solar cells)
- Electronic silicon (semiconductors, the purest silicon, "9N" = 99.9999999 of purity)

The reaction (5.1) takes place in reduction furnaces similar to those used for calcium carbide, iron and some others (see Figure 5.1 for a sketch of the overall process). More specifically, silicon is obtained in submerged "arc" furnaces that use three-phase alternating current (see Figure 5.3). A submerged arc furnace is a large furnace loaded with selected raw materials. During the process, these raw materials melt and chemically react. The liquid silicon leaving the furnace is cooled and further processed into an appropriate size depending on different applications. A reference book for silicon metallurgy is [13].

Electrodes are the main components of reduction furnaces. The typical diameter of the electrodes is one or two meters while their length is more than ten meters, their weight being greater than 10 Mt. Electric current enters each electrode through copper contact clamps which completely embrace the column approximately one meter above the charge level. Current goes down, crossing the column length comprised between the contact clamps and the lower end of the column generating heat by the Joule effect. At the tip of the electrode an electric arc is produced, reaching temperatures of about 2500 °C.

Classical electrodes extensively used in industry include *pure graphite*, *prebaked* and *Søderberg* electrodes. The latter are used for ferrosilicon production. Its advantages with respect to pure graphite or prebaked electrodes are that they are built in larger sizes and cost less. However, as the electrode is consumed it has to be slipped, typically 0.5 m per day. New sections of casing are welded at the top when needed and Søderberg paste is replenished in the electrode center. In this way, the steel casing moves with the carbon body so it melts and pollutes silicon: the iron contribution of the electrode casing, in addition to the iron rendered from the carbon ash and the impurities in the quartz, give in all cases an iron content in excess of 1%. This is why they cannot be used to obtain silicon with metallurgical quality (for short, *silicon metal*) to be used, for instance, to produce aluminum-silicon alloys. Accordingly, prebaked electrodes were for many years the only alternative for commercial silicon metal production until the arrival of the ELSA electrode into the market. The ELSA electrode (see Figure 5.2) consists of a central column of baked carbonaceous material, graphite or similar, surrounded by a Søderberg-like paste. There is a steel casing that contains the paste until it is baked.

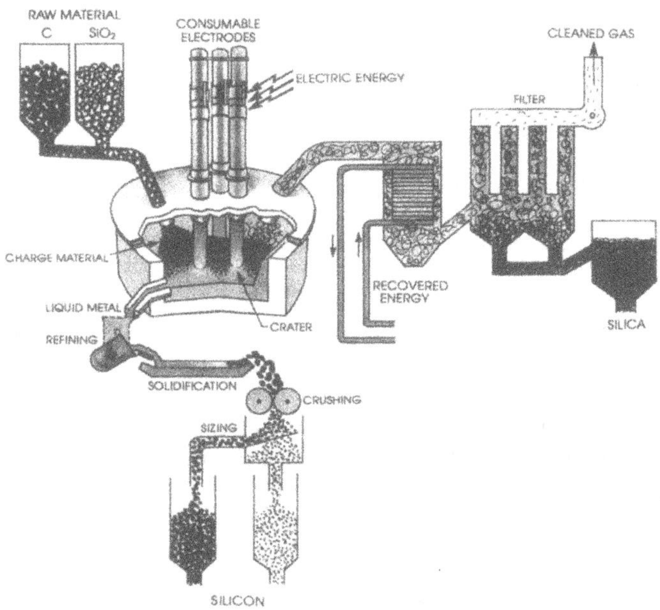

Figure 5.1 Silicon production.

The furnace has two different slipping systems: one for the casing and another one for the central column. The combination of both systems is necessary so as to slip the casing as little as possible and also to carry out the correct extrusion of the carbon electrode with the central column slipping rings. Contrary to Søderberg electrodes, the casing is not necessarily consumed and then it is possible to produce non-polluted silicon with so-called metallurgical quality, which can be used, for instance, to produce aluminum-silicon alloys.

The result is that the furnace operation is similar to prebaked electrodes, but the compound electrode is less expensive. Moreover, unlike Søderberg electrodes, the inside of the casing is absolutely smooth so as to allow the slippage of the electrode. Thus, the casing only acts as an extrusion sleeve.

Another important advantage of the ELSA electrode is the supply of raw materials because there are many more factories in the world making graphite and Søderberg paste than making prebaked electrodes whose transport costs

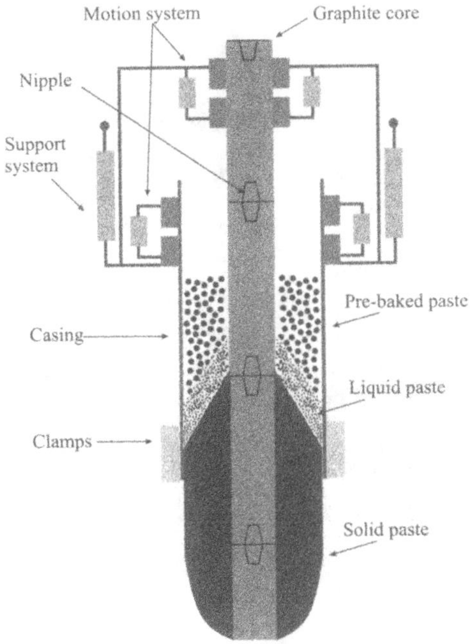

Figure 5.2 The ELSA electrode.

have always been very important. The disadvantage is that slipping velocity is not free as in prebaked electrodes, because the paste has to be baked before leaving the casing so it is necessary to have a minimum period of time between two consecutive slippings. Thus, as we noticed for Søderberg electrodes, baking of the paste is a crucial point in the working of this type of electrode.

More precisely, the paste baking should be a continuous process for a successful operation. The paste is baked in the contact clamps zone between 100°C and 500°C, where it suffers several changes of state. The baking of the paste is closely related to the electrode current and slipping rate. Indeed, the electrode has to be slipped downwards to compensate its consumption on the tip, but slipping has to be done in small increments in order to avoid breakages in the soft paste. During normal operation, the baking zone is stabilized within the lower part of the holder. If the baking zone comes below the clamps it can cause a leakage of soft paste. Then, the lower part of the electrode may slide into the furnace pot, causing a volatile species and the paste to catch fire. Therefore it is important to know the relation between electric current and slipping rate, and other parameters like temperature in the surroundings, temperature of the cooling-water, properties of the paste, etc., which can affect the position of the baking zone.

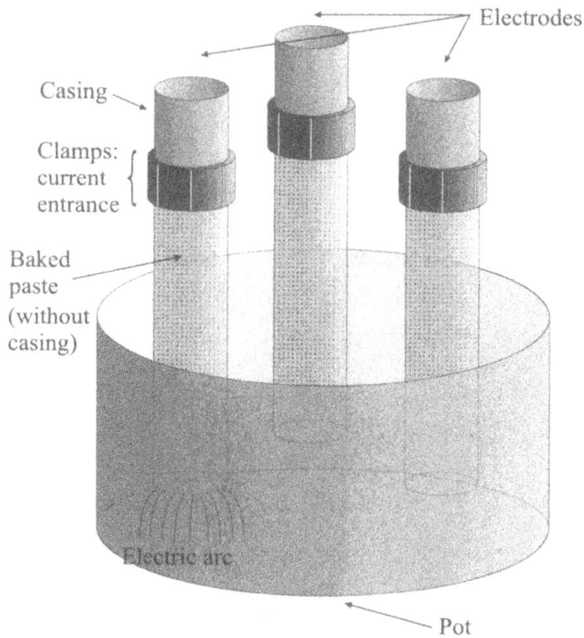

Figure 5.3 Sketch of a reduction furnace.

5.2.2 Mathematical Modeling

Since we are dealing with electromagnetic and heat transfer problems, mathematical modeling of the ELSA electrode needs two coupled submodels. Firstly, we have to compute the electromagnetic field and, more specifically, the distribution of electric current represented by the current density field. This can be done by solving the quasi-static Maxwell equations, which are also called the eddy current model. From this field it is possible to determine the heat released by the electric current at each volume element of the electrode and then to compute the temperature field by solving a heat transfer model. Notice that several nonlinearities come to place arising from the change of state of the paste (Stefan's problem), from the dependency of thermo-physical parameters on temperature and from the radiative heat transfer. Moreover, the electromagnetic and thermal models are coupled together because electric current is the heat source and, conversely, the electric conductivity of materials depends on temperature.

Neglecting the proximity effect, we can consider axisymmetric models as a first step, in order to reduce the computational cost for solving the involved partial differential equations.

Once the temperature is known, it could be also used in a thermo-elasticity model in order to determine the mechanical stresses which give us information on the structural behavior of the electrodes in order to prevent breakages.

Each furnace has three cylindrical electrodes. Electric current enters through halfway each electrode through eight clamps made of copper that completely embrace the column (see Figure 5.3). Inside the clamps, water is flowing for cooling purposes. As electric current goes down, crossing the lower half of the electrode, it generates heat by the Joule effect. At the tip of the electrode an electric arc is produced, reaching temperatures of about 2500 °C.

The paste surrounding the graphite starts baking at 350 °C in the contact clamp zone. From the top of the electrode to this level, the paste is held by a casing made of steel with a thickness of several millimeters.

5.2.2.1 The Electromagnetic Submodel. Neglecting the effects due to the proximity of the other two electrodes, we can assume axisymmetry in the problem and write the equations in cylindrical coordinates on a vertical section Ω of the electrode (see Figure 5.4). For this purpose it is assumed that the input current does not depend on the meridian section and that it remains on this section. Moreover we also include the casing and the contact clamps in the domain Ω.

To calculate the electromagnetic field, we must solve the Maxwell equations. Industrial current alternates with frequency $f = 50$ Hz. In this situation, the time scale for variation of the electromagnetic field is much smaller than the one for variation of temperature. Thus, we may consider the eddy current model to compute the electromagnetic field in the frequency domain, and then the heat source due to the Joule effect is determined by taking the mean value on a cycle.

The eddy current model is obtained from Maxwell's equations (see, for instance, [9]) by neglecting the term involving the electric displacement in Ampère's equation. This can be done because the geometrical length scale in our situation is much lower than the typical wavelength of the current source.

Moreover, assuming the current source is harmonic and all materials have linear electromagnetic behavior, all fields can be written in the form

$$\mathcal{G}(x,t) = \mathrm{Re}(e^{\mathrm{i}\omega t} G(x)), \tag{5.2}$$

where G is a complex field (vector or scalar, depending on the nature of $\mathcal{G}$) called *complex amplitude or phasor* and ω is the angular frequency.

Using this expression in the time-domain eddy current model we get the frequency-domain eddy current model,

$$\mathrm{i}\omega\mathbf{B} + \mathbf{curl}\,\mathbf{E} = 0, \tag{5.3}$$

$$\mathbf{curl}\,\mathbf{H} = \mathbf{J}, \tag{5.4}$$

$$\mathrm{div}\,\mathbf{B} = 0, \tag{5.5}$$

$$\mathbf{B} = \mu\mathbf{H}, \tag{5.6}$$

$$\mathbf{J} = \sigma\mathbf{E}, \tag{5.7}$$

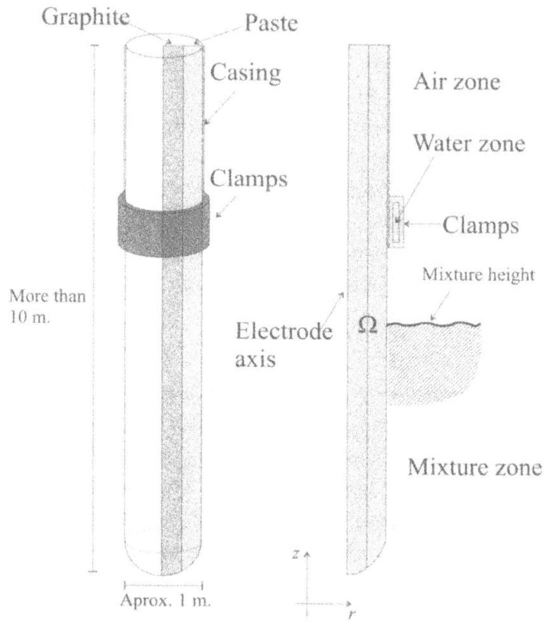

Figure 5.4 Sketch of domain Ω.

where μ is the magnetic permeability, σ is the temperature dependent electrical conductivity, and $\mathbf{B}$, $\mathbf{E}$ and $\mathbf{H}$ are the complex amplitudes associated with the magnetic induction, the electric field and the magnetic field, respectively.

Straightforward computations from (5.3)–(5.7) allow us to obtain a single equation for the magnetic field in conductors, namely,

$$i\omega\mu\mathbf{H} + \mathbf{curl}\Big(\frac{1}{\sigma(x,T)}\,\mathbf{curl}\,\mathbf{H}\Big) = 0, \tag{5.8}$$

where T denotes temperature.

The assumption of axisymmetry of the input current leads to the independency on the angular variable θ of all fields and to a null azimuthal component of the current density, namely,

$$\mathbf{J}(r,\theta,z) = J_r(r,z)\mathbf{e}_r + J_z(r,z)\mathbf{e}_z,$$

where (r,z) belongs to the two-dimensional domain Ω (see Figure 5.4). It is useful to recall the expression of the **curl** of a vector field in cylindrical coordinates:

$$\mathbf{curl}\,\mathbf{F}(r,\theta,z) = \frac{1}{r}\begin{vmatrix} \mathbf{e}_r & \mathbf{e}_\theta & \mathbf{e}_z \\ \frac{\partial}{\partial r} & \frac{\partial}{\partial \theta} & \frac{\partial}{\partial z} \\ F_r & rF_\theta & F_z \end{vmatrix}. \tag{5.9}$$

The above expression for $\mathbf{J}$, together with (5.3), (5.6) and (5.7), imply that the magnetic field only has tangential component:

$$\mathbf{H}(r,\theta,z) = H_\theta(r,z)\mathbf{e}_\theta. \tag{5.10}$$

Using (5.10), Eq. (5.8) reduces to

$$i\omega\mu H_\theta - \frac{\partial}{\partial z}\Big(\frac{1}{\sigma}\frac{\partial H_\theta}{\partial z}\Big) - \frac{\partial}{\partial r}\Big(\frac{1}{\sigma r}\frac{\partial (rH_\theta)}{\partial r}\Big) = 0. \tag{5.11}$$

In the computational domain, cooling water tubes are also included. Taking into account that water is a dielectric material, $\mathbf{J} = 0$ in it. Then Eq. (5.7) must be rewritten in the form

$$\mathbf{J} = \sigma\mathbf{E} \quad \text{on } C, \tag{5.12}$$

$$\mathbf{J} = 0 \quad \text{on } D, \tag{5.13}$$

where D is the region corresponding to water in the domain Ω and $C = \Omega\backslash D$. Hence, Eq. (5.4) becomes

$$\mathbf{curl}\,\mathbf{H} = 0 \quad \text{on } D \tag{5.14}$$

in the water. In order to handle this equation, we can use a so-called penalty method. The idea is to consider water as a conductor but with an electrical conductivity much smaller that the ones for the other conductors involved in the model. In other words, Eq. (5.14) is approximated by

$$-\delta\,\mathbf{E} + \mathbf{curl}\,\mathbf{H} = 0 \text{ on } D,$$

for δ very small compared with σ. In this way, equation (5.11) also holds in the water but with σ replaced by δ.

5.2.2.2 Electromagnetic Boundary Conditions According to the differential operators involved in the model, so-called *essential* and *natural* boundary conditions for the above problem consists of prescribing the values of

a) $\mathbf{H}\times\mathbf{n}$,

b) $\mathbf{J}\times\mathbf{n} = \mathbf{curl}\,\mathbf{H}\times\mathbf{n}$,

respectively, where $\mathbf{n}$ denotes an outward unit normal vector to the boundary. In our case, due to cylindrical symmetry, the first condition means that H_θ is given, which is a Dirichlet boundary condition for partial differential equation (5.11). In order to determine H_θ on the boundary we proceed as follows: by the axial symmetry we know that

$$\mathbf{J}\cdot\mathbf{n} = \mathbf{curl}\,\mathbf{H}\cdot\mathbf{n} = \frac{1}{r}\frac{\partial (r\,H_\theta)}{\partial r},$$

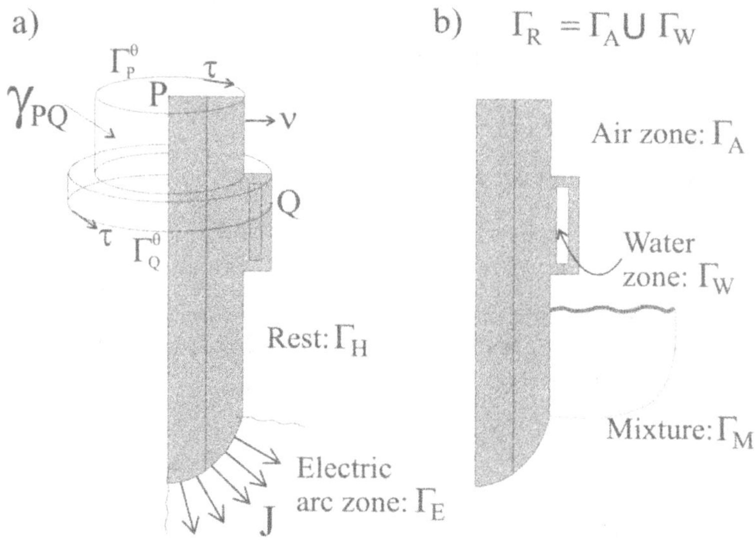

Figure 5.5 Boundary conditions: a) electromagnetic; b) thermal.

where τ is a unit vector tangent to the boundary (see Fig. 5.5).

If we take a piece of boundary Γ_H of the two-dimensional domain, between P and Q (see Figure 5.5) and denote γ_{PQ} the surface generated by its rotation, it is possible to determine $\mathbf{H}$ at a point Q if the intensity of current crossing γ_{PQ}, to be called $i(\gamma_{PQ})$, is given. Indeed, by Stokes' theorem we have

$$\begin{aligned} i(\gamma_{PQ}) = \int_{\gamma_{PQ}} \mathbf{J} \cdot \mathbf{n} \, d\gamma = \int_{\gamma_{PQ}} \mathbf{curl}\,\mathbf{H} \cdot \mathbf{n} \, d\gamma = \int_{\Gamma_P^\theta \cup \Gamma_Q^\theta} \mathbf{H} \cdot \tau \, d\Gamma \\ = \int_{\Gamma_Q^\theta} H_\theta \, r \, d\theta - \int_{\Gamma_P^\theta} H_\theta \, r \, d\theta, \end{aligned}$$

Γ_P^θ and Γ_Q^θ being the boundary curves of γ_{PQ}. Since H_θ does not depend on θ we finally get

$$i(\gamma_{PQ}) = 2\pi \Big((r\,H_\theta)(Q) - (r\,H_\theta)(P) \Big)$$

and therefore

$$H_\theta(Q) = \frac{1}{r(Q)} \left(r(P)\,H_\theta(P) + \frac{i(\gamma_{PQ})}{2\pi} \right).$$

In its turn, the *natural* boundary condition **b)** is suitable for the part of the boundary in contact with the electric arc, Γ_E, because we may assume that the current gets out of the electrode perpendicular to this surface.

5.2.2.3 The Thermal Submodel As previously said, electric current dissipates heat inside conductors that is responsible for increasing the temperature of the electrode. This temperature can be obtained by solving the following partial differential equation (it is the *energy conservation* equation which is also called "heat equation")

$$\rho(x,T)\,c(x,T)\left(\frac{\partial T}{\partial t}+\mathbf{v}(x,t)\cdot\operatorname{grad}T\right)-\operatorname{div}\left(k(x,T)\operatorname{grad}T\right)=\mathcal{Q}(x), \quad (5.15)$$

where ρ is density, c is specific heat, $\mathbf{v}$ is velocity, k is thermal conductivity and $\mathcal{Q}$ is the heat source, in our case due to the Joule effect. Temperature T depends on point x and time t.

Electrode slipping is vertical and serves to make up for the wear suffered at its bottom due to high temperatures and chemical reactions. While this movement is very slow (several centimeters per hour) it is enough to determine that electrodes never are in a thermal steady state. The constitutive materials are restored in the upper zone. All of this allows us to suppose that $\mathbf{v}(x,t)=V(t)\mathbf{e}_z$, where $\mathbf{e}_z$ denotes the unit vector along the axial direction, and that domain Ω does not depend on time.

The heat source is the time average of Joule effect along a cycle, namely

$$\mathcal{Q}(x)=\frac{\omega}{2\pi}\int_0^{\frac{2\pi}{\omega}}\mathcal{J}(x,t)\cdot\mathcal{E}(x,t)\mathrm{d}t.$$

Taking into account the expression of a harmonic field (5.2) and Eq. (5.7) we get (see Exercise 5.4)

$$\mathcal{Q}(x)=\operatorname{Re}(\mathbf{J}(\mathbf{x})\cdot\overline{\mathbf{E}(x)})=\frac{1}{\sigma}|\mathbf{J}(x)|^2=\frac{1}{\sigma}\frac{|\mathbf{curl}H_\theta(x)|^2}{2}.$$

Paste undergoes a change of state (generically denominated as "baking") at a temperature between 100°C and 550°C to be denoted by T_b. Then, Eq. (5.15) must be corrected to take into account the latent heat involved in this process by introducing an enthalpy function. More precisely, the heat transfer equation is rewritten in the form

$$\frac{\partial e}{\partial t}+V(t)\frac{\partial e}{\partial z}-\operatorname{div}\left(k\operatorname{grad}T\right)=\mathcal{Q}, \quad (5.16)$$

where e denotes the enthalpy density which is expressed as a function of temperature by

$$e\in\mathcal{H}(T),$$

where

$$\mathcal{H}(T)=\begin{cases}\int_0^T\rho(s)\,c(s)\,ds, & T<T_b,\\ \left[\int_0^{T_b}\rho(s)\,c(s)\,ds,\int_0^{T_b}\rho(s)\,c(s)\,ds+\rho(T_b)\,L\right], & T=T_b,\\ \int_0^T\rho(s)\,c(s)\,ds+\rho(T_b)\,L, & T>T_b.\end{cases} \quad (5.17)$$

Taking into account axisymmetry, Eq. (5.16) becomes

$$\dot{e} - \frac{1}{r}\frac{\partial}{\partial r}\left(k\frac{\partial T}{\partial r}\right) - \frac{\partial}{\partial z}\left(k\frac{\partial T}{\partial z}\right) = \mathcal{Q}, \tag{5.18}$$

where $\dot{e} =: \frac{\partial e}{\partial t} + V(t)\frac{\partial e}{\partial z}$ denotes the material time derivative of enthalpy.

5.2.2.4 Thermal Boundary Conditions Let us assume that the temperature on the surface of the electrode submerged in the mixture is given and equal to T_M. This means that we have a Dirichlet boundary condition on Γ_M. Besides, on Γ_R, a radiation-convection boundary condition is prescribed. More precisely, we write

$$k\frac{\partial T}{\partial \mathbf{n}} = g(T_c, T_r, T) \quad \text{on } \Gamma_R,$$

where the nonlinear function g is defined by

$$g(T_c, T_r, T) := h(T_c - T) + \gamma(T_r^4 - T^4). \tag{5.19}$$

In (5.19), h is the coefficient of convective heat transfer, T_c and T_r are the external convection and radiation temperatures, respectively, and coefficient γ is the product of the Stefan-Boltzman constant, 5.669×10^{-8} $\mathrm{W/m^2K^4}$, times a parameter related to the emissivity of the surface of the electrode.

With respect to convective heat transfer, let us introduce the *Nusselt number* which is the nondimensional number defined by

$$\mathrm{Nu} = \frac{h\mathcal{L}}{k_a},$$

$\mathcal{L}$ being the characteristic length of the cooled surface and k_a is the thermal conductivity of air.

The convection coefficient for air has been obtained by using standard formulas. For free convection on a vertical plane surface the following correlations for the Nusselt number can be found in the literature (see, for instance, [11]):

$$\mathrm{Nu} = 0.59\,(\mathrm{Gr\,Pr})^{1/4}, \qquad 10^4 \leq \mathrm{Gr\,Pr} \leq 10^9,$$
$$\mathrm{Nu} = 0.13\,(\mathrm{Gr\,Pr})^{1/3}, \qquad 10^9 \leq \mathrm{Gr\,Pr} \leq 10^{12}.$$

where Pr and Gr are the Prandtl and Grashoff nondimensional numbers.

For free convection on horizontal plates, Ref. [6] collects several formulas of various authors. For a heated plate facing up (or a cooled plate facing down) we have

$$\mathrm{Nu} = 0.54\,(\mathrm{Gr\,Pr})^{1/4}, \qquad 10^5 \leq \mathrm{Gr\,Pr} \leq 2 \times 10^7,$$
$$\mathrm{Nu} = 0.15\,(\mathrm{Gr\,Pr})^{1/3}, \qquad 2 \times 10^7 \leq \mathrm{Gr\,Pr} \leq 3 \times 10^{10}.$$

For a heated plate facing down (or a cooled plate facing up)

$$\mathrm{Nu} = 0.27\,(\mathrm{Gr}\,\mathrm{Pr})^{1/4}, \qquad 3 \times 10^5 \leq \mathrm{Gr}\,\mathrm{Pr} \leq 3 \times 10^{10}.$$

On the internal surface in contact with cooling water, only convective heat transfer is considered. The corresponding coefficient of convective transfer has been obtained by evaluating the heat transfer rate from the electrode to water. An average value of the latter quantity is given by

$$H = \rho_w\, c_w\, q(T_o - T_i), \tag{5.20}$$

where ρ_w and c_w are, respectively, the mean density and specific heat of water in the range of temperatures under consideration; T_i and T_o are the input and output temperatures of water in the cooling pipes and q is the water flow rate. Then a mean value for h can be calculated as

$$h = \frac{H}{S(T_s - T_o)}, \tag{5.21}$$

T_s being an average temperature of the cooled surface inside the contact clamps and S the area of this inner surface.

In fact, since T_s is unknown, we cannot use this expression to obtain h directly. Therefore, we have adjusted h using numerical results. More precisely, we have first solved for different values of h the coupled problem; then we have calculated the total heat exchange between electrode and water, that is,

$$\int_{\Gamma_W} h(T - T_0)\, \mathrm{d}\Gamma,$$

and finally we have selected the value of h such that the corresponding total heat is closest to the theoretical value according to (5.20).

5.2.3 Numerical Solution

Since the model is time dependent, for numerical solution we perform two successive discretizations in time and in space.

5.2.3.1 Time Discretization. Let $[0, t_f]$ be the time interval for simulation and $\Pi = \{t_0, \ldots, t_I\}$ a mesh consisting of I equally spaced points. Let us denote by Δt, the corresponding time step.

To integrate the equation in time, we use an Euler implicit scheme. We approximate the value of the total derivative of enthalpy $\dot{e}((r, z), t)$ at $(r, z) \in \Omega$ and $t = t_{n+1}$ by the two-point finite difference formula:

$$\dot{e}((r, z), t_{n+1}) \approx \frac{e^{n+1}(r, z) - e^n\left(X^n(r, z)\right)}{\Delta t},$$

where $X^n(r, z)$ represents the position occupied at time t^n by the material point which is at position (r, z) at time t_{n+1}. Since the velocity field is space independent, X^n is simply given by

$$X^n(r, z) = (r, z - \int_{t_n}^{t_{n+1}} V_n(t)\, dt).$$

Multiplying Eq. (5.18) discretized in time by a test function and integrating in the meridian section Ω, we obtain, after using a Green's formula and taking boundary conditions into account, the following *weak formulation* of the discretized thermal problem:

(WTP): *For each $t_{n+1} \in \Pi$, find a function T^{n+1} such that $T^{n+1} = T_M$ on Γ_M and furthermore*

$$\int_\Omega \frac{1}{\Delta t} e^{n+1}\, W\, r\, drdz + \int_\Omega k(r, z, T^{n+1}) \left(\frac{\partial T^{n+1}}{\partial z} \frac{\partial W}{\partial z} + \frac{\partial T^{n+1}}{\partial r} \frac{\partial W}{\partial r} \right) r\, drdz$$
$$- \int_{\Gamma_R} g(T_c, T_r, T^{n+1})\, W\, r\, d\Gamma = \int_\Omega \frac{1}{\sigma(r, z, T^{n+1})} |\mathbf{curl}\ H_\theta^{n+1}|^2 W\, r\, drdz \tag{5.22}$$

for all test function W which is null on Γ_M.

Similarly, an approximation of the magnetic field at time t_{n+1}, H_θ^{n+1} is determined as the solution of the following *weak formulation* of the electromagnetic problem

(WEP): *Find a function H_θ^{n+1} such that $H_\theta^{n+1} = g$ on Γ_H and*

$$\int_\Omega i\omega\mu H_\theta^{n+1} \bar{G}\, r\, drdz + \int_\Omega \frac{1}{\sigma(T^{n+1})} \left(\frac{\partial H_\theta^{n+1}}{\partial z} \frac{\partial \bar{G}}{\partial z} + \frac{1}{r^2} \frac{\partial (r H_\theta^{n+1})}{\partial r} \frac{\partial (r\bar{G})}{\partial r} \right) r\, drdz = 0, \tag{5.23}$$

for all test function G null on Γ_M.

5.2.3.2 Space Discretization. Problems (5.22)and (5.23) can be spatially discretized by a standard finite element method. More precisely, let us consider a mesh of domain Ω consisting of triangles. Then we approximate both temperature and magnetic field by continuous piecewise linear finite elements defined on this mesh.

5.2.3.3 Solving the Discrete Problem: An Iterative Algorithm. We notice that a coupled nonlinear system must be solved at each time step, because heat source depends on the solution of the electromagnetic problem and parameters k and σ depend on temperature. A fixed point algorithm has been used to solve this system which is described below. Moreover, enthalpy has been defined as a multivalued function of temperature preventing the use of Newton-like methods. Fortunately, $\mathcal{H}$ as defined in (5.17) is monotonous and it is maximal in that its graph cannot be strictly included into another monotonous graph. Thus, in order to solve the nonlinear problem at each time step, we can use an iterative algorithm introduced in a general setting in Ref. [4]. This algorithm is based on the following result:

Lemma 5.2.1 *Let $\mathcal{H}$ be a maximal monotone operator. Then the following statements are equivalent:*

- $e \in \mathcal{H}(T)$,
- $p = \mathcal{H}_\lambda^\alpha(T + \lambda p)$, *with* α, $\lambda > 0$ *such that* $\lambda\alpha < 1$,

where $\mathcal{H}_\lambda^\alpha$ is the so-called Yosida approximation of the shifted operator $\mathcal{H} - \alpha I$ that is defined as follows,

$$\mathcal{H}_\lambda^\alpha(s) = \frac{[I - (I + \lambda(\mathcal{H} - \alpha I))^{-1}](s)}{\lambda}, \quad s \in \mathbb{R}, \quad q \in \Omega.$$

This result leads us to use the following algorithm, sketched in Fig. 5.6.

Initial step.— Let T^0 be given. H^0 is calculated as the solution of the linear equation,

$$\int_\Omega i\omega\mu H^0 \bar{G}\, r\, dr\, dz + \int_\Omega \frac{1}{\sigma(T^0)} \left(\frac{1}{r}\frac{\partial}{\partial r}(rH^0) \frac{1}{r}\frac{\partial}{\partial r}(r\bar{G}) + \frac{\partial H^0}{\partial z}\frac{\partial \bar{G}}{\partial z} \right) r\, dr\, dz = 0, \tag{5.24}$$

for all test function G null on Γ_M.

Step n+1.— Let us suppose T^n and H^n are known. Then, at time t_{n+1}, functions T^{n+1} and H^{n+1} are obtained as the limit of sequences T_s^{n+1}, H_s^{n+1} constructed with the following iterative algorithm:

1. *Initialization.* Let T_0^{n+1}, H_0^{n+1} be given by, for instance, $T_0^{n+1} = T^n$ and $H_0^{n+1} = H^n$.

2. *Iteration s.* Let us suppose $T_{s-1}^{n+1}, H_{s-1}^{n+1}$ are known. We successively determine H_s^{n+1} and T_s^{n+1} as follows:

- H_s^{n+1} is the solution of

$$\int_\Omega i\omega\mu H_s^{n+1}\bar{G}\, r\, dr\, dz + \int_\Omega \frac{1}{\sigma\left(T_{s-1}^{n+1}\right)}\left(\frac{1}{r}\frac{\partial}{\partial r}(rH_s^{n+1})\frac{1}{r}\frac{\partial}{\partial r}(r\bar{G}) + \frac{\partial H_s^{n+1}}{\partial z}\frac{\partial \bar{G}}{\partial z}\right) r\, dr\, dz = 0, \tag{5.25}$$

for all test function G null on Γ_M.

- T_s^{n+1} is the solution of the problem

$$\frac{\alpha}{\Delta t}\int_\Omega T_s^{n+1}\, W\, r\, dr\, dz + \int_\Omega k\left(T_{s-1}^{n+1}\right) \mathrm{grad}\left(T_s^{n+1}\right)\cdot \mathrm{grad}\, W\, r\, dra\, dz$$
$$= -\int_{\Gamma_{RT}} g(T_C, T_r, T_s^{n+1}\, W\, r\, d\Gamma + \frac{1}{\Delta t}\int_\Omega \left(e^n \circ X^n - p_s^{n+1}\right) W\, r\, dr\, dz$$
$$+ \int_\Omega \frac{1}{\sigma\left(T_{s-1}^{n+1}\right)} \left|\mathbf{curl}\,(H_s^{n+1})\right|^2 W\, r\, dr\, dz, \tag{5.26}$$

for all test function W null on Γ_M.

- p_s^{n+1} is calculated by

$$p_s^{n+1} = \mathcal{H}_\lambda^\alpha\left(T_s^{n+1} + \lambda p_s^{n+1}\right).$$

5.2.4 Numerical Results

A computer code implementing the above numerical methods has been written. It has been applied to simulate the evolution of temperature in a single ELSA electrode under industrial-like working conditions. In particular, slipping and switching off have been considered.

As an initial temperature we have taking the one corresponding to the steady state which has also been determined numerically.

It can be seen in Figure 5.7 that, in a real situation in the factory, an electrode never reaches its thermal steady state. Moreover, a global look at this figure shows that temperature does not change too much along the time.

In Figure 18.26, one can see how temperature decreases near the boundary due to water and air cooling and increases again about one meter below the contact clamps, where the mixture of quartz and carbonaceous materials protect the column from heat losses.

Electric current enters the electrode through a very small area in the lower part of the clamps. Current intensity is a datum for the model. The real part of the current density is shown in Figure 5.9 which serves to check that the major part of it goes to the graphite through the lower zone of the clamps.

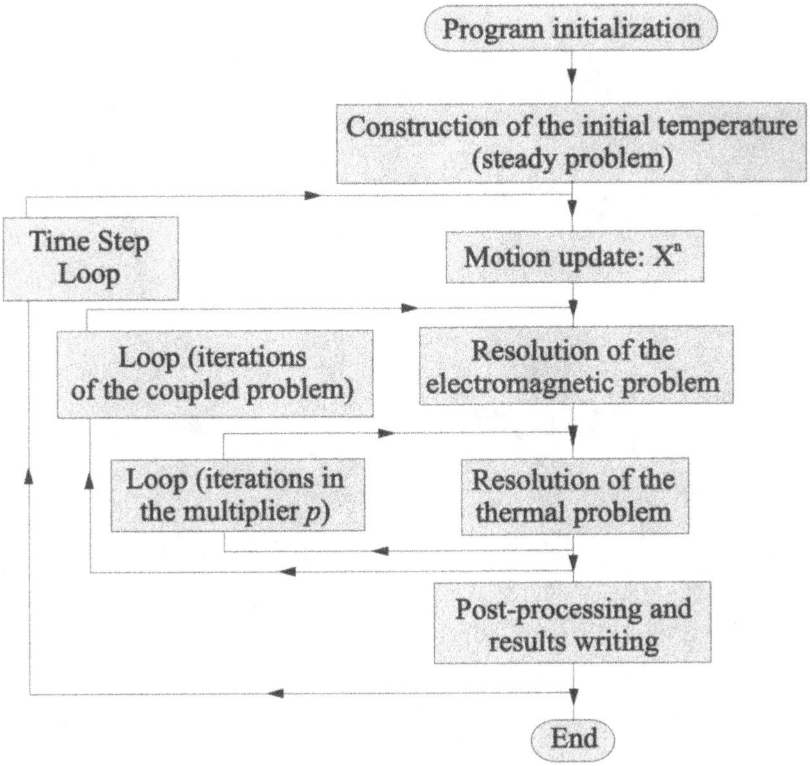

Figure 5.6 Flow chart of the algorithm.

Accordingly, the production of heat is more important in this zone (see the Joule heating in Figure 5.10).

The heat necessary for paste baking comes from the graphite, which is a better heat conductor, and it is also produced by the Joule effect at the lower end of the clamps. The interaction between this heat and the refrigeration due to water produces high gradients of temperature in this zone.

5.3 NUMERICAL SIMULATION OF PIT LAKE WATER QUALITY

5.3.1 Introduction to the Problem

In the last years, one common strategy for the environmental and landscape recuperation of open pit mines after their closure consists of filling the mining void with water. Diverting a neighboring river to fill the pit has been a frequently used technique to accomplish this task. Since mining lakes are generally in contact with natural rivers or streams, they must accomplish certain water standards set by the corresponding authorities.

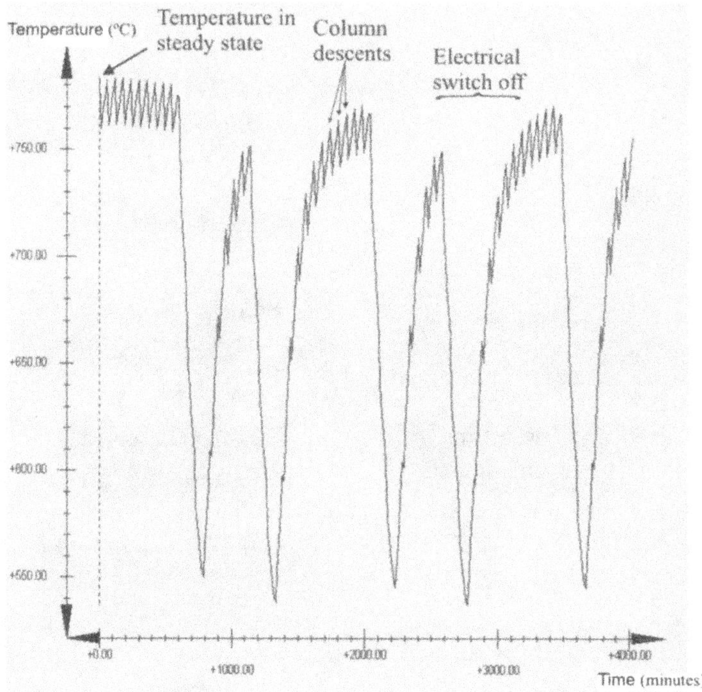

Figure 5.7 Temperature evolution.

In the context of this problem, numerical modeling constitutes a unique tool to predict the water quality of a pit lake, not only at its maximum level, but also at each stage of the flooding period. Thus, it allows for the application in advance of any remediation strategy aimed at obtaining acceptable water quality conditions at the end of the flooding period.

Modeling the water quality of a future pit lake requires, at a first stage, knowing which are the main factors that affect it. In this sense, the most important environmental problem associated with mining lakes is related to the presence of iron sulfides at the pit walls. The oxidation of these materials release acidity and heavy metals that might constitute a source of pollution of the pit lake (see [5], [8]). Waters of this type are very hazardous for the environment. They are also highly reactive, generally triggering a chain of chemical reactions whose relative importance will determine the final lake water quality.

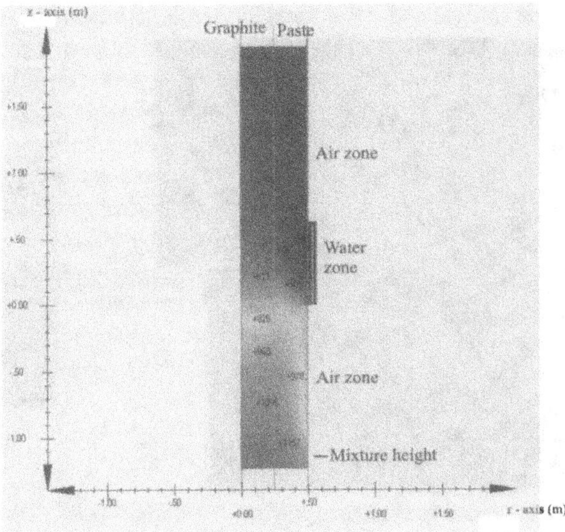

Figure 5.8 Temperature in clamp's zone.

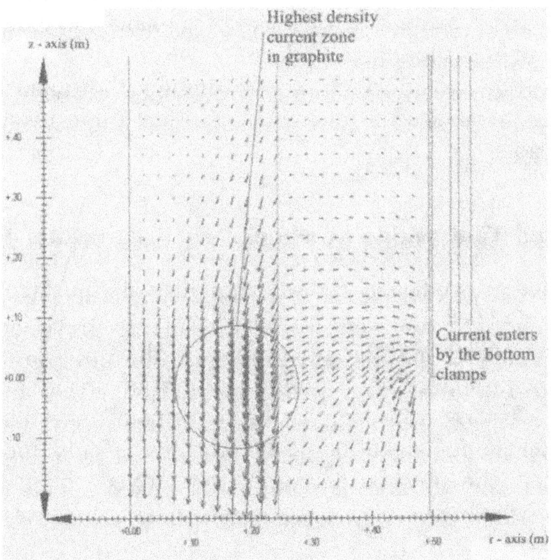

Figure 5.9 Real part of current density.

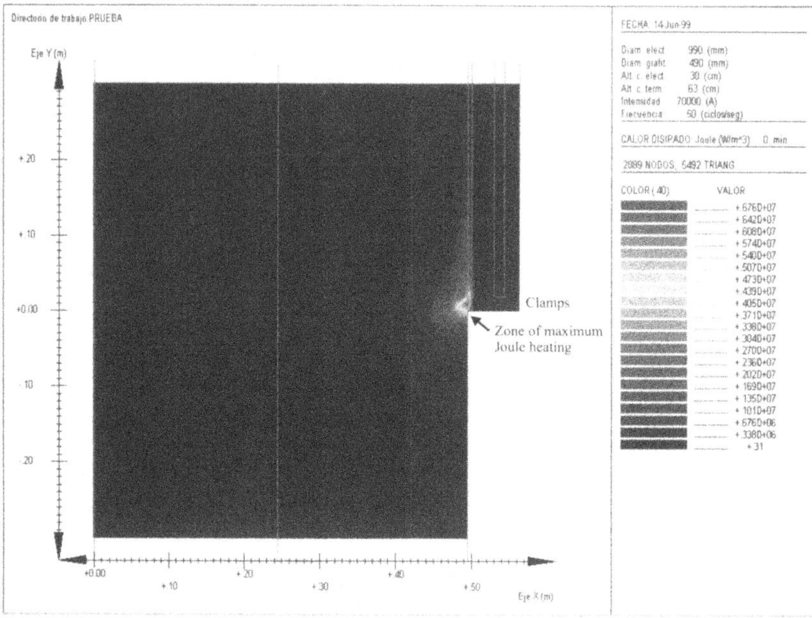

Figure 5.10 Heat released by the Joule effect.

The input of water sources of different natures (different degrees of pollution) and the vertical circulation patterns of the lake contribute to an increase the complexity of this modeling work.

Some hints to the way in which an environmental problem as the one introduced here must be treated from a mathematical point of view are provided in the next sections.

5.3.2 A Stirred Tank Model to Predict Pit Lake Water Quality

Any model aimed at predicting the water quality of a pit lake should take into account all the factors that were introduced in the previous section. There exist conceptual models of different complexity that integrate all these factors trying to provide a reliable water quality estimation, either analyzing extreme situations (conservative approach) or keeping them more realistic.

In this section the focus will be posed on a *stirred tank* model that consists of assuming that the pit lake is completely mixed. This is equivalent to considering the whole lake as a water mass whose concentration is the same at each point.

Complete mixing is a rather unlikely hypothesis in the particular case of pit lakes, either due to their characteristic morphology (the ratio lake depth/lake diameter is higher than in natural lakes) that is not prone to experiment mixing events that affect the whole water column (see [5]), or as a consequence or their geochemistry. Sometimes a denser and more polluted water layer occupies the lake bottom and does not mix with the upper ones. If complete mixing occurred, we would be considering the most pessimistic situation in terms of water quality: the more polluted water layers at the lake bottom would mix up with the cleaner ones at the surface and then the last ones, which are generally in contact with rivers, streams, etc., would be more polluted. In this sense, the stirred tank model, based on the hypothesis of complete mixing, constitutes a valuable tool for decision makers.

The following sections will be devoted to formulate mathematically a stirred tank model—a zero-dimensional model—whose water quality will depend on the geochemistry of the water sources that feed the lake as well as the chemical reactions that were mentioned in Section 5.3.1. Therefore, its development mainly consists of setting the suitable volume and mass conservation equations.

5.3.2.1 Notation

Related to the Lake Morphology. The conceptualization of the stirred tank model is represented in Figure 5.11.

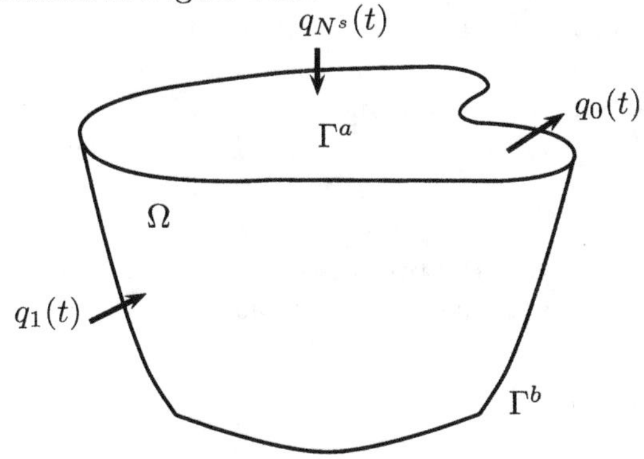

Figure 5.11 Stirred tank conceptual model.

Frequently, the water quality of pit lakes is studied from the beginning of the lake flooding to its end but, for the sake of simplicity, in this presentation we will focus on an already flooded lake. In this way, Ω denotes the region of the space that is occupied by the lake and V is its (constant in time) volume.

The boundary of Ω, named Γ, is such that

$$\Gamma = \Gamma^b \cup \Gamma^a, \tag{5.27}$$

with Γ^b and Γ^a the parts of the boundary that represent the pit wall/water interface and the air/water interface, respectively.

At each time, the lake receives water from different sources: river water, rain, subterranean water, infiltration water, etc. The number of these water sources is represented by N^s and $q_j(t)$, $j = 1, \ldots, N^s$ is used to refer to their respective flow rate in m^3/s. Since the lake volume remains constant, all the water inputs must be compensated by an equivalent output. In this sense, if we take into account that the water inputs intercept Ω at certain places of the boundary Γ, it can also be defined as

$$\Gamma = \Gamma_1 \cup \cdots \cup \Gamma_{N^s} \cup \Gamma^O \cup \Gamma^I, \tag{5.28}$$

with Γ_j, $j = 1, \ldots, N^s$ the part of Γ through which the j-th water source enters Ω, Γ^O, the part of Γ through which the completely mixed waters of the lake leave and Γ^I the impervious part of Γ.

It must be mentioned that the lake morphology data, such as its volume V, the area of the air/water interface S^a, the area of the pit wall/water interface S^b, etc., are generally known, as they are usually provided by the mining companies.

Related to the Water Quality. The water quality of a pit lake is determined by the presence of certain chemical species E_i, $i = 1, \ldots, N$ whose concentration at each time are expressed by $y_i(t)$, $i = 1, \ldots, N$. As it was already mentioned, the time evolution of the concentration of each chemical species will depend on the geochemistry of the water sources that enter the pit but also on the chemical reactions that occur in its interior. Focusing on the chemical reactions, the notation that will be considered to represent their effect is as follows:

- $r_i^a(t, y_1(t), \ldots, y_N(t))$ is the production rate in mol/(s m^2) of species E_i from the chemical reactions that take place at Γ^a.

- $r_i^c(t, y_1(t), \ldots, y_N(t))$ is the production rate in mol/(s m^3) of species E_i from the chemical reactions that take place in the water column.

- $r_{ij}^b(t, y_1(t), \ldots, y_N(t))$ is the production rate in mol/(s m^2) of species E_i from the chemical reactions that occur at Γ_j^b, $j = 1, \ldots, M^r$. Since each mineral presents its own reaction rate, notation Γ_j^b, $j = 1, \ldots, M^r$ is used to refer to the part of the boundary Γ^b that the j-th mineral occupies, M^r being the total number of reacting minerals at this surface whose effect on the pit lake water quality will be considered.

5.3.2.2 *Conservation Equations*

Volume Conservation Equations. If we denote by $\mathbf{n}$ the normal unit vector pointing outwards the lake domain Ω, by $\mathbf{v}_j$, $j = 1, \ldots, N^s$ the velocity of the water sources that enter the pit and by $\mathbf{v}$ the velocity of the pit lake outflow, it follows that

$$\frac{dV(t)}{dt} = -\sum_{j=1}^{N^s} \int_{\Gamma_j} \mathbf{v}_j(x,t).\mathbf{n}(x)dS_x + \int_{\Gamma^O} \mathbf{v}(x,t).\mathbf{n}(x)dS_x = \sum_{j=1}^{N^s} q_j(t) - q_0(t). \tag{5.29}$$

Notice that the sign of the first integral is negative because we are dealing with inputs and $\mathbf{n}$ is defined to be pointing outwards. A similar argument is applied to explain the positive sign of the second integral.

Since our model assumes that the lake is already flooded, $dV(t)/dt = 0$. Therefore, $q_0(t)$, which is the water flow rate outside the lake, verifies

$$q_0(t) = \sum_{j=1}^{N^s} q_j(t). \tag{5.30}$$

Mass Conservation Equations. The amount of substance in moles $m_i(t)$, $i = 1, \ldots, N$ of the ith chemical species in Ω is given by

$$m_i(t) = \int_{\Omega} y_i(t)dV_x = y_i(t)V, \tag{5.31}$$

being its time evolution calculated as

$$\frac{dm_i(t)}{dt} = \int_{\Omega} r_i^c(t,\mathbf{y}(t))dV_x + \sum_{j=1}^{M^r} \int_{\Gamma_j^b} r_{ij}^b(t,\mathbf{y}(t))d\Gamma_x + \int_{\Gamma^a} r_i^a(t,\mathbf{y}(t))d\Gamma_x$$
$$-\sum_{j=1}^{N^s} \int_{\Gamma_j} a_{ij}(t)\mathbf{v}_j(x,t)\cdot\mathbf{n}(x)d\Gamma_x + \int_{\Gamma^o} y_i(t)\mathbf{v}(x,t)\cdot\mathbf{n}(x)d\Gamma_x =$$
$$Vr_i^c(t,\mathbf{y}(t)) + S^a r_i^a(t,\mathbf{y}(t)) + \sum_{j=1}^{M^r} S_j^b r_{ij}^b(t,\mathbf{y}(t)) + \sum_{j=1}^{N^s} a_{ij}(t)q_j(t) -$$
$$y_i(t)q_0(t), \tag{5.32}$$

where $\mathbf{y}(t) = (y_1(t), \ldots, y_N(t))$, $a_{ij}(t)$, $i = 1, \ldots, N$, $j = 1, \ldots, N^s$ is the concentration of the i-th chemical species in the jth water entrance to the lake at time t, S^a is the area of the air/water interface and S_j^b, $j = 1, \ldots, M^r$ is the area occupied by the jth mineral at the pit wall.

If we take the time derivative of Eq. (5.31), we get

$$\frac{dm_i(t)}{dt} = V\frac{dy_i(t)}{dt}. \tag{5.33}$$

The time evolution of the concentration of each chemical species can be easily obtained by combining equations (5.30), (5.32) and (5.33). Namely,

$$\frac{dy_i(t)}{dt} = r_i^c(t, \mathbf{y}(t))$$

$$+\frac{1}{V}\left[S^a r_i^a(t, \mathbf{y}(t)) + \sum_{j=1}^{M^r} S_j^b r_{ij}^b(t, \mathbf{y}(t)) + \sum_{j=1}^{N^s}(a_{ij}(t) - y_i(t))q_j(t)\right]. \quad (5.34)$$

By now we have just mentioned the way in which the chemical reactions must be taken into consideration but nothing has been said about the kinetics of these reactions, i.e., about terms r_i^c, r_i^a and r_{ij}^b. This will be the objective of the next section.

5.3.3 Mathematical Models for Chemical Reaction Systems

The chemical species whose concentrations determine the pit lake water quality are involved in a set of chemical reactions that are occurring at different places within the lake, as it was explained in Section 5.3.2.2. Depending on the velocity at which these reactions proceed with respect to the time scale of the problem, two basic theories are applied: finite rate chemical kinetics and equilibrium. If the velocity of the chemical reactions is fast when compared to the time scale of the problem, the system could be considered to reach equilibrium. On the other hand, if the reaction rate is similar to the selected time scale, reactions should be kinetically described (see Ref. [12]).

In this section we will introduce the formal mathematical framework to describe the evolution of the concentration of a set of chemical species according to the theories of chemical kinetics and equilibrium. Notation is based on Ref. [2].

Let us start by considering that the chemical reactions in which the N chemical species of interest, E_i, $i = 1, \ldots, N$ are involved, can be represented by

$$\nu_1^l E_1 + ... + \nu_N^l E_N \to \lambda_1^l E_1 + ... + \lambda_N^l E_N, \qquad 1 \leq l \leq L, \quad (5.35)$$

with ν_i^l and λ_i^l the stoichiometric coefficients, that are always positive. In the following, they will also be assumed to be integers.

5.3.3.1 Finite Rate Chemical Kinetics. According to this theory, the evolution of the concentration of the ith chemical species is governed by an ordinary differential equation system (ODE) of the form

$$\frac{dy_i(t)}{dt} = \sum_{l=1}^{L}(\lambda_i^l - \nu_i^l)\delta_l(y_1, \ldots, y_N), \quad (5.36)$$

where δ_l is the velocity of the l-th chemical reaction, its expression being dependent on the complexity of the reaction. In this sense,

- If the reactions are elementary, meaning that they proceed in a single step, δ_l is written as

$$\delta_l = k_l \prod_{i=1}^{N} y_i(t)^{\nu_i^l}, \tag{5.37}$$

  with k_l the rate constant, that depends on the temperature through the Arrhenius law (see, for instance, Ref. [12]).

- In most of the literature sources

$$\delta_l = k_l \prod_{i=1}^{N} y_i(t)^{\nu_i^{l'}}, \tag{5.38}$$

  where $\nu_i^{l'}$ is the partial reaction order with respect to the i-th reactant and $\sum_i^N \nu_i^{l'}$ is the total reaction order. In general, $\nu_i^{l'}$ has nothing to do with the stoichiometric coefficient of the ith reactant.

- In complex reactions, the mathematical expression for δ_l does not follow any rule, as the ones above, its formula being empirically obtained in most of the cases. Therefore, in general, δ_l will be a function

$$\delta_l = h(y_1, \dots, y_N). \tag{5.39}$$

If we assume, for simplicity, that all the chemical reactions are elementary, the problem that must be solved to obtain the water quality consists of the following system of ODEs

$$\begin{cases} \frac{dy_i(t)}{dt} = \sum_{l=1}^{L} (\lambda_i^l - \nu_i^l) k_l \prod_{j=1}^{N} y_j(t)^{\nu_j^l}, & i = 1, \dots, N, \\ y_i(0) = y_{i,init}, \end{cases} \tag{5.40}$$

which is obtained by using (5.37) in Ref. (5.36).

A proof on the existence and uniqueness of solution to problem (5.40) can be found in [3].

5.3.3.2 Chemical Equilibrium. From a thermodynamic point of view, a system in chemical equilibrium is a steady state characterized by a minimum in the Gibbs free energy. Consequently, one of the available methods to obtain the equilibrium concentration of a set of chemical species consists of minimizing the Gibbs free energy subjected to certain restrictions (see Refs. [2], [12], [12], etc.). However, in this chapter we will exploit the existing relationship between the kinetic theory and equilibrium. In this sense, we will keep the same notation for the set of chemical species E_i, $i = 1, \dots, N$, being involved in a set of chemical reactions such as those in Eq. (5.35), but this time $l = 1, \dots, 2J$ representing a set of J reversible couples of reactions that are characterized by the fact that the stoichiometric coefficients satisfy

$$\nu_i^{2j-1} = \lambda_i^{2j}, \qquad j = 1, \dots, J. \tag{5.41}$$

Before going into further details of this method, let us introduce the following definition.

Definition 5.3.1 *We will denote by n_i, $i = 1, \dots, N$ the molinity of species E_i, i.e., the number of kmol of the ith chemical species per kg of solution. Thus,*

$$n_i = \frac{y_i}{\rho}, \tag{5.42}$$

where ρ is the solution mass density in kg/m^3.

In order to keep the explanation of this method as simple as possible, only one of the reversible reactions in (5.35) will be considered, namely,

$$\nu_1 E_1 + \dots + \nu_N E_N \underset{2}{\overset{1}{\rightleftharpoons}} \lambda_1 E_1 + \dots + \lambda_N E_N. \tag{5.43}$$

We define

$$\lambda_i := \nu_i^2 = \lambda_i^1, \tag{5.44}$$

$$\nu_i := \lambda_i^2 = \nu_i^1. \tag{5.45}$$

By assuming that both the forward and backward reactions in (5.43) are elementary, the reaction velocities δ_1 and δ_2 can be written according to Eq. (5.37) as

$$\delta_1 = k_1 \prod_{i=1}^{N} y_i^{\nu_i} \qquad \text{and} \qquad \delta_2 = k_2 \prod_{i=1}^{N} y_i^{\lambda_i}. \tag{5.46}$$

Thus, the ODE system in Eq. (5.40) becomes

$$\frac{dy_i(t)}{dt} = (\lambda_i - \nu_i)\delta^*(t), \tag{5.47}$$

with

$$\delta^* = \left(k_1 \prod_{i=1}^{N} y_i^{\nu_i} - k_2 \prod_{i=1}^{N} y_i^{\lambda_i} \right). \tag{5.48}$$

Since $y_i = n_i \rho$ (after Definition 5.3.1), equation (5.47) can be rearranged in such a way that we obtain

$$\frac{1}{(\lambda_i - \nu_i)} \rho \frac{dn_i(t)}{dt} = \delta^*(t). \tag{5.49}$$

By integrating this system of equations

$$\frac{n_1(t) - n_{1,init}}{(\lambda_1 - \nu_1)} = \cdots = \frac{n_N(t) - n_{N,init}}{(\lambda_N - \nu_N)} = \xi(t), \tag{5.50}$$

with $\xi(t)$, the *reaction extent*, given by

$$\xi(t) = \frac{1}{\rho} \int_0^t \delta^*(s) ds. \tag{5.51}$$

From Eq. (5.50), n_i, $i = 1, \ldots, N$ can be written in terms of the reaction extent as

$$n_i(t) = n_{i,init} + (\lambda_i - \nu_i)\xi(t). \tag{5.52}$$

Let us go back for a moment to the definition of equilibrium as a steady state. In this sense, it is characterized by the fact that $dy_i(t)/dt = 0$, which is equivalent to $\delta^* = 0$ in Eq. (5.48) and, hence,

$$\frac{k_1}{k_2} = \prod_{i=1}^{N} y_i^{\lambda_i - \nu_i} = K^e, \tag{5.53}$$

where K^e is called the equilibrium constant for reaction (5.43).

Equation (5.53) expresses the relationship between the equilibrium constant and the forward and backward rate constants.

By taking into account in Eq. (5.53) that $y_i = \rho n_i$ and replacing n_i by its expression as a function of the reaction extent in (5.52), we obtain

$$K^e(\theta) = \rho^{\sum_{i=1}^{N}(\lambda_i - \nu_i)} \prod_{i=1}^{N} (n_{i,init} + (\lambda_i - \nu_i)\xi)^{\lambda_i - \nu_i}, \tag{5.54}$$

which is an algebraic equation that must be solved for ξ to obtain the equilibrium concentration or, in other words, the water quality when equilibrium conditions are applicable.

5.3.3.3 Coexistence of Slow and Fast Chemical Reactions. There exist problems in which the chemical species E_i, $i = 1, \ldots, N$ are involved in chemical reactions that proceed at different time scales, some of them being very fast in the time frame of interest, some others being slower.

In this section, the notation for the chemical reactions in Eq. (5.35) is conserved, although $l = 1, \ldots, L+2J$, with the first L slow chemical reactions and the last $2J$, J couples of reversible reactions that satisfy

$$\nu_i^{L+2j-1} = \lambda_i^{L+2j}, \tag{5.55}$$

$$\lambda_i^{L+2j-1} = \nu_i^{L+2j}, \qquad j = 1, \ldots, J. \tag{5.56}$$

From Eq. (5.36) and by considering a kinetic description of the equilibrium reactions [see Eqs. (5.47) and (5.48)] we obtain

$$\frac{dy_i(t)}{dt} = \sum_{l=1}^{L}(\lambda_i^l - \nu_i^l)\delta_l(t,\mathbf{y}(t)) + \sum_{j=1}^{J}(\lambda_i^{L+2j-1} - \nu_i^{L+2j-1})\Big(\delta_{L+2j-1}(t,\mathbf{y}(t)) - \delta_{L+2j}(t,\mathbf{y}(t))\Big), \quad (5.57)$$

where δ_l, $l = 1, \ldots, L$ are the velocities of the slow chemical reactions and δ_{L+2j-1} and δ_{L+2j}, $j = 1, \ldots, J$ are the velocities of the fast forward and fast backward reactions, respectively.

Under the assumption of elementary reactions [see Eq. (5.37)], (5.57) is transformed into

$$\frac{dy_i(t)}{dt} = \sum_{l=1}^{L}(\lambda_i^l - \nu_i^l)k_l\prod_{i=1}^{N} y_i^{\nu_i^l} + \sum_{j=1}^{J}(\lambda_i^{L+2j-1} - \nu_i^{L+2j-1})\left[k_{L+2j-1}\prod_{i=1}^{N} y_i^{\nu_i^{L+2j-1}} - k_{L+2j}\prod_{i=1}^{N} y_i^{\lambda_i^{L+2j-1}}\right]. \quad (5.58)$$

In Eq. (5.58), all the rate constants present different units. In this sense, it is not possible to compare them in a homogeneous way and, hence, it is difficult to define by means of a numerical value what is a slow reaction or what is a fast one. In order to overcome this problem, it is necessary to scale (5.58); thus, all the magnitudes would be in the same units and, hence, they could be compared. For this purpose, the following dimensionless variables are defined

$$\hat{y}_i(\hat{t}) = \frac{y_i(t)}{Y_i} \quad \text{and} \quad \hat{t} = \frac{t}{T}, \quad (5.59)$$

where Y_i and T are the typical scales for concentration and time in the problem. The scaled problem is obtained by writing the original one in terms of the dimensionless variables (5.59) . Namely,

$$\frac{Y_i}{T}\frac{d\hat{y}_i}{d\hat{t}} = \sum_{l=1}^{L+2J}(\lambda_i^l - \nu_i^l)\hat{k}_l\prod_{i=1}^{N} \hat{y}_i^{\nu_i^l} + \sum_{j=1}^{J}(\lambda_i^{L+2j-1} - \nu_i^{L+2j-1})\left[\hat{k}_{L+2j-1}\prod_{i=1}^{N} \hat{y}_i^{\nu_i^{L+2j-1}} - \hat{k}_{L+2j}\prod_{i=1}^{N} \hat{y}_i^{\lambda_i^{L+2j-1}}\right], \quad (5.60)$$

where

$$\hat{k}_l = k_l \prod_{i=1}^{N} Y_i^{\nu_i^l}, \qquad l = 1, \dots, L, \tag{5.61}$$

$$\hat{k}_{L+2j-1} = k_{L+2j-1} \prod_{i=1}^{N} Y_i^{\nu_i^{L+2j-1}}, \tag{5.62}$$

$$\hat{k}_{L+2j} = k_{L+2j} \prod_{i=1}^{N} Y_i^{\lambda_i^{L+2j-1}}. \tag{5.63}$$

Now, in Eq. (5.60) all the rate constants are given in the same units, so we are able to compare them in order to reach to the limit.

Let us assume that $\varepsilon > 0$ is a small positive parameter describing the ratio of fast time scales to slow ones. More precisely, let us assume the following properties:

$$\hat{k}_l = \mathcal{O}(1), \quad l = 1, \dots, L, \tag{5.64}$$

$$C_1\varepsilon^{-1} \leq \hat{k}_{L+2j-1} \leq C_2\varepsilon^{-1}, \tag{5.65}$$

$$C_1\varepsilon^{-1} \leq \hat{k}_{L+2j} \leq C_2\varepsilon^{-1}, \tag{5.66}$$

for $j = 1, \dots, J,$ for small enough ε, and for some positive constants C_1 and C_2, and

$$\hat{K}_j^e = \frac{\hat{k}_{L+2j-1}}{\hat{k}_{L+2j}}, \quad j = 1, \dots, J \qquad \text{are independent of } \varepsilon. \tag{5.67}$$

Under assumptions (5.65) to (5.67), the model can be written as

$$\frac{d\hat{\mathbf{y}}(t)}{dt} = \hat{\mathbf{f}}(t, \hat{\mathbf{y}}(t)) + \frac{1}{\varepsilon}\mathcal{A}\hat{\mathbf{g}}^{\varepsilon}(t, \hat{\mathbf{y}}(t)), \tag{5.68}$$

with $\hat{\mathbf{f}} = (\hat{f}_1, \dots, \hat{f}_N)$ being the function that involves the slow chemical reactions, $\hat{f}_i(t, \hat{\mathbf{y}}(t)) = \frac{T}{Y_i}\sum_{l=1}^{L}(\lambda_i^l - \nu_i^l)\hat{k}_l \prod_{i=1}^{N} \hat{y}_i^{\nu_i^l}$, and $\frac{1}{\varepsilon}\mathcal{A}\hat{\mathbf{g}}^{\varepsilon}(t, \hat{\mathbf{y}}(t))$ the term that represents the contribution of the fast reversible reactions. In this last term, $\mathcal{A}$ is a $N \times J$ matrix of components $\mathcal{A}_{ij} = \frac{T}{Y_i}(\lambda_i^{L+2j-1} - \nu_i^{L+2j-1})$ and $\hat{\mathbf{g}}_{\varepsilon} = (g_1^{\varepsilon}, \dots, g_N^{\varepsilon})$, with $\hat{g}_j^{\varepsilon}(t, \hat{\mathbf{y}}(t)) = \varepsilon\hat{k}_{L+2j}g_j(t, \hat{\mathbf{y}}(t))$, being $g_j(\hat{\mathbf{y}}) = \hat{K}_j^e(\theta)\prod_{i=1}^{N} \hat{y}_i^{\nu_i^{L+2j-1}} - \prod_{i=1}^{N} \hat{y}_i^{\lambda_i^{L+2j-1}}$.

Passing to the limit when $\varepsilon \to 0$ in the nondimensional scaled model (5.68) yields the dimensional limit model,

$$\left.\begin{aligned} &\mathbf{y}'(t) = \mathbf{f}(t, \mathbf{y}(t)) + \mathcal{A}^e\mathbf{p}^e(t), \\ &\mathbf{g}(t, \mathbf{y}(t)) = 0, \\ &\mathbf{y}(0) = \mathbf{y}_{init}, \end{aligned}\right\} \tag{5.69}$$

where $\mathcal{A}^e_{ij} = \lambda_i^{L+2j-1} - \nu_i^{L+2j-1}$ (the complete development of this limit model can be found in Ref. [3]).

Functions $\mathbf{p}^e = (p^e_1, \ldots, p^e_J)$ can be considered as Lagrange multipliers associated with restrictions $\mathbf{g}(t, \mathbf{y}(t)) = (g_1, \ldots, g_J)$. Since p^e_j, $j = 1, \ldots, J$ can be either positive, negative or zero, these restrictions are equivalent to $p^e_j \in \mathcal{H}(g_j)$, $\forall j = 1, \ldots, J$, with $\mathcal{H}$ the multi-valued maximal monotone function,

$$\mathcal{H}(x) = \begin{cases} \emptyset & \text{if } \quad x < 0 \quad \text{and} \quad x > 0, \\ (-\infty, \infty) & \text{if } \qquad\qquad x = 0, \end{cases} \tag{5.70}$$

where $\emptyset$ denotes the empty set. In addition, $p^e_j \in \mathcal{H}(g_j) \Leftrightarrow p^e_j = \mathcal{H}_\lambda(g_j + \lambda p^e_j) = p^e_j + \frac{1}{\lambda} g_j$, $\forall \lambda > 0$ and not necessarily a small number. This equivalence which is trivial in the present case, is a general result for maximal monotone operators such as $\mathcal{H}$, with $\mathcal{H}_\lambda$ its Yosida approximation (see Lemma 5.2.1).

Functions $\mathcal{H}(x)$ and $\mathcal{H}_\lambda(x)$ are shown in Fig. 5.12.

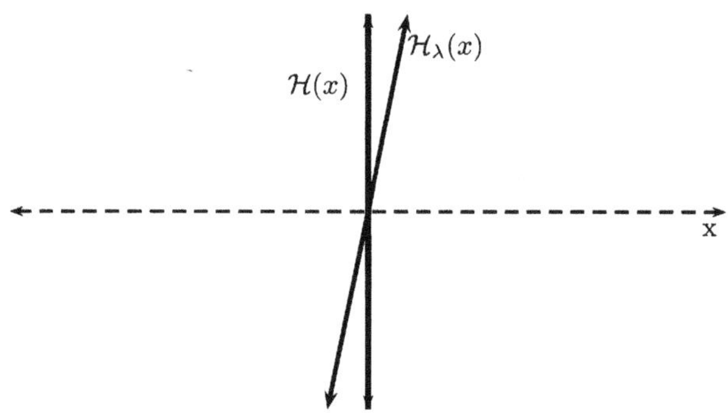

Figure 5.12 Functions $\mathcal{H}(x)$ and $\mathcal{H}_\lambda(x)$

The Particular Case of Solubility Equilibria. Solubility equilibria are particular cases of equilibrium reactions in which certain dissolved chemical species are in equilibrium with a solid that contains them. These reactions may advance in the sense of precipitation, meaning solid formation, which occurs when the concentration of the dissolved species exceeds a certain threshold, or in the sense of dissolution, implying a decrease in the concentration of the solid and an increase in the concentration of the dissolved chemical species.

In this section we will deal with problems in which slow chemical reactions coexist with solubility equilibria. Moreover, solubility equilibria will be treated only in the sense of precipitation in such a way that, when a solid has formed, it will irreversibly disappear from the dissolved phase.

Again, the general set of reactions in Eq. (5.35) will be considered, although this time $l = 1, \ldots, L+M$, L being the number of slow chemical reactions and M the number of solubility equilibria. Solubility equilibria increase by M the number of chemical species, thus E_i, $i = 1, \ldots, N+M$, with $E_{N+1}, \ldots, E_{N+M}$ the precipitated solids .

Since the solubility reactions are only considered in the sense of precipitation (a kind of unilateral equilibrium) and precipitates are generally represented as reactants (the species for which $\nu - \lambda > 0$), a solubility reaction will be written as

$$\begin{aligned}
&\nu_1^{L+m} E_1 + \ldots + \nu_N^{L+m} E_N + \nu_{N+1}^{L+m} E_{N+1} + \ldots + \nu_{N+M}^{L+m} E_{N+M} \\
&\leftharpoondown\!\!\rightharpoonup \lambda_1^{L+m} E_1 + \ldots + \lambda_N^{L+m} E_N + \lambda_{N+1}^{L+m} E_{N+1} + \ldots + \lambda_{N+M}^{L+m} E_{N+M}, \; 1 \leq m \leq M,
\end{aligned} \tag{5.71}$$

where the bigger arrow points to the left, indicating that just the precipitation way is important, although a smaller arrow pointing to the right still exist to represent that solubility is an equilibrium and thus requires information both from reactants and products.

On the basis of the information above, the time evolution of a chemical species that is involved both in slow and solubility equilibrium reactions is written as

$$\begin{aligned}
\frac{dy_i(t)}{dt} = f_i(t, \mathbf{y}(t)) + \sum_{m=1}^{M} (-\lambda_i^{L+m} + \nu_i^{L+m}) k_{L+m} \Bigg[&- K_m^s \prod_{j=1}^{N} y_j(t)^{\nu_j^{L+m}} \\
&+ \prod_{j=1}^{N} y_j(t)^{\lambda_j^{L+m}} \Bigg]^+, \quad i = 1, \ldots, N+M,
\end{aligned} \tag{5.72}$$

where $f_i(t, \mathbf{y}(t))$ denotes the contribution of the slow chemical reactions, k_{L+m}, $m = 1, \ldots, M$ is the rate constant in the sense of precipitation and K_m^s, $m = 1, \ldots, M$ is the solubility equilibrium constant for the mth precipitation reaction. The solubility equilibrium constants include the concentration of the precipitated solids because they are considered to be constant.

Notice that just the positive part of the term in squared brackets in Eq. (5.72) is taken into account, meaning that just the displacement of the chemical reaction (5.72) from right to left is accounted for. Recall that $a^+ = 0$ if $a < 0$ and $a^+ = a$ otherwise.

The limit of (5.72) is obtained by applying a similar procedure as for Eq. (5.58), which is

$$\begin{cases}
\dfrac{d\mathbf{y}(t)}{dt} = \mathbf{f}(t, \mathbf{y}(t)) + \mathcal{A}^s \mathbf{p}^s(t), & (5.73) \\
\mathbf{g}^s(t, \mathbf{y}(t)) \leq 0, & (5.74) \\
\mathbf{p}^s(t) \geq 0, & (5.75) \\
\mathbf{g}^s(t, \mathbf{y}(t)) \mathbf{p}^s(t) = 0, & (5.76) \\
\mathbf{y}(0) = \mathbf{y}_{init}, & (5.77)
\end{cases}$$

where $\mathcal{A}^s$ is the $(N+M)\times M$ matrix of components $\mathcal{A}^s_{im} = (-\lambda_i^{L+m} + \nu_i^{L+m})$, $\mathbf{p}^s(t) = (p_1^s(t), \ldots, p_M^s(t))$ is the vector of Lagrange multipliers associated with the inequality restriction functions $\mathbf{g}^s(t, \mathbf{y}(t)) \leq 0$ and the product $\mathbf{g}^s(t,\mathbf{y}(t))\mathbf{p}^s(t)$ refers to the vector $(g_1^s p_1^s, \ldots, g_M^s p_M^s)$. The m-th component of $\mathbf{g}^s(t,\mathbf{y}(t))$ is given by

$$g^s_m(t, \mathbf{y}(t)) = -K^s_m \prod_{i=1}^{N} y_i(t)^{\nu_i^{L+m}} + \prod_{i=1}^{N} y_i(t)^{\lambda_i^{L+m}}, \qquad m = 1, \ldots, M. \quad (5.78)$$

Conditions (5.74)–(5.76) are equivalent to $p^s_m \in \mathcal{G}(g^s_m)$, $\forall m = 1, \ldots, M$, with $\mathcal{G}$ the multi-valued function

$$\mathcal{G}(x) = \begin{cases} 0, & x < 0, \\ \emptyset, & x > 0, \\ [0, \infty), & x = 0, \end{cases} \quad (5.79)$$

where $\emptyset$ denotes the empty set. Moreover, $p^s_m \in \mathcal{G}(g^s_m) \iff p^s_m = \mathcal{G}_\lambda(g^s_m + \lambda p^s_m) = max\{0, p^s_m + \frac{1}{\lambda} g^s_m\}$ $\forall \lambda > 0$ and not necessarily a small number (see again Lemma 5.2.1). We recall that function $\mathcal{G}_\lambda$ is the Yosida approximation of the maximal monotone operator $\mathcal{G}$, both of them being represented in Figure 5.13.

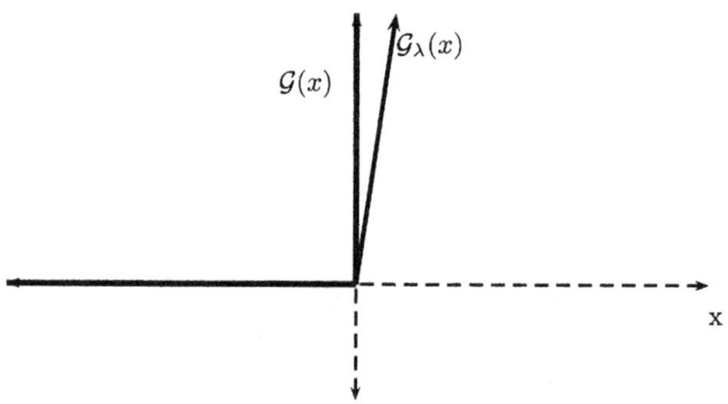

Figure 5.13 Functions $\mathcal{G}(x)$ and $\mathcal{G}_\lambda(x)$

5.3.3.4 The Complete Stirred Tank Problem. Once the mathematical treatment of chemical reactions that proceed at different velocities has been explained in detail, we can go back to Section 5.3.2 to complete the formal mathematical writing of the stirred tank model. In particular, we will focus on Eq. (5.34). First of all, it will be assumed that the contributions of the chemical reactions that occur at the air/water $[r^a_i(t, \mathbf{y}(t))]$ and at the pit wall/water interfaces $[r^b_{ij}(t, \mathbf{y}(t))]$ are slow, as it normally occurs. On the other hand, the chemical reactions that take place in the water column may be slow or fast, the latter considered to be in equilibrium. Solubility equilibria are also allowed.

On the basis of the notation in (5.69) we consider that:

- $r_i^c(t,\mathbf{y}(t)) = f_i^c(t,\mathbf{y}(t)) + \sum_{j=1}^J \mathcal{A}_{ij}^e p_j^e(t,\mathbf{y}(t)) + \sum_{k=1}^M \mathcal{A}_{ik} p_k^s(t,\mathbf{y}(t))$, $i = 1,\dots,N+M$, with f_i^c the slow chemical reactions occurring at the water column, p_j^e the Lagrange multiplier associated with the jth equilibrium reaction and p_k^s the Lagrangre multiplier associated with the kth solubility reaction.

- $S^a r^a(t,\mathbf{y}(t)) = f_i^a(t,\mathbf{y}(t))$.

- $\sum_{j=1}^{M^r} S^b r_{ij}^b(t,\mathbf{y}(t)) = f_i^b(t,\mathbf{y}(t))$.

Therefore, the complete stirred tank model would be written as

$$
\left\{
\begin{array}{ll}
\dfrac{dy_i(t)}{dt} = f_i^c(t,\mathbf{y}(t)) + \displaystyle\sum_{j=1}^{J} \mathcal{A}_{ij}^e p_j^e(t) + \sum_{k=1}^{M} \mathcal{A}_{ik}^s p_k^s(t) + \frac{1}{V}\Bigg[f_i^a(t,\mathbf{y}(t)) + f_i^b(t,\mathbf{y}(t)) & \\
\qquad + \displaystyle\sum_{j=1}^{N^s} (a_{ij}(t) - y_i(t)) q_j(t) \Bigg], \quad i = 1,\dots,N+M, & (5.80) \\
g_j^e(t,\mathbf{y}(t)) = 0, \quad j = 1,\dots,J, & (5.81) \\
g_k^s(t,\mathbf{y}(t)) \le 0, \quad k = 1,\dots,M, & (5.82) \\
p_k^s(t) \ge 0, & (5.83) \\
g_k^s(t) p_k^s(t) = 0, & (5.84) \\
y_i(0) = y_{i,init}. & (5.85)
\end{array}
\right.
$$

5.3.4 Numerical Solution of the Model

5.3.4.1 Numerical methods. This section concerns the numerical solution of the chemical model involving Lagrange multipliers (5.69).

Two main elements are proposed in order to carry out this task:

1. A Euler implicit scheme for the time discretization of the problem.

2. An iterative algorithm to solve the discrete problem.

Time discretization of the problem. The time interval of interest $(0, t_f)$ is uniformly partitioned, the approximate solution of the problem being calculated at each $t_n = n\Delta t$, with $\Delta t = t_f/N$ and $0 \le n \le N$.

The approximate solution of the problem at each t_n is obtained by considering an Euler implicit scheme. The discrete version of (5.69) is as follows

$$\begin{cases} \text{If} \quad n = 0, \\ \quad \mathbf{y}^0 = \mathbf{y}_{init}, & (5.86) \\ \text{If} \quad n \geq 0, \\ \quad \mathbf{y}^{n+1} = \mathbf{y}^n + \Delta t\big[\mathbf{f}(t_{n+1}, \mathbf{y}^{n+1}) + \mathcal{A}\mathbf{p}^{n+1}\big], & (5.87) \\ \quad \mathbf{g}(t_{n+1}, \mathbf{y}^{n+1}) = \mathbf{0}, & (5.88) \end{cases}$$

where $\mathbf{y}^n$ denotes the approximation to $\mathbf{y}(t_n)$, as it is usually done.

Thus, the nonlinear system of numerical equations (5.87) and (5.88) must be solved at each time step in order to obtain $\mathbf{y}^{n+1}$ and $\mathbf{p}^{n+1}$ from $\mathbf{y}^n$.

The Iterative Algorithm. The proposed iterative algorithm consists of two nested loops:

- An external loop to create the sequence $\mathbf{p}_r^{n+1}$ that converges to the multiplier solution of the problem.
- An internal loop to solve the nonlinear system regarding $\mathbf{y}^{n+1}$.

Figure 5.14 shows the flow diagram of the algorithm. Notice that the curved-edge rectangles are not themselves steps of the algorithm, but the task that will be carried out in the loop afterwards. It mainly comprises four steps:

1. *Initial guess* for the Lagrange multipliers: $\mathbf{p}_o^{n+1} = \mathbf{p}^n$.
2. *Calculation of* $\mathbf{y}_r^{n+1}$, by solving the non-linear system in (5.87), where the Lagrange multiplier provided in the previous step is used. The numerical solution of this nonlinear system must be obtained iteratively by applying, for example, Newton's method. This iterative algorithm constitutes the "internal loop."
3. *Updating* $\mathbf{p}_r^{n+1}$ once $\mathbf{y}_r^{n+1}$ has been obtained as [see the comments after Eq. (5.70)]
$$\mathbf{p}_r^{n+1} = \mathbf{p}_{r-1}^{n+1} + \frac{1}{\lambda}\mathbf{g}(t_{n+1}, \mathbf{y}_r^{n+1}). \tag{5.89}$$
4. *Applying* a suitable *convergence test* for the Lagrange multipliers.

The successive repetition of the external loop (and the nested internal loop) provides the approximate solution of the problem at time t_{n+1}.

5.3.5 Numerical Results: A Simplified Chemical Problem.

The above numerical methods have been applied to simulate the water quality of the pit lake that is now being flooded in the open pit coal mine of the

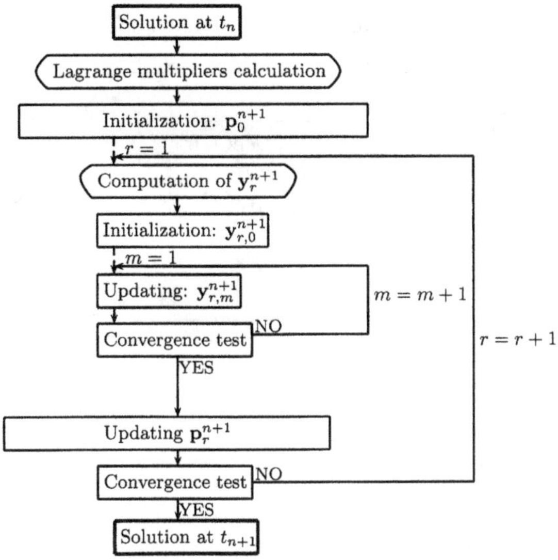

Figure 5.14 Flow diagram for the iterative algorithm.

Spanish company LIMEISA, in Cerceda (A Coruña, Spain). The geochemical model retained is very complex and can be found in Ref. [7] where numerical results are extensively shown.

A simplification of this model has been chosen to illustrate the way in which a water quality problem can be stated and subsequently solved. For this purpose, we will consider that the lake has no water inputs/outputs and that no chemical reactions at the air/water or pit wall/water interfaces are taking place. These assumptions imply that only the first three terms after the equal sign in Eq. (5.80) are non-null. In other words, we will deal with a fictitious example in which a set of chemical reactions are occurring at the water column. These reactions, although real, are treated in a fictitious way because the reaction velocities, equilibrium constants and solubility constants do not have to do with the ones that can be found in the literature for the same chemical reactions.

Problem statement. The water quality of our stirred tank model will be defined by six chemical species, five of them dissolved ($N = 5$, $i = 1, \dots, 5$ in Table 5.1) and a solid one ($M = 1$, $i = 8$ in Table 5.1). These species are involved in four chemical reactions: one is slow ($L = 1$, $l = 1$ in Table 5.2), two homogeneous equilibria ($J = 2$, $l = 2, 3$ in Table 5.2) and a solubility equilibrium ($M = 1$, $l = 4$ in Table 5.2).

Notice that δ_1 in Table 5.2 denotes the velocity of the slow chemical reactions, K_1 and K_2 the equilibrium constants and K^s the solubility constant. It must also be mentioned that, in reaction $l = 1$ the concentration of dissolved

Table 5.1 Chemical species.

i	E_i
1	$FeOH^{2+}$
2	OH^-
3	Fe^{3+}
4	H^+
5	Fe^{2+}
6	$Fe(OH)_{3(s)}$

Table 5.2 Chemical reactions.

l	Chemical reaction	Reaction velocity or equilibrium constant
1	$Fe^{2+} + H^+ + \frac{1}{4}O_{2(aq)} \rightarrow Fe^{3+} + \frac{1}{2}H_2O$	δ_1
2	$Fe^{3+} + H_2O \rightleftharpoons FeOH^{2+} + H^+$	K_1
3	$H_20 \rightleftharpoons H^+ + OH^-$	K_2
4	$Fe(OH)_{3(s)} + 3H^+ \rightleftharpoons Fe^{3+} + 3H_2O$	K^s

oxygen is considered to be enough for the reaction to proceed. In general, the reaction velocity δ_1 depends on the oxygen and Fe^{2+} concentrations, but for simplicity we will assume that it is constant.

The homogeneous equilibria $l = 2, 3$ impose restrictions on the concentration of the chemical species. Namely,

$$g_1 = -y_1 + \frac{K_1 y_3}{y_4} = 0, \quad (5.90) \qquad g_2 = -y_2 + \frac{K_2}{y_4} = 0, \quad (5.91)$$

where $y_1, \ldots, y_6$ are the concentrations of the chemical species in Table 5.1. The concentration of Fe^{3+} (species E_3 in Table 5.1) is also constrained by the solubility equilibrium $l = 4$, which is the associated restriction function given by

$$g^s = y_3 - \frac{K^s}{y_4^3} \leq 0. \tag{5.92}$$

Thus, the problem that must be solved to obtain the concentration of the chemical species along time in the stirred tank model is

$$\left\{\begin{array}{l} y_1'(t) = p_1^e(t), \\ y_2'(t) = p_2^e(t), \\ y_3'(t) = \delta_1 - p_1^e(t) - p^s(t), \\ y_4'(t) = -\delta_1 + p_1^e(t) + p_2^e(t) + 3p^s, \\ y_5'(t) = -\delta_1, \\ y_6'(t) = p^s, \\ g_1 = -y_1 + \frac{K_1 y_3}{y_4} = 0, \\ g_2 = -y_2 + \frac{K_2}{y_4} = 0, \\ g^s = y_3 - \frac{K^s}{y_4^3} \leq 0, \\ p^s \geq 0, \\ p^s g^s = 0, \\ \mathbf{y}(0) = \mathbf{y}_{init}. \end{array}\right. \tag{5.93}$$

where p_1^e, p_2^e are the Lagrange multipliers associated with equilibria $l = 2, 3$ and p^s is the Lagrange multiplier associated with solubility equilibrium $l = 4$.

Numerical results. The proposed example is solved by considering the data summarized in Table 5.3, regarding reaction velocities and equilibrium constants, and Table 5.4 that compiles the initial conditions for the problem. The initial conditions are obtained by considering that the slow and solubility reactions are frozen at $t = 0$ and that the initial $pH = 6.5$ ($pH = -log(y_4)$). The algorithm in 5.3.4.1 has been considered to carry out the numerical solution of the example, although in this case it will not be necessary to solve a nonlinear system of equations at each iteration of the external loop r to obtain $\mathbf{y}_r^{n+1}$ because δ_1 is constant.

Table 5.3 Data for the example.

Reac. vel or eq. cte.	Value
δ_1	10^{-8}
K_1	10^{-7}
K_2	10^{-14}
K^s	10^{12}

Table 5.4 Initial conditions.

i	$y_i(0)$ (mol/l)
1	0
2	$10^{-7.5}$
3	0
4	$10^{-6.5}$
5	10^{-4}
6	0

The λ value in Eq. (5.89) was selected to be 0.1 both for the Lagrange multipliers associated with homogeneous equilibria and the ones related to solubility reactions. The convergence criterium for the external loop is $\|\mathbf{p}_r^{n+1} - \mathbf{p}_{r-1}^{n+1}\| \leq max(10^{-3}\|\mathbf{p}_r^{n+1}\|, (10^{-3})^2)$.

Figure 5.15 shows the obtained results for Fe^{2+}, Fe^{3+} and pH both when solubility equilibrium is considered (purple line) and when it is not (green line). The precipitation of $Fe(OH)_3$, which starts less than five seconds after the beginning of the simulation, brings as a consequence a decrease in the pH (see Figure 5.15 C) because H^+ is released in reaction $l = 4$. The solubility reaction also constrains the concentration of Fe^{3+}, as it can be seen in Figure 5.15 B. Finally, Figure 5.15 A shows that the solubility equilibrium has no influence on the concentration of Fe^{2+}, as it is only involved in reaction $l = 1$ that proceeds at a constant rate.

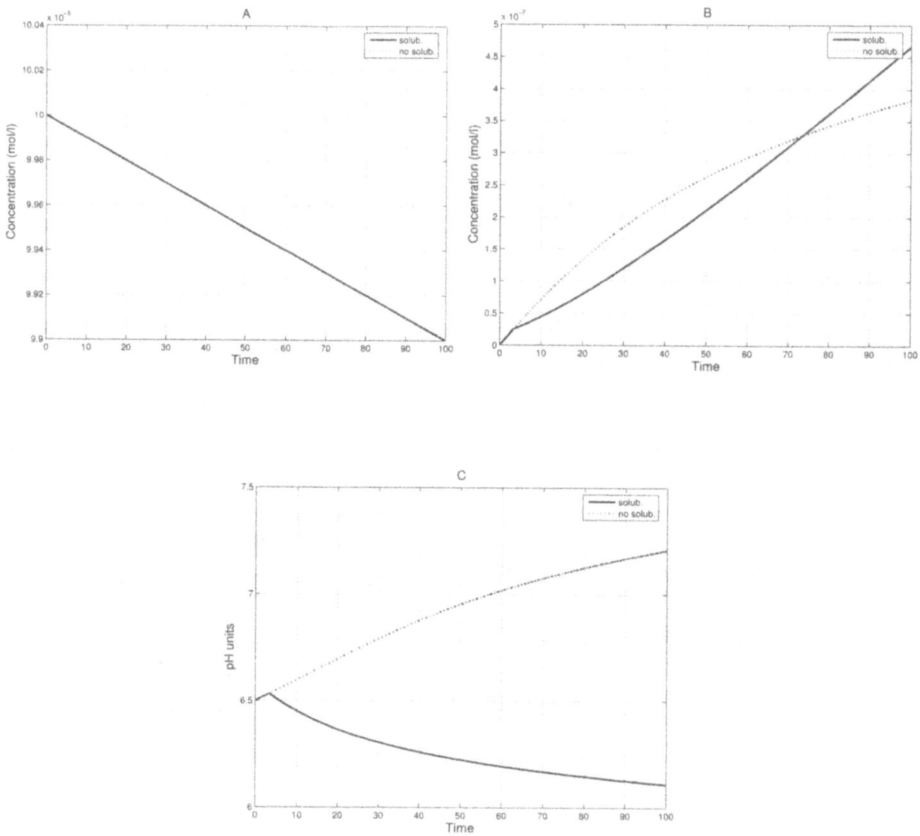

Figure 5.15 Time evolution of: **A:** concentration of Fe^{2+}, **B:** concentration of Fe^{3+} and **C:** pH.

EXERCISES

5.1 Obtain Eq. (5.8).

5.2 From the expression of the curl operator in cylindrical coordinates (5.9) and assuming that

$$\mathbf{A} = A_\theta(r, z)\mathbf{e}_\theta$$

obtain successively,

- **curlA**,
- **curlcurlA**.

5.3 Obtain Eq. (5.11).

5.4 Assume that that $\mathcal{A}$ and $\mathcal{B}$ are harmonic fields with angular frequency ω and complex amplitudes **A** and **C**, respectively [see (5.2)]. Prove the equality,

$$\frac{\omega}{2\pi}\int_0^{\frac{2\pi}{\omega}} \mathcal{A}(\mathbf{x}, t) \cdot \mathcal{C}(\mathbf{x}, t)\, dt = \frac{1}{2}\mathrm{Re}(\mathbf{A}(\mathbf{x}) \cdot \overline{\mathbf{C}(\mathbf{x})}).$$

5.5 In Section 5.3.2, the equation that allows for the calculation of the water quality of a pit lake according to the stirred tank model was derived. Extend this calculation to the period that ranges from the beginning of flooding to the moment in which it is full.

5.6 The calculation of the water quality of a system that is considered to be in equilibrium was done in Section 5.3.3.2 for a unique equilibrium reaction. Extend this calculation to the full set of equilibrium reactions.

REFERENCES

1. Bermúdez, A., *Continuum Thermomechanics*, Birkhäuser, (2005).
2. Bermúdez, A., Bullón, J., Pena, F. and Salgado, P., A numerical method for transient simulation of metallurgical compound electrodes, *Finite Elements in Analysis and Design*, 39, 283–299 (2003).
3. Bermúdez, A. and García-García, L. M., Mathematical modeling in chemistry. Application to water quality problems, *Applied Numerical Mathematics* (DOI: 10.1016/j.apnum.2011.05.002) (2011).
4. Bermúdez, A. and C. Moreno, C., Duality methods for solving variational inequalities, *Computers and Mathematics with Applications*, 7, 43–58 (1981).
5. Castro, J. M. and Moore, J. N., Pit lakes: their characteristics and the potential for their remediation. *Environmental Geology*, 39, n. 11, 1254–1260 (2000).
6. Chapman, A. J., *Fundamentals of Heat Transfer*, Collier McMillan, London (1987).
7. García García, Luz M., *Numerical resolution of water quality models: application to the closure of open pit mines*, Ph.D. thesis, University of Santiago de Compostela, Spain (2010).

8. Davis, A., Lyons, W. B. and Miller, G. C., Understanding the water quality of pit lakes, *Environmental Science and Technology,* 30, n. 3, 118A–123A (1996).
9. Johnk, C. T. A., *Engineering Electromagnetic Fields and Waves*, Springer, Berlin (2001).
10. Lery, T., Primicerio, M., Esteban, M.J., Fontes, M., Maday, Y., Mehrmann, V., Quadros, G., Schilders, W., Schuppert, A., Tewkesbury, H. (Eds.) *European Success Stories in Industrial Mathematics*, Springer, Heidelberg (2011).
11. McAdams, W. H., *Heat transmission*, McGraw-Hill, New York (1954).
12. Morel, F. M. M. and Hering, J. G., *Principles and applications of Aquatic Chemistry*, John Wiley and Sons, New York (1993).
13. Schei, A., Tuset, J.K., and Tveit, H., *Production of High Silicon Alloys.* Tapir Forlag, Trondheim (1998).
14. Smith, W. R. and Missen, R. W., *Chemical Reaction Equilibrium Analysis: Theory and Algorithms*, Wiley, New York (1982).

CHAPTER 6

BINARY AND ORDINAL DATA ANALYSIS IN ECONOMICS: MODELING AND ESTIMATION

IVAN JELIAZKOV AND MOHAMMAD ARSHAD RAHMAN

Department of Economics, University of California, Irvine

6.1 INTRODUCTION

This chapter is concerned with the analysis of statistical models for binary and ordinal outcomes. Binary data arise when a particular response variable of interest y_i can take only two values, i.e., $y_i \in \{0, 1\}$, where the index $i = 1, \ldots, n$ refers to units in the sample such as individuals, families, firms, and so on. Such dichotomous outcomes are widespread in the social and natural sciences. For example, to understand socio-economic processes, economists often need to analyze individuals' binary decisions such as whether to make a particular purchase, participate in the labor force, obtain a college degree, see a doctor, migrate to a different country, or vote in an election. By convention, $y_i = 1$ typically indicates the occurrence of the event of interest, whereas the occurrence of its complement is denoted by $y_i = 0$.

Mathematical Modeling with Multidisciplinary Applications.
Edited by Xin-She Yang

We also examine modeling and estimation issues related to another type of data, called ordinal data, where y_i can take one of J ordered values, $j = 1, \ldots, J$. The defining feature of ordinal data is that even though the outcomes are monotone, the scale on which they are measured is not assumed to be cardinal and differences between categories are not directly comparable. For instance, in quantifying survey responses on consumer satisfaction, 1 could be assigned to "very unhappy," 2 to "not too happy," 3 to "happy," and 4 to "very happy," but even though the scale tells us that 4 implies more happiness than 2, this does not mean that 4 implies twice as much happiness as 2, or that the difference in happiness between 1 and 3 is the same as that between 2 and 4. Even though ordinal data models were developed primarily for the analysis of data on rankings, they offer a flexible modeling framework that can also be very useful in the analysis of certain types of count data.

In this chapter we pursue several goals. We briefly review relevant results from the theory of choice which formalize the link between economic theory and empirical practice in binary and ordinal data analysis. We then turn our attention to the topic of estimation and highlight the identification issues that arise in binary and ordinal models. We review both classical and Bayesian approaches to estimation, and introduce a new simulation-based estimation algorithm for logit models based on data augmentation. Even though the theoretical foundations for this algorithm have been available for decades, the approach has remained unexploited until now. Our estimation approach removes important obstacles that have hindered extensions of logistic regression to multivariate and hierarchical model settings.

Another topic that we examine here is covariate effect estimation, which allows us to evaluate the impact of particular covariates on the outcome of interest and gives concrete practical meaning to the parameters of the model. The techniques are illustrated in two applications in economics including women's labor force participation and educational attainment. The methods discussed here form a foundation for studying other more complex recent developments in the literature such as extensions to panel data, multivariate and multinomial outcomes, dynamics, mixed models, and copula models.

6.2 THEORETICAL FOUNDATIONS

There exist a number of statistical models for binary and ordinal data, but they share a common foundation in which the observed discrete outcomes can be represented by the crossing of particular thresholds by an underlying continuous latent variable. This latent variable threshold-crossing formulation can in turn be related to the theory of choice in economics to form an elegant link between behavioral and statistical models. Because of this link, models for discrete data in econometrics are also frequently referred to as discrete choice models. The derivations are important because the latent variable representation turns out to be particularly useful not only in theory, but also in estimation. It also helps clarify the relationship between empirical models based on different distributional assumptions and provides a basis for the calculation of important quantities in economics, such as consumer surplus or willingness to pay. Note, however, that the econometric techniques are fully general and can be used to represent various phenomena that do not necessarily entail references to utility or choice (e.g., weather patterns, accident probabilities, volcanic eruptions, etc.).

In order for the decision problem to be well-posed, the set of available alternatives, or *choice set*, must be defined so that alternatives are (i) *mutually exclusive*, i.e., they represent distinct non-overlapping outcomes, and (ii) *exhaustive*, so that all possible outcomes are fully accounted for. These criteria are easily satisfied in the context of binary and ordinal data where the dependent variable y_i is simply an indicator variable for the occurrence of a particular event. One should keep in mind, that while in some contexts the dichotomy can be a natural feature of the data (e.g., medical tests, welfare program participation, home ownership, criminal recidivism, etc.), in other cases it can be introduced subjectively by the researcher to study a particular socio-economic phenomenon. For example, in studying market participation, a researcher may set $y_i = 1$ for producers whose sales in a given market are positive and $y_i = 0$ for all others. At first glance this discretization may seem unreasonable as it leads to loss of information on magnitudes (since both small and large sellers are treated alike). However, economic theory suggests that the presence of fixed costs leads firms to treat market entry and exit differently than the problem of how much to produce conditionally on being in the market. For this reason, the delineation of firms into market participants (regardless of sales volume) and non-participants (those with zero sales) can be an important first step in studying market outcomes. In the case of ordinal data, the outcomes will easily satisfy the first criterion if the dependent variable $y_i \in \{1, \ldots, J\}$ is defined as the sum of indicator variables over a particular monotone set of events. The second criterion, on the other hand, can either be satisfied naturally if outcomes are measured on a finite scale (as in surveys, or bond and stock ratings) or may have to be imposed by specifying a composite category that captures all possible outcomes beyond a certain value (as is common in the analysis of count data). Therefore, the nature of the choice set in binary and ordinal data models is in sharp contrast with standard models for continuous dependent variables, such as consumption or growth.

6.2.1 Binary Outcomes

The roots of the random utility framework that underlies discrete choice models in econometrics can be traced back to the pioneering work of Refs. [15], [16], and [17]. A detailed recent review with applications to problems in modern econometrics is given in Ref. [25]. The basic setup involves utility maximizing decision makers, who choose among competing alternatives associated with certain levels of utility. The theory is quite general and can handle a variety of possible choices; the same ideas apply in our binary data context where there are only two possible alternatives. Specifically, individual i has two levels of utility, U_{i1} and U_{i0}, that are associated with $y_i = 1$ or $y_i = 0$, respectively. The utility maximizing agent then selects the option providing the higher of the two utilities:

$$y_i = \begin{cases} 1, & \text{if } U_{i1} > U_{i0}, \\ 0, & \text{otherwise.} \end{cases}$$

The utilities U_{i1} and U_{i0} are known to the decision maker but are unknown to the researcher, who can only observe a vector x_i of characteristics of the decision maker that can be related to utility through $U_{ij} = x_i'\beta_j + \varepsilon_{ij}$ for $j = 0, 1$. The term $x_i'\beta_j$ is sometimes referred to as *representative utility*, whereas ε_{ij} captures unobserved factors that affect utility but are not included in $x_i'\beta_j$. In essence, $x_i'\beta_j$ is a systematic component and $\varepsilon_{i,j}$ is a stochastic (from the point of view of the researcher) part of individual utility.

This theoretical setup will be used to make probabilistic statements about the observed choices y_i conditionally on x_i. In the remainder of this chapter, conditioning of one variable on another will be denoted by a vertical bar '$|$', for example, $\Pr(A|B)$ will represent the conditional probability of A given B. Similarly, if s is a continuous random variable $f(s|t)$ will be used to denote the conditional density of s given t. In some contexts, when it is important to make clear the link between a random variable and its density, we may use notation such as $s|t$ to emphasize that we are interested in a random variable with density $f(s|t)$, i.e., $s|t \sim f(s|t)$, as opposed to a random variable s with density $f(s)$, i.e., $s \sim f(s)$.

To develop a model for the observed choices, note that given x_i and the parameters β_0 and β_1, the conditional probability of observing $y_i = 1$ can be expressed as an exceedance probability between the two utility levels

$$\begin{aligned} \Pr(y_i = 1|x_i, \beta_0, \beta_1) &= \Pr(U_{i1} > U_{i0}) \\ &= \Pr(x_i'\beta_1 + \varepsilon_{i1} > x_i'\beta_0 + \varepsilon_{i0}) \\ &= \Pr[(\varepsilon_{i0} - \varepsilon_{i1}) < x_i'(\beta_1 - \beta_0)]. \end{aligned} \tag{6.1}$$

The model is operationalized by specifying a density for the random variable $(\varepsilon_{i0} - \varepsilon_{i1})$, but before we consider specific cases, we need to address the important topic of parameter identification. From Eq. (6.1) we see that the choice probability depends only on the differences in utilities between alternatives, not on the absolute level of utilities. Specifically, because the probability in (6.1) depends on the difference $(\beta_1 - \beta_0)$, it will not change if we add an arbitrary constant c to both β_0 and β_1, i.e., $x_i'(\beta_1 - \beta_0) = x_i'(\tilde{\beta}_1 - \tilde{\beta}_0)$, where $\tilde{\beta}_1 = \beta_1 + c$ and $\tilde{\beta}_0 = \beta_0 + c$. Second, the scale of utility is not identified because the probability is unchanged if both sides of (6.1) are multiplied by an arbitrary constant $c > 0$, i.e., $\Pr[(\varepsilon_{i0} - \varepsilon_{i1}) < x_i'(\beta_1 - \beta_0)] = \Pr[c(\varepsilon_{i0} - \varepsilon_{i1}) < cx_i'(\beta_1 - \beta_0)]$.

To deal with these problems, we need to fix both the location and scale of utility. The location is fixed by measuring utility relative to that of the baseline category, U_{i0}. In other words, we work with the differenced form

$$z_i = x_i'\beta + \nu_i, \qquad i = 1, \ldots, n, \tag{6.2}$$

where $z_i = U_{i1} - U_{i0}$, $\beta = \beta_1 - \beta_0$, and $\nu_i = \varepsilon_{i1} - \varepsilon_{i0}$. As a result, the relationship between the observed outcome y_i and the latent z_i is given by

$$y_i = \begin{cases} 1, & \text{if } z_i > 0, \\ 0, & \text{otherwise,} \end{cases} \tag{6.3}$$

which can alternatively be written as $y_i = 1\{z_i > 0\}$ using the indicator function $1\{\cdot\}$ that takes the value 1 if its argument is true and 0 otherwise. The scale of utility is normalized by fixing the variance of ν_i and treating it as given rather than as a parameter to be estimated; doing so is only a normalization that does not restrict the underlying flexibility of the model. (One should keep in mind that the value at which the variance of ν_i is fixed will be model specific.) In the following examples, we review the three most common model specifications used in empirical analysis — probit, logit, and t-link or robit.

■ EXAMPLE 6.1

The probit model is obtained by assuming that the errors in (6.2) follow a standard normal distribution $\nu_i \sim N(0, 1)$ with probability density function (pdf) and cumulative distribution function (cdf) given by

$$\phi(\nu_i) = (2\pi)^{-1/2} e^{-\nu_i^2/2} \quad \text{and} \quad \Phi(\nu_i) = \int_{-\infty}^{\nu_i} \phi(t) dt.$$

Note that the pdf $\phi(\cdot)$ is symmetric and the variance of ν_i is fixed at 1 as a normalization. In addition, even though the expression for the Gaussian cdf $\Phi(\cdot)$ does not have a closed form solution, it is readily available in most statistical software packages.

■ EXAMPLE 6.2

The logit model is obtained by assuming that the errors in (6.2) follow a logistic distribution whose cdf $F_L(\cdot)$ and pdf $f_L(\cdot)$ are explicitly available (see Exercise 6.1 for a derivation of $f_L(\cdot)$ from $F_L(\cdot)$):

$$F_L(\nu_i) = \left(1 + e^{-\nu_i}\right)^{-1} \quad \text{and} \quad f_L(\nu_i) = F_L(\nu_i)[1 - F_L(\nu_i)].$$

The logistic distribution is symmetric with mean 0, variance $\pi^2/3$, and heavier tails than the normal distribution. The tail mass makes it more likely to observe "nonconforming" behavior such as choosing $y_i = 0$ for large positive $x_i'\beta$ or $y_i = 1$ for large negative $x_i'\beta$. In Exercise 6.2, we derive another well known result (see Refs. [16] and [18]) that the logit choice probabilities are obtained if the errors ε_{i0} and ε_{i1} in (6.1) follow an extreme value type I distribution.

EXAMPLE 6.3

The t-link or "robit" model is obtained by assuming that the errors in (6.2) follow a standard Student's t distribution with τ degrees of freedom. The distribution is symmetric around 0, has variance $\tau/(\tau-2)$ for $\tau > 2$, and its pdf $f_{T_\tau}(\cdot)$ and cdf $F_{T_\tau}(\cdot)$ are given by

$$f_{T_\tau}(\nu_i) = \frac{\Gamma(\frac{\tau+1}{2})}{\Gamma(\frac{\tau}{2})\sqrt{\tau\pi}}\left(1 + \frac{\nu_i^2}{\tau}\right)^{-\frac{\tau+1}{2}} \quad \text{and} \quad F_{T_\tau}(\nu_i) = \int_0^{\nu_i} f_{T_\tau}(s)ds,$$

where $\Gamma(s) = \int_0^\infty t^{s-1}e^{-t}dt$ denotes the gamma function (which equals $(s-1)!$ for positive integer values of s). Note that the variance of the t distribution is larger than in the probit case but approaches 1 for $\tau \to \infty$. Also, the cdf $F_{T_\tau}(\cdot)$ does not have a closed form solution, but is readily available in most statistical software packages.

An appealing feature of the t-link model is its flexibility: low values of τ produce heavier tails than the logistic distribution, setting $\tau \approx 8$ approximates the logit model, and as $\tau \to \infty$, the t distribution approximates the standard normal. Figure 6.1 shows the log-densities for the standard normal, scaled logistic and scaled t with 4 degrees of freedom (the scaling is done so that all three variances are 1). Because the t-link offers a modeling approach that is robust to variations in the tail behavior of the latent z_i, it has also been referred to by the portmanteau word "robit" ("robust" + the suffix "-it" to resemble probit and logit).

Given the three specifications we have just considered, we can now obtain the outcome probabilities $\Pr(y_i = 1|\beta)$ and $\Pr(y_i = 0|\beta) = 1 - \Pr(y_i = 1|\beta)$ (we suppress the dependence of these probabilities on x_i for notational convenience). In particular, from (6.2) and (6.3) and under the assumption

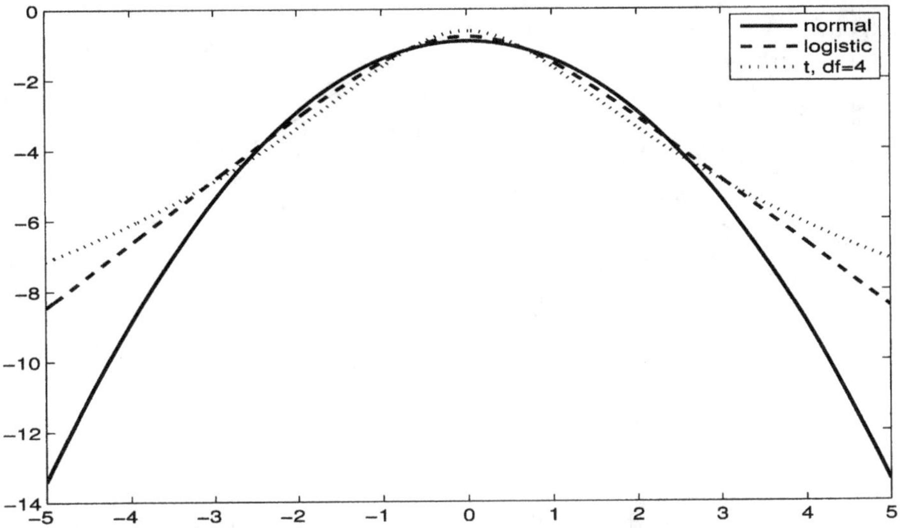

Figure 6.1 Log-densities for the standard normal, scaled logistic and Student's t with 4 degrees of freedom.

that the density of ν_i is symmetric, we have that

$$\begin{aligned}\Pr(y_i = 1|\beta) &= \Pr(z_i > 0)\\ &= \Pr(x_i'\beta + \nu_i > 0)\\ &= 1 - \Pr(\nu_i < -x_i'\beta)\\ &= 1 - [1 - \Pr(\nu_i < x_i'\beta)]\\ &= \Pr(\nu_i < x_i'\beta) = F(x_i'\beta),\end{aligned}$$

where $F(\cdot)$ is the assumed cdf of ν_i – as before, $F(\cdot) = \Phi(\cdot)$ produces the probit model, $F(\cdot) = F_L(\cdot)$ leads to logit, and $F(\cdot) = F_{T_\tau}(\cdot)$ gives the t-link model. Symmetry is used in obtaining the second to last line, and while all models considered here involve symmetric distributions, readers are cautioned to be careful in general.

6.2.2 Ordinal Outcomes

We now turn attention to models for ordinal data, where the alternatives are inherently ordered or ranked. Common applications that involve ordered outcomes include sentiment or opinion surveys, quality tests, health assessment studies, the level of employment (unemployed, part-time, full-time), the level and usage of insurance, and others.

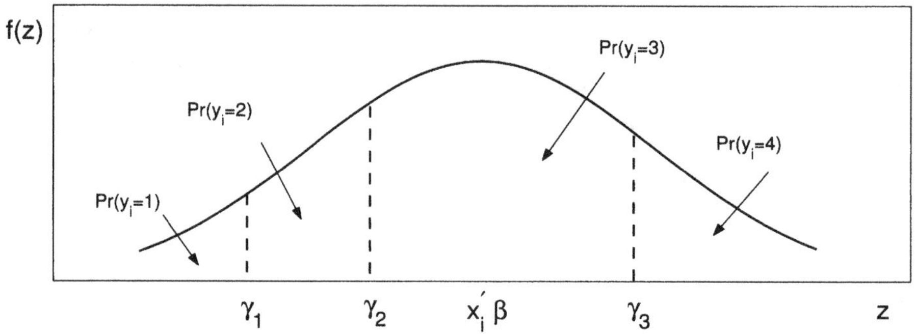

Figure 6.2 Outcome probabilities in an ordinal data model.

Similarly to the models studied in Section 6.2.1, ordinal data models can be motivated by an underlying latent variable threshold-crossing framework. In particular, as in (6.2) we assume that a continuous latent random variable z_i depends on a k-vector of covariates x_i through the relationship $z_i = x_i'\beta + \nu_i$, $i = 1, \ldots, n$, but with the difference that the observed outcomes $y_i \in \{1, \ldots, J\}$ arise according to

$$y_i = j \quad \text{if} \quad \gamma_{j-1} < z_i \leq \gamma_j, \tag{6.4}$$

where $-\infty = \gamma_0 < \gamma_1 < \ldots < \gamma_{J-1} < \gamma_J = \infty$ are cutpoint parameters that determine the discretization of the data into J ordered categories. An alternative way of writing (6.4) is to let $y_i = \sum_{j=1}^{J} 1\{z_i > \gamma_{j-1}\}$. Given this representation and a particular cdf $F(\nu_i)$, the probability of observing $y_i = j$, conditional on β and $\gamma = (\gamma_1, \ldots, \gamma_{J-1})'$, is given by

$$\Pr(y_i = j|\beta, \gamma) = F(\gamma_j - x_i'\beta) - F(\gamma_{j-1} - x_i'\beta). \tag{6.5}$$

Figure 6.2 depicts the probabilities of y_i falling in category j as determined by (6.5) in a four-category setting. As before, various choices of the cdf $F(\cdot)$ are possible—e.g., $F(\cdot) = \Phi(\cdot)$, $F(\cdot) = F_L(\cdot)$, $F(\cdot) = F_{T_\tau}(\cdot)$, and so on—but the ordinal probit model is one of the most practical because it is tractable in univariate cases and can be easily generalized to flexible multivariate and hierarchical settings. In contrast, the logistic distribution can not handle correlations in multivariate settings.

As with models for binary data, we require both location and scale restrictions in order to identify the parameters. To see the need for doing so, note that the probabilities in (6.5) are invariant to shifting and rescaling the parameters by some arbitrary constants c and $d > 0$ because

$$F(\gamma_j - x_i'\beta) = F(\gamma_j + c - (x_i'\beta + c))$$

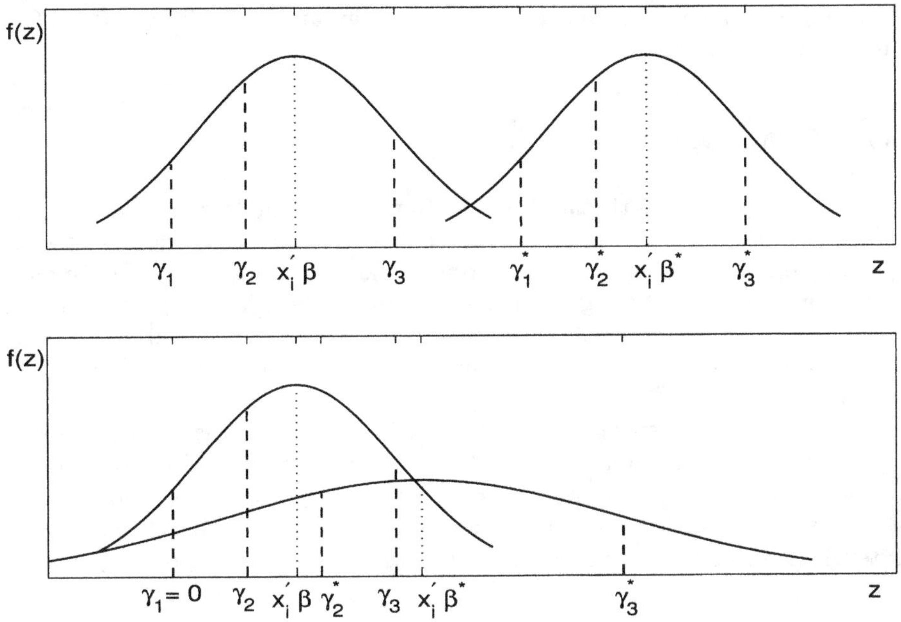

Figure 6.3 Parameter identification in ordinal data models.

and

$$F(\gamma_j - x_i'\beta) = F\left(\frac{\gamma_{j-1}d - x_i'\beta d}{d}\right),$$

which can be applied to both terms in (6.5) without affecting $\Pr(y_i = j|\beta, \gamma)$. The first identification problem is easily corrected by fixing a cutpoint – in particular, letting $\gamma_1 = 0$ removes the possibility for shifting the distribution without changing the probability of observing y_i. As in binary data models, we resolve the possibility for rescaling (our second identification problem) by fixing the variance of ν_i. The variance equals 1 in the probit case, $\pi^2/3$ in the logit case, and $\tau/(\tau - 2)$ in the t-link model.

Figure 6.3 illustrates these identification considerations. The first panel in the figure illustrates that shifting the density and all cutpoints leaves the probability unaffected; the second panel shows that even if one sets $\gamma_1 = 0$, in the absence of a scale restriction, one can simultaneously rescale $F(\cdot)$, the mean, and the remaining cutpoints without affecting $\Pr(y_i = j|\beta, \gamma)$.

In addition to fixing one cutpoint and the variance of ν_i, there are other possible ways to achieve parameter identification. For example, as an alternative to letting $\gamma_1 = 0$, it is possible to identify the parameters by dropping the intercept term from $x_i'\beta$. Moreover, instead of fixing the variance of ν_i, one can impose a scale restriction by fixing two cutpoints (e.g., $\gamma_1 = 0$ and $\gamma_{J-1} = 1$). The presence and effectiveness of these alternative approaches has been examined in Ref. [13] and the references therein, however, these alternatives will not be examined here because they are primarily of interest in multivariate models.

6.3 ESTIMATION

This section reviews both classical and Bayesian methods for estimating the models considered in Section 6.2. Classical estimation in this class of models typically employs the method of maximum likelihood, which requires numerical optimization of the log-likelihood function. Bayesian estimates, on the other hand, are generally obtained by Markov chain Monte Carlo (MCMC) simulation methods such as Gibbs sampling or the Metropolis-Hastings algorithm.

In addition to reviewing existing estimation methods, this chapter also introduces a new estimation algorithm for logit models that has been overlooked in the literature. The method not only supplements our toolkit for dealing with logistic regression, but also lays a foundation for estimating important extensions of the logit model to multivariate and hierarchical settings.

6.3.1 Maximum Likelihood Estimation

Consider a set of observations $y = (y_1, \ldots, y_n)'$ that comes from some statistical model with sampling density $f(y|\theta)$ written in terms of a parameter vector θ. Because $f(y|\theta)$ provides a mathematical description of the probabilistic phenomenon that generates the observed data sample y given θ, it is called the data generating process. Note that the data generating process is a function of the data conditionally on the parameters, and indeed we can think of it as the mathematical model by which, given θ, nature generates y. In practice, empirical researchers see the sample y generated from $f(y|\theta)$, but do not know the value of θ. When $f(y|\theta)$ is viewed as a function of the parameter vector θ given the sample y, it is called the likelihood function. Although the two functions refer to the same object, $f(y|\theta)$, they emphasize (and take as arguments) its two different components. A thorough review of likelihood inference can be found in standard statistics and econometrics references such as [11].

The maximum likelihood estimator (or MLE) is defined as the value of θ that maximizes the log-likelihood function

$$\hat{\theta}_{MLE} = \arg\max_{\theta} \ln f(y|\theta), \tag{6.6}$$

or heuristically, it is the value of θ that makes the observed sample y as "likely" as possible within the confines of the assumed data generating process. Note that because the logarithmic transformation is monotone, the value $\hat{\theta}_{MLE}$ that maximizes $\ln f(y|\theta)$ also maximizes $f(y|\theta)$, however, it is common to work with $\ln f(y|\theta)$ because it is more stable and easier to evaluate than

$f(y|\theta)$, and also because the most important statistical properties of $\hat{\theta}_{MLE}$ are associated with features of $\ln f(y|\theta)$. Specifically, it is known that under mild regularity conditions, the maximum likelihood estimator $\hat{\theta}_{MLE}$ defined in (6.6) is consistent and asymptotically normally distributed. Consistency means that as the sample size $n \to \infty$, the probability limit (or plim) of $\hat{\theta}_{MLE}$ is the true value θ_0, i.e., $\text{plim}\,\hat{\theta}_{\text{MLE}} = \theta_0$. Asymptotic normality means that in large samples, as $n \to \infty$,

$$\hat{\theta}_{MLE} \sim N\left(\theta_0, V^{-1}\right),$$

where V is the Fisher information defined as the negative of the expected value of the second derivative (or Hessian) matrix of the log-likelihood

$$V = -E\left[\frac{\partial^2 \ln f(y|\theta)}{\partial\theta\partial\theta'}\right]$$

evaluated at θ_0 and the expectation is taken with respect to $f(y|\theta_0)$. Because it is typically impossible to evaluate this expectation, it is common to approximate V by the observed Hessian

$$V = -\frac{\partial^2 \ln f(y|\theta)}{\partial\theta\partial\theta'},$$

which is evaluated at the maximum likelihood value $\theta = \hat{\theta}_{MLE}$. The standard errors of the individual elements of $\hat{\theta}_{MLE}$ are given by the square root of the diagonal entries of V^{-1}, and those can be used in testing and constructing confidence intervals. Next, we consider the likelihood functions for the models studied in Section 6.2.

For the binary data models that we considered in Section 6.2.1, the likelihood function can be written as

$$\begin{aligned} f(y|\beta) &= \Pr(y_1, y_2, \ldots, y_n|\beta) \\ &= \prod_{i=1}^{n} \Pr(y_i|\beta) \\ &= \left\{\prod_{i:y_i=1} F(x_i'\beta)\right\}\left\{\prod_{i:y_i=0} [1 - F(x_i'\beta)]\right\} \\ &= \prod_{i=1}^{n} [F(x_i'\beta)]^{y_i} [1 - F(x_i'\beta)]^{(1-y_i)}, \end{aligned} \tag{6.7}$$

where the second line follows by assuming independence among the observations and the last line is simply a convenient expression for picking the relevant probability. This likelihood function captures all three binary data models discussed in Section 6.2.1 — probit, logit, and t-link — which could be obtained by using the appropriate cdf in place of $F(\cdot)$ as discussed in Section 6.2.1.

In order to find the maximum likelihood estimate $\hat{\beta}_{MLE}$, we maximize the log-likelihood function

$$\ln f(y|\beta) = \sum_{i=1}^{n} \{y_i F(x_i'\beta) + (1 - y_i)[1 - F(x_i'\beta)]\},$$

which is typically done iteratively using standard hill climbing algorithms such as Newton-Raphson or BHHH (see Ref. [3]) because the first-order condition for maximization

$$\frac{\partial \ln f(y|\beta)}{\partial \beta} \equiv \sum_{i=n}^{n} \left[\frac{y_i f(x_i'\beta)}{F(x_i'\beta)} - (1 - y_i) \frac{f(x_i'\beta)}{1 - F(x_i'\beta)} \right] x_i = 0$$

does not admit an explicit analytical solution even though the log-likelihood is typically well behaved (unimodal and concave) in this class of models.

Turning attention to ordinal outcomes, Eq. (6.5) and the assumption of independent sampling give the following likelihood function for the ordinal data model

$$\begin{aligned} f(y|\beta, \gamma) &= \prod_{i=1}^{n} \Pr(y_i|\beta, \gamma) \\ &= \prod_{i=1}^{n} \left[F(\gamma_j - x_i'\beta) - F(\gamma_{j-1} - x_i'\beta) \right], \end{aligned} \tag{6.8}$$

where the index j on the cutpoints in the second line is determined by the realization of y_i (recall that because y_i takes values in $\{1, \ldots, J\}$, it can be used for indexing the cutpoints, i.e. $\gamma_j = \gamma_{y_i}$ and $\gamma_{j-1} = \gamma_{y_i - 1}$).

A minor complication arises in maximizing $\ln f(y|\beta, \gamma)$ because the values of the free cutpoints must satisfy an ordering constraint: $\gamma_1 = 0 < \gamma_2 < \ldots < \gamma_{J-1}$. In order to avoid the computational complexities associated with constrained optimization, it is useful to reparameterize the problem in order to remove those constraints. For example, optimization can be simplified by transforming the cutpoints γ so as to remove the ordering constraint by the one-to-one map

$$\delta_j = \ln(\gamma_j - \gamma_{j-1}),\ 2 \leq j \leq J - 1, \tag{6.9}$$

and rewriting the likelihood as a function of β and $\delta = (\delta_2, \ldots, \delta_{J-1})'$, i.e. drawing inferences from $f(y|\beta, \delta)$. Other transformations have been considered in Ref. [4] and comparisons have been drawn in Ref. [13], but these transformations relate to alternative identification restrictions of the scale of the model and will not be examined here.

6.3.2 Bayesian Estimation

In contrast to classical (or frequentist) inference, which only involves the likelihood function $f(y|\theta)$, Bayesian analysis rests on Bayes' theorem

$$\pi(\theta|y) = \frac{f(y|\theta)\pi(\theta)}{\int f(y|\theta)\pi(\theta)d\theta},$$

and inference is based on the posterior density $\pi(\theta|y)$, which is proportional to the product of the likelihood and the prior density $\pi(\theta)$. There are important theoretical advantages of Bayesian analysis over classical inference, which have been carefully reviewed in [10], [14], [21], and [23]. For example, the posterior density allows for finite sample inferences about the unknown parameter vector θ that incorporates information from the observed sample (which enters through the likelihood) and non-sample information (e.g., from previous studies, theoretical considerations, the researcher's experience, etc.), which enters through the prior. In addition to their finite sample properties, Bayesian estimators also have desirable asymptotic properties (as $n \to \infty$).

An important practical benefit of Bayesian estimation is that inference is possible even in models where the likelihood $f(y|\theta)$ is difficult to evaluate and hence maximum likelihood estimation is infeasible. In those cases, progress has been made possible by recent advances in simulation-based estimation and data augmentation which allow sampling from $\pi(\theta|y)$ without requiring evaluation of $f(y|\theta)$. Such simulation methods, based on MCMC theory, have enabled inference in previously intractable applications. Once a sample of draws $\{\theta\}$ from $\pi(\theta|y)$ is available, those draws can be used to summarize features of the posterior (such as mean, variance, quantiles, etc.) and construct point and interval estimates.

For the binary and ordinal data models we have examined in this chapter, Bayes' theorem will lead to a posterior density

$$\pi(\theta|y) \propto f(y|\theta)\pi(\theta)$$

that typically does not belong to a known family of distributions and cannot be sampled directly. This is because even if the prior $\pi(\theta)$ is selected from a well-known class of distributions (e.g., Gaussian), the parameters enter the likelihood $f(y|\theta)$ in such a way [note the nonlinearity in Eqs. (6.7) and (6.8)] that the posterior $\pi(\theta|y)$ does not have a recognizable analytical representation.

In this section we present tools for dealing with this problem in two ways. First, we discuss a general MCMC simulation technique, called the Metropolis-Hastings algorithm, that can be employed to produce draws from intractable distributions. Second, we review a method that circumvents the problem by augmenting the sampling scheme with an additional vector of variables in a way that restores tractability. The benefit of this approach, called data augmentation, is that it enables estimation by Gibbs sampling (another MCMC simulation technique). In the remainder of this Section, we review all of these methods and propose a new data augmentation algorithm for the logit model which has not appeared elsewhere in the literature.

6.3.2.1 Metropolis-Hastings Algorithm. The Metropolis-Hastings (MH) algorithm [19], [12], [24], [5] is a versatile Markov chain simulation method for non-standard distributions. Denoting the current value of θ by θ^c, it proceeds by generating a proposed value $\theta^p \sim q(\theta|y)$ from the proposal density $q(\cdot)$. In principle $q(\cdot)$ can depend on θ^c (e.g., in random walk proposal densities), but in this chapter we examine a version of the MH algorithm, called independence chain MH, in which the proposal density does not vary with θ^c. The proposed draw θ^p is accepted with probability

$$\alpha_{MH}(\theta^c, \theta^p|y) = \min\left\{1, \frac{f(y|\theta^p)\pi(\theta^p)q(\theta^c|y)}{f(y|\theta^c)\pi(\theta^c)q(\theta^p|y)}\right\},$$

and if θ^p is rejected, θ^c is repeated as the next value of θ in the Markov chain. As shown by [12] (also see [24], [5]), the limiting distribution of the draws of θ coming from the MH algorithm is $\pi(\theta|y)$. In practice, this means that after a transient phase (called the burn-in period), draws obtained by MH simulation can be viewed as coming from $\pi(\theta|y)$.

To apply the independence chain MH algorithm in our context, we note that a suitable proposal density can be obtained by employing the MLE results from Section 6.3.1. In particular, for any of the models studied in Sections 6.2.1 and 6.2.2, the proposal density can be constructed as a multivariate t density

$$q(\theta|y) = f_{T_\omega}(\theta|\hat{\theta}, a\Psi),$$

with mean $\hat{\theta} = \hat{\theta}_{MLE}$ and scale matrix $a\Psi$, where Ψ is given by the inverse of the negative Hessian of the log-likelihood

$$\Psi = -\left[\frac{\partial^2 \ln f(y|\theta)}{\partial\theta\partial\theta'}\right]^{-1}.$$

evaluated at $\theta = \hat{\theta}_{MLE}$, a is a scalar tuning parameter, and ω is the degrees of freedom of the proposal density. The tuning parameter a is typically taken to be $a \geq 1$ and ω is usually set at a small value, both of which are intended to ensure that the proposal has sufficiently heavy tails to explore the space more thoroughly. In the examples in this chapter, we use $\omega = 10$ and $a = 1.25$.

The independence chain MH algorithm can then be employed to estimate probit, logit and robit models for binary data using the likelihood in (6.7) with parameter vector $\theta = \beta$, or ordinal models using likelihood (6.8) written in terms of the transformed cutpoints in (6.9) whereby the parameter vector θ is given by $\theta = (\beta', \delta')'$.

6.3.2.2 Gibbs Sampling and Data Augmentation. Gibbs sampling (see Ref. [9]) is an MCMC method for simulation from a distribution when its full conditional densities have known form. To review the approach, suppose there are three parameter blocks θ_1, θ_2, and θ_3 with joint density $\pi(\theta_1, \theta_2, \theta_3|y)$. The Gibbs sampler produces draws $\{\theta_1, \theta_2, \theta_3\} \sim \pi(\theta_1, \theta_2, \theta_3|y)$ by sequentially drawing from the set of full conditional densities $\pi(\theta_1|y, \theta_2, \theta_3)$, $\pi(\theta_2|y, \theta_1, \theta_3)$ and $\pi(\theta_3|y, \theta_1, \theta_2)$. Under mild conditions, it can be shown that the Markov chain formed by the Gibbs sampler has a limiting invariant distribution that is the distribution of interest $\pi(\theta_1, \theta_2, \theta_3|y)$. This means that draws obtained by Gibbs sampling after the initial burn-in period, can be viewed as coming from $\pi(\theta_1, \theta_2, \theta_3|y)$. Some authors have likened the way in which the Gibbs sampler traverses the parameter space to the way a rook moves in chess. In addition, the particular order in which the full conditional densities are sampled does not affect the limiting distribution. A thorough review of the method and its applications in econometrics is offered in [6].

The application of Gibbs sampling to models for binary and ordinal data is complicated by the fact that the posterior and its full conditional densities are not of known form. However, a method known as data augmentation can be used to overcome this problem.

The idea behind data augmentation is simple. Instead of focusing on the intractable posterior density

$$\pi(\theta|y) \propto f(y|\theta)\pi(\theta),$$

we choose to work with $\pi(\theta, w|y)$, a density judiciously augmented with w in such a way that the full-conditionals $\pi(\theta|y, w)$ and $\pi(w|y, \theta)$ are tractable and can be sampled directly. As a result, a Gibbs sampler constructed using sequential sampling from $\pi(\theta|y, w)$ and $\pi(w|y, \theta)$ will produce draws $\{\theta, w\} \sim \pi(\theta, w|y)$.

But how do we relate the draws $\{\theta, w\}$ from $\pi(\theta, w|y)$ to our original goal of sampling $\theta \sim \pi(\theta|y)$? This is easily done by only collecting the draws of θ and simply ignoring w. The approach works because by the properties of cdfs, given two vectors of constants a_θ and a_w conformable with θ and w, respectively, the marginal cdf is obtained from the joint cdf as

$$F(a_\theta) \equiv \Pr(\theta \ll a_\theta) = \lim_{a_w \to \infty} F(a_\theta, a_w) \equiv \Pr(\theta \ll a_\theta, w \ll \infty),$$

where "$\ll$" is used to denote element-by-element weak inequality comparison. The condition $w \ll \infty$ always holds and in this sense we "simply ignore" w to obtain draws $\theta \sim \pi(\theta|y)$ from $\{\theta, w\} \sim \pi(\theta, w|y)$.

Having presented the theory behind data augmentation, we now discuss its specific application to the models considered in this chapter.

■ EXAMPLE 6.4

The binary probit model can be estimated easily, as shown in [1], if we were to introduce the latent $z = (z_1, \ldots, z_n)'$ from Eq. (6.2) into our MCMC simulation algorithm. Specifically, instead of working with $\pi(\beta|y)$, we specify a Gibbs sampler to simulate the augmented posterior $\pi(\beta, z|y)$, which can be written as

$$\begin{aligned}\pi(\beta, z|y) &\propto f(y|\beta, z)\, f(\beta, z) \\ &= f(y|\beta, z)\, f(z|\beta)\, \pi(\beta) \\ &= \left\{\prod_{i=1}^{n} f(y_i|z_i)\right\} f(z|\beta)\, \pi(\beta).\end{aligned} \tag{6.10}$$

Note that the last line of (6.10) involves terms that are easy to evaluate. In particular, $f(y_i|z_i) = 1\{z_i \in \mathcal{B}_i\}$, where

$$\mathcal{B}_i = \begin{cases} (0, \infty) & \text{if} \quad y_i = 1, \\ (-\infty, 0] & \text{if} \quad y_i = 0, \end{cases} \tag{6.11}$$

which follows from the relationship between y_i and z_i in binary data models. Note that conditionally on z_i, y_i does not depend on β. In addition, $f(z|\beta) = f_N(z|X\beta, I_n)$, where $X = (x_1', \ldots, x_n')'$ is the matrix of covariates and I_n denotes the $n \times n$ identity matrix; this follows from the latent variable representation of the probit model, namely $z_i = x_i'\beta + \nu_i$ with $\nu_i \sim N(0,1)$ for $i = 1, \ldots, n$. Finally, $\pi(\beta)$ is the prior distribution on β which we assume to be $f_N(\beta|\beta_0, B_0)$, i.e., β is assumed to be *a priori* normally distributed, i.e., $\beta \sim N(\beta_0, B_0)$.

A Gibbs sampler now can be easily constructed to explore $\pi(\beta, z|y)$ because the full conditional densities $\pi(\beta|y, z)$ and $\pi(z|y, \beta)$ are of known form. Specifically, $\pi(\beta|y, z)$ is proportional to the terms in (6.10) that involve β so that $\pi(\beta|y, z) \propto f(z|\beta)\, \pi(\beta)$, which technically does not depend on y. Because both $f(z|\beta)$ and $\pi(\beta)$ are normal, the full conditional is also normal and therefore we draw

$$\beta|y, z \sim N(\hat{\beta}, \hat{B}),$$

where $\hat{B} = (B_0^{-1} + X'X)^{-1}$ and $\hat{\beta} = \hat{B}(B_0^{-1}\beta_0 + X'z)$. Details of the derivation are considered in Exercise 6.3.

The density $\pi(z|y, \beta)$ is proportional to the terms in (6.10) that involve z so that

$$\begin{aligned}\pi(z|y, \beta) &\propto \left\{\prod_{i=1}^{n} 1\{z_i \in \mathcal{B}_i\}\right\} f_N(z|X\beta, I_n) \\ &= \prod_{i=1}^{n} \left[1\{z_i \in \mathcal{B}_i\} f_N(z_i|x_i'\beta, 1)\right],\end{aligned}$$

whereby $z|y, \beta$ is easily sampled by drawing z_i, $i = 1, \ldots, n$, from appropriately truncated normal densities

$$z_i|y_i, \beta \sim TN_{\mathcal{B}_i}(x_i'\beta, 1),$$

where the region of truncation $\mathcal{B}_i$ is defined in (6.11).

■ EXAMPLE 6.5

The t-link (robit) model can be estimated by extending the data augmentation approach presented in Example 6.4. The discussion follows [1] and rests on the result (see, e.g., [2]) that the t distribution can be represented as a scale mixture of normals. Specifically, if for $i = 1, \ldots, n$, λ_i has a gamma distribution

$$\lambda_i \sim G(\tau/2, \tau/2), \tag{6.12}$$

and conditionally on λ_i, we have

$$z_i|\lambda_i \sim N(x_i'\beta, 1/\lambda_i), \tag{6.13}$$

then marginally of λ_i, z_i is distributed

$$z_i \sim T_\tau(x_i'\beta, 1).$$

Therefore, letting $\lambda = (\lambda_1, \ldots, \lambda_n)'$, we can consider the augmented posterior

$$\begin{aligned}\pi(\beta, z, \lambda|y) &\propto f(y|\beta, z, \lambda)\, f(\beta, z, \lambda) \\ &= f(y|\beta, z, \lambda)\, f(z|\beta, \lambda)\, \pi(\beta)\pi(\lambda) \\ &= \left\{\prod_{i=1}^{n} f(y_i|z_i)\right\} f(z|\beta, \lambda)\, \pi(\beta)\pi(\lambda),\end{aligned} \tag{6.14}$$

where $f(y_i|z_i) = 1\{z_i \in \mathcal{B}_i\}$ as before, $f(z|\beta, \lambda) = f_N(z|X\beta, \Lambda^{-1})$ with $\Lambda = \text{diag}(\lambda)$ which follows from (6.13), $\pi(\beta) = f_N(\beta|\beta_0, B_0)$ is the prior on β, and $\pi(\lambda)$ is given by the product of n independent gamma densities stemming from (6.12)

$$\pi(\lambda) = \prod_{i=1}^{n} f_G(\lambda_i|\tau/2, \tau/2).$$

It is then quite straightforward to show that the Gibbs sampler for simulating from $\pi(\beta, z, \lambda|y)$ can be constructed by sequentially drawing from the following full conditionals

$$\beta|z, \lambda \sim N(\hat{\beta}, \hat{B}),$$

with $\hat{B} = (B_0^{-1} + X'\Lambda z)^{-1}$ and $\hat{\beta} = \hat{B}(B_0^{-1}\beta_0 + X'\Lambda z)$,

$$z_i|y, \beta, \lambda_i \sim TN_{\mathcal{B}_i}(x_i'\beta, \lambda_i^{-1}), \qquad i = 1, \ldots, n,$$

and

$$\lambda_i|y, \beta, z \sim G\left(\frac{\tau+1}{2}, \frac{\tau + (z_i - x_i'\beta)^2}{2}\right), \qquad i = 1, \ldots, n.$$

■ EXAMPLE 6.6

The logit model can be estimated by pursuing a new data augmentation scheme that has not been exploited in the literature. Because the logistic distribution can be written as a scale mixture of normals with respect to the Kolmogorov distribution [2, 22], we can write that

$$f_L(s|\mu) = \int f_N(s|\mu, 4\kappa^2) f_K(\kappa) d\kappa, \tag{6.15}$$

where $f_L(s|\mu)$ denotes the density of a random variable that has a logistic distribution around μ and variance $\pi^2/3$, and $f_K(\kappa)$ represents the Kolmogorov density $f_K(\kappa) = 8\kappa \sum_{j=1}^{\infty}(-1)^{j+1} j^2 e^{-2j^2\kappa^2}$. This implies, analogously to Example 6.5, that if κ_i has a Kolmogorov distribution and conditionally on κ_i, $z_i|\kappa_i \sim N(x_i'\beta, 4\kappa_i^2)$, then marginally of κ_i, z_i has logistic density $f_L(z_i|x_i'\beta)$.

Therefore, letting $\kappa = (\kappa_1, \ldots, \kappa_n)'$, we can consider the augmented posterior

$$\begin{aligned}\pi(\beta, z, \kappa|y) &\propto f(y|\beta, z, \kappa) f(\beta, z, \kappa) \\ &= f(y|\beta, z, \kappa) f(z|\beta, \kappa) \pi(\beta)\pi(\kappa) \\ &= \left\{\prod_{i=1}^{n} f(y_i|z_i)\right\} f(z|\beta, \kappa) \pi(\beta)\pi(\kappa).\end{aligned} \tag{6.16}$$

where $f(y_i|z_i) = 1\{z_i \in \mathcal{B}_i\}$, $f(z|\beta, \kappa) = f_N(z|X\beta, K)$ with $K = \text{diag}(4\kappa^2)$, $\pi(\beta) = f_N(\beta|\beta_0, B_0)$, and $\pi(\kappa) = \prod_{i=1}^{n} f_K(\kappa_i)$.

The resulting Gibbs sampler for simulating from $\pi(\beta, z, \kappa|y)$ is constructed by sequentially drawing from the following full conditionals

$$\beta|z, \kappa \sim N(\hat{\beta}, \hat{B}),$$

with $\hat{B} = (B_0^{-1} + X'K^{-1}z)^{-1}$ and $\hat{\beta} = \hat{B}(B_0^{-1}\beta_0 + X'K^{-1}z)$,

$$z_i|y, \beta, \kappa_i \sim TN_{\mathcal{B}_i}(x_i'\beta, 4\kappa_i^2), \qquad i = 1, \ldots, n,$$

and

$$\kappa_i|y, \beta, z_i \sim f(\kappa_i|z_i, \beta), \qquad i = 1, \ldots, n, \tag{6.17}$$

where $f(\kappa_i|z_i, \beta)$ does not belong to a known family of distributions. However, a very convenient result can be obtained by representing this distribution in terms of Bayes' formula as

$$\begin{aligned} f(\kappa_i|z_i, \beta) &= \frac{f(z_i|\beta, \kappa_i)f(\kappa_i)}{\int f(z_i|\beta, \kappa_i)f(\kappa_i)d\kappa_i} \\ &= \frac{f_N(z_i|x_i'\beta, 4\kappa_i^2)f_K(\kappa_i)}{f_L(z_i|x_i'\beta)}. \end{aligned} \tag{6.18}$$

The last line in (6.18) follows by recognizing that the numerator densities are Gaussian and Kolmogorov, and the denominator, by Eq. (6.15), is simply the logistic density. Therefore, the unknown $f(\kappa_i|z_i, \beta)$ can now be represented very simply in terms of other well-known densities.

The fact that $f(\kappa_i|z_i, \beta)$ can be evaluated explicitly means that one can also evaluate the corresponding cdf

$$F_{\kappa|z,\beta}(\kappa_i|z_i, \beta) = \int_0^{\kappa_i} f(s|z_i, \beta)ds.$$

In turn, $F_{\kappa|z,\beta}(\kappa_i|z_i, \beta)$ can be utilized to produce the draws needed in (6.17) by solving $\kappa_i = F_{\kappa|z,\beta}^{-1}(u)$, where $u \sim U(0, 1)$ is a uniform random variable on the unit interval. The latter technique is known as the inverse cdf method and follows because

$$\Pr(\kappa_i \leq a) = \Pr(F_{\kappa|z,\beta}^{-1}(u) \leq a) = \Pr(u \leq F_{\kappa|z,\beta}(a)) = F_{\kappa|z,\beta}(a).$$

This completes the proposed Gibbs sampling scheme for logit models. However, to provide additional intuition about the behavior of $f(\kappa_i|z_i, \beta)$ and compare it to the Kolmogorov distribution $f_K(\kappa_i)$, Figure 6.4 plots $f(\kappa_i|z_i, \beta)$ for two settings of $z_i - x_i'\beta$. The figure reveals that when z_i is close to the mean $x_i'\beta$ the mass of the distribution is closer to the origin than when z_i is far (in absolute terms) from $x_i'\beta$. This is to be expected because κ_i enters the conditional variance of z_i.

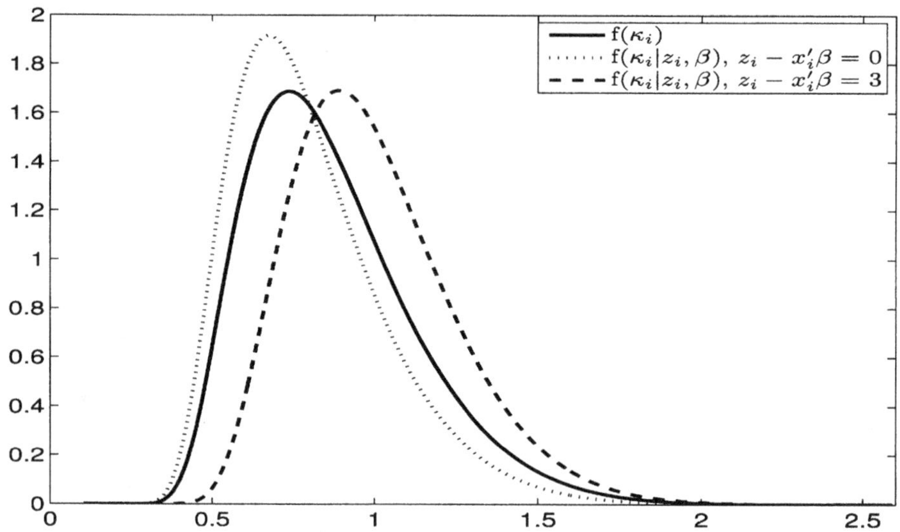

Figure 6.4 Behavior of the density $f(\kappa_i|z_i, \beta)$ relative to $f_K(\kappa_i)$.

■ EXAMPLE 6.7

The analysis of the ordinal probit model is similar to the cases considered in the preceding examples. In particular, given the priors $\beta \sim N(\beta_0, B_0)$ and $\delta \sim N(d_0, D_0)$, the augmented posterior distribution is given by

$$\begin{aligned}\pi(\beta, \delta, z|y) &\propto f(y|\beta, \delta, z) f(z|\beta)\pi(\beta)\pi(\delta) \\ &= \left\{\prod_{i=1}^{n} f(y|\delta, z_i)\right\} f(z|\beta)\pi(\beta)\pi(\delta),\end{aligned} \tag{6.19}$$

where $f(y_i|\delta, z_i) = 1\{\gamma_{j-1} < z_i \leq \gamma_j\}$, the correspondence between γ and δ is determined by (6.9), and the cutpoint index j is given by the realization of y_i. Furthermore, $f(z|\beta) = f_N(z|X\beta, I_n)$, $\pi(\beta) = f_N(\beta|\beta_0, B_0)$, and $\pi(\delta) = f_N(\delta|d_0, D_0)$.

It has been noted in the literature that in order to design an efficient MCMC sampler for the ordinal probit model, δ and z must be simulated jointly, not conditionally on each other. The reason that conditional sampling does not mix well is that z and δ (which determines the values of γ) constrain each other through the restrictions $\{\gamma_{y[i]-1} < z_i < \gamma_{y[i]}\}$, whereby the sampler can only slowly explore the posterior distribution. Several alternatives for joint sampling are reviewed in [13], and the following simulation scheme is suggested.

1. Sample $\delta, z|y, \beta$ in one block as follows:

(a) Sample $\delta|y, \beta$ marginally of z by drawing $\delta^p \sim q(\delta|y, \beta)$ from a proposal density $q(\delta|y, \beta) = f_{T_\omega}(\delta|\hat{\delta}, \hat{D})$, where

$$\hat{\delta} = \arg\max_{\delta} \ln f(y|\beta, \delta) \quad \text{and} \quad \hat{D} = -\left[\frac{\partial^2 \ln f(y|\beta, \delta)}{\partial\delta\partial\delta'}\right]^{-1}\Bigg|_{\delta=\hat{\delta}}.$$

Accept δ^p with probability

$$\alpha_{MH}(\delta, \delta^p) = \min\left\{1, \frac{f(y|\beta, \delta^p)\pi(\delta^p)}{f(y|\beta, \delta^c)\pi(\delta^c)} \frac{q(\delta^c|y, \beta)}{q(\delta^p|y, \beta)}\right\},$$

otherwise repeat the current value δ^c.

(b) Sample $z_i|y, \beta, \delta \sim TN_{(\gamma_{j-1}, \gamma_j)}(x_i'\beta, 1)$ for $i = 1, \ldots, n$, where γ is obtained by the one-to-one mapping relating γ and δ.

2. Sample $\beta|z \sim N(\hat{\beta}, \hat{B})$ with

$$\hat{B} = (B_0 + X'X)^{-1} \quad \text{and} \quad \hat{\beta} = \hat{B}(B_0^{-1}\beta_0 + X'z).$$

In Step 1 of this algorithm, the degrees of freedom parameter ω is taken to be a low number such as 5 or 10 to ensure that the proposal density has sufficiently heavy tails. Grouping δ and z into a single sampling block dramatically improves the mixing of the Markov chain.

We complete the discussion of data augmentation by emphasizing its practical appeal. For instance, data augmentation is often the only viable estimation approach in a variety of multivariate and hierarchical models. Maximum likelihood estimation becomes infeasible in those settings owing to the intractability of the likelihood function. However, data augmentation allows us to circumvent this difficulty by simulating from well-known distributions without having to evaluate the likelihood. This has enabled inference in difficult settings such as multivariate and multinomial probit, mixed logit, multivariate ordinal probit, copula models, panel data models, models with incidental truncation, treatment models, and many others.

6.3.3 Marginal Effects

Having estimated the parameters of a model, one is typically interested in the practical implications of those estimates. However, interpretation of the parameter estimates is complicated by the nonlinearity of the models we have considered. In binary data models, for example, $E(y_i|x_i, \beta) = \Pr(y_i = 1|x_i, \beta) = F(x_i'\beta)$. Therefore, the marginal effect of changing some continuous covariate in x_i, say x_h, is not simply given by β_h. This can be easily seen by taking the derivative of $\Pr(y_i = 1|x_i, \beta)$ with respect to x_h

$$\frac{\partial \Pr(y_i = 1|x_i, \beta)}{\partial x_h} = \frac{\partial F(x_i'\beta)}{\partial x_h} = f(x_i'\beta)\beta_h,$$

and hence the marginal effect of x_h depends on β_h, but also on all of the covariates in x_i, all of the parameters in β, and will differ with the choice of cdf $F(\cdot)$ and respective pdf $f(\cdot)$. Table 6.1 gives the choice probability and marginal effects for the three commonly used binary data models.

Table 6.1 Marginal effects in binary data models.

Model	Probability $P(y_i = 1\|x_i, \beta))$	Marginal Effect of x_h
Logit	$F_L(x_i'\beta) = \frac{e^{x_i'\beta}}{1+e^{x_i'\beta}}$	$f_L(x_i'\beta)\beta_h$
Probit	$\Phi(x_i'\beta) = \int_{-\infty}^{x_i'\beta} \phi(z)dz$	$\phi(x_i'\beta)\beta_h$
t-link	$F_{T_\tau}(x_i'\beta) = \int_{-\infty}^{x_i'\beta} f_{T_\tau}(t)dt$	$f_{T_\tau}(x_i'\beta)\beta_h$

Given a specific model, there are several approaches to compute the average marginal effect of covariate x_h. One approach is to evaluate the marginal effect using the sample average of the regressors $\bar{x}_i$ and the point estimate $\hat{\beta}$, i.e. $f(\bar{x}_i'\hat{\beta})\hat{\beta}_h$. However, this average effect may not represent the effect in the population well because $f(\cdot)$ is a nonlinear function and by Jensen's inequality $f(\bar{x}_i'\hat{\beta}) \neq \overline{f(x_i'\hat{\beta})}$. Therefore, a more reasonable approach would be to calculate the sample average of the marginal effects

$$\overline{f(x_i'\hat{\beta})\hat{\beta}_h} = n^{-1}\sum_{i=1}^{n} f(x_i'\hat{\beta})\hat{\beta}_h.$$

Even though this quantity is better than computing $f(\bar{x}_i'\hat{\beta})\hat{\beta}_h$ as suggested in Ref. [26], it has an important drawback: it does not account for the variability in β. For this reason, Refs. [8] and [13] suggest that the average covariate effect should be computed by averaging over both the covariates and parameters. If estimation is done by MCMC simulation, one can use draws $\beta^{(m)} \sim \pi(\beta|y)$ to construct the average covariate effect as follows

$$\overline{f(x_i'\beta)\beta_h} = \frac{1}{nM}\sum_{i=1}^{n}\sum_{m=1}^{M} f\left(x_i'\beta^{(m)}\right)\beta_h^{(m)}.$$

Note that unlike the earlier quantities we considered, $\overline{f(x_i'\beta)\beta_h}$ produces an estimate of the average effect that accounts for variability in both x_i and β.

6.4 APPLICATIONS

6.4.1 Women's Labor Force Participation

We apply the techniques of this chapter to study the determinants of women's labor force participation, a topic that has been extensively studied because of the large increases in women's participation and hours of work in the post-war period. For instance, there has been a sevenfold increase in the participation rate of married women since the 1920's. Understanding labor force participation and entry and exit decisions is a fundamental prerequisite for understanding wages because wages are not observed for women who do not work.

The data set used in this application has been studied in [20] and [7]. The sample consists of 753 married women, 428 of whom were employed. The variables in the data set are summarized in Table 6.2.

Table 6.2 Covariates in the women's labor supply example.

Covariate	Explanation	Mean	SD
KLT6	Number of kids under 6 years old	0.28	0.52
KGE6	Number of kids 6–18 years old	1.35	1.32
NWINC	Estimated nonwife income (1975, in $10,000)	2.01	1.16
MEDU	Mother's years of schooling	9.25	3.37
FEDU	Father's years of schooling	8.81	3.57
HEDU	Husband's years of schooling	12.49	3.02
AGE	Woman's age in years	42.54	8.07
EXPER	Actual labor market experience in years	10.63	8.07

We implemented the techniques developed in this chapter to estimate probit, logit, and t-link models of the binary participation decision. Estimation was carried out by the MCMC simulation methods discussed in Section 6.3.2 and our results are summarized in Table 6.3.

The estimates in Table 6.3 are consistent with the predictions of economic theory. For example, having young children reduces labor force participation as evidenced by the negative mean and a 95% credibility interval that lies below zero, but older children have little impact on the mother's decision to work. Again, consistent with economic theory, higher nonwife income and lower parents' and husband's schooling reduce participation. The table also shows that age has a strong negative effect, which is consistent with cohort and life-cycle effects, whereas experience has a strong positive effect on probability of working, which is consistent with increases in productivity as experience grows.

Table 6.3 Parameter estimates in the women's labor force participation application.

	Probit		t-link		Logit	
Covariate	Mean	SD	Mean	SD	Mean	SD
1	1.1758	0.4358	1.1737	0.4586	1.3931	0.6188
KLT6	-0.7964	0.1115	-0.8285	0.1210	-1.2476	0.1847
KGE6	0.0346	0.0415	0.0362	0.0443	0.0763	0.0695
NWINC	-0.0773	0.0484	-0.0817	0.0531	-0.1384	0.0825
MEDU	0.0320	0.0184	0.0339	0.0197	0.0580	0.0306
FEDU	0.0143	0.0175	0.0158	0.0189	0.0250	0.0300
HEDU	0.0251	0.0188	0.0265	0.0207	0.0476	0.0326
AGE	-0.0517	0.0078	-0.0534	0.0083	-0.0769	0.0117
EXPER	0.0745	0.0074	0.0796	0.0084	0.1270	0.0138

6.4.2 An Ordinal Model of Educational Attainment

We now consider the ordinal probit model of educational attainment studied in [13]. Educational attainment has been the subject of a large literature because of its implications for earnings, economic growth, and social well-being. The setting is suitable for ordinal modeling because the dependent variable is naturally categorized by measurable thresholds into a number of distinct groups. This application considers the following four ordered outcomes: (1) less than a high school education, (2) high school degree, (3) some college or associate's degree, and (4) college or graduate degree. The data are obtained from the National Longitudinal Survey of Youth (NLSY79).

In this study it is of interest to examine the effect of family background and individual variables on educational attainment. The family background variables included in the data set are: the highest grade completed by the individual's father and mother, whether the mother worked, square root of family income, an indicator for whether the youth lived in an urban area, and an indicator for whether the youth lived in the South. The individual variables include gender and race, as well as three indicator variables that control for age cohort affects. The sample is restricted to those cohorts that were between 14 and 17 years old in 1979. The sample is restricted to include only individuals whose records have all relevant variables. Additionally, the sample excludes disabled individuals and those who report more than 11 years of education at age 15. The resulting sample consists of 3923 individuals.

The model was estimated by the MCMC simulation techniques discussed in Section 6.3.2. The results are presented in Table 6.4. The coefficient estimates in the table are consistent with other findings in the literature. Parental education and income have a positive effect on educational attainment, as might be expected. *A priori*, the effect of mother's labor force participation is theoretically ambiguous — on the one hand, a mother's work force participation could be detrimental due to reduced parental supervision, but on the other, it provides a positive example for her children to follow. The empirical findings in Table 6.4 indicate that the net effect is positive, although it is not precisely estimated. Conditionally on the remaining covariates, we also see that blacks and individuals from the South have higher educational attainment.

Parameter	Covariate	Mean	SD
β	Intercept	-1.34	0.09
	Family income (sq. rt.)	0.14	0.01
	Mother's education	0.05	0.01
	Father's education	0.07	0.01
	Mother worked	0.03	0.04
	Female	0.16	0.04
	Black	0.15	0.04
	Urban	-0.05	0.04
	South	0.05	0.04
	Age cohort 2	-0.03	0.05
	Age cohort 3	0.00	0.06
	Age cohort 4	0.23	0.06
δ	(transformed cutpoint)	0.08	0.02
	(transformed cutpoint)	-0.28	0.03

Table 6.4 Parameters estimates in the educational attainment application.

Following [13], we computed the effect of an increase in family income on educational outcomes following the discussion in Section 6.3.3. For the overall sample, the effect of a \$1000 increase in family income is to lower the probability of dropping out of high school by approximately 0.0050, lower the probability of only obtaining a high school degree by 0.0006, but increase the probability of having some college or associate's degree by 0.0020 and increase the probability of getting a college or graduate degree by 0.0036. We also computed these effects for specific subsamples that are of interest. For the subsample of females, the effects of an income increase on the four outcome probabilities were comparable at approximately −0.0048, −0.0009, 0.0019, and 0.0038, respectively. For the subsample of blacks, the effects of income change were somewhat stronger – in that subsample, an increase of \$1000 in family income changed the four educational outcome probabilities by −0.0060, −0.0009, 0.0026, and 0.0043, respectively.

6.5 CONCLUSIONS

This chapter has introduced the theory behind binary and ordinal models in economics, and has examined their estimation by both maximum likelihood and Bayesian simulation methods. We have reviewed several existing

MCMC algorithms and have proposed a new data augmentation method for the estimation of logit models. The ability to implement data augmentation techniques makes it possible to extend these techniques and estimate models in which the likelihood function is intractable.

The methods are examined in two applications dealing with labor force participation and educational attainment. The applications illustrate that the models and estimation methods are practical and can uncover interesting features in the data.

EXERCISES

6.1 The cdf of the logistic distribution is given by

$$F_L(\nu) = \left(1 + e^{-\nu}\right)^{-1} = \frac{e^{\nu}}{1 + e^{\nu}}.$$

Show that the logistic pdf, $f_L(\nu)$, can be written as

$$f_L(\nu) = F_L(\nu)\left[1 - F_(\nu)\right].$$

6.2 Suppose the random utility model is given by

$$U_{ij} = x_i'\beta_j + \varepsilon_{ij}, \quad i = 1, \ldots, n, \quad j = 0, 1.$$

Starting with Eq. (6.1), show that if ε_{i0} and ε_{i1} are independent and identically distributed as extreme value type I with density

$$f_{EV}(\varepsilon) = e^{-\varepsilon}e^{-e^{-\varepsilon}}$$

and cumulative distribution function

$$F_{EV}(\varepsilon) = e^{-e^{-\varepsilon}},$$

then (6.1) gives rise to the logistic outcome probability

$$\Pr(y_i = 1|\beta) = \frac{1}{1 + e^{-x_i'\beta}},$$

where $\beta = \beta_1 - \beta_0$.

6.3 Consider the probit model and assume the prior $\beta \sim N(\beta_0, B_0)$. Show that given the latent data z, the full conditional distribution for β is

$$\beta|y, z \sim N(\hat{\beta}, \hat{B}),$$

where $\hat{B} = (B_0^{-1} + X'X)^{-1}$ and $\hat{\beta} = \hat{B}(B_0^{-1}\beta_0 + X'z)$.

REFERENCES

1. Albert, J. and Chib, S., "Bayesian Analysis of Binary and Polychotomous Response Data," *Journal of American Statistical Association*, Vol. 88, No. 422, pp. 669–679, (1993).

2. Andrews, D. F. and Mallows, C. L., "Scale Mixtures of Normal Distributions," *Journal of the Royal Statistical Society, Series B*, Vol. 36, No. 1, pp. 99–102, (1974).

3. Berndt, E., Hall, B., Hall, R., and Hausman J., "Estimation and Inference in Nonlinear Structural Models," *Annals of Economic and Social Measurement*, Vol. 3, pp. 653–665 (1974).

4. Chen, M.-H. and Dey, D. K., "Bayesian Analysis for Correlated Ordinal Data Models," in D. Dey, S. Ghosh, and B. Mallick (eds.), *Generalized Linear Models: A Bayesian Perspective*, pp. 133–157. New York: Marcel-Dekker.

5. Chib, S. and Greenberg, E., "Understanding the Metropolis-Hastings Algorithm," *The American Statistician*, Vol. 49, No. 4, pp. 327–335 (1995).

6. Chib, S. and Greenberg, E., "Markov Chain Monte Carlo Simulation Methods in Econometrics," *Econometric Theory*, Vol. 12, No. 3, pp. 409–431 (1996).

7. Chib, S., Greenberg, E., and Jeliazkov, I., "Estimation of Semiparametric Models in the Presence of Endogeneity and Sample Selection," *Journal of Computational and Graphical Statistics*, Vol. 18, pp. 321–348 (2009).

8. Chib, S. and Jeliazkov, I., "Inference in Semiparametric Dynamic Models for Binary Longitudinal Data," *Journal of the American Statistical Association*, Vol. 101, pp. 685–700 (2006).

9. Gelfand, A. E. and Smith, A. F. M., "Sampling Based Approaches to Calculating Marginal Densities," *Journal of the American Statistical Association*, 85, 398–409 (1990).

10. Greenberg, E., *Introduction to Bayesian Econometrics*, Cambridge University Press, Cambridge (2007).

11. Greene, W. H., *Econometric Analysis*, 7th edition, Prentice Hall, New Jersey (2011).

12. Hastings, W. K., "Monte Carlo Sampling Methods using Markov Chains and Their Applications," *Biometrika*, Vol. 57, pp. 97–109 (1970).

13. Jeliazkov, I., Graves, J., and Kutzbach, M., "Fitting and Comparison of Models for Multivariate Ordinal Outcomes," *Advances in Econometrics: Bayesian Econometrics*, Vol. 23, pp. 115–156 (2008).

14. Koop, G. and Poirier, D.J. and Tobias, J.L., *Bayesian Econometric Methods*, Cambridge University Press, Cambridge (2007).

15. Luce, R. D., *Individual Choice Behavior*. John Wiley & Sons, New York (1959).

16. Luce, D. and Suppes, P., "Preferences, Utility and Subjective Probability," *Handbook of Mathematical Psychology*, R. D. Luce, R. Bush, and E. Galanter (eds.), John Wiley & Sons, New York (1965).

17. Marschak, J., "Binary-Choice Constraints and Random Utility Indicators," *Mathematical Methods in the Social Sciences*, K. J. Arrow, S. Karlin, and P. Suppes (eds.), Stanford University Press, pp. 312–329 (1960).

18. McFadden, D., "Conditional Logit Analysis of Qualitative Choice Behavior," *Frontiers in Econometrics*, P. Zarembka (ed.), pp. 105–142, Academic Press, New York (1974).

19. Metropolis, N., Rosenbluth, A. W., Rosenbluth, M. N., Teller, A. H., and Teller, E., "Equations of State Calculations by Fast Computing Machines," *Journal of Chemical Physics*, Vol. 21, pp. 1087–1092 (1953).

20. Mroz, T. A., "The Sensitivity of an Empirical Model of Married Women's Hours of Work to Economic and Statistical Assumptions," *Econometrica*, Vol. 55, pp. 765–799 (1987).

21. O'Hagan, A., *Kendalls Advanced Theory of Statistics: Bayesian Inference*. John Wiley & Sons, New York (1994).

22. Poirier, D. J., "A Curious Relationship between Probit and Logit Models," *Southern Economic Journal*, Vol. 40, pp. 640–641 (1978).

23. Poirier, D. J., *Intermediate Statistics and Econometrics: A Comparative Approach*, Cambridge, MA: MIT Press (1995).

24. Tierney, L., "Markov Chains for Exploring Posterior Distributions," (with discussion), *Annals of Statistics*, Vol. 22, pp. 1701–1762 (1994).

25. Train, K., *Discrete Choice Methods with Simulation*, Cambridge University Press, Cambridge (2009).

26. Verlinda, J. A., "A comparison of two common approaches for estimating marginal effects in binary choice models," *Applied Economics Letters*, Vol. 13, pp. 77–80 (2006).

CHAPTER 7

INVERSE PROBLEMS IN ODEs

H. KUNZE[1] AND D. LA TORRE[2]

[1]Department of Mathematics and Statistics, University of Guelph, Canada
[2] Department of Economics, Business and Statistics, University of Milan, Italy

Early undergraduate courses in differential equations typically focus on solution methods, mathematical modeling, and interpretation of solutions and models. Seeking the solution of a given differential equation or system is often called the "direct problem." On the other hand, the "inverse problem" asks us to find an appropriate model (differential equation or system), the solution of which agrees well with some experimental or real-world observations.

For example, we might believe that a particular predator-prey system of differential equations models the interactions between the rabbits and foxes in a given region, and we can gather some population data over some period of time. From this data, the inverse problem might ask us to estimate the parameters (the coefficients) in the model. Many questions may come to mind. Some are:

Mathematical Modeling with Multidisciplinary Applications.
Edited by Xin-She Yang

1. Since the differential equations are nonlinear, we cannot solve them explicitly, and we can only solve them numerically if we pick values for the coefficients. So, how do we solve the inverse problem?

2. If we only gather population data every week, say, and there are surely measurement errors, our method for solving the inverse problem needs to be pretty robust. Is it possible to construct such a method?

3. The method should be mathematically sound. Can we build the theory as well as a practically useful method?

In this chapter, we will formulate one approach to solving this sort of inverse problem. The mathematical theory will be based on Banach's fixed point theorem, and, in particular, a simple corollary named the collage theorem (because of its original usefulness in fractal imaging). We will see many practical examples and develop an appreciation for the robustness of the general method. This approach to solving inverse problems in differential equations first saw light in 2000: as we learn how some older mathematical theory can be used to build new theory and tools, we reinforce how mathematics is vibrant and always growing.

7.1 BANACH'S FIXED POINT THEOREM & THE COLLAGE THEOREM

Banach's Fixed Point Theorem, also called the Contraction Mapping Principle, is a central component of most first courses in real analysis. (For greater details on the ideas presented in this section, see Ref. [4], a standard text used in such a course.) The two key ingredients of the theorem are a "complete metric space" and a "contraction map" that send the space into itself.

The first ingredient can be quite challenging. A space consists of the same type of elements: the space of real numbers contains real numbers, a vector space contains vectors, an image space contains images, and a function space contains functions, for example. Loosely, a metric space is a space along with a way to measure distance between elements in the space. For example, the distance between real numbers might be taken to be the absolute value of their difference; and we might use the Euclidean metric to measure distance between (finite) vectors, square rooting the sum of the squares of the differences of the components of the vectors. There are many ways to measure the distance between functions in a function a space. We will see some choices in the other sections of this chapter. A metric must satisfy these properties:

1. $d(x, y) \geq 0$ for all $x, y \in X$, and $d(x, y) = 0$ if and only if $x = y$;

2. $d(x, y) = d(y, x)$ for all $x, y \in X$;

3. $d(x, z) \leq d(x, y) + d(y, z)$ for all $x, y, z \in X$.

The final property is the familiar triangle inequality. All of the properties make sense when you thing of d as a distance between familiar objects.

In general, we denote a space by the letter X, a metric by the letter d, and the corresponding metric space by the ordered pair (X, d).

A complete metric space has a special property that only rears its head in this section, where we present proofs of the fundamental results we use in subsequent sections. We capture this property in the following definition.

Definition 1 *Let (X, d) be a metric space. The sequence $\{x_n\}_{n=1}^{\infty}$ is Cauchy if for every $\varepsilon > 0$ there is an M such that*

$$d(x_m, x_n) < \varepsilon \quad \textit{whenever} \quad n > M \textit{ and } m > M.$$

The metric space (X, d) is called complete when every Cauchy sequence converges to some element $x \in X$. That is,

$$d(x_n, x) \to 0 \textit{ as } n \to \infty.$$

The second ingredient also has its complications, but an easy interpretation brings comfort.

Definition 2 *A map $T : X \to X$ is a contraction map (with respect to the metric d) if there exists a $c \in [0, 1)$ such that*

$$d(Tx, Ty) \leq c\, d(x, y) \textit{ for all } x, y \in X.$$

The smallest value of c for which this inequality holds is called the contraction factor of the contraction map T.

In order to understand Definition 2, first observe that x and y get sent to Tx and Ty, respectively, by the map T. As a result, the inequality in the definition says that T moves points in X closer together: since $c \in [0, 1)$, the distance between Tx and Ty is smaller than the distance between the original points, x and y. In order for T to be contractive on (X, d), the preceding statement must be true for *every* choice of x and y. See Figure 7.1.

■ EXAMPLE 7.1

Let $X = [0, 1]$, the unit interval of real numbers, and $d(x, y) = |x - y|$ for $x, y \in X$. We consider some choices for $T : X \to X$.

1. If $Tx = \frac{1}{2}x$, then we see that for $x, y \in X$

$$d(Tx, Ty) = \left|\frac{x}{2} - \frac{y}{2}\right| = \frac{1}{2}|x - y| = \frac{1}{2}d(x, y).$$

The map T is contractive on (X, d) with contraction factor $\frac{1}{2}$.

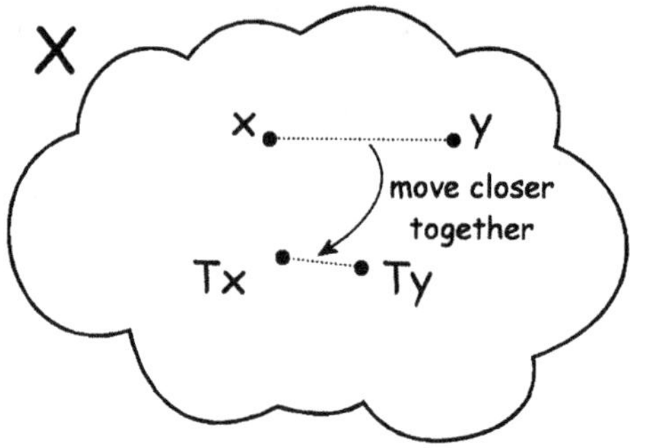

Figure 7.1 A contractive map T moves points closer together.

2. If $Tx = \frac{1}{3}x + \frac{2}{3}$, note that T does send X to itself and for $x, y \in X$

$$d(Tx, Ty) = \left|\frac{x}{3} + \frac{2}{3} - \frac{y}{3} - \frac{2}{3}\right| = \frac{1}{3}|x - y| = \frac{1}{3}d(x, y).$$

The map T is contractive on (X, d) with contraction factor $\frac{1}{3}$.

3. If $Tx = \alpha x + \beta$, then provided $0 \le \beta \le 1$ and $0 \le \alpha + \beta \le 1$, we see $T : X \to X$. We calculate that

$$d(Tx, Ty) = |\alpha x + \beta - (\alpha y + \beta)| = |\alpha||x - y| = |\alpha| d(x, y).$$

The map T is contractive provided that $|\alpha| < 1$.

Banach's Fixed Point Theorem states the remarkable fact that when a contraction map acts on a complete metric space there is exactly one element of the space that does not move when the map acts upon it. This special point that does not move is rather appropriately called the fixed point of the map. We state and prove the theorem.

Theorem 1 *(Banach's Fixed Point Theorem) Let (X, d) be a complete metric space and $T : X \to X$ be a contraction map with contraction factor $c \in [0, 1)$. Then*

1. *there is a unique $\bar{x} \in X$ such that $T\bar{x} = \bar{x}$;*

2. *for any $x_0 \in X$, the sequence defined by $x_{n+1} = Tx_n$ converges (in the metric d) to $\bar{x}$; that is, $d(x_n, \bar{x}) \to 0$ as $n \to \infty$.*

Proof: To prove existence, pick an arbitrary $x_0 \in X$. Define the sequence $x_{n+1} = Tx_n$, $n = 0, 1, 2, 3, \ldots$. We will prove that the sequence is Cauchy in (X, d). By the completeness of the space this will mean that the sequence converges to a point $\bar{x} \in X$; that is, $d(x_n, \bar{x}) \to 0$ as $n \to \infty$. We are then able to conclude that

$$\begin{aligned} \lim_{n\to\infty} x_{n+1} &= \lim_{n\to\infty} Tx_n \\ &= T \lim_{n\to\infty} x_n, \end{aligned}$$

because contraction maps are continuous (see Exercise 7.1). Taking the limits, we arrive at $\bar{x} = T\bar{x}$, and we see that $\bar{x}$ is a fixed point of T.

To prove that the sequence is Cauchy, based on Definition 1, for $n > m \geq 2$, we consider

$$\begin{aligned} d(x_m, x_n) &= d(Tx_{m-1}, Tx_{n-1}) \\ &\leq c\, d(x_{m-1}, x_{n-1}) \\ &= c\, d(Tx_{m-2}, Tx_{n-2}) \\ &\leq c^2\, d(x_{m-2}, x_{n-2}). \end{aligned}$$

We see the pattern. We can continue the process m times, until the first argument in d is x_0. We conclude that

$$d_(x_m, x_n) \leq c^m\, d(x_0, (x_{n-m}) \text{ for } n > m \geq 0.$$

This result motivates us to consider

$$\begin{aligned} d(x_0, x_k) &\leq d(x_0, x_1) + d(x_1, x_2) + d(x_2, x_3) + \cdots + d(x_{k-1}, x_k) \\ &\leq d(x_0, x_1) + c\, d(x_0, x_1) + c^2\, d(x_0, x_1) + \cdots + c^{k-1} d(x_0, x_1) \\ &\leq \left(1 + c + c^2 + \cdots + c^{k-1}\right) d(x_0, x_1) \\ &\leq \frac{1 - c^k}{1 - c} d(x_0, x_1) \\ &\leq \frac{1}{1 - c} d(x_0, x_1), \end{aligned}$$

where we sum the geometric series in the second-last line. Putting things together, we conclude that

$$d_(x_m, x_n) \leq \frac{c^m}{1 - c} d(x_0, x_1) \text{ for } n > m \geq 0.$$

Since $c \in [0, 1)$, when m (and hence n) become large enough, the bound on the right-hand side gets arbitrarily small. We conclude that the sequence is Cauchy in (X, d). As presented earlier, this means that the first conclusion of the theorem holds.

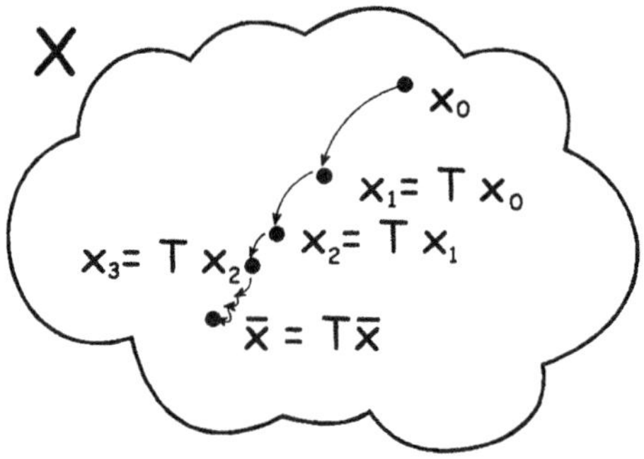

Figure 7.2 Banach's Theorem: Repeated iteration of T takes us to its fixed point.

To prove uniqueness, assume the opposite. Suppose that there are two distinct points $\bar{x}, \bar{y} \in X$ with $T\bar{x} = \bar{x}$ and $T\bar{y} = \bar{y}$. Then

$$d(\bar{x}, \bar{y}) = d(T\bar{x}, T\bar{y}) \leq c\, d(\bar{x}, \bar{y}).$$

Since $\bar{x} \neq \bar{y}$, we know that $d(\bar{x}, \bar{y}) \neq 0$, using a property of metrics. So we can divide by this quantity to obtain $1 \leq c$, a contraction. We conclude that the fixed point is unique. ■

The second conclusion of Banach's fixed point theorem is illustrated in Figure 7.2. Note that a map can have a fixed point without being contractivity, but that none of the theory in this chapter will apply.

■ EXAMPLE 7.2

Let $X = [0, 1]$, the unit interval of real numbers, and $d(x, y) = |x - y|$ for $x, y \in X$. Each of the maps in Example 7.1 is contractive. We can find the unique fixed point by solving $x = Tx$.

1. When $Tx = \frac{1}{2}x$, we find that $x = Tx = \frac{1}{2}x$ has solution $\bar{x} = 0$. This makes sense, as we see that $T^n x = \frac{1}{2^n}x \to 0$ as $n \to \infty$ regardless of the choice of x.
2. Solving $Tx = \frac{1}{3}x + \frac{2}{3} = x$, we get $\frac{2}{3}x = \frac{2}{3}$, with solution $\bar{x} = 1$.
3. Solving $Tx = \alpha x + \beta = x$, we find that $\beta = (1 - \alpha)x$ with solution $\bar{x} = \frac{\beta}{1-\alpha}$. Recall from Example 7.1 that $|\alpha| < 1$ and, combining the other inequalities, $0 \leq \beta \leq 1 - \alpha$. Together these inequalities tell us that $\bar{x} \in [0, 1]$.

Banach's Fixed Point Theorem has a simple corollary called the Collage Theorem. The name of this result comes from its usefulness in fractal imaging, a topic removed from the current discussion.

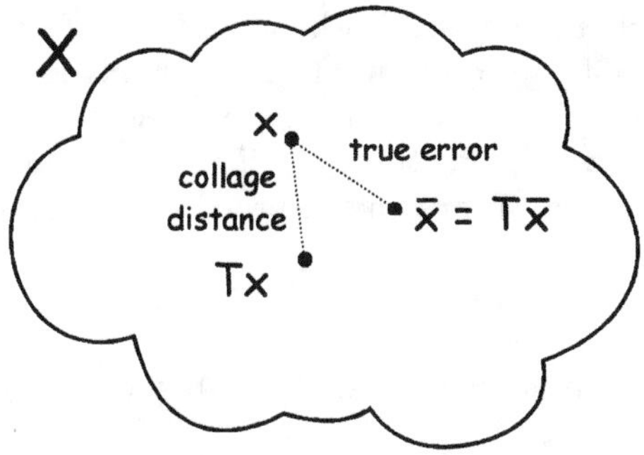

Figure 7.3 Collage Theorem: The true error can be controlled by the collage distance.

Theorem 2 *(Collage theorem) Let (X, d) be a complete metric space and $T : X \to X$ be contractive with contractivity factor $c \in [0, 1)$. Denote by $\bar{x}$ the unique fixed point of T, as guaranteed by Banach's Fixed Point Theorem. Then, for any $x \in X$,*

$$d(x, \bar{x}) \leq \frac{1}{1-c} d(x, Tx).$$

Proof: Using the triangle inequality, we have

$$\begin{aligned} d(x, \bar{x}) &\leq d(x, Tx) + d(Tx, \bar{x}) \\ &= d(x, Tx) + d(Tx, T\bar{x}) \\ &\leq d(x, Tx) + c\, d(x, \bar{x}) \\ \Rightarrow (1-c)\, d(x, \bar{x}) &\leq d(x, Tx). \end{aligned}$$

Dividing by $1 - c$, which is nonzero, gives the desired result. ∎

If we think of $d(x, \bar{x})$ as the *error* in approximating x by the fixed point $\bar{x}$, we see that the inequality in the Collage Theorem says that this approximation error can be controlled by minimizing the *collage distance* $d(x, Tx)$. We illustrate the theorem in Figure 7.3.

7.2 EXISTENCE-UNIQUENESS OF SOLUTIONS TO INITIAL VALUE PROBLEMS

We use Banach's fixed point theorem to prove that under suitable conditions the ODE initial value problem (IVP)

$$x'(t) = f(x, t), \tag{7.1}$$

$$x(t_0) = x_0, \tag{7.2}$$

has a unique solution in some neighborhood of $t = t_0$. Proofs of this result can be found in [1] and [3], two often-used texts for introductory and advanced differential equations courses, respectively, but the proofs in most references do not use Banach's fixed point theorem. The nice (and accessible) book on metric spaces by Copson [2] presents a proof similar to the one we present here.

Notice that the solution to (7.1) and (7.2) may well not exist for all real t values. In general, we do not have *global* existence of a solution, but we can prove that we do have *local* existence. With this in mind, we define

$$\begin{aligned} I &= [t_0 - a, t_0 + a] \text{ for some } a > 0, \\ X = C(I) &= \{\text{space of continuous functions } x(t) \text{ on } I.\}, \\ \text{and } d_\infty(x, y) &= \|x - y\|_\infty = \max_{t \in I} |x(t) - y(t)|. \end{aligned}$$

Note that the max in the definition of d_∞ exists because $|x(t) - y(t)|$ is a continuous function. By the Extreme Value Theorem from calculus, a continuous function on a closed interval achieves an absolute maximum and an absolute minimum.

Now, we merely state that the space $(X, d) = (C(I), d_\infty)$ is a complete metric space. This statement says that d_∞ is in fact a metric on $C(I)$. This claim is not hard to prove, since it mostly takes advantage of properties of absolute value. But the statement also says that the metric space is complete. This is a harder claim to prove, requiring an argument involving Cauchy sequences of functions in $C(I)$. The proof is a standard part of a real analysis course. It involves first proving that the sequence converges pointwise and then proving that the convergence is actually uniform, so that the limit of the sequence inherits the continuity of the sequence terms.

We want to introduce a contraction map on $(C(I), d_\infty)$. Integrating the ODE with respect to t from $t = 0$ to $t = t$, we get

$$\begin{aligned} \int_{t_0}^{t} x'(s)\, ds &= \int_{t_0}^{t} f(x(s), s)\, ds \\ \Rightarrow x(t) - x(0) &= \int_{t_0}^{t} f(x(s), s)\, ds \\ \Rightarrow \qquad x(t) &= x_0 + \int_{t_0}^{t} f(x(s), s)\, ds. \end{aligned} \tag{7.3}$$

Here, we use the Fundamental Theorem of Calculus to get to line two, and then we rearrange and use the initial condition (7.2) to get the final line. The resulting integral equation features the function $x(t)$ on both sides. It sits on the left-hand side, but it also appears inside f on the right-hand side. In Exercise 7.3, you are asked to prove that $x(t)$ is a solution to the IVP if and only if it satisfies the integral equation. Based upon the integral equation

(7.3), we define the *Picard operator*

$$(Tx)(t) = x_0 + \int_{t_0}^{t} f(x(s), s)\, ds. \tag{7.4}$$

Using the result of Exercise 7.3, we see that $x(t)$ is a fixed point of T if and only if it is a solution to the IVP. It would be nice if T was a contraction map on (X, d), since that is the focus of our discussion.

First, for convenience, we note that if $t_0 \neq 0$ or $x_0 \neq 0$, we can let $s = t - t_0$ and $y(s) = x(s) - x_0$ so that $y(s)$ satisfies $y(0) = 0$ and the transformed ODE. Hence, without loss of generality, in the remainder of this section, we assume that our IVP takes the form

$$x'(t) = f(x, t), \tag{7.5}$$
$$x(0) = 0, \tag{7.6}$$

Note that this means that $I = [-a, a]$. We define

$$\begin{aligned} D &= \{(x,t) \,|\, |x| \leq b,\ |t| \leq a\} \\ \bar{C}(I) &= \{x \in C(I) \,|\, \|x\|_\infty \leq b\}. \end{aligned}$$

The metric space $(\bar{C}(I), d_\infty)$ is complete. We also assume that f is continuous and satisfies $\max_{(x,t)\in D} |f(x,t)| \leq b/a$, and a *Lipschitz* condition on D: there exists a constant $K > 0$ such that

$$|f(x_1, t) - f(x_2, t)| \leq K|x_1 - x_2| \text{ for all } (x_i, t) \in D,$$

such that $c = Ka < 1$. If necessary, we can adjust the value of $a > 0$ to make it small enough so that $Ka < 1$.

Now we prove two results.

Theorem 3 *With the preceding definitions,* $T : \bar{C}(I) \to \bar{C}(I)$.

Proof: We must prove that $Tx \in \bar{C}(I)$ when $x \in \bar{C}(I)$. Since f is continuous and x is continuous, the composition in the integral is continuous. That means that the integral produces a continuous function. We only need to show that $\|Tx\|_\infty \leq b$. Since $t_0 = 0$ and $x_0 = 0$, we find that

$$\begin{aligned} \|Tx\|_\infty &= \max_{t\in I} \left| \int_0^t f(x(s), s)\, ds \right| \\ &= \max_{s\in I} |f(x(s), s)| \cdot \max_{t\in I} \left| \int_0^t ds \right| \\ &\leq \max_{(x,t)\in D} |f(x,t)| \cdot \max_{t\in I} |t| \\ &\leq \frac{b}{a} \cdot a = b, \end{aligned}$$

proving the result. ■

Theorem 4 *With the preceding definitions,*

$$d_\infty(Tx, Ty) \le c\, d_\infty(x, y), \text{ for } x, y \in \bar{C}(I),$$

where $c = Ka < 1$.

Proof: For $x, y \in \bar{C}(I)$, we calculate that

$$\begin{aligned}
d_\infty(Tx, Ty) &= \max_{t\in I}\left|\int_0^t f(x(s), s)\, ds - \int_0^t f(y(s), s)\, ds\right| \\
&= \max_{t\in I}\left|\int_0^t (f(x(s), s) - f(y(s), s))\, ds\right| \\
&\le \max_{t\in I}\left|\int_0^t |f(x(s), s) - f(y(s), s)|\, ds\right| \\
&\le \max_{t\in I}\left|\int_0^t K\,|x(s) - y(s)|\, ds\right| \\
&\le K \max_{s\in I}|x(s) - y(s)| \cdot \max_{t\in I}\left|\int_0^t ds\right| \\
&= K\, d_\infty(x(s), y(s)) \cdot \max_{t\in I}|t| \\
&\le Ka\, d_\infty(x(s), y(s)),
\end{aligned}$$

as required. ■

The contractivity of T on $(\bar{C}(I), d_\infty)$ implies the existence of a unique element $\bar{x} \in \bar{C}(I)$ such that $T\bar{x} = \bar{x}$, using Banach's Fixed Point Theorem. By Exercise 7.3, this fixed point is therefor the unique solution to the IVP (7.5) and (7.6).

7.3 SOLVING INVERSE PROBLEMS FOR ODEs

The inverse problem of interest to us is: Given a solution curve $x(t)$ for $t \in [0, 1]$, perhaps the interpolation of experimental data points, find an ODE $x'(t) = f(x, t)$ that admits $x(t)$ as an approximate solution, where f may be restricted to a particular family of functions based on understanding of the underlying model.

Based on the discussion in Section 7.2, we can rephrase the inverse problem as: Given a solution curve $x(t)$ for $t \in [0, 1]$, perhaps the interpolation of experimental data points, find a Picard operator that admits $x(t)$ as an approximate fixed point, where f may be restricted to a particular family of functions based on understanding of the underlying model.

We call $x(t)$ the *target* solution. The second formulation of the inverse problem makes it clear that we are seeking to approximation $x(t)$ by a fixed point $\bar{x}(t)$ of some Picard operator T. By restricting the form of $f(x,t)$, typically defining it in terms of some coefficients, we similarly restrict the possible Picard operators. In general, we cannot express the possible fixed points in terms of these coefficients, so direct minimization of the true approximation error $d_\infty(x, \bar{x})$ (with $\bar{x}$ defined by the coefficients) is not possible. Instead, we call on the Collage Theorem, minimizing the collage distance $d_\infty(x, Tx)$ in order to control the true approximation. But it is troublesome to work with the d_∞ metric. We switch to a different metric for functions, called the $\mathcal{L}^2$ metric, given by

$$d_2(x,y) = \left(\int_I (x(t) - y(t))^2 \right)^{\frac{1}{2}}.$$

There is a technical catch: $(\bar{C}(I), d_2)$ is *not* complete. The functions in $\mathcal{L}^2(I)$ are those that satisfy $d_2(x, 0) < \infty$. It can be shown that

$$\bar{C}(I) \subset C(I) \subset \mathcal{L}^2(I).$$

We know that the fixed point of T lies in $\bar{C}(I)$, so it will be in $\mathcal{L}^2(I)$. We still have to prove that T is contractive with respect to d_2. We state the result as a theorem.

Theorem 5 *With the preceding definitions,*

$$d_2(Tx, Ty) \leq \frac{c}{\sqrt{2}} d_2(x,y), \text{ for } x, y \in \bar{C}(I),$$

where $c = Ka < 1$.

Proof: For $x, y \in \bar{C}(I)$, we calculate that

$$\begin{aligned} d_2^2(Tx, Ty) &= \int_I \left[\int_0^t f(x(s), s)\, ds - \int_0^t f(y(s), s)\, ds \right]^2 dt \\ &= \int_I \left[\int_0^t (f(x(s), s) - f(y(s), s))\, ds \right]^2 dt \\ &\leq \int_I \left[\int_0^t |f(x(s), s) - f(y(s), s)|\, ds \right]^2 dt \\ &\leq \int_I \left[\int_0^t K\, |x(s) - y(s)|\, ds \right]^2 dt. \end{aligned} \tag{7.7}$$

For convenience, consider $t > 0$ only; a similar argument applies for $t < 0$. The Cauchy-Schwarz inequality in this setting says that

$$\int_0^t g(s)h(s)\, ds \leq \left[\int_0^t (g(s))^2\, ds \right]^{\frac{1}{2}} \left[\int_0^t (h(s))^2\, ds \right]^{\frac{1}{2}}.$$

We apply the inequality with $g(s) = 1$ and $h(s) = |x(s) - y(s)|$. Then

$$\begin{aligned} \int_0^t 1 \cdot |x(s) - y(s)|\, ds &\le \left[\int_0^t 1\, ds\right]^{\frac{1}{2}} \left[\int_0^t |x(s) - y(s)|^2\, ds\right]^{\frac{1}{2}} \\ &\le t^{\frac{1}{2}} \left[\int_0^t |x(s) - y(s)|^2\, ds\right]^{\frac{1}{2}}. \end{aligned} \tag{7.8}$$

Plugging (7.8) into (7.7) gives

$$\begin{aligned} d_2^2(Tx, Ty) &\le K^2 \int_0^a \left[t^{\frac{1}{2}} \left[\int_0^t |x(s) - y(s)|^2\, ds\right]^{\frac{1}{2}}\right]^2 dt \\ &= K^2 \int_0^a \int_0^t t|x(s) - y(s)|^2\, dsdt \\ &= K^2 \int_0^a \int_s^a t|x(s) - y(s)|^2\, dtds \\ &= K^2 \int_s^a t\, dt \int_0^a |x(s) - y(s)|^2\, ds \\ &= K^2 \left(\frac{s^2}{2} - \frac{a^2}{2}\right) \int_0^a |x(s) - y(s)|^2\, ds \\ &\le \frac{K^2 a^2}{2} \int_0^a |x(s) - y(s)|^2\, ds, \end{aligned}$$

and square rooting gives the result. Upon combining with the argument for $t < 0$, the theorem is proved. ■

■ EXAMPLE 7.3

Let $x(t) = Ae^{Bt} + C$ be the target solution, where A, B, and C are real numbers, with $B \neq 0$. This is the exact solution of the linear IVP

$$\begin{aligned} x'(t) &= -BC + Bx, \\ x(0) &= A + C. \end{aligned}$$

Given the target solution $x(t)$, we seek an IVP of the form

$$\begin{aligned} x'(t) &= c_0 + c_1 x, \\ x(0) &= x_0, \end{aligned}$$

where c_0, c_1, and x_0 are parameters. We construct the Picard operator

$$\begin{aligned}(Tx)(t) &= x_0 + \int_0^t (c_0 + c_1 x(s))\,ds \\ &= x_0 + \int_0^t (c_0 + c_1 A e^{Bs} + c_1 C)\,ds \\ &= x_0 + (c_0 + c_1 C)t + \frac{c_1 A}{B}\left(e^{Bt} - 1\right) \\ &= x_0 - \frac{c_1 A}{B} + (c_0 + c_1 C)t + \frac{c_1 A}{B} e^{Bt}.\end{aligned}$$

The squared collage distance on $[0, 1]$ is

$$d_2^2(x, Tx) = \int_0^1 \left(Ae^{Bt} + C - \left(x_0 - \frac{c_1 A}{B} + (c_0 + c_1 C)t + \frac{c_1 A}{B} e^{Bt}\right)\right)^2 dt.$$

Notice that the integrand is a linear function of x_0, c_0, and c_1. Squaring and integrating leads to the expression

$$\begin{aligned}d_2^2(x, Tx) = \frac{1}{6B^3}\Big(&6x_0^2B^3 + 9c_1^2A^2 + 3A^2e^{2B}B^2 + 3c_1^2A^2e^{2B} \\ &-12AB^2C - 3A^2B^2 + 6C^2B^3 + 12Ae^BB^2C \\ &-12x_0Bc_1A - 12ABc_0 + 12c_0c_1A + 12c_1^2AC \\ &+12AB^2x_0 - 6A^2Bc_1 - 6C^2B^3c_1 - 12CB^3x_0 - 6CB^3c_0 \\ &+6x_0B^3c_0 + 6c_1^2A^2B + 2c_1^2C^2B^3 + 12CB^2c_1A \\ &-12x_0B^2c_1A + 6x_0B^3c_1C - 6c_1Ac_0B^2 - 6c_1^2ACB^2 \\ &+4c_0B^3c_1C + 12x_0Bc_1Ae^B + 12c_1^2ACe^BB \\ &+12c_0c_1Ae^BB - 12CB^2c_1Ae^B + 2c_0^2B^3 - 12c_1^2A^2e^B \\ &-12Ae^BB^2x_0 + 12A^2e^BBc_1 - 12AB^2c_0e^B + 12ABc_0e^B \\ &-12c_1^2ACe^B - 12c_0c_1Ae^B - 6A^2e^{2B}Bc_1\Big).\end{aligned}$$

We display the expression to stress that solving such problems is really not work for pencil and paper. Using calculus, we differentiate with respect to x_0, c_0, and c_1, setting the first partial derivatives equal to zero. We implement these steps using mathematical software on a computer. As we would hope, the result is that the squared collage distance is minimized when

$$x_0 = A + C, \; c_0 = -BC, \; c_1 = B.$$

EXAMPLE 7.4

Let $x(t) = t^2$ be the target solution on $I = [0,1]$. We see that $x'(t) = 2t = 2\sqrt{x}$ for ≥ 0, so the true DE satisfied by $x(t)$ is not linear. However, looking for a linear IVP

$$\begin{aligned} x'(t) &= c_0 + c_1 x, \\ x(0) &= x_0, \end{aligned}$$

we follow the same process: define the Picard operator and calculate the square collage distance $d_2^2(x, Tx)$. Minimizing this distance with computer software, we obtain the IVP

$$\begin{aligned} x'(t) &= \frac{5}{12} + \frac{35}{18}x, \\ x(0) &= -\frac{1}{27}, \end{aligned}$$

with corresponding (minimized) collage distance $d_2(x, Tx) = 0.0124$. The solution to this IVP is

$$\bar{x}(t) = \frac{67}{378}e^{\frac{35}{18}t - \frac{3}{14}}.$$

We can calculate that $d_2(x, \bar{x}) = 0.0123$. Note that $\bar{x}(0) \neq x(0)$. We can impose the condition that $x_0 = x(0) = 0$, to find the minimal-collage IVP

$$\begin{aligned} x'(t) &= \frac{5}{12} + \frac{35}{16}x, \\ x(0) &= 0, \end{aligned}$$

with corresponding collage distance $d_2(x, Tx) = 0.0186$ and solution

$$\bar{x}(t) = \frac{1}{7}e^{\frac{35}{16}t - \frac{1}{7}}.$$

As expected, the distance $d_2(x, \bar{x}) = 0.0463$ is larger than in the previous case where x_0 was not constrained.

If we allow f to be quadratic and leave x_0 unconstrained, we obtain the IVP

$$\begin{aligned} x'(t) &= \frac{35}{128} + \frac{105}{32}x - \frac{231}{128}x^2, \\ x(0) &= -\frac{1}{60}, \end{aligned}$$

with corresponding collage distance $d_2(x, Tx) = 0.0047$ and true error $d_2(x, \bar{x}) = 0.0049$.

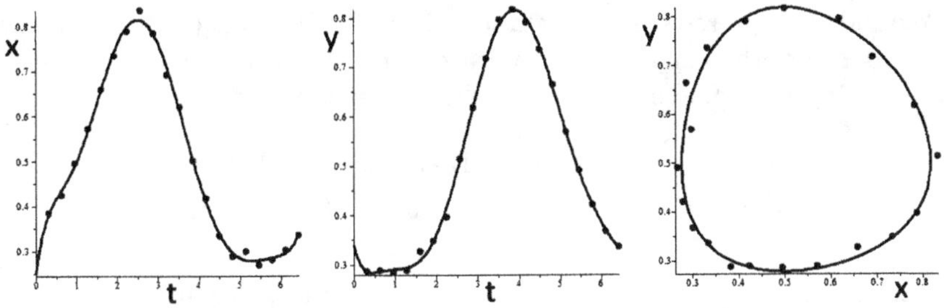

Figure 7.4 Target solutions for the Lotka-Volterra system

EXAMPLE 7.5

Given the parametric representation of a curve $x = x(t)$, $y = y(t)$, $t \geq 0$, we can look for a two-dimensional system of ODEs of the form

$$\dot{x}(t) = f(x, y),\ x(0) = x_0, \tag{7.9}$$
$$\dot{y}(t) = g(x, y),\ y(0) = y_0, \tag{7.10}$$

with conditions on f and g to be specified below. We consider the Lotka-Volterra system

$$\dot{x}(t) = x - 2xy,\ x(0) = \frac{1}{3}, \tag{7.11}$$
$$\dot{y}(t) = 2xy - y,\ y(0) = \frac{1}{3}, \tag{7.12}$$

the solution of which is a periodic cycle. We solve the system numerically. We can determine that the periodic cycle has period 6.35. Sampling the numerical solution with $\Delta t = 0.3$, we gather 21 data points for both x and y. We fit eighth-degree polynomials to each data set. To four decimal places, the result is

$$\begin{aligned} x(t) &= 0.2497 + 0.6565t - 1.0696t^2 + 1.0191t^3 - 0.4384t^4 + 0.0847t^5 \\ &\quad - 0.0050t^6 - 0.0007t^7 + 0.0001t^8, \\ y(t) &= 0.3401 - 0.3477t + 0.7872t^2 - 0.8188t^3 + 0.4183t^4 - 0.0950t^5 \\ &\quad + 0.0054t^6 + 0.0014t^7 - 0.0002t^8. \end{aligned}$$

These polynomials are our target solution to the system of ODEs. See Figure 7.4. We look for a system of the form

$$\begin{aligned} \dot{x}(t) &= c_0 x + c_1 xy,\ x(0) = x_0, \\ \dot{y}(t) &= c_2 xy + c_3 y,\ y(0) = y_0. \end{aligned}$$

We build the squared collage distances $d_2^2(x, T_x X)$ and $d_2^2(y, T_y Y)$, where T_x and T_y are the usual Picard operators, for the respective component. Minimizing the collage distances yields the system

$$\begin{aligned} \dot{x}(t) &= 1.0349x - 2.0471xy, \; x(0) = 0.3241, \\ \dot{y}(t) &= -0.9943xy + 1.9839y, \; y(0) = 0.3388. \end{aligned}$$

This algorithm for solving ODE inverse problems is robust. When low-amplitude noise is added to the sample values, we see small changes in the minimal-collage ODE. This effect occurs because of a continuity result for contractive maps and their fixed points.

As an exercise, use Maple, Mathematica, or other mathematical software to program the algorithm for the preceding examples.

EXERCISES

7.1 Prove that any contraction map is continuous.

7.2 Prove that d_∞ is a metric on $C(I)$.

7.3 Prove that $x(t)$ is a solution to the IVP (7.1) and (7.2) if and only if it satisfies the integral equation (7.3).

7.4 Consider the IVP

$$\begin{aligned} x'(t) &= x, \\ x(0) &= 1. \end{aligned}$$

(a) Verify that the IVP satisfies the properties required for the existence of a unique (local) solution.

(b) Verify that $x(t) = e^t$ is a solution of the IVP (and therefore the unique solution).

(c) Starting with $x_0 = 1$, use the Picard operator T to construct the sequence $x_{n+1} = Tx_n$, for $n = 1, 2, 3$, and 4. Are you convinced that the sequence approaches the Taylor series of e^t centered at $t = 0$?

7.5 Consider the IVP

$$\begin{aligned} x'(t) &= x^2 \\ x(0) &= 1. \end{aligned}$$

(a) Verify that $x(t) = \frac{1}{1-t}$ is a solution of the IVP.

(b) Starting with $x_0 = 1$, use the Picard operator T to construct the sequence $x_{n+1} = Tx_n$, for $n = 1, 2, 3$, and 4. Are you convinced that the sequence approaches the Taylor series of $\frac{1}{1-t}$ centered at $t = 0$?

7.6 Give the argument for the proof of Theorem 5 in the case $t < 0$.

7.7 Prove that the Picard operator T is contractive with respect to the metric $d_1(x, y) = \int_I |x(t) - y(t)|\, dt$.

REFERENCES

1. Boyce, W. E. and DiPrima, R. C., *Elementary Differential Equations and Boundary Value Problems*, Wiley, Ninth Edition (2008).
2. Copson E. T., *Metric Spaces*, Cambridge University Press, Cambridge (1988).
3. Perko, L., *Differential Equations and Dynamical Systems*, Springer, Third Edition (2006).
4. Rudin, W., *Real and Complex Analysis*, McGraw-Hill, Third Edition, New York (1986).

CHAPTER 8

ESTIMATION OF MODEL PARAMETERS

ROBERT PICHÉ

Tampere University of Technology, Tampere, Finland

8.1 ESTIMATION IS AN INVERSE PROBLEM

Mathematical models are created to help understand real-world data. *Estimation* is the process of determining the values of parameters of a mathematical model by comparing real-world observations with the corresponding results predicted by the model. Estimation might be done to improve the model's usability as a tool for prediction or for "what if" scenarios; this is sometimes called *model calibration*. The estimated values of the parameters are typically of interest too, because they represent important quantities, for example, velocity, population, market volatility.

In other chapters of this book, mathematical modeling procedures are presented for setting up and solving questions of the form: "with given parameters, what observations would result?" In this phase of modeling, we seek to set up a mathematical problem whose solutions exist, are unique, and are continuous with respect to the parameters. In estimation, the question is dif-

Mathematical Modeling with Multidisciplinary Applications.
Edited by Xin-She Yang

ferent: "what do the observations tell me about the parameters?" Estimation is, in this sense, an *inverse* problem (Figure 8.1).

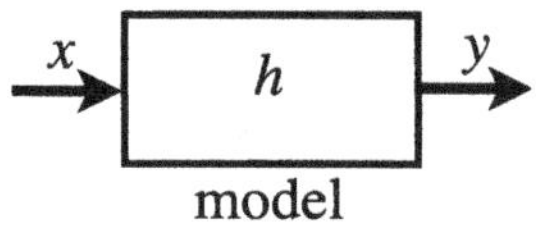

Figure 8.1 Mathematical model seen as a system with inputs (parameters) and outputs (observations).

This chapter introduces a probability-theory based approach to estimation known as Bayesian estimation. This approach is nowadays widely applied in many areas of science and technology [1, 2]. This chapter presents some basic versions of the method that could, in principle, be applied to many mathematical models. To give the discussion a concrete and familiar setting, the method is illustrated by using the problem of estimating your position using observations such as angles between landmarks (triangulation) or ranging signals from satellites (trilateration).

The chapter is organized as follows. We start with a very condensed review of facts and formulas about the multivariate normal distribution that will be needed in this chapter. In Section 8.3, the notation for the observation model is introduced, and it is shown how linearization is used to obtain models that are more tractable for estimation computations. Basic models for positioning via triangulation and via trilateration are presented. In Section 8.4, the Bayesian estimation problem with nonlinear observations and multivariate normal distributions is set up. Two solutions methods for computing key features of the Bayesian estimate are presented. The first method is computation of the posterior distribution's mean and covariance by using the moment matching approximation. The second method is computation of the posterior distribution's mode by using a numerical optimization algorithm. The two methods are illustrated by the detailed solution of a triangulation problem.

Notation: In this chapter, real vectors are denoted with lowercase italic font, real matrices with uppercase italic, and random vectors with lowercase bold, The symbol 0 denotes a zero scalar, vector, or matrix, according to context. The expectation operator is denoted $\mathbb{E}$. (In this chapter, it is understood that whenever this operator is used, the random variable's probability distribution is such that the expectation exists.) Abbreviations used include pdf (probability density function), and spd (symmetric positive definite). A dot is used to indicate an approximate relation: $\doteq$ for equality and $\dot{\sim}$ for probability distribution specification.

8.2 THE MULTIVARIATE NORMAL DISTRIBUTION

Definition 8.1 *The* mean *of a random vector* $\mathbf{x}$ *is the vector* $\mathbb{E}\mathbf{x}$. *The* covariance *of* $\mathbf{x}$, *denoted* $\mathrm{var}\mathbf{x}$, *is the square matrix*

$$\mathrm{var}\mathbf{x} = \mathbb{E}((\mathbf{x} - \mathbb{E}\mathbf{x})(\mathbf{x} - \mathbb{E}\mathbf{x})^T).$$

When $\mathbf{x}$ *is a scalar,* $\mathrm{var}\mathbf{x}$ *is called its variance.*

Fact 8.1 $\mathrm{var}\mathbf{x}$ *is symmetric and non-negative definite, and* $\mathrm{var}\mathbf{x} = \mathbb{E}(\mathbf{x}\mathbf{x}^T) - (\mathbb{E}\mathbf{x})(\mathbb{E}\mathbf{x})^T$.

A linear transformation is a mapping of the form $x \mapsto Ax$, and an affine transformation is a mapping of the form $x \mapsto b + Ax$.

Fact 8.2 *The mean and covariance of an affine transformation of a random variable are* $\mathbb{E}(b + A\mathbf{x}) = b + A\,\mathbb{E}\mathbf{x}$ *and* $\mathrm{var}(b + A\mathbf{x}) = A(\mathrm{var}\mathbf{x})A^T$.

A standard normal random vector is characterized by the fact that its components are independent standard normal random variables, as follows:

Definition 8.2 *An n-variate random vector* $\mathbf{u}$ *is said to have a* standard normal *distribution, denoted* $\mathbf{u} \sim \mathrm{N}(0, I)$, *when its probability density function (pdf) is*

$$\begin{aligned} p_{\mathbf{u}}(u) &= (2\pi)^{-n/2} \mathrm{e}^{-\frac{1}{2}u^T u} \\ &= \left(\frac{1}{\sqrt{2\pi}}\mathrm{e}^{-\frac{1}{2}u_1^2}\right) \cdots \left(\frac{1}{\sqrt{2\pi}}\mathrm{e}^{-\frac{1}{2}u_n^2}\right). \end{aligned}$$

Fact 8.3 *The mean of a standard normal random vector* $\mathbf{u}$ *is* $\mathbb{E}\mathbf{u} = 0$ *and its covariance is* $\mathrm{var}\mathbf{u} = I$.

A normal random vector can be characterized as an affine transformation of a standard normal random vector:

Definition 8.3 *A random vector* $\mathbf{x}$ *is said to have a* normal *distribution, denoted* $\mathbf{x} \sim \mathrm{N}(b, AA^T)$, *if* $\mathbf{x} = b + A\mathbf{u}$ *for some standard normal random vector* $\mathbf{u}$, *vector* b, *and matrix* A.

The following facts from matrix algebra relate the normal distribution's second parameter and the affine transformation's matrix.

Fact 8.4 *For any matrix* A, *the product* AA^T *is symmetric and non-negative definite. If the columns of* A *are linearly independent, the product* AA^T *is symmetric positive definite (spd). For any symmetric non-negative matrix* C, *there exists a matrix* A *such that* $C = AA^T$. *For any spd matrix* C, *there exists a matrix* A *having linearly independent columns such that* $C = AA^T$; *such a matrix* A *can be computed using the Cholesky factorization algorithm.*

The parameters of the normal probability distribution are the mean and covariance of the random vector:

Theorem 8.1 *If* $\mathbf{x} \sim \mathrm{N}(b, C)$ *then* $\mathbb{E}\mathbf{x} = b$ *and* $\mathrm{var}\mathbf{x} = C$.

Proof. Let $C = AA^T$. Then $\mathbb{E}\mathbf{x} = \mathbb{E}(b + A\mathbf{u}) = b + A\mathbb{E}(\mathbf{u}) = b + A \cdot 0 = b$ and $\mathrm{var}\mathbf{x} = \mathrm{var}(b + A\mathbf{u}) = \mathrm{var}(A\mathbf{u}) = A(\mathrm{var}\mathbf{u})A^T = AIA^T = AA^T = C$. ■

In the characterization of the normal distribution in Definition 8.3, the covariance matrix is not required to be invertible. When the covariance matrix is invertible, the distribution is nondegenerate, that is, the pdf exists:

Fact 8.5 *If* $\mathbf{x} \sim \mathrm{N}(b, C)$ *is an n-variate random vector with invertible C, then its pdf is*

$$p_{\mathbf{x}}(x) = (2\pi)^{-n/2}(\det C)^{-1/2}\mathrm{e}^{-\frac{1}{2}(x-b)^T C^{-1}(x-b)}.$$

The following theorem states that normally distributed random vectors remain normal under affine transformations. In particular, the marginal distributions of $\mathbf{x}$ are normal.

Theorem 8.2 *If* $\mathbf{x} \sim \mathrm{N}(b, C)$ *and* $\mathbf{y} = d + B\mathbf{x}$ *then* $\mathbf{y} \sim \mathrm{N}(d + Bb, BCB^T)$ *and* $\mathbf{x}_{1:k} \sim \mathrm{N}(b_{1:k}, C_{1:k,1:k})$.

Proof. Let $C = AA^T$. Then $\mathbf{y} = d + B(b + A\mathbf{u}) = d + Bb + BA\mathbf{u}$, so (by Definition 8.3) we have $\mathbf{y} \sim \mathrm{N}(d + Bb, (BA)(BA)^T)$. The first assertion then follows from the fact that $(BA)(BA)^T = BAA^TB^T = BCB^T$. The second assertion follows from the fact that the subvector is a linear transformation of the full vector, $\mathbf{x}_{1:k} = [I_k, 0]\mathbf{x}$, and so $\mathbf{x}_{1:k} \sim \mathrm{N}([I_k, 0]b, [I_k, 0]C[I_k, 0]^T) = \mathrm{N}(b_{1:k}, C_{1:k,1:k})$. ■

The following Definition and two Facts apply to a random vector having any distribution, provided only that the covariance exists.

Definition 8.4 *The off-diagonal blocks of the covariance matrix*

$$\mathrm{var}\left(\begin{bmatrix} \mathbf{y} \\ \mathbf{z} \end{bmatrix}\right) = \begin{bmatrix} C_{yy} & C_{yz} \\ C_{zy} & C_{zz} \end{bmatrix}$$

are called cross-covariances *and are denoted* $\mathrm{cov}(\mathbf{y}, \mathbf{z}) = C_{yz}$ *and* $\mathrm{cov}(\mathbf{z}, \mathbf{y}) = C_{zy}$. *The random vectors* $\mathbf{y}$ *and* $\mathbf{z}$ *are said to be* uncorrelated *if* $\mathrm{cov}(\mathbf{y}, \mathbf{z}) = 0$.

Fact 8.6 (Chebyshev inequality) *For any* $\lambda > 0$,

$$\mathbb{P}(\mathbf{x}^T\mathbf{x} \geq \lambda) \leq \frac{\mathbb{E}\mathbf{x}^T\mathbf{x}}{\lambda}.$$

Consequently, for any $\alpha \in [0,1)$*, the sphere centered at* $\mathbb{E}\mathbf{x}$ *and having radius* $\sqrt{\frac{\text{trace var}\mathbf{x}}{1-\alpha}}$ *contains at least* α *of the probability of* $\mathbf{x}$*, that is,*

$$\mathbb{P}(\|\mathbf{x} - \mathbb{E}\mathbf{x}\|^2 \le \frac{\text{trace var}\mathbf{x}}{1-\alpha}) \ge \alpha.$$

Fact 8.7 *If* $\mathbf{y}, \mathbf{z}$ *are statistically independent then* $\mathbf{y}, \mathbf{z}$ *are uncorrelated.*

The converse of Fact 8.7 is true for random vectors that are jointly normal:

Fact 8.8 *If the random vector* $\begin{bmatrix} \mathbf{y} \\ \mathbf{z} \end{bmatrix}$ *is normally distributed with uncorrelated* $\mathbf{y}, \mathbf{z}$ *then* $\mathbf{y}, \mathbf{z}$ *are statistically independent.*

The equal-density contours of the pdf of a nondegenerate normal random vector are ellipsoids:

Fact 8.9 *Let* $\text{chi2inv}(\cdot, n)$ *denote the inverse cumulative distribution function of the chi-squared distribution with* n *degrees of freedom. If* $\mathbf{x} \sim \text{N}(b, C)$ *is an* n*-variate non-degenerate normal random vector and* $\alpha \in (0,1)$*, then the ellipsoid*

$$\mathcal{E} = \left\{x : (x-b)^T C^{-1} (x-b) < \text{chi2inv}(\alpha, n)\right\}.$$

contains α *of the probability of* $\mathbf{x}$*, that is,* $\mathbb{P}(\mathbf{x} \in \mathcal{E}) = \alpha$*. In particular, because* $\text{chi2inv}(0.95, 1) = 1.96^2$*, an interval containing* 95% *of the probability of a univariate non-degenerate normal* $\mathbf{x}$ *is*

$$\mathcal{E} = \left\{x : \frac{(x-b)^2}{C} < 1.96^2\right\} = b \pm 1.96\sqrt{C}.$$

Also, because $\text{chi2inv}(0.95, 2) = 5.99$, 95% *of the probability of a bivariate non-degenerate normal* $\mathbf{x}$ *is contained in the ellipse*

$$\mathcal{E} = \left\{x : (x-b)^T C^{-1} (x-b) < 5.99\right\}.$$

In MATLAB *or Octave, this ellipse can be plotted with the code*

```
[u,s,v]=svd(5.99*C); t=0:0.02:2*pi;
e=u*sqrt(s)*[cos(t);sin(t)];
plot(b(1)+e(1,:),b(2)+e(2,:)); axis equal
```

If $\begin{bmatrix} \mathbf{x} \\ \mathbf{y} \end{bmatrix}$ is a random vector and y is a real vector such that $p_{\mathbf{y}}(y) > 0$, then $\mathbf{x} \,|\, (\mathbf{y} = y)$ denotes the random vector, called the conditional distribution of $\mathbf{x}$ given $\mathbf{y} = y$, whose pdf is

$$p_{\mathbf{x}\,|\,\mathbf{y}}(x \,|\, y) := \frac{p_{\begin{bmatrix} \mathbf{x} \\ \mathbf{y} \end{bmatrix}}\left(\begin{bmatrix} x \\ y \end{bmatrix}\right)}{p_{\mathbf{y}}(y)},$$

For jointly normal random vectors, conditional distributions are normal:

Theorem 8.3 *If*

$$\begin{bmatrix} \mathbf{x} \\ \mathbf{y} \end{bmatrix} \sim \mathrm{N}\left(\begin{bmatrix} m_x \\ m_y \end{bmatrix}, \begin{bmatrix} C_x & C_{xy} \\ C_{yx} & C_y \end{bmatrix} \right)$$

with nonsingular C_y, *then*

$$\mathbf{x} \,|\, (\mathbf{y} = y) \sim \mathrm{N}\left(m_x + C_{xy}C_y^{-1}(y - m_y),\; C_x - C_{xy}C_y^{-1}C_{yx} \right).$$

Proof. Let

$$\mathbf{z} = \mathbf{x} - m_x - C_{xy}C_y^{-1}(\mathbf{y} - m_y) = C_{xy}C_y^{-1}m_y - m_x + \begin{bmatrix} I, & -C_{xy}C_y^{-1} \end{bmatrix} \begin{bmatrix} \mathbf{x} \\ \mathbf{y} \end{bmatrix}.$$

This random vector is an affine transformation of a normal random vector and so is normal. Its mean is $\mathbb{E}\mathbf{z} = \mathbb{E}\big(\mathbf{x} - m_y - C_{xy}C_y^{-1}(\mathbf{y} - m_y)\big) = 0$ and its covariance is

$$C_z = \begin{bmatrix} I, & -C_{xy}C_y^{-1} \end{bmatrix} \begin{bmatrix} C_x & C_{xy} \\ C_{yx} & C_y \end{bmatrix} \begin{bmatrix} I \\ -C_y^{-1}C_{yx} \end{bmatrix} = C_x - C_{xy}C_y^{-1}C_{yx}.$$

The random vectors $\mathbf{y}$ and $\mathbf{z}$ are independent, because they are jointly normal and their cross-covariance is

$$\mathbb{E}(\mathbf{y} - \mathbb{E}\mathbf{y})(\mathbf{z} - \mathbb{E}\mathbf{z})^T = \mathbb{E}(\mathbf{y} - m_y)(\mathbf{x} - m_x - C_{xy}C_y^{-1}(\mathbf{y} - m_y))^T = 0.$$

Thus

$$\begin{aligned} \mathbf{x} \,|\, (\mathbf{y} = y) &= \big(\mathbf{z} + m_x + C_{xy}C_y^{-1}(\mathbf{y} - m_y)\big) \,|\, (\mathbf{y} = y) \\ &= m_x + C_{xy}C_y^{-1}(y - m_y) + \underbrace{\mathbf{z} \,|\, (\mathbf{y} = y)}_{\sim \mathrm{N}(0, C_z)}, \end{aligned}$$

that is, the conditional random vector is the sum of a constant and a zero-mean normal random vector whose covariance is C_z. ■

8.3 MODEL OF OBSERVATIONS

8.3.1 Deterministic Model and its Linearization

A deterministic model for how an observation (n_y-vector y) depends on input data or model parameters (n_x-vector x) has the form

$$y = h(x), \tag{8.1}$$

where $h : \mathbb{R}^{n_x} \longrightarrow \mathbb{R}^{n_y}$ is a known function. This model, illustrated in Figure (8.1), is very general, describing any mathematical model of a deterministic system having a finite number of inputs and outputs. The deterministic model will be the foundation on which the probabilistic model, used for estimation, will be constructed in the next section.

The computational methods that will be introduced later in the chapter all make use of an approximation of the deterministic model; this approximation is obtained as follows. In the neighborhood of a given fixed vector $\hat{x}$, the deterministic model (8.1) can be linearized using a first-order Taylor polynomial

$$h(x) \doteq h(\hat{x}) + H\,(x - \hat{x}), \tag{8.2}$$

where the matrix H is the Jacobian of h, evaluated at the reference point:

$$H_{ij} = \left.\frac{\partial h_i(x)}{\partial x_j}\right|_{x=\hat{x}}. \tag{8.3}$$

Denoting $b = h(\hat{x}) - H\hat{x}$, the linearized model can be written in the form

$$y \doteq b + Hx. \tag{8.4}$$

When there are n_k observation vectors, the models are denoted

$$y[k] = h[k](x) \doteq b[k] + H[k]x \quad (k = 1, \ldots, n_k). \tag{8.5}$$

Here are two examples of observation models drawn from the field of surveying and navigation.

■ EXAMPLE 8.1 Triangulation

During a cross-country ski trek in Lapland, you take out your map and recognize some landmarks around you: a ski lift, a cellphone network mast, a lake in the distance, etc. With your magnetic compass, you measure the headings of the landmarks. Denoting y as the heading (in radians, clockwise from north) to a landmark, x as your position (x_1 is the northing, x_2 is the easting), and s as the landmark's position, the geometry (Figure 8.2) can be modeled as

$$s_1 - x_1 = r\cos y, \quad s_2 - x_2 = r\sin y, \tag{8.6}$$

where $r = \|s - x\|$.

Using the two-argument arctangent function, the observation model can be written in the form of (8.1) with

$$h(x) = \operatorname{atan}(s_2 - x_2,\, s_1 - x_1) \in (-\pi, \pi].$$

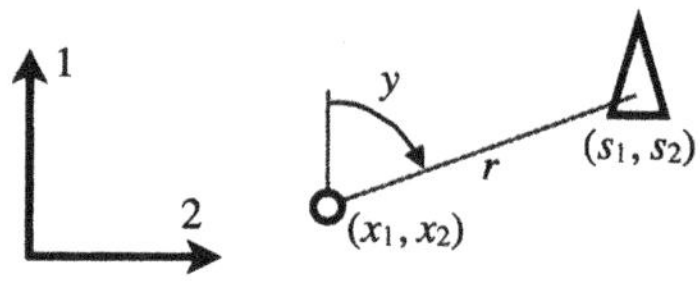

Figure 8.2 Geometry of positioning by triangulation

Differentiating (8.6) gives

$$-\,\mathrm{d}x_1 = \mathrm{d}r \cdot \cos y - r \sin y \cdot \mathrm{d}y, \qquad -\,\mathrm{d}x_2 = \mathrm{d}r \cdot \sin y + r \cos y \cdot \mathrm{d}y$$

Solving for the differentials $\mathrm{d}y$ and $\mathrm{d}r$ gives

$$\mathrm{d}y = r^{-1} \sin y \cdot \mathrm{d}x_1 - r^{-1} \cos y \cdot \mathrm{d}x_2, \quad \mathrm{d}r = \text{(formula not needed)}$$

Comparing terms with the total derivative formula $\mathrm{d}y = \frac{\partial y}{\partial x_1}\,\mathrm{d}x_1 + \frac{\partial y}{\partial x_2}\,\mathrm{d}x_2$, it follows that

$$\frac{\partial h}{\partial x_1} = r^{-1} \sin y, \qquad \frac{\partial h}{\partial x_2} = -r^{-1} \cos y$$

Substituting $r = \|s - x\|$ and (8.6), the linearized model is obtained in the form of (8.4) with the 1×2 Jacobian matrix

$$H = \left[\frac{s_2 - \hat{x}_2}{\|s - \hat{x}\|^2}, \; -\frac{s_1 - \hat{x}_1}{\|s - \hat{x}\|^2} \right].$$

■ EXAMPLE 8.2 GNSS Pseudo-range

The previous example might seem rather dated to the modern reader, because positioning nowadays is mostly done using devices that process data from global navigation satellite systems (GNSS) such as GPS and GLONASS. These satellites periodically transmit coded messages from which a receiver can recover the satellite's current position and the time difference between the satellite's atomic clock and the receiver's clock. This time difference, multiplied by c, the speed of radio waves, is called the *pseudo-range* and is denoted y. If the 3-vectors s and $x_{1:3}$ denote the position of the satellite and of the receiver, x_4 denotes the amount added to y due to the lack of synchronization, and d denotes the part of y that is due to other effects (including atmospheric delays and relativity), then a simple deterministic model for the pseudo-range is

$$y = \|x_{1:3} - s\| + x_4 + d,$$

This is a model of the form of (8.1), with $h(x)$ being the expression on the right-hand side of the equation. The linearized model is (8.4) with the 1×4 Jacobian matrix

$$H = \left[\frac{(\hat{x}_{1:3} - s)^T}{\|\hat{x}_{1:3} - s\|}, \ 1 \right].$$

8.3.2 Probabilistic Model

Because of the various idealizations and approximations that are made during the modeling process, we don't expect real-world observations to agree exactly with the model. Indeed, observations differ even when the same experiments are repeated with all the conditions, as far as we can determine, identical. In probabilistic (statistical) approaches to estimation, this variability is taken into account by modeling the observation as a random variable.

In Bayesian estimation, the quantities being estimated are also modeled as random variables. This approach is based on the idea that the laws of probability serve as a logically consistent way of modeling one's state of knowledge (and ignorance) about these values.

A general probabilistic version of the observation model (8.1) is a specification of a conditional probability distribution for $\mathbf{y}$ given the parameters, that is, a specification of the random variable $\mathbf{y} \mid (\mathbf{x} = x)$. In the case of continuous random variables, the specification of $\mathbf{y} \mid (\mathbf{x} = x)$ can be in the form of a pdf, and this pdf is denoted

$$p_{\mathbf{y}|\mathbf{x}}(y \mid x), \tag{8.7}$$

In this chapter, we consider a specific instance of (8.7) that is obtained by adding a zero-mean multivariate normally distributed (i.e. Gaussian) term $\mathbf{v} \sim \mathrm{N}(0, R)$, independent of $\mathbf{x}$, to the deterministic observation function h evaluated with random argument:

$$\mathbf{y} = h(\mathbf{x}) + \mathbf{v}.$$

The term $\mathbf{v}$ is sometimes called "observation error" or "noise"; the amount of uncertainty of the observation is represented by the covariance matrix R. In the case where the observation is scalar-valued, R is a scalar and is called the variance.

Because $\mathbf{v}$ is independent of $\mathbf{x}$, we have

$$\mathbf{y} \mid (\mathbf{x} = x) = h(\mathbf{x}) \mid (\mathbf{x} = x) + \mathbf{v} \mid (\mathbf{x} = x) = h(x) + \mathbf{v},$$

and so using Theorem 8.2 our "additive Gaussian noise" observation model can be written

$$\mathbf{y} \mid (\mathbf{x} = x) \sim \mathrm{N}(h(x), R). \tag{8.8}$$

8.4 ESTIMATION

8.4.1 Bayesian Inference

You could think of the probabilistic observation model (8.7) as a computer simulation program. Given certain parameter values, the program uses a random number generator to produce simulated observations that follow a certain probability distribution. From this point of view, estimation is the inverse problem: determine (the probability distribution of) the parameters, given the realized observation values.

In principle, the estimation problem is entirely solved by Bayes' formula. If the pdf $p_{\mathbf{x}}(x)$ models your state of knowledge about $\mathbf{x}$ before being informed of the observation values, then the pdf $p_{\mathbf{x}|\mathbf{y}}(x \mid y)$ is a model of your state of knowledge that incorporates this additional information. This pdf is given by Bayes' formula

$$p_{\mathbf{x}|\mathbf{y}}(x|y) \propto p_{\mathbf{y}|\mathbf{x}}(y \mid x) p_{\mathbf{x}}(x) \tag{8.9}$$

and is the pdf of the random variable $\mathbf{x} \mid (\mathbf{y} = y)$, which is known as the *posterior*. The pdf $p_{\mathbf{y}|\mathbf{x}}$ is the observation model (8.7) and the pdf $p_{\mathbf{x}}$ is known as the *prior* pdf. The proportionality symbol in Bayes' formula is used to represent the fact that the expression on the right-hand side needs to be scaled by a constant (not depending on x) in order to be a proper pdf; the constant is determined by the fact that the integral with respect to x should be unity.

With Bayes' formula we can compute (up to a scaling factor) the values of the posterior density at any point x, but this is not particularly useful by itself. One is typically interested in *summarizing statistics* of the distribution, such as the mean, the mode, the covariance matrix, or a sphere containing 95% of the probability. Because of the nonlinearity of the observation function h, there are usually no formulas for these summarizing statistics, and some numerical methods and approximations are needed.

In the next two subsections, two methods are presented to compute summarizing statistics of the posterior distribution: moment matching and optimization.

8.4.2 Moment Matching

The moment-matching approach is as follows: compute (approximately) the mean and covariance of the joint distribution of $(\mathbf{x}, \mathbf{y})$, then approximate the joint distribution by a normal distribution having that mean and covariance. (Notice that two approximations are introduced!) The mean and covariance of the posterior that corresponds to this normal joint distribution are then given by the formulas of Theorem 8.3.

The moment-matching method's first approximation arises in the computation of the moments μ, S, and C, defined as follows.

Fact 8.10 *If* $\mathbf{x} \sim \mathrm{N}(m, P)$ *and* $\mathbf{v} \sim \mathrm{N}(0, R)$ *are independent and* $\mathbf{y} = h(\mathbf{x}) + \mathbf{v}$ *then*

$$\begin{aligned} \mu &:= \mathbb{E}\mathbf{y} = \mathbb{E}h(\mathbf{x}), \\ S &:= \mathrm{var}\mathbf{y} = R + \mathbb{E}(h(\mathbf{x}) - \mu)(h(\mathbf{x}) - \mu)^T, \\ C &:= \mathrm{cov}(\mathbf{x}, \mathbf{y}) = \mathbb{E}(\mathbf{x} - m)(h(\mathbf{x}) - \mu)^T. \end{aligned}$$

The method's second approximation arises when the joint distribution is assumed to be a multivariate normal:

$$\begin{bmatrix} \mathbf{x} \\ \mathbf{y} \end{bmatrix} \dot{\sim} \mathrm{N}\left(\begin{bmatrix} m \\ \mu \end{bmatrix}, \begin{bmatrix} P & C \\ C^T & S \end{bmatrix} \right).$$

Theorem 8.3 then gives the approximation of the posterior as $\mathbf{x} \,|\, (\mathbf{y} = y) \dot{\sim} \mathrm{N}(q, Q)$ with

$$q = m + CS^{-1}(y - \mu), \quad Q = P - CS^{-1}C^T \tag{8.10}$$

provided that S is invertible.

There still remains the question of how to compute the moments in Fact 8.10. One alternative is to make use of the linearization approximation of the observation model, Eq. (8.4), with $\hat{x} = m$. When $h(x) = b + Hx$, the moments are

$$\mu = h(m), \quad S = R + HPH^T, \quad C = PH^T.$$

Substituting these into (8.10), and introducing

$$K = PH^T(R + HPH^T)^{-1}, \tag{8.11}$$

the posterior mean and covariance are obtained as

$$q = m + K(y - h(m)), \quad Q = P - KHP. \tag{8.12}$$

If P is invertible, then by applying the Woodbury matrix identity[2] the update formulas (8.12) can be written as

$$Q = (H^T R^{-1} H + P^{-1})^{-1}, \quad q = QH^T R^{-1}(y - b) + QP^{-1}m.$$

The limiting case $P^{-1} \to 0$ can be interpreted as a model of "infinite" uncertainty about the prior $\mathbf{x}$. Even though the prior with $P^{-1} \to 0$ is not a proper probability distribution, the corresponding limiting value of the approximate posterior is a proper distribution, namely $\mathbf{x} \,|\, (\mathbf{y} = y) \dot{\sim} \mathrm{N}(q, Q)$ with

$$Q = (H^T R^{-1} H)^{-1}, \quad q = QH^T R^{-1}(y - b), \tag{8.13}$$

[2] $(A + BDC)^{-1} = A^{-1} - A^{-1}B(D^{-1} + CA^{-1}B)^{-1}CA^{-1}$.

provided that the inverses exist.

A convenient feature of moment matching is that the posterior is (approximated to be) a normal distribution. Then, if further observations are obtained that are conditionally independent of the earlier observations given x, they can be assimilated into the posterior by updating it. That is, if $m[k-1]$ and $P[k-1]$ denote the mean and covariance after observations $y[1{:}k-1]$ have been obtained, and $\mathbf{y}[k] \mid (\mathbf{x} = x) \sim \mathrm{N}(h[k](x), R[k])$ with $h[k](x) \doteq b[k] + H[k]x$, then $\mathbf{x} \mid (\mathbf{y}[1{:}k] = y[1{:}k]) \mathrel{\dot\sim} \mathrm{N}(m[k], P[k])$ with

$$\begin{aligned} K[k] &= P[k-1]H[k]^T(R[k] + H[k]P[k-1]H[k]^T)^{-1}, \\ m[k] &= m[k-1] + K[k](y[k] - h[k](m[k-1])), \\ P[k] &= P[k-1] - K[k]H[k]P[k-1]. \end{aligned} \tag{8.14}$$

This recursive one-observation-at-a-time updating procedure can be justified using Bayes' formula, as follows:

$$\begin{aligned} p_{\mathbf{x}|\mathbf{y}[1:2]}(x \mid y[1{:}2]) &\propto p_{\mathbf{y}[1:2]|\mathbf{x}}(y[1{:}2] \mid x)\, p_{\mathbf{x}}(x) \\ &= p_{\mathbf{y}[2]|\mathbf{x}}(y[2] \mid x)\, p_{\mathbf{y}[1]|\mathbf{x}}(y[1] \mid x)\, p_{\mathbf{x}}(x) \\ &\propto p_{\mathbf{y}[2]|\mathbf{x}}(y[2] \mid x)\, p_{\mathbf{x}|\mathbf{y}[1]}(x \mid y[1]), \end{aligned}$$

and similarly for $y[1{:}3]$, $y[1{:}4]$, etc.

To summarize, given

- independent observation vectors $y[k]$ for $k = 1, 2, \ldots, n$
- observation functions $x \mapsto h[k](x)$
- additive zero-mean normal observation noises with covariance $R[k]$
- a normal prior distribution with mean $m[0]$ and covariance $P[0]$

then the mean $m[n]$ and covariance $P[n]$ of the normal approximation of the posterior distribution $\mathbf{x} \mid (\mathbf{y}[1{:}n] = y[1{:}n])$ can be computed as follows:

1. Initialize $k \leftarrow 1$
2. Find the mapping $x \mapsto b[k] + H[k]x$ by linearizing h about $m[k-1]$.
3. Compute $m[k]$ and $P[k]$ using (8.14).
4. If $k = n$, stop, otherwise increment $k \leftarrow k + 1$ and go to 2.

If $P[0]^{-1} = 0$, then the update in step 3 for $k = 1$ is computed using

$$\begin{aligned} P[1] &\leftarrow (H[1]^T R[1]^{-1} H[1])^{-1} \\ m[1] &\leftarrow P[1]H[1]^T R[1]^{-1}(y[1] - b[1]) \end{aligned}$$

provided that the inverses exist.

■ **EXAMPLE 8.3 The Deep Chasm**

Frodo and his companions come to the edge of a chasm. Frodo fearfully peers over the edge and gasps: the chasm appears to be between 100 and 200 m deep! He drops a pebble and hears it hit the bottom 5 to 6 seconds later. How does Frodo revise his belief about the depth of the chasm in light of the result of the pebble-dropping experiment?

Let us model the prior distribution as a normal distribution that has 95% of the probability lying in the interval [100, 200]. Using Fact 8.9, a prior of the form $\mathbf{x} \sim \mathrm{N}(m, P)$ has 95% of its probability in the interval $m \pm 1.96\sqrt{P}$, so we set $m = 150$ and $P = (\frac{50}{1.96})^2 = 651$.

Assuming zero initial velocity and constant acceleration $g = 9.81\,\mathrm{m/s}^2$, the distance fallen by the pebble in time y is $\frac{1}{2}gy^2$. Assuming instantaneous sound travel, a deterministic model of the time observation is therefore

$$y = \sqrt{\frac{2x}{g}} =: h(x).$$

The linearization of this model about $x = \hat{x}$ is (8.4) with

$$H = \left.\frac{\mathrm{d}h(x)}{\mathrm{d}x}\right|_{x=\hat{x}} = \frac{1}{\sqrt{2g\hat{x}}}.$$

The observation $[5, 6] = 5.5 \pm 0.5$ is interpreted as a realization $y = 5.5$ of an observation of the form (8.8) with variance $R = (\frac{0.5}{1.96})^2 = 0.0651$.

Using (8.11 and 8.12), we compute

$$\begin{aligned}
H &= \frac{1}{\sqrt{2gm}} = \frac{1}{\sqrt{2 \cdot 9.81 \cdot 150}} = 0.01843,\\
K &= PH^T(R + HPH^T)^{-1} = \frac{651 \cdot 0.01843}{0.0651 + 651(0.01843)^2} = 41.91,\\
q &= m + K(y - h(m)) = 150 + 41.91\left(5.5 - \sqrt{\frac{2 \cdot 150}{9.81}}\right) = 149.91,\\
Q &= P - KHP = 147.97.
\end{aligned}$$

The posterior distribution is thus approximated to be $\mathbf{x} \,|\, (\mathbf{y} = y) \sim \mathrm{N}(149.91, 147.97)$. Frodo is 95% certain that the chasm depth (in meters) is a number in the interval $149.91 \pm 1.96\sqrt{147.97} = [126, 174]$.

■ EXAMPLE 8.4 Triangulation (Continued)

Given a prior estimate $\hat{x} = [500, 1000]^T$, and the data in Table 8.1 and Figure 8.3, what is your location?

In order to illustrate the use of recursive formulas, in this solution the first two headings will be treated as a single observation, and then the estimate will be updated using the third heading as a new observation.

Table 8.1 Data for the triangulation problem.

landmark no.	northing (m)	easting (m)	angle (deg)	std. dev. (deg)
1	1126	370	-45	1
2	1643	2478	52	2
3	443	2173	91	1

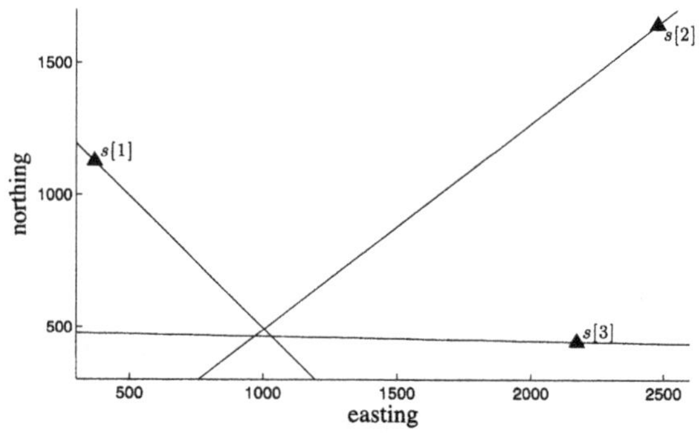

Figure 8.3 Geometry of the triangulation problem

The observation model for the first observation is $\mathbf{y}[1]|(\mathbf{x} = x) \sim \mathrm{N}(h[1](x), R[1])$ with

$$h[1](x) = \text{atan}\left(\begin{bmatrix} 370 \\ 2478 \end{bmatrix} - x_2, \begin{bmatrix} 1126 \\ 1643 \end{bmatrix} - x_1\right), \quad R[1] = \sigma_0^2 \begin{bmatrix} 1^2 & 0 \\ 0 & 2^2 \end{bmatrix},$$

and $\sigma_0 = \frac{\pi}{180}$ (this is 1 degree in radians). The atan function with vector arguments is evaluated element-wise.

The realized value of the observation is

$$y[1] = \frac{\pi}{180}\begin{bmatrix} -45 \\ 52 \end{bmatrix} = \begin{bmatrix} -0.7854 \\ 0.9076 \end{bmatrix}.$$

The linearized model is $h[1](x) \doteq b[1] + H[1]x$ with

$$\begin{aligned} H[1] &= \begin{bmatrix} \frac{370-1000}{(1126-500)^2+(370-1000)^2} & -\frac{1126-500}{(1126-500)^2+(370-1000)^2} \\ \frac{2478-1000}{(1643-500)^2+(2478-1000)^2} & -\frac{1643-500}{(1643-500)^2+(2478-1000)^2} \end{bmatrix} \\ &\doteq 10^{-6}\begin{bmatrix} -798.71 & -793.63 \\ 423.38 & -327.42 \end{bmatrix} \end{aligned}$$

and

$$\begin{aligned} b[1] &= h[1](\hat{x}) - H[1]\hat{x} \\ &= \operatorname{atan}\left(\begin{bmatrix} 370 \\ 2478 \end{bmatrix} - 1000, \begin{bmatrix} 1126 \\ 1643 \end{bmatrix} - 500\right) \\ &\quad - 10^{-6}\begin{bmatrix} -798.71 & -793.63 \\ 423.38 & -327.42 \end{bmatrix}\begin{bmatrix} 500 \\ 1000 \end{bmatrix} = \begin{bmatrix} 0.404 \\ 1.0283 \end{bmatrix}. \end{aligned}$$

Assuming a prior with mean $\hat{x}$ and "infinite" covariance, we apply the formula (8.13) and obtain the approximate posterior $\mathbf{x}|(\mathbf{y}[1] = y[1]) \dot{\sim} \mathrm{N}(m[1], P[1])$ with

$$P[1] = (H[1]^T R[1]^{-1} H[1])^{-1} = \begin{bmatrix} 2241 & -2045 \\ -2045 & 2330 \end{bmatrix},$$

$$m[1] = P[1]H[1]^T R[1]^{-1}(y[1] - b[1]) = \begin{bmatrix} 491.7 \\ 1004.4 \end{bmatrix}.$$

The approximate posterior distribution's mean is shown in Figure 8.4; it lies on the intersection of the lines of sight to the landmarks. Also shown is an ellipse containing 95% of the probability; the ellipse is computed using Fact 8.9.

Next, we update this estimate using the third heading as the new observation. The observation model for the observation is $\mathbf{y}[2]|(\mathbf{x} = x) \sim \mathrm{N}(h[2](\mathbf{x}), R[2])$ with

$$h[2](x) = \operatorname{atan}(2173 - x_2, 443 - x_1), \quad R[2] = \sigma_0^2.$$

The realized value of the observation is $y[2] = 91 \cdot \frac{\pi}{180} = 1.5882$.

The model linearized about $m[1]$ is $h[2](x) \doteq b[2] + H[2]x$ with

$$H[2] = \begin{bmatrix} \frac{2173-1004.4}{(2173-1004.4)^2+(443-491.7)^2} & -\frac{443-491.7}{(2173-1004.4)^2+(443-491.7)^2} \end{bmatrix}$$

$$\doteq 10^{-6}\begin{bmatrix} 854.24 & 35.60 \end{bmatrix}.$$

Applying the updating formulas (8.14), we obtain the new posterior $\mathbf{x}|(\mathbf{y}[1{:}2] = y[1{:}2]) \dot{\sim} \mathrm{N}(m[2], P[2])$ with

$$K[2] = P[1]H[2]^T(R[2] + H[2]P[1]H[2]^T)^{-1} = \begin{bmatrix} 1012.7 \\ -915.0 \end{bmatrix},$$

$$P[2] = P[1] - K[2]H[2]P[1] = \begin{bmatrix} 376.1 & -359.9 \\ -359.9 & 807.5 \end{bmatrix},$$

$$m[2] = m[1] + K[2](y[2] - h[2](m[1])) = \begin{bmatrix} 467.2 \\ 1026.5 \end{bmatrix}.$$

The posterior mean and an ellipse containing 95% of its probability are shown in Figure 8.4. An interval for the northing that contains 95% of the posterior probability is

$$467.2 \pm 1.96\sqrt{376.1} = [429., 505.2].$$

An interval for the easting is

$$1026.5 \pm 1.96\sqrt{807.5} = [970.8, 1082.2].$$

In this solution, the observations were processed in two batches, and the linearization reference point was updated between batches. If all three observations had been processed in a single batch, with a single linearization about the prior mean, the answer would be (slightly) different.

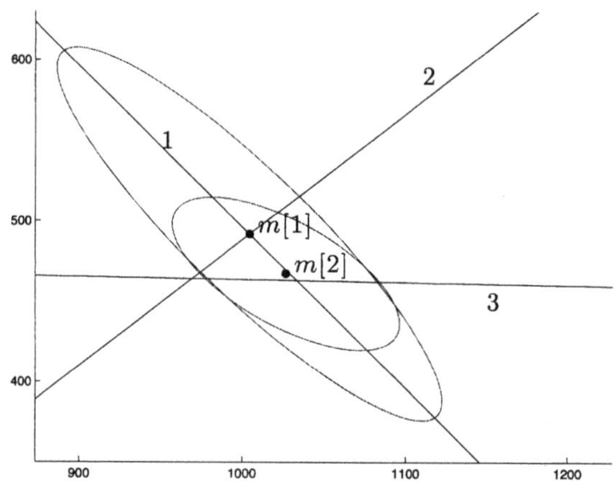

Figure 8.4 Position estimated by triangulation. The headings are shown as lines from the (distant) landmarks; the large ellipse centered at $m[1]$ is the 95% ellipse for the position estimate based on the first two headings; the small ellipse centered at $m[2]$ is the 95% ellipse for the position estimate based on all three headings.

8.4.3 Estimation by Optimization

Up to this point we have assumed a normal prior $\mathbf{x} \sim \mathrm{N}(m, P)$ with spd P and normal conditional observation $\mathbf{y} \,|\, \mathbf{x} \sim \mathrm{N}(h(x), R)$. Here we further assume that R is spd. By Bayes' formula (8.9), the posterior density is

$$\begin{aligned} p_{\mathbf{x}\,|\,\mathbf{y}}(x\,|\,y) &\propto p_{\mathbf{y}|\mathbf{x}}(y\,|\,x)p_{\mathbf{x}}(x) \\ &\propto \mathrm{e}^{-\frac{1}{2}(h(x)-y)^T R^{-1}(h(x)-y)}\mathrm{e}^{-\frac{1}{2}(x-m)^T P^{-1}(x-m)} \\ &\propto \mathrm{e}^{-\phi(x)}, \end{aligned}$$

where

$$\phi(x) = \tfrac{1}{2}(h(x)-y)^T R^{-1}(h(x)-y) + \tfrac{1}{2}(x-m)^T P^{-1}(x-m) \tag{8.15}$$

Instead of seeking the mean of the posterior, as in the previous section, let us seek the *mode* of the posterior, that is, the point of maximum density. This is called the maximum a-posteriori (MAP) estimate, and it can be computed by using numerical optimization algorithms to find the minimizer of the "cost function" ϕ.

According to Fact 8.4, the covariance matrices can be factorized as $P = AA^T$ and $R = BB^T$. Then the cost function can be written as a sum of squares

$$\phi(x) = \tfrac{1}{2}\|f(x)\|^2, \tag{8.16}$$

with the (weighted) residual

$$f(x) = \begin{bmatrix} A^{-1}(x-m) \\ B^{-1}(h(x)-y) \end{bmatrix}.$$

Minimization of a cost function of the form (8.16) is called a nonlinear least squares problem. There exist several algorithms for this class of optimization problems; in the following we present the algorithm known as the Gauss-Newton method.

The "Gauss" part of the method's name comes from the use of the theory of least squares, which is summarised as follows.

Fact 8.11 (Least squares) *For any matrix A, there exists a unique solution to the equation set*

$$AXA = A,\ XAX = X,\ (AX)^T = AX,\ (XA)^T = XA.$$

This solution is called the Moore-Penrose pseudoinverse and is denoted A^+. For any compatible-sized vectors b and x, there holds

$$\|b - Ax\|^2 = \|b - AA^+b\|^2 + \|A(x - A^+b)\|^2.$$

Consequently, $x = A^+b$ is a minimizer of $\|b - Ax\|^2$. If A has linearly independent columns then $A^+ = (A^TA)^{-1}A^T$ and the minimizer $x = (A^TA)^{-1}A^Tb$ is unique.

The "Newton" part of the method's name comes from the linearization of the residual function, similarly to what is done in the Newton method for finding the zero of a function. The linearization about a given reference point is

$$f(\hat{x} + d) \doteq f(\hat{x}) + Jd,$$

where J is the Jacobian of f, evaluated at the reference point:

$$J = \left.\frac{\partial f(x)}{\partial x}\right|_{x=\hat{x}} = \begin{bmatrix} A^{-1} \\ B^{-1}H \end{bmatrix}.$$

The linearization of the residual function corresponds to the approximation of the cost function as

$$\phi(\hat{x} + d) = \tfrac{1}{2}\|f(\hat{x} + d)\|^2 \doteq \tfrac{1}{2}\|f(\hat{x}) + Jd\|^2.$$

Invoking Fact 8.11, and noting that J has linearly independent columns, the approximated cost function is minimized by taking d to be

$$\begin{aligned} d_{\text{gn}} &= -(J^T J)^{-1} J^T f(\hat{x}) \\ &= -(H^T R^{-1} H + P^{-1})^{-1}(P^{-1}(\hat{x} - m) + H^T R^{-1}(h(\hat{x}) - y)) \\ &= m - \hat{x} + K\big(y - h(\hat{x}) - H\,(m - \hat{x})\big), \end{aligned} \tag{8.17}$$

where

$$K = PH^T\big(R + HPH^T\big)^{-1}.$$

The Gauss-Newton algorithm for finding the minimizer of (8.15) consists of assigning the initial value $\hat{x}$, then repeatedly updating $\hat{x} \leftarrow \hat{x} + d_{\text{gn}}$ using and (8.3) and (8.17) until $\|d_{\text{gn}}\|$ is sufficiently small.

Notice that if the initial value is taken to be $\hat{x} = m$, then the first Gauss-Newton update $m + d_{\text{gn}}$ coincides exactly with the posterior mean found by the method of moment matching with linearization that was described in Section 8.4.2.

In the case of a prior with $P^{-1} \to 0$, and provided that H has linearly independent columns, in place of (8.17) one can use the formula

$$d_{\text{gn}} = -\big(H^T R^{-1} H\big)^{-1} H^T R^{-1}(h(\hat{x}) - y)). \tag{8.18}$$

Sometimes the Gauss-Newton step d_{gn} may be so large that the cost function actually increases. In order to avoid this, one can take a smaller step by scaling d_{gn} if necessary to ensure that the cost function is decreasing. The following theorem establishes that this is always possible, because the Gauss-Newton step d_{gn} is a "descent direction."

Theorem 8.4 *There exists a positive $\underline{\alpha}$ such that the inequality $\phi(\hat{x} + \alpha d_{\text{gn}}) < \phi(\hat{x})$ holds for all $\alpha \in (0, \underline{\alpha}]$.*

Proof. The Taylor expansion of the objective function is

$$\phi(\hat{x} + \alpha d) = \phi(\hat{x}) + \alpha \phi'(\hat{x}) d + O(\alpha^2),$$

so d is a descent direction if $\phi'(\hat{x})d < 0$. For $\phi(x) = \frac{1}{2}\|f(x)\|^2$ we have

$$\frac{\partial \phi}{\partial x_j} = \sum_i f_i(x) \frac{\partial f_i}{\partial x_j},$$

and for the Gauss-Newton step we have

$$\phi'(\hat{x}) d_{\text{gn}} = f(\hat{x})^T J d_{\text{gn}} = -d_{\text{gn}}^T (J^T J)^{-1} d_{\text{gn}},$$

which is < 0 because $J^T J$ is spd. ■

In each iteration of the "descending" Gauss-Newton method, initialize $\alpha \leftarrow 1$, repeat the "line search"

$$\text{while } \phi(\hat{x} + \alpha d_{\text{gn}}) \geq \phi(\hat{x}) : \quad \alpha \leftarrow \frac{\alpha}{2}$$

to determine a suitable scale factor α, and update $\hat{x} \leftarrow \hat{x} + \alpha d_{\text{gn}}$. (A computer implementation in floating point arithmetic should of course include a limit on the number of line search steps, to avoid an infinite loop.)

To summarize, given

- a vector of observations y
- an observation function $x \mapsto h(x)$
- additive zero-mean normal observation noises with covariance R
- a normal prior distribution with mean m and covariance P

then the MAP estimate $\hat{x}$, that is, the mode of the posterior distribution $\mathbf{x} \,|\, (\mathbf{y} = y)$, can be computed using the descending Gauss-Newton algorithm as follows:

1. Initialize $\hat{x} \leftarrow m$.
2. Find the mapping $x \mapsto b + Hx$ by linearizing h about $\hat{x}$.
3. Compute d_{gn} using (8.17) [or (8.18) if $P^{-1} = 0$] and set $\alpha \leftarrow 1$.
4. Compute $\phi(\hat{x})$ using (8.15) with $x \leftarrow \hat{x}$.
5. Compute $\phi(\hat{x} + \alpha d_{\text{gn}})$ using (8.15) with $x \leftarrow \hat{x} + \alpha d_{\text{gn}}$.
6. If $\phi(\hat{x} + \alpha d_{\text{gn}}) \geq \phi(\hat{x})$, set $\alpha \leftarrow \frac{1}{2}\alpha$ and go to 5.
7. Update $\hat{x} \leftarrow \hat{x} + \alpha d_{\text{gn}}$.
8. If $\|d_{\text{gn}}\|$ is small enough, stop; otherwise, go to 2.

The ordinary Gauss-Newton algorithm is obtained by omitting steps 4–6.

■ EXAMPLE 8.5 Triangulation (Continued)

Let's find the MAP estimate of position for the data in Example 8.4. The Gauss-Newton iteration may be started from the posterior mean found earlier:

$$\hat{x} = \begin{bmatrix} 467.2 \\ 1026.5 \end{bmatrix}.$$

The observation function evaluated at this point is

$$h(\hat{x}) = \operatorname{atan}\left(\begin{bmatrix} 370 \\ 2478 \\ 2173 \end{bmatrix} - 1026.5, \begin{bmatrix} 1126 \\ 1643 \\ 443 \end{bmatrix} - 467.2 \right) = \begin{bmatrix} -0.7836 \\ 0.8900 \\ 1.5919 \end{bmatrix}.$$

This is very close to the realized observations

$$y = \begin{bmatrix} -0.7854 \\ 0.9076 \\ 1.5882 \end{bmatrix}$$

and so it can be expected that further iterations will not change the answer much. However, for the sake of demonstration of the algorithm, let's carry out the computations.

The observation function's Jacobian is

$$H(\hat{x}) = \begin{bmatrix} \frac{370-1026.5}{(1126-467.2)^2+(370-1026.5)^2} & -\frac{1126-467.2}{(1126-467.2)^2+(370-1026.5)^2} \\ \frac{2478-1026.5}{(1643-467.2)^2+(2478-1026.5)^2} & -\frac{1643-467.2}{(1643-467.2)^2+(2478-1026.5)^2} \\ \frac{2173-1026.5}{(2173-1026.5)^2+(443-467.2)^2} & -\frac{443-467.2}{(2173-1026.5)^2+(443-467.2)^2} \end{bmatrix}$$

$$\doteq 10^{-6} \begin{bmatrix} -758.95 & -761.61 \\ 415.98 & -336.97 \\ 871.83 & 18.40 \end{bmatrix}.$$

Applying the formula (8.18), we obtain the correction step

$$d_{\mathrm{gn}} = 10^{-12} \begin{bmatrix} -0.1557 \\ -0.1471 \end{bmatrix}.$$

This step is very small (less than a millionth of a micron!), so the iteration was indeed unnecessary.

8.5 CONCLUSION

This chapter has presented a basic version of Bayesian estimation in which normally distributed noise is added to a nonlinear observation model. This basic framework can be used to estimate parameters for a wide variety of models, and it can be extended further:

- Besides the Gauss-Newton method presented here, optimization literature and software packages offer many alternative methods, of which the Levenberg-Marquardt algorithm is one of the most popular.
- In this chapter the noise covariance was assumed to be known. It is possible to treat the covariance matrix as an additional unknown parameter that is then estimated from the data using Bayes' rule, along with the other parameters. For linear observation models, closed-form formulas are given for example in Ref. [3].
- The moment matching method described in Section 8.4.2 can be extended to estimate parameters of models of dynamic (time-varying) phenomena. This estimator is called the Extended Kalman Filter, and is described in detail for example in Refs. [4, 5].
- Outliers tend to be much more common in real-world data than in random samples from a normal distribution. Instead of normally distributed noise, one could use a heavy-tailed probability distribution such as the Student-t. For a linear (or linearized) observation model, the posterior mean and covariance can be computed using the MATLAB code provided in Ref. [6].

EXERCISES

8.1 This exercise is a continuation of Example 8.3. Suppose now that one of Frodo's companions, the sharp-eared Legolas, drops a pebble and says that the sound of its striking the ground came back 5.45–5.60 s later. Update the probability distribution that models Frodo's belief about the depth.

8.2 A hall has a floor plan that is an equilateral triangle (Figure 8.5).

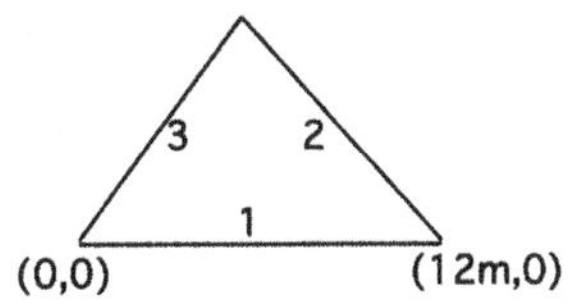

Figure 8.5 Floor plan for Exercise 8.2.

Find the observation function $\mathbf{y}_{1:3} = h(\mathbf{x}) + \mathbf{v}$ when $\mathbf{y}_i$ is the in-plane distance between the ith wall and the in-plane position $\mathbf{x}_{1:2}$ of a point in the hall. Assume the observations have independent zero-mean errors.

Standing in the hall, you measure your distance to walls 1, 2, and 3 to be 2 m, 4 m, and 6 m, respectively. Where are you?

Suppose each observation has a standard deviation of 50 cm. Using the Chebyshev inequality, find the radius of a disk that, with 95% probability, contains your position. Also, plot the smallest ellipse that, with 95% probability, contains your location.

8.3 The in-plane position of a robot has prior $\mathbf{x} \sim \mathrm{N}([0,0]^T, 30^2 I)$. The observations $y = [108,\ 46]^T$ are the distances from the robot to two infrared ranging devices located at $s[1] = [100,\ 0]^T$ and $s[2] = [0,\ 50]^T$. Each range observation has a zero-mean error with standard deviation 10.

Plot some level curves of $\phi(x) = -\log p_{\mathbf{x}\,|\,\mathbf{y}}(x \,|\, y)$ in the region $[-50, 70] \times [-20, 100]$.

Find the posterior mode using the Gauss-Newton method.

REFERENCES

1. Bertsch McGrayne, S., *The Theory That Would Not Die: How Bayes' Rule Cracked the Enigma Code, Hunted Down Russian Submarines, and Emerged Triumphant from Two Centuries of Controversy*, Yale University Press, 2011.
2. Jaynes, E. T., *Probability Theory: The Logic of Science*, Cambridge University Press, 2003.
3. Koch, K-R., *Introduction to Bayesian Statistics*, 2nd ed., Springer, 2007.
4. Maybeck, P. S., *Stochastic Models, Estimation, and Control*, Vol. 1, Academic Press, 1979; Navtech, 1994.
5. Challa, S., Morelande, M. R., Mušicki, D., and Evans, R. J., *Fundamentals of Object Tracking*, Cambridge University Press, 2011.
6. Piché, R., MVSREGRESS — Robust multivariate linear regression based on the Student-t distribution, `http://www.mathworks.com/matlabcentral/fileexchange/31230-mvsregress`

CHAPTER 9

LINEAR AND NONLINEAR PARABOLIC PARTIAL DIFFERENTIAL EQUATIONS IN FINANCIAL ENGINEERING

L. A. Boukas[1], K. I. Vasileiadis[2], S. Z. Xanthopoulos[2], A. N. Yannacopoulos[3]

[1] Department of Information and Communication Systems Engineering, University of the Aegean, Greece
[2] Laboratory for Financial and Actuarial Mathematics, Department of Statistics and Actuarial - Financial Mathematics, University of the Aegean, Greece
[3] Department of Statistics, Athens University of Economics and Business, Greece

9.1 FINANCIAL DERIVATIVES

Financial markets are mechanisms that facilitate the trading (buying and selling) of various assets, as, for example, financial securities (e.g., stocks and bonds), commodities (e.g., precious metals or agricultural products), etc. Among the various types of financial markets (e.g., capital markets, money markets, commodities markets, etc.), the market for derivative securities has seen an enormous growth during the last 30 years. One could say that the main role of the derivatives market is to facilitate the transfer of the financial risk that is inherent in other assets.

Mathematical Modeling with Multidisciplinary Applications.
Edited by Xin-She Yang

Generally speaking, a derivative security (or derivative contract or simply derivative) is a contract, the value of which depends on a reference value of some other underlying asset(s).

Usually, the underlying asset, to which a derivative refers, is a tradable financial security or a commodity, and the reference value is the price of this underlying asset. However, the reference value could be more complex, like, for example, the average price of the underlying asset during the life of the derivative contract or the maximum price of a basket of underlying assets, etc. On top of this, a derivative may refer to more "exotic" underlying assets, like, for example, the height of snow in a skiing resort, the number of CO_2 molecules in the city center or the value of other derivatives, etc.

Each derivative contract has two counter parties, usually referred to as the "buyer" and the "seller" of the contract. Accordingly there are two opposite positions in a derivative, the one of the buyer (the long position) and the one of the seller (the short position).

The most simple examples of derivative securities are the so-called Forwards and Options.

■ EXAMPLE 9.1 Forwards

A forward contract (or simply forward) is an agreement according to which the two counter parties agree that they will perform a trade in the future as follows: the buyer of the contract is *obliged* to buy from the seller of the contract (and the seller of the contract is *obliged* to sell to the buyer of the contract), a prespecified quantity of an underlying asset at a prespecified future time and at a prespecified price.

The prespecified future time at which the future trade will take place is called *time of maturity* of the forward contract, while the prespecified price at which the future trade will take place is called *delivery price* of the forward contract.

Obviously, a variable that affects decisively the value of the forward contract is the current price (spot price) of the underlying asset. Usually at the time that the two counter parties enter the forward contract, the delivery price of the underlying asset is selected in a way that the forward contract has zero value to the two counter parties. After this time, the value of the forward contract changes, becoming positive for one of the counter parties and equally negative for the other counter party. The delivery price for which a forward contract has zero value is called the *forward price* of the forward contract.

One can see directly that at maturity, the value (to the buyer) of a forward with delivery price K is given by $f_T = S(T) - K$, where $S(T)$ denotes the price of the underlying asset at the time of maturity of the forward. Obviously, the value (to the seller) of the forward with delivery price K is given as $-f_T = K - S(T)$.

■ EXAMPLE 9.2 Options

Options are contracts that give to their holder the *right* to perform a future trade according to prespecified terms. In contrast to forwards the holder of an option is not obliged to perform this future trade. Therefore an option is exercised only if it is profitable to its holder.

Depending on the flexibility to exercise an option, there are two main types of options, European and American. The options of European type can be exercised only at the end of the life of the option (expiry) while the options of American type can be exercised at any time until the end of the life of the option.

Depending on the type of the future transaction (buying or selling the underlying asset) that an option offers to its holder, we distinguish two basic kinds of options: the call option and the put option (sometimes referred to as the plain vanilla options).

► A call option (of European type), on an underlying asset A, with exercise price K and expiry T, is a contract between two counter parties (the buyer and the seller) that gives to the buyer of the option, the right to buy the underlying asset A at time T and price K.

► A put option (of European type), on an underlying asset A, with exercise price K and expiry T, is a contract between two counter parties (the buyer and the seller) that gives to the buyer of the option, the right to sell the underlying asset A at time T and price K.

The buyer of an option pays a premium to the seller of the option and acquires the right to exercise the option (without being obliged to do so) during the exercise time. We say that the buyer of an option has a long position on the option. The seller of the option receives the premium of the option in advance and has the obligation to satisfy the buyer according to the contractual terms of the option in the case that the buyer decides to exercise the option. We say that the seller of the option has a short position on the option.

Therefore we notice that there exist four different basic positions on options, namely: long call, short call, long put, short put.

One can see directly that the value of a call option at expiration is given as $C_T = \max(S(T) - K,\ 0)$, while that of a put option is given as $P_T = \max(K - S(T),\ 0)$.

However, the value of an option at a time before expiration is not so easy to determine, as we will see in the following sections.

9.2 MOTIVATION FOR A MODEL FOR THE PRICE OF STOCKS

Before we can say anything of some importance about the valuation of a derivative security, we need a model describing the behavior of the price of the underlying asset. This is equivalent to assuming a model that describes the returns of the underlying asset in various periods of time.

So we will start by explaining briefly why it is plausible to assume that the return of an asset is a normally distributed random variable.

It is very convenient in finance to work with the geometric (or logarithmic) return of an asset instead of the arithmetic return.[3]

From now on (and without loss of generality) we will assume that the underlying asset is a (nondividend paying) stock. The simplest "realistic" model describing the behavior of the price of stocks is the geometric Brownian motion.

The geometric (or logarithmic) return of the stock during some time period $[t_1,\ t_2]$ is defined as $R_{[t_1,t_2]} = \ln\left(\frac{S(t_2)}{S(t_1)}\right)$. This is equivalent to $S(t_2) = S(t_1)\exp(R_{[t_1,t_2]})$, which amounts to continuous compounding of the return. Notice the nice property $R_{[t_1,t_2]} + R_{[t_2,t_3]} = R_{[t_1,t_3]}$.

Let us consider a very small time period Δt measured in years (for example, $\Delta t = 1/10^{1000}$ of a year). Let us consider the times $t_0 = 0$ and $t_i = i \cdot \Delta t$ for $i \in \mathbb{N}$. We denote by $S(t_i)$ the random variable that represents the price of the stock at time t_i as it is perceived at time 0. For each $i = 1, 2, \ldots$ let $R_{\Delta t_i} \equiv R_{[t_{i-1},t_i]} \equiv \ln\left(\frac{S(t_i)}{S(t_{i-1})}\right)$ denote the random variables representing the return of the asset during the time period $[t_{i-1}, t_i]$.

The only assumption that we will make is that the returns $R_{\Delta t_i}$ are independent and identically distributed random variables with mean value $\mu_{\Delta t}$ and variance $\sigma^2_{\Delta t}$.

Let us consider now the time $T = 1$ year (i.e., $t_n = 1$ for some n or equivalently $\Delta t = 1/n$). Let $\mu = \mathbb{E}[R_{[0,1]}]$ denote the expected return and $\sigma^2 = Var(R_{[0,1]})$ denote the variance of the return of the stock during the forthcoming year.

Clearly, $R_{[0,1]} = R_{\Delta t_1} + \ldots + R_{\Delta t_n}$. But then $\mu = n \cdot \mu_{\Delta t} \Rightarrow \mu_{\Delta t} = \mu \cdot 1/n \Rightarrow \mu_{\Delta t} = \mu \cdot \Delta t$ and

$$\sigma^2 = n \cdot \sigma^2_{\Delta t} \Rightarrow \sigma^2_{\Delta t} = \sigma^2 \cdot 1/n \Rightarrow \sigma^2_{\Delta t} = \sigma^2 \cdot \Delta t.$$

Let us consider now an arbitrary time $T = t_N = N \cdot \Delta t$ for some large enough $N \in \mathbb{N}$ and denote by $R_T = R_{[0,T]} = \ln\left(\frac{S(T)}{S(0)}\right)$ the random variable representing the return of the stock during the time period $[0, T]$.

[3] Let $S(t)$ denote the price of a stock at some time t. The arithmetic return of the stock during some time period $t_1,\ t_2$ is defined as $r_{[t_1,t_2]} = \frac{S(t_2)-S(t_1)}{S(t_1)}$. This is equivalent to $S(t_2) = S(t_1)(1 + r_{[t_1,t_2]})$. Notice however that $r_{[t_1,t_2]} + r_{[t_2,t_3]} \neq r_{[t_1,t_3]}$.

Obviously, $R_T = R_{\Delta t_1} + ... + R_{\Delta t_N}$. But then $\mathbb{E}[R_T] = N \cdot \mu_{\Delta t} = \mu \cdot N \cdot \Delta t = \mu \cdot T$ and

$$Var(R_T) = N \cdot \sigma_{\Delta t}^2 = \sigma^2 \cdot N \cdot \Delta t = \sigma^2 \cdot T.$$

Then, the Central Limit Theorem implies that $R_T \to N(\mu T,\ \sigma^2 T)$, i.e., R_T is asymptotically distributed as a normal with mean μT and variance $\sigma^2 T$.

Therefore, if μ is the expected one-year return of the stock and σ^2 is the variance of the one-year return of the stock, then the random variable $S(T)$ that represents the price of the stock at some time T is lognormally distributed since

$$\ln\left(\frac{S(T)}{S(0)}\right) \sim \mathcal{N}(\mu T,\ \sigma^2 T).$$

Thus, in order to model the price of the stock with a stochastic process we would like a process that would result into $S(t) = S(0) \cdot \exp(Z)$ where Z is a normal distribution with mean μt and variance $\sigma^2 t$. Moreover, we would like the returns that are produced by this process at non-overlapping intervals to be independent random variables. By the properties of the lognormal distribution it follows that

$$\mathbb{E}[S(T)|S(0)] = S(0) \exp\left(\mu T + \frac{\sigma^2 T}{2}\right).$$

Such a class of processes, which also allows to describe the model in a dynamic fashion, can be effectively produced by using the properties of the so-called Brownian motion or Wiener process. This is the topic of the next section.

9.3 STOCK PRICES INVOLVING THE WIENER PROCESS

As argued in Section 9.2 above, the simplest "realistic" model for the price of stocks is geometric Brownian motion. According to this model the logarithmic returns of the stocks follow the normal distribution. Let $S(t)$ be the price of the stock at time t. This is a random variable, depending on the state of the economy, i.e. the contingencies that have occurred till time t. These contingencies are assumed to have formed the prices of the stock via the market forces of supply and demand (if you believe the fable of general equilibrium) or by oligopolistic processes (if you do not). Being mathematicians and leaving theoretical economics aside, we assume that

$$Y := \ln\left(\frac{S(t)}{S(0)}\right) \sim \mathcal{N}(\hat{\mu}t, \sigma^2 t),$$

or in other words

$$P(Y < y) = \frac{1}{\sqrt{2\pi\sigma^2 t}} \int_{-\infty}^{y} \exp\left(-\frac{(z - \hat{\mu}t)^2}{2\sigma^2 t}\right) dz.$$

This model, simple as it may be, is still a credible model for the price of stocks and gives rise to the lognormal distribution. Under certain circumstances this model serves as a good model for the time series of stock prices observed in some markets.

An alternative way of seeing this model is in a dynamic fashion. To do this we need to introduce a stochastic process, which is called the Brownian motion or the Wiener process. So as not to get into the intricacies of stochastic process theory, we will content ourselves with the working definition of a stochastic process as a collection of random variables (indexed by $t \in \mathbb{R}_+$). For an excellent mathematical introduction to the Wiener process and its applications in finance, see Refs. [5] and [6].

Definition 3 *The Wiener process $\{W(t)\}_{t\in\mathbb{R}_+}$ is a collection of random variables with the following properties:*

1. *The random variables $W(t_i) - W(t_{i-1})$ and $W(t_{i-1}) - W(t_{i-2})$ are independent random variables for all $t_1, \ldots, t_n$, $t_{i-1} < t_i$, $i = 2, \ldots, n$.*
2. *When considered as a (random function) of time t, the function $W(t)$ is a continuous function (almost surely[4]) such that $W(0) = 0$.*
3. $W(t) - W(s) \sim \mathcal{N}(0, t-s)$

By the properties of the normal distribution and the above definition it may be seen that

$$\hat{\mu}t + \sigma W(t) \sim \mathcal{N}(\hat{\mu}t, \sigma^2 t)$$

so that our model for the logarithmic returns can be expressed equivalently as

$$S(t) = S(0)\exp(\hat{\mu}t + \sigma W(t)).$$

To obtain a dynamic version of this model we need to associate the change of the value of the stock in the time interval $[t, t+dt)$ with the observed value of the stock at time t. This reminds us of connecting the temporal derivative of the random function $S(t)$ with its value $S(t)$ at time t, i.e., setting up a differential equation for the evolution of the stock value. However, life is not all that rosy! One of the intriguing properties of the Wiener process is that it is a function which may be continuous but is nowhere differentiable. In fact it is a function presenting variation in all scales, a property inherited either by its construction as a scaled random walk or as a random Fourier series, and this shows up in the observation that it is a function of infinite variation, i.e.,

$$\sup \sum_{i=1}^{n} |W(t_i) - W(t_{i-1})| = \infty,$$

[4] i.e., for all realizations of the random process apart from some of measure zero.

where $\{t_1, t_2, \cdots, t_n\}$ is a partition of $[0, t]$ and the supremum is taken over all partitions of $[0, t]$. This tells us that when we observe the absolute value of the variation of the Wiener process at the finest scale we may take, and then add it all up, the variation is infinite.

All seems to be lost, but the following observation comes to rescue. If we take the quadratic variation of the Wiener process, then this is finite and in particular

$$\sup \sum_{i=1}^{n} |W(t_i) - W(t_{i-1})|^2 = t,$$

where again the supremum is taken over all possible partitions of $[0, t]$. The above observations, i.e., the connection of a finite square variation with an infinite variation is very deep and intimately related with the random walk nature of the Wiener process.[5]

The infinite variation of the Wiener process does not allow us to interpret the integral $\int_0^t f(s, \omega) dW(s)$ as a Stieltjes integral. We should rather interpret it as a "new" integral called the Itō integral which is defined as the limit as $n \to \infty$, in the $L^2(\Omega, \mathcal{F}, P)$ sense of the sequence of random variables

$$\xi_n := \sum_{i=1}^{n} f(t_i, \omega)(W(t_{i+1}) - W(t_i)),$$

where $\{t_i\}$ is a partition of $[0, t]$. Recall that as sequence of random variables $\{\xi_n\}$ converges to a random variable ξ in the $L^2(\Omega, \mathcal{F}, P)$ sense, if $\mathbb{E}[(\xi_n - \xi)^2] \to 0$ as $n \to \infty$. It can be shown that this limit exists, giving rise to a new random variable denoted by $\int_0^t f(s, \omega)\, dW(s)$ with the properties

$$\mathbb{E}\left[\int_0^t f(s, \omega)\, dW(s)\right] = 0,$$

$$\mathbb{E}\left[\left(\int_0^t f(s, \omega)\, dW(s)\right)^2\right] = \mathbb{E}\left[\int_0^t (f(s, \omega))^2\, ds\right].$$

It may further be shown that the second property is crucial in developing this theory as the class of stochastic processes for which the Itō integral is defined as simply those for which the right-hand side of this equality (called the Itō isometry) makes sense.

Based on this new concept of integration, which was proposed by Kyoshi Itō, a new calculus was developed, bearing his name, that allows us to bypass the absence of derivatives of the Wiener process and understand and model the changes of $S(t)$ in a generalized fashion under an integral sign. Absence

[5] Any square integrable continuous martingale can be shown to have this property.

of space does not allow us to give full credit to this elegant theory so we only summarize Itō's lemma that allows us to monitor changes in the value of a function calculated on the Wiener process.

Proposition 1 *Let $f : \mathbb{R}_+ \times \mathbb{R} \to \mathbb{R}$ be a $C^{1,2}$ function. Define the process $Y(t) := S(0) + \mu t + \sigma W(t)$. Then,*

$$f(t, Y(t)) - f(0, Y(0)) \\ = \int_0^t \left\{ \frac{\partial f}{\partial t} + \frac{\partial f}{\partial x}\mu + \frac{1}{2}\frac{\partial^2 f}{\partial x^2}\sigma^2 \right\} (s, Y(s))ds + \int_0^t \mu \frac{\partial f}{\partial x}(s, Y(s))\, dW(s).$$

There are versions of Itō's lemma covering more complicated stochastic processes than the process $Y(\cdot)$ considered above. For instance, it may be useful to consider a stochastic process $X(\cdot)$ such that at any time $t \in [0, T]$ it holds that

$$X(t) = X_0 + \int_0^t \mu(s, X(s))ds + \int_0^t \sigma(s, X(s))dW(s), \tag{9.1}$$

where the last integral is understood as an Itō integral and the functions μ and σ are known functions. Such a process is called an Itō process. Equation (9.1) is an integral equation which involves the Itō integral of the unknown process $X(\cdot)$. It is also common to encounter this equation in a (shorthand) differential form as

$$dX(t) = \mu(t, X(t))dt + \sigma(t, X(t))dW(t), \tag{9.2}$$

with initial condition $X(0) = X_0$. Such differential equations are called stochastic differential equations, and there is a rich mathematical theory covering their existence and uniqueness properties as well as their qualitative behavior. We content here to say that the above equation is well posed under, e.g., Lipschitz type conditions for μ and σ.

A version of Itō's lemma for solution of stochastic differential equations is given in the following:

Proposition 2 *Let $f : \mathbb{R}_+ \times \mathbb{R} \to \mathbb{R}$ be a $C^{1,2}$ function. Define the process $X(\cdot)$ as the solution of the stochastic differential equation [9.1) (equivalently (9.2)]. Then,*

$$f(t, X(t)) - f(0, X(0)) \\ = \int_0^t \left\{ \frac{\partial f}{\partial t} + \frac{\partial f}{\partial x}\mu + \frac{1}{2}\frac{\partial^2 f}{\partial x^2}\sigma^2 \right\} (s, X(s))ds + \int_0^t \mu \frac{\partial f}{\partial x}(s, X(s))\, dW(s),$$

where of course now μ and σ are functions of x.

9.4 CONNECTION BETWEEN THE WIENER PROCESS AND PDEs

There is an intimate connection between the Wiener process and functionals of the Wiener process with the solutions of linear partial differential equations of the parabolic type. This connection can already be made apparent by the version of Itō's lemma presented in Proposition 2 which involves a partial differential operator in the drift term, but we would like to devote some more time to this important part of the theory.

By definition the Wiener process is distributed by the normal (Gaussian) distribution. This means that if one wishes to calculate the moments of the random variable $X = f(W(t))$ where f is a Borel function defined on the reals, then

$$\mathbb{E}[f(W(t))] = \int_{-\infty}^{\infty} f(y)\frac{1}{\sqrt{2\pi t}}\exp\left(-\frac{y^2}{2t}\right)dy.$$

If on the other hand we assume that we start our Wiener process at the initial point x rather than 0 and then we wish to calculate the moments of the random variable $X = f(x + W(t))$ then a similar calculation will give us

$$\begin{aligned}\mathbb{E}[f(x+W(t))] &= \int_{-\infty}^{\infty} f(x+y)\frac{1}{\sqrt{2\pi t}}\exp\left(-\frac{x^2}{2t}\right)dy \\ &= \int_{-\infty}^{\infty} f(y)\frac{1}{\sqrt{2\pi t}}\exp\left(-\frac{(x-y)^2}{2t}\right)dy.\end{aligned}$$

If we thus define the function $U : \mathbb{R}_+ \times \mathbb{R} \to \mathbb{R}$ by $U(t,x) := \mathbb{E}[f(x + W(t))]$ we see by the above calculation that

$$U(t,x) = \int_{-\infty}^{\infty} f(y)\frac{1}{\sqrt{2\pi t}}\exp\left(-\frac{(x-y)^2}{2t}\right)dy. \tag{9.3}$$

The last integral rings a familiar bell. It is nothing else but the integral representation of the solution of the heat equation

$$\frac{\partial U}{\partial t} = \frac{1}{2}\frac{\partial^2 U}{\partial t^2} \tag{9.4}$$

subject to the initial condition $U(0,x) = f(x)$. Therefore the interpretation of the integral in Eq. (9.3) is as the solution of the initial value problem (Cauchy problem) (9.4) with the stated initial condition. This is very important as it leads to the fundamental relation between the solutions of the heat equation and the Wiener process that can be stated as

> The solution of the heat equation (9.4) with the initial condition $U(0,x) = f(x)$ can be expressed in terms of the expectation
>
> $$U(t,x) = \mathbb{E}_x[f(\bar{W}(t))], \tag{9.5}$$

where by $\mathbb{E}_x[\cdot]$ we denote the expectation over the paths of the (generalized) Wiener process[6] $\bar{W}(t)$ which is a Wiener process that has started its path at point x (rather than at point 0).

The interpretation of this formula is very clear. If you want to find the solution of the heat equation at time t and at the position x, then you have to start a path of the Wiener process at x and run it for time t. Then calculate your initial condition (the function f at the point that this Wiener process has reached by time t. This gives you a random variable $f(\bar{W}(t)) = f(x_W(t))$. The expectation of this random variable is the function we seek. This can be approximated as follows: Repeat the above procedure for as many paths as you can and then approximate the expectation of the random variable $f(\bar{W}(t)) = f(x_W(t))$ by the standard estimator for the expectation, the sample mean. This will give us an approximation of the solution of the heat equation at (t, x).

The above connection between the heat equation and the Wiener process is not just a coincidence. It is a fact that holds in general for any Itō process and is related to properties of these processes such as the Markov property and the semi-martingale property, which unfortunately we may not introduce here at length due to space limitations. At any rate this connection, known as the celebrated Feynman-Kac representation formula, connects solutions of general linear parabolic partial differential equations with expectations of functionals of Itō processes. The basic quantity connecting these two notions is the generator operator of the Itō process which for a general process of the form (9.2) is the operator $\mathcal{L} : C^2(\mathbb{R}) \to C(\mathbb{R})$ defined by

$$\mathcal{L}f(x) := \mu(x)\frac{\partial f}{\partial x} + \sigma(x)^2 \frac{1}{2}\frac{\partial^2 f}{\partial x^2}.$$

Then the general result (which is of course connected with Itō's lemma) is that if we start the Itō process defined by Eq. (9.2) at point x and leave the process running for time t, and calculate the expectation of the functional $f(X(t)$, then

$$\mathbb{E}[f(X(t))] = U(t, x),$$

where $U(t, x)$ is the solution of the linear parabolic partial differential equation

$$\frac{\partial U}{\partial t} = \mathcal{L}U$$

with initial condition $U(0, x) = f(x)$. This result can be extended for Itō processes in $\mathbb{R}^n$ (see the forthcoming section on dynamic programming) or even for Itō processes taking values in infinite dimensional spaces.

[6]By the properties of the Wiener process we can see that $\bar{W}(t) = x + W(t)$ where $W(t)$ is the standard Wiener process. Furthermore, $\bar{W}(t) \sim \mathcal{N}(x, t)$ whereas $\bar{W}(t+s) - \bar{W}(t) \sim \mathcal{N}(0, s)$.

9.5 THE BLACK-SCHOLES-MERTON EQUATION

Consider a financial option with payoff $\Phi(S(T))$ at expiry T, where $\Phi : \mathbb{R} \to \mathbb{R}$ is a given function. If the option is a European call, then $\Phi(x) = (x - K)^+$, whereas if it is a European put then $\Phi(x) = (K - x)^+$. The value of the financial option at time t is considered to be a function of the value of the underlying asset at time t. Therefore, we assume[7] the existence of a function $V : [0, T] \times \mathbb{R} \to \mathbb{R}_+$ such that $V(t, S(t))$ provides the price of the option in the market at time t. This function must have the property $V(T, S(T)) = \Phi(S(T))$.

If we manage to specify this function V then we have a nice way to price the option. All we need is to know the price of the stock at time t, $S(t)$, this is easy to obtain from the market, and we substitute this value in the function to provide the price an investor will pay at time t to acquire this option and thus guarantee the payoff $\Phi(S(T))$ upon expiry T. It is the aim of this section to show that if such a function exists, then it must be the solution of a linear parabolic PDE.

We now consider a portfolio consisting of the option and a position of Δ units in the underlying asset (the stock). The total value of this portfolio at time t, given that the underlying asset has value $S(t)$ will be

$$\Pi(t, S(t)) = \Delta S(t) - V(t, S(t)).$$

This portfolio corresponds to the writer of the option (short position in the option), i.e. to the agent that guarantees $\Phi(S(T))$ upon expiry.

The value of this portfolio [being a function of $S(t)$] is a random function that fluctuates with the price of the underlying asset. Can we choose its composition, in other words Δ, so that the total value of this portfolio is unaffected by the fluctuations of the underlying asset. This effectively means that Δ is chosen so that if, e.g., the fluctuations of $S(t)$ are such that the value of the option rises, then the position in the underlying asset suppresses this rise. The choice of Δ can be made with the aid of Itō's lemma. Since

$$dS(t) = \mu S(t)dt + \sigma S(t)dW(t),$$

a straightforward application of Itō's lemma (see Proposition 2) gives us that

$$d\Pi(t, S(t)) = \Delta(\mu S(t)dt + \sigma S(t)dW(t))$$
$$-\frac{\partial V}{\partial t}dt + \frac{\partial V}{\partial s}(\mu S(t)dt + \sigma S(t)dW(t)) + \frac{1}{2}\frac{\partial^2 V}{\partial s^2}\sigma^2 S(t)^2 dt,$$

[7]This assumption can be proved using the Markov property for the Wiener process.

where the function V and its derivatives are calculated at $(t, S(t))$. Collecting alike terms together we rewrite the above as

$$d\Pi(t,S(t)) = \underbrace{\left(\Delta - \frac{\partial V}{\partial s}\right)\sigma S(t) dW(t)}$$

$$-\left(-\Delta\mu S(t) + \frac{\partial V}{\partial t} + \frac{\partial V}{\partial s}\mu S(t) + \frac{1}{2}\frac{\partial^2 V}{\partial s^2}\sigma^2 S(t)^2\right) dt.$$

The underbraced terms are those giving rise to the fluctuations, therefore they may be suppressed by choosing

$$\Delta = \frac{\partial V}{\partial s}(t, S(t))$$

for every pair $(t, S(t))$. These pairs may be interpreted as alternative scenaria for the behavior of the market. Therefore, the composition of this portfolio may be given by a function $\Delta : \mathbb{R}_+ \times \mathbb{R} \to \mathbb{R}$, which is defined by the function V as $\Delta := \frac{\partial V}{\partial s}$. With this choice the value of the portfolio changes as

$$d\Pi(t,S(t)) = -\left(\frac{\partial V}{\partial t} + \frac{1}{2}\frac{\partial^2 V}{\partial s^2}\sigma^2 S(t)^2\right) dt$$

and is not subject to the fluctuations of the underlying asset.

Such a portfolio must therefore appreciate in value with the return of the riskless asset r. If this were not true then there would be opportunities for riskless profit in the market. Such opportunities (called arbitrage) are not compatible with the theory of general equilibrium. Therefore, absence of arbitrage gives us that

$$d\Pi = r\Pi dt$$

and upon substituting in this fundamental equation the expression for Π, $d\Pi$ and Δ we obtain that the function V must satisfy the following equation:

$$\frac{\partial V}{\partial t} + rs\frac{\partial V}{\partial s} + \frac{1}{2}\sigma^2 s^2 \frac{\partial^2 V}{\partial s^2} - rV = 0. \tag{9.6}$$

This is a partial differential equation of the parabolic type for the unknown function V, called the Black-Scholes-Merton (BSM) equation. If we manage to solve this equation with the final value $V(T,s) = \Phi(s)$ (and show that the solution has the required regularity properties), then we have completed our program for pricing the option. At the same time we have also managed to find the necessary position Δ in the underlying asset, needed by the writer of the option in order to make her portfolio insensitive to the fluctuations of the underlying assets. This is called hedging in the financial world and is a very important step toward the management of the risks incurring due to the various financial assets.

9.6 SOLUTION OF THE BLACK-SCHOLES-MERTON EQUATION

Since its derivation in the 1970's many alternative ways for the resolution of the BSM equation have been proposed. In practice all possible methods for writing down explicit solutions of this equation have been tried, from integral transforms to semigroup methods or transformation techniques.

It is an easy exercise in intermediate calculus to see that if V is to be understood as a function of τ, x instead of t, s where

$$S = K\exp(x), \quad t = T - \frac{2\tau}{\sigma^2}, \quad q = \frac{2r}{\sigma^2}, \tag{9.7}$$

then $U(t,x) = K^{-1}\exp(\frac{1}{2}(q-1)x + (\frac{1}{4}(q-1)^2 + q)\tau)V(\tau,x)$ solves the heat equation (see, e.g., Ref. [8])

$$\frac{\partial U}{\partial t} = \frac{\partial^2 U}{\partial x^2}$$

with initial condition $U(0,x) = K^{-1}\exp(\frac{1}{2}(q-1)x)\Phi(K\exp(x))$. Note that in the new coordinates $t = T$ corresponds to $\tau = 0$, therefore, the final condition of the original equation is turned to an initial condition of the transformed equation. This equation is well known to be solvable in terms of the integral formula

$$U(t,x) = \frac{1}{2\sqrt{\pi t}}\int_{-\infty}^{\infty} U(0,y)\exp\left(-\frac{(x-y)^2}{4t}\right)dy,$$

which when transformed back into the original coordinates gives

$$V(t,s) = \frac{1}{\sqrt{2\pi t}}\exp(-r(T-t))$$

$$\times\int_{-\infty}^{\infty}\Phi\left(s\exp\left(\left(yr - \frac{\sigma^2}{2}\right)(T-t) + \sigma y\right)\right) \times \exp\left(-\frac{(s-y)^2}{2(T-t)}\right)dy.$$

Since Φ is known that is needed is to complete this integration and obtain the function V which is the pricing function. For certain choices of Φ this calculation is feasible in closed form. For example, if we wish to price a call option then $\Phi(x) = (x-K)^+$ and by elementary calculus we obtain the value of the call option in the following form

$$\begin{aligned}
V(t,s) &= s\,N(d_1) - K\exp(-r(T-t))\,N(d_2),\\
d_1 &= \frac{\ln(s/K) + (r+\sigma^2/2)(T-t))}{\sigma\sqrt{T-t}},\\
d_2 &= d_1 - \sigma\sqrt{T-t},
\end{aligned} \tag{9.8}$$

where N is the cumulative distribution function for the standard normal.

Similarly, for a put option we have

$$\begin{aligned}
V(t,S) &= -s\,N(-d_1) + K\exp(-r(T-t))\,N(-d_2),\\
d_1 &= \frac{\ln(s/K) + (r+\sigma^2/2)(T-t))}{\sigma\sqrt{T-t}},\\
d_2 &= d_1 - \sigma\sqrt{T-t}.
\end{aligned} \tag{9.9}$$

The above formulas may be used for providing a benchmark price for these financial options. Futhermore, observe the connection with the Feynman-Kac representation.

9.7 FREE BOUNDARY-VALUE PROBLEMS AND VALUATION OF AMERICAN OPTIONS

An important class of options are American options which differ from the European type that we have encountered so far by their feature that allows them to be exercised not only at expiry but whenever it is considered as suitable by the holder. Suppose that we have an American option whose payoff when exercised at time τ is a given function of the value of the underlying at this time $S(\tau)$, and let us call it $\Phi(\tau, S(\tau))$. The exact form of the function Φ depends on the type of the option, e.g., for a call option exercised at τ the payoff will be $\exp(-r\tau)\,(S(\tau)-K)^+$, where K is the strike of the call and the factor $\exp(-r\,\tau)$ corresponds to a discounting factor which transfers the value of the payoff which will be obtained at time τ to its value at time $t=0$.

An important observation is that the time τ is a random time, the holder of the option does not know its value beforehand and also its value depends on the contingencies of the market. However, it is not just any random time. It must have the property that the holder of the option may know whether this time has arrived by time t, only by her knowledge of the market by this time. In other words, the holder of the option may decide of when it is the best time to exercise the option only by her prior knowledge of what has passed (the history of the market) and not from any future insights concerning the market. This introduces the notion of the stopping time which is a very important notion in stochastic processes.

The optimal time at which you should exercise the option must be a stopping time, otherwise you must have access to some inside information concerning the market. This is not allowed in the theory since it may lead to arbitrage opportunities. Therefore the time τ must be decided only by the history of the market. The answer to the question $\tau \leq t$ must be decided for only by knowledge of $\{S(u),\, u \leq t\}$ (or rather by having access to the smallest σ-algebra that makes these random variables measurable).

Proposition 3 *The price of the American option is given by a function $V : \mathbb{R}_+ \times \mathbb{R} \to \mathbb{R}_+$ that solves the variational inequality*

$$\max\left\{-\frac{\partial V}{\partial t} + rs\frac{\partial V}{\partial s} + \frac{1}{2}\sigma^2 s^2 \frac{\partial^2 V}{\partial s^2} - rV, \Phi(s) - V\right\} = 0. \tag{9.10}$$

An alternative (and more compact way) of writing this, is as a free boundary-value problem. Define the operator $\mathcal{L}$ by

$$(\mathcal{L}u)(t,x) := -\frac{\partial u}{\partial t} + rs\frac{\partial u}{\partial s} + \frac{1}{2}\sigma^2 s^2 \frac{\partial^2 u}{\partial s^2} - ru, \tag{9.11}$$

in terms of this operator.

Proposition 4 *Consider the solution of the free boundary-value problem*

$$\mathcal{L}V = 0,\ \ V \geq \Phi,\ \ on\, \mathcal{D}$$
$$\mathcal{L}V < 0,\ \ V = \Phi,\ \ on\, \mathbb{R}_+^2 \setminus \mathcal{D}$$

where $\mathcal{D} := \{(t,s) \in \mathbb{R}_+^2 \,:\, 0 \leq s \leq d(t)\}$ where d is to be determined by the solution of the problem. Then the solution V is the pricing function and the optimal exercise time is the first time that the price process leaves $\mathcal{D}$.

The above problem is called a free boundary-value problem since it is a boundary-value problem on a domain which is not prescribed *a priori* but is determined by the solution of the problem. It is a version of the famous Stefan problem which among other things determines the boundary between, e.g., ice and water while the ice is melting in the water.

Free boundary-value problems cannot be solved in closed form except in certain very simple cases. In general they may be solved only numerically. Of course there is a well-developed analytical theory which provides existence and uniqueness results as well as important qualitative results.

An important version of the American option is the perpetual American option which does not expire. Its expiration time T can be considered as $T \to \infty$. For such an option the price function V will not depend on t, therefore we obtain the following version of the variational inequality

Proposition 5 *The price of the perpetual American option is given by a function $V : \mathbb{R}_+ \times \mathbb{R} \to \mathbb{R}_+$ that solves the variational inequality.*

$$\max\left\{rs\frac{\partial V}{\partial s} + \frac{1}{2}\sigma^2 s^2 \frac{\partial^2 V}{\partial s^2} - rV, \Phi(s) - V\right\} = 0. \tag{9.12}$$

The following comments may clarify why the above free boundary-value problems may provide us with the pricing methodology for American options.

While the American option is kept and not exercised it is equivalent to a European option, therefore its value is given by the same equation as that of a European option. Thus, in the continuation region (when the option is not to be exercised) we have that

$$\frac{\partial V}{\partial t} + \mathcal{L}V = 0, \text{ continuation region.}$$

What happens when we enter the exercise region (stopping region)? Then in that case we expect that the value of holding the option is smaller than the value of exercising it, therefore, we should no longer treat the option as a European one since its value is now lower than that of a European option. How would this show in terms of the differential operator $\mathcal{L}$? To understand that we need to invoke a general property of parabolic equations which gives rise to a very useful comparison principle.

Lemma 9.7.1 *Suppose that we have two functions ϕ_1 and ϕ_2, where the first one solves the differential equation*

$$\frac{\partial \phi_1}{\partial t} + \mathcal{L}\phi_1 = 0,$$

whereas the second one solves

$$\frac{\partial \phi_2}{\partial t} + \mathcal{L}\phi_2 < 0.$$

Then $\phi_2 < \phi_1$.

Using this lemma we see that in the stopping region the value of the option must satisfy the differential inequality

$$\frac{\partial V}{\partial t} + \mathcal{L}V < 0, \text{ stopping region.}$$

Indeed, consider for example the American put option. When we are in the stopping region, we exercise the option and V becomes $V = (K - S)^+$. Substituting this function into the Black-Scholes operator we see that $\frac{\partial V}{\partial t} + \mathcal{L}V = -rK < 0$. However, the above argument holds for more general options.

To summarize, we see that in general the value of the American option satisfies the differential inequality

$$\frac{\partial V}{\partial t} + \mathcal{L}V \leq 0,$$

where the equality holds in the continuation region and the inequality holds in the stopping region.

The form of the stopping region depends on the particular type of option. For example if we consider an American put option,[8] then we expect the holder to exercise the option whenever the price of the underlying falls below a given threshold s^* (since then it is of benefit to the holder of the option to exercise and sell the underlying for K). Therefore, the continuation region for the American put will be the region $s > s^*$, where of course s^* is to be determined. For the American put its price satisfies

$$\begin{aligned} &\frac{\partial V}{\partial t} + \mathcal{L}V = 0, \quad s > s^*, \\ &V = K - s, \qquad s \leq s^*. \end{aligned}$$

How can we obtain s^*? This may be obtained by the continuity property of the first derivative $\frac{\partial V}{\partial s}$ at $s = s^*$ (smooth pasting condition).

We now formulate this problem as a linear complementarity problem. Rewrite the above as

$$\begin{aligned} &\text{if } V(t,s) > \Phi(s) \quad \frac{\partial V}{\partial t} + \mathcal{L}V = 0, \\ &\text{if } V(t,s) = \Phi(s) \quad \frac{\partial V}{\partial t} + \mathcal{L}V < 0, \end{aligned}$$

where $\Phi(s)$ is the payoff the holder of the option will acquire if she chooses to exercise at time t when the value of the underlying is equal to s. We note that we may rewrite the above as an equality of the form

$$\begin{aligned} &\left(\frac{\partial V}{\partial t} + \mathcal{L}V\right)(V - \Phi) = 0, \\ -&\left(\frac{\partial V}{\partial t} + \mathcal{L}V\right) \geq 0, \quad V - \Phi \geq 0. \end{aligned}$$

This formulation is very useful for the numerical solution of free boundary value problems.

Further, an alternative formulation is as

$$\max\{\Phi - V, \frac{\partial V}{\partial t} + \mathcal{L}V\} = 0,$$

or equivalently

$$\min\{V - \Phi, -\frac{\partial V}{\partial t} - \mathcal{L}V\} = 0,$$

and $V = \Phi$ on ∂D.

[8]The American call on a non-divident paying asset is never optimal to exercise early.

Consider the set of functions

$$\mathcal{K} := \{v \in C^0, \ \ v \geq \Phi\}$$

This is a convex set. Consider any $v \in \mathcal{K}$ so that $v - \Phi \geq 0$. Since $-\frac{\partial V}{\partial t} - \mathcal{L}V \geq 0$ we have upon integrating that

$$\int \left(-\frac{\partial V}{\partial t} - \mathcal{L}V \right) (v - \Phi) \geq 0$$

Furthermore, since $\left(-\frac{\partial V}{\partial t} - \mathcal{L}V\right)(V - \Phi) = 0$ for all t, s and integrating and subtracting we obtain that

$$\int \left(-\frac{\partial V}{\partial t} - \mathcal{L}V \right) (v - V) \geq 0, \ \ \forall v \in \mathcal{K}.$$

Interpreting the integral as an inner product we rewrite the problem as a variational inequality

$$\left\langle -\frac{\partial V}{\partial t} - \mathcal{L}V, v - V \right\rangle \geq 0, \ \ \forall v \in \mathcal{K}.$$

9.8 THE HAMILTON-JACOBI-BELLMAN EQUATION

We now consider another important class of nonlinear equations in financial engineering, the Hamilton-Jacobi-Bellman equations. These are fully nonlinear equations in the sense that the higher derivatives of the unknown function appear in a highly nonlinear function. This type of equations is one of the most difficult partial differential equations (PDEs), however, they appear in a variety of applications which are related to stochastic optimal control. It is the aim of this section to introduce the Hamilton-Jacobi-Bellman equation and motivate its use in a variety of applications in mathematical finance (for a more thorough discussion, see Refs. [6] or [3]).

Consider the following example to motivate the discussion.

■ EXAMPLE 9.3

Assume that an investor creates a portfolio which consists of the riskless asset B with return r, i.e., $dB = rBdt$ and the risky asset S with drift μ and volatility σ, i.e., $dS = \mu Sdt + \sigma SdW$. The investor's relative weights between these two titles at time t are denoted by $\bar{u}_t^0$ and $\bar{u}_t^1$ for the riskless and risky asset, respectively, and her consumption rate by $c(t)$. The value $X(t)$ of this self-financing portfolio at time t will be given by the solution of the stochastic differential equation (SDE)

$$dX\,(t) = X(t)(\bar{u}_t^0 r + \bar{u}_t^1 \mu)dt - c(t)dt + \sigma \bar{u}_t^1 X(t)dW(t).$$

We see that the portfolio's value at time t depends on the choice of portfolio weights and the consumption process, therefore we must recall at all times that $X(t) = X(t, \bar{u}^0, \bar{u}^1, c)$. The portfolio weights and the consumption process will be called the control variables of the problem, while the portfolio value will be called the state variable. There is also a number of *control constraints*, which we will see below.

Investors will choose the control processes in order to achieve a specific purpose. In this particular case, this purpose will be to maximize a utility function of intertemporary consumption and a utility function of the final wealth,

$$J(c,x) := \mathbb{E}\left[\int_0^T U_1(t, c(t))dt + U_2(X(T))\right],$$

where U_1 is the instantaneous utility function for consumption, whereas U_2 is a "legacy" function which measures the utility of having a portion of the wealth left at the end of the period. An example of $U_1(t, c(t))$ could be $U_1(t, c(t)) = e^{-\delta t}\ln(c(t))$ and of $U_2 i(X(t)) = -(X(T) - A)^2$, which means that the investor wants to maximize a logarithmic utility function of consumption discounted during the whole time period $[0, T]$ and to minimize the mean quadratic distance of final wealth from a target A that she has set. The selection of the control variables $\bar{u}^0$, $\bar{u}^1$, c will be made so that the selected target is reached. Control variables will be subject to the natural constraint $\bar{u}_t^0 + \bar{u}_t^1 = 1$, $\forall t \geq 0$ and consumption must follow the condition $c(t) \geq 0$, $\forall t \geq 0$. These constraints can be modeled geometrically, considering that the process $\bar{u}_t = (\bar{u}_t^0, \bar{u}_t^1, c_t)$ will belong to a set $G \subset \mathbb{R}^3$.

The portfolio problem thus becomes

$$\max_{\bar{u}^0, \bar{u}^1, c} \mathbb{E}\left[\int_0^T U_1(t, c(t))dt + U_2(X(T))\right],$$

subject to

$$dX(t) = X(t)(\bar{u}_t^0 r + \bar{u}_t^1 \mu)dt - c(t)dt + \sigma \bar{u}_t^1 X(t)dW(t),$$

and $X(0) = x_0$, $\bar{u}_0^t + \bar{u}_1^t = 1$, $c(t) \geq 0$ for all $t \geq 0$.

A problem of this kind is called a *stochastic optimal control problem*. In the next sections we will study a fairly general class of stochastic optimal control problems.

Now, we will set the formal problem, stating a fairly general class of optimal control problems. To this end, consider the functions

$$\begin{aligned} \mu &: \mathbb{R}_+ \times \mathbb{R}^n \times \mathbb{R}^k \to \mathbb{R}^n, \\ \sigma &: \mathbb{R}_+ \times \mathbb{R}^n \times \mathbb{R}^k \to \mathbb{R}^{n\times d}. \end{aligned}$$

where $\mu(t,x,u)$ and $\sigma(t,x,u)$ denote the drift and diffusion coefficient of a stochastic process, given that the control variable u is exerted on the system while being at state x and time t. For a given point $x_0 \in \mathbb{R}^n$, we will consider the following *controlled* stochastic differential equation:

$$dX(t) = \mu(t, X(t), u(t))dt + \sigma(t, X(t), u(t))dW(t), \tag{9.13}$$

$$X(0) = x_0. \tag{9.14}$$

We view the n-dimensional process X as a state process, which we are trying to "control." We can (partly) control the state process X, by choosing the k-dimensional *control process* u in a suitable way. W is a d-dimensional Wiener process, and we must now try to give a precise meaning to the formal expressions (9.13) and (9.14).

Our first modeling problem concerns the class of admissible control processes. In most concrete cases it is natural to require that the control process u is adapted to the X process. One way to guarantee that is to assume the control process to be a functional of the observed values of the state process up to the time under consideration. A particular case of that is to assume that $u(t) = g(t, X(t))$, where $g : \mathbb{R}_+ \times \mathbb{R}^n \to \mathbb{R}^k$ is a deterministic Borel function that remains to be specified.

Such a control procedure is called a *feedback control law.* In the sequel we will restrict our attention to only such control laws.

Suppose that we have chosen a fixed control law u, specified by $u(t) = g(t, X(t))$. Then we can insert u into (9.13) to obtain the standard SDE:

$$dX(t) = \mu(t, X(t), g(t, X(t)))dt + \sigma(t, X(t), g(t, X(t)))dW(t). \tag{9.15}$$

Therefore, once an appropriate feedback control law has been established, then the optimal path is obtained by the solution of the standard SDE.

Definition 4 *A control law called* ***admissible*** *if*

- $u(t) = g(t, X(t)) \in U$ *for all* $t \in \mathbb{R}_+$ *and* $x \in \mathbb{R}^n$, *where* $U \subset \mathbb{R}^k$, *the geometric set modeling the constraints.*
- *For any given initial point* (t, x), *and for any* $s \in [t, T]$, *the SDE*

  $$\begin{aligned} &dX(s) = \mu(s, X(s), g(s, X(s)))ds + \sigma(s, X(s), g(s, X(s)))dW(s), \\ &X(t) = x, \end{aligned}$$

  has a unique solution.

The class of admissible control laws is denoted by $\mathcal{U}$.

For a given control law u we will denote the corresponding solution of (9.13) by X^u.

We now go on to the objective function of the control problem, and therefore we consider as given a pair of functions

$$\begin{aligned} U_1 &: \mathbb{R}_+ \times \mathbb{R}^n \times \mathbb{R}^k \to \mathbb{R}, \\ U_2 &: \mathbb{R}^n \to \mathbb{R}. \end{aligned}$$

Now we define the *objective function* of our problem as the functional $J : \mathcal{U} \to \mathbb{R}$, defined by

$$J(u) = \mathbb{E}\left[\int_0^T U_1(t, X^u(t), u(t))dt + U_2(X^u(T))\right],$$

where X^u is the solution to (9.13) with the given initial condition $X_0 = x_0$.

Thus, our formal problem can be written as that of maximizing J over all $u \in \mathcal{U}$. We define the *optimal value* $\hat{J}$ by

$$\hat{J} = \sup_{u \in \mathcal{U}} J(u).$$

This is an optimization problem in infinite dimensions since $\mathcal{U}$ is a set of stochastic processes. If there exists an admissible control law $\hat{u}$ with the property that

$$J(\hat{u}) = \hat{J},$$

then we say that $\hat{u}$ is an *optimal control law* for the given problem. Note that, as for any optimization problem, the optimal law may not exist, i.e. the supremum may not necessarily be achieved. For a given concrete control problem our main objective is of course to find the optimal control law (if it exists), show the existence of an optimal control law and then find explicitly or approximate its form.

9.8.1 The Hamilton-Jacobi-Bellman Equation

The Hamilton-Jacobi-Bellman (HJB) equation is based on the principle of dynamic programming that allows us to determine the feedback law for the optimal control by solving an appropriate finite dimensional optimization problem in lieu of the original infinite dimensional optimization problem. This leads to a fully nonlinear PDE, the HJB equation, the solution of which provides of both the value function and the optimal control law.

To obtain a HJB equation we need to employ the *embedding procedure* which is described as follows. We choose a point t (fixed) in time, with $0 \leq t \leq T$. We also choose a point x (fixed) in the state space, i.e., $x \in \mathbb{R}^n$.

Definition 5 *For a fixed pair* (t, x) *as above, define the control problem* $\mathfrak{P}(t, x)$*:*

$$\max_{u \in \mathcal{U}} \left(\mathbb{E}_{t,x} \left[\int_t^T U_1(s, X^u(s), u(s)) ds + U_2(X^u(T)) \right] \right) \tag{9.16}$$

subject to the constraints

$$\begin{aligned} dX^u(s) &= \mu\left(s, X^u(s), u(s, X^u(s))\right) ds + \sigma\left(s, X^u(s), u(s, X^u(s))\right) dW(s), \\ X^u(s) &= x, \end{aligned} \tag{9.17}$$

where by $\mathbb{E}_{t,x}$ *we will denote the expectation with respect to the probability law defined by the process (9.17).*

Clearly the problem $\mathfrak{P}(0, x_0)$ corresponds to the original problem under consideration. If we manage to solve the general problem $\mathfrak{P}(t, x)$ for every $(t, x) \in \mathbb{R}_+ \times \mathbb{R}^n$, then by substituting $t = 0$ and $x = x_0$ we obtain the solution of our original problem. We have therefore "embedded" our original problem into a general class of related problems. As it turns out it is simpler to solve the general class of problems than the specific one.

We note that in terms of the definition above, our original problem is the problem $\mathfrak{P}(0, x_0)$. A somewhat drastic interpretation of the problem $\mathfrak{P}(t, x)$ is that the investor stopped keeping track of her portfolio at time zero. Suddenly she starts keeping track again of her portfolio, noticing that the time now is t and that her state process has moved to the point x. She now tries to do as well as possible under the circumstances, so she wants to maximize her utility over the remaining time, given the fact that she starts at time t in the state x.

Now we define the *optimal value function.*

Definition 6 *The optimal value function* $V : \mathbb{R}_+ \times \mathbb{R}^n \to \mathbb{R}$ *is defined by*

$$V(t, x) = \sup_{u \in \mathcal{U}} J(u),$$

on the solution of problem $\mathfrak{P}(t, x)$.

Therefore, the optimal value function gives us the optimal expected utility, given that we start at state x at time t.

Our main objective is to derive a PDE for the value function V. For the sake of simplicity we present the derivation of the PDE in a heuristic way which of course can be turned into a fully rigorous mathematical argument under the necessary technical assumptions.

To this end, assume the existence of an optimal control law and sufficient regularity (at least $C^{1,2}$) of the optimal function V:

We perturb the optimal control law $\hat{u}$, to u^*:

$$u^*(s,y) = \begin{cases} u(s,y), & (s,y) \in [t, t+h) \times \mathbb{R}^n, \\ \hat{u}(s,y), & (s,y) \in [t+h, T] \times \mathbb{R}^n, \end{cases}$$

where h is sufficiently small and u is any control.

Clearly u^* is suboptimal leading to smaller values for the expected utility than $\hat{u}$. In full analogy with Fermat's rule for maximization of real valued functions, as taught in first year calculus classes, in this infinite dimensional problem we expect the control law $\hat{u}$ to be optimal if the variation of the expected utility for the control laws $\hat{u}$ and any "small"[9] perturbations of $\hat{u}$, u^*, vanishes. This simple observation leads to a generalization of the derivative, the Gâteaux derivative, which allows us to define first-order conditions for the solution of infinite dimensional optimization problems in full analogy with what happens for real valued functions.

The expected utility for the time interval $[t, T]$, if the control procedure u^* is adopted is given by

$$\mathbb{E}_{t,x}\left[\int_t^T U_1(s, X^{u^*}(s), u^*(s))ds\right].$$

We divide the time interval $[t, T]$ into two parts, the intervals $[t, t+h)$ and $(t+h, T]$. Clearly the control u^* coincides with u in the first interval and with the optimal control $\hat{u}$ in the second. The expected utility, for the interval $[t, t+h)$ is given by

$$\mathbb{E}_{t,x}\left[\int_t^{t+h} U_1(s, X^u(s), u(s))ds\right].$$

In the interval $[t+h, T]$ we observe that at time $t+h$ we will be in the (stochastic) state $X^u(t+h)$. Since, by definition, we will use the optimal strategy during the entire interval $[t+h, T]$ we see that the remaining expected utility at time $t+h$ is given by $V(t+h, X^u(t+h))$. This observation follows by the fact that the controlled process has the markov property. Thus the expected utility over the interval $[t+h, T]$, conditional on the fact that at time t we are in state x, is given by

$$\mathbb{E}_{t,x}[V(t+h, X^u(t+h))].$$

[9]In terms of the appropriate norm.

Thus the total expected utility for the control procedure u^* is

$$\mathbb{E}_{t,x}\left[\int_t^{t+h} U_1(s, X^u(s), u(s))ds + V\left(t+h, X^u(t+h)\right)\right].$$

Clearly if u^* consisted solely of $\hat{u}$ the above expression would coincide with $V(t,x)$, and since by definition the control law $\hat{u}$ is the optimal one, we have the inequality

$$V(t,x) \geq \mathbb{E}_{t,x}\left[\int_t^{t+h} U_1(s, X^u(s), u(s))ds + V\left(t+h, X^u(t+h)\right)\right], \quad (9.18)$$

with equality sign if and only if $u^* = \hat{u}$ for all $s \in [t,T]$.

Definition 7 *In what follows, we will use the notation ξ^u for the generator operator with action defined by*

$$\xi^u U_1 := \sum_{i=1}^{n} \mu_i(x,u)\frac{\partial}{\partial x_i}U_1 + \sum_{i,j=1}^{n}\sum_{k=1}^{d} \sigma_{ik}(x,u)\sigma_{kj}(x,u)\frac{\partial^2}{\partial x_i \partial x_j}U_1,$$

where μ_i are the coordinates of the vector μ and σ_{ij} are the elements of the matrix σ.

Since, by assumption, V is smooth, we now use the Itō formula to obtain

$$V\left(t+h, X^u(t+h)\right) - V(t,x) = \int_t^{t+h} \nabla_x V\left(s, X^u(s)\right)\sigma^u dW(s)$$
$$+ \int_t^{t+h} \left\{\frac{\partial V}{\partial t}\left(s, X^u(s)\right) + \xi^u V\left(s, X^u(s)\right)\right\}ds,$$

from which the first integral vanishes upon application of the expectation operator $\mathbb{E}_{t,x}$ (given the assumed smoothness of the value function). We can then insert the resulting equation into the inequality (9.18). The term $V(t,x)$ will cancel, leaving us with the inequality

$$\mathbb{E}_{t,x}\left[\int_t^{t+h}\left[U_1(s, X^u(s), u(s)) + \frac{\partial V}{\partial t}\left(s, X^u(s)\right) + \xi^u V\left(s, X^u(s)\right)\right]ds\right] \leq 0. \quad (9.19)$$

Now we divide by h, move h within the expectation and let h tend to zero. We further assume that $u(t)$ can be expressed in feedback form as $u(t) = g(t, X(t))$ for a suitable deterministic function g. Assuming enough regularity to allow us to take the limit within the expectation, using the fundamental theorem of integral calculus, and recalling that $X(t) = x$, we get

$$U_1(t, x, \bar{u}) + \frac{\partial V}{\partial t}(t, x) + \xi^{\bar{u}} V(t, x) \leq 0, \tag{9.20}$$

where $\bar{u}$ denotes the value of the law u evaluated at (t, x), i.e., $\bar{u} = g(t, x)$. Since the control law $u(t) = g(t, X(t))$ was arbitrary, this inequality will hold for all choices of $\bar{u} \in U$, and we will have equality if and only if $\bar{u} = \hat{g}(t, x)$ where $\hat{g}$ corresponds to the optimal feedback law. We thus have the following equation:

$$\frac{\partial V}{\partial t}(t, x) + \sup_{\bar{u} \in U} \left\{ U_1(t, x, \bar{u}) + \xi^{\bar{u}} V(t, x) \right\} = 0.$$

During the discussion the point (t, x) was fixed, but since it was chosen as an arbitrary point we see that the equation holds in fact for all $(t, x) \in (0, T) \times \mathbb{R}^n$. Thus the above equation can be considered as a PDE the solution of which will give us the value function. We obviously need some boundary conditions. One such condition is easily obtained, since we have $V(t, x) = \Phi(x)$ for all $x \in \mathbb{R}^n$. We have now arrived at, our goal, namely the derivation of the Hamilton-Jacobi-Bellman equation.

Theorem 6 (Hamilton-Jacobi-Bellman equation) *Assuming solvability of the optimal control problem and under sufficient regularity assumptions the value function V satisfies the HJB equation*

$$\frac{\partial V}{\partial t}(t, x) + \sup_{\bar{u} \in U} \left\{ U_1(t, x, \bar{u}) + \xi^{\bar{u}} V(t, x) \right\} = 0, \tag{9.21}$$

with final condition $V(T, x) = U_2(x)$.

Remark 9.8.1 *Note that the optimization problem in (9.21) is considered as a static, finite dimensional optimization problem*

We now proceed as follows:

1. Consider the HJB equation as a PDE for an unknown function V.

2. Fix an arbitrary point $((t, x) \in [0, T] \times \mathbb{R}^n$ and solve, for this fixed choice of (t, x), the static optimization problem

$$\max_{\bar{u} \in U} \left\{ U_1(t, x, \bar{u}) + \xi^{\bar{u}} V(t, x) \right\}.$$

   In this problem $\bar{u}$ is the only variable, whereas t and x are considered to be fixed parameters. The functions U_1, μ, σ and V are considered as given.

3. The optimal choice of $\bar{u}$, denoted by $\hat{u}$, will of course depend on our choice of t and x, but it will also depend on the function V and its

various partial derivatives (which are hiding under the sign $\xi^{\bar{u}}$). To highlight these dependencies we write $\hat{u}$ as:

$$\hat{u} = \hat{u}(t, x, V, V_x, V_{xx}). \tag{9.22}$$

4. The function $\hat{u}(t, x, V, V_x, V_{xx})$ is our candidate for the optimal control law, but since we do not know V this description is incomplete. Therefore we substitute the expression for $\hat{u}$ in (9.22) into the PDE (9.21), giving us the PDE

$$\frac{\partial V}{\partial t}(t, x) + U_1^{\hat{u}}(t, x) + \xi^{\hat{u}} V(t, x) = 0, \tag{9.23}$$

which is to be solved with the final condition $V(T, x) = \Phi(x)$.

5. If we can solve the above PDE we can put the solution V into expression (9.22) and identify V as the optimal value function, and $\hat{u}$ as the optimal control law.

The above procedure was based on very heuristic arguments, however, it may be shown that under conditions it holds in a rigorous manner. An important problem concerning the HJB equation is the lack of solutions which are smooth enough for the above arguments to hold, except for very simple models. However, this difficulty can be surpassed with the notion of weak solutions for the HJB equation, called viscosity solutions. Most of the above steps can be generalized using the notion of viscosity solutions in lieu of the notion of classical solution.

9.8.2 An Explicitly Worked Example

We will now analyze the problem introduced in Example 9.3. We first notice that we can get rid of the constraint $\bar{u}_t^0 + \bar{u}_t^1 = 1$ by defining a new control variable w as $w = \bar{u}^1$, and then substituting $1 - w$ for $\bar{u}^0$. This gives us the state dynamics

$$dX(t) = w(t)\left[\mu - r\right] X(t)dt + \left(rX(t) - c(t)\right) dt + w(t)\sigma X(t)dW(t), \tag{9.24}$$

and the corresponding HJB equation is

$$\frac{\partial V}{\partial t} + \sup_{c \geq 0, w \in \mathbb{R}} \left\{ U_1(t, c) + wx(\mu - r)\frac{\partial V}{\partial x} + (rx - c)\frac{\partial V}{\partial x} + \frac{1}{2}x^2 w^2 \sigma^2 \frac{\partial^2 V}{\partial x^2} \right\} = 0.$$

We assume for simplicity that $U_2 = 0$ therefore this provides us with the final condition $V(T, x) = 0$.

We now consider the case where U_1 is of the form

$$U_1(t, c) = e^{-\delta t} c^\gamma,$$

where $0 < \gamma < 1$ and δ is a discount factor. The economic reasoning behind this is that we now have an infinite marginal utility at $c = 0$. This will force the optimal consumption plan to be positive throughout the planning period, a fact which will facilitate the analytical treatment of the problem.

The static optimization problem to be solved with respect to c and w is thus that of maximizing

$$e^{-\delta t}c^{\gamma} + wx(\mu - r)\frac{\partial V}{\partial x} + (rx - c)\frac{\partial V}{\partial x} + \frac{1}{2}x^2w^2\sigma^2\frac{\partial^2 V}{\partial x^2},$$

and, assuming an interior solution, the first order conditions are

$$\gamma c^{\gamma-1} = e^{-\delta t}V_x, \tag{9.25}$$

$$w = \frac{-V_x}{x \cdot V_{xx}} \cdot \frac{\mu - r}{\sigma^2}. \tag{9.26}$$

We again see that in order to implement the optimal consumption–investment plan (9.25) and (9.26) we need to know the optimal value function V. We therefore suggest a trial solution, and in view of the shape of the instantaneous utility function it is natural to try a solution of the form

$$V(t, x) = e^{-\delta t}h(t)x^{\gamma}, \tag{9.27}$$

where, because of the final conditions, we must demand that

$$h(T) = 0. \tag{9.28}$$

Given a V of this form we have (using $\cdot$ to denote the time derivative)

$$\frac{\partial V}{\partial t} = e^{-\delta t}\dot{h}x^{\gamma} - \delta e^{-\delta t}hx^{\gamma}, \tag{9.29}$$

$$\frac{\partial V}{\partial x} = \gamma e^{-\delta t}hx^{\gamma-1}, \tag{9.30}$$

$$\frac{\partial^2 V}{\partial x^2} = \gamma(\gamma - 1)e^{-\delta t}hx^{\gamma-2}. \tag{9.31}$$

Inserting these expressions into (9.25) and (9.26) we get

$$\hat{w}(t, x) = \frac{\mu - r}{\sigma^2(1 - \gamma)}, \tag{9.32}$$

$$\hat{c}(t, x) = xh(t)^{-1/(1-\gamma)}. \tag{9.33}$$

This looks very promising: we see that the candidate optimal portfolio is constant and that the candidate optimal consumption rule is linear in the wealth variable. We now want to show that a function of the form (9.27) actually solves the HJB equation. We therefore substitute the expressions (9.29)–(9.33) into the HJB equation. This gives us the equation

$$x^\gamma \left\{ \dot{h}(t) + Ah(t) + Bh(t)^{-\gamma/(1-\gamma)} \right\} = 0,$$

where the constants A and B are given by

$$\begin{aligned} A &= \frac{\gamma(\mu - r)^2}{\sigma^2(1-\gamma)} + r\gamma - \frac{1}{2}\frac{\gamma(\mu - r)^2}{\sigma^2(1-\gamma)} - \delta, \\ B &= 1 - \gamma. \end{aligned}$$

If this equation is to hold for all x and all t, then we see that h must solve the Ordinary Differential Equation (ODE)

$$\dot{h}(t) + Ah(t) + Bh(t)^{-\gamma/(1-\gamma)} = 0, \tag{9.34}$$

with final condition $h(T) = 0$. An equation of this kind is known as a Bernoulli equation, and it can be solved explicitly.

Summing up, we have shown that if we define V as in (9.27) with h defined as the solution to (9.34), and if we define $\hat{w}$ and $\hat{c}$ by (9.32) and (9.33), then V satisfies the HJB equation, and $\hat{w}$, $\hat{c}$ attain the supremum in the equation.

9.8.3 Viscosity Solutions

The HJB equation very seldom has classical solutions and even more seldom we may find closed-form analytic solutions. In fact the above examples are probably the only available examples for which a closed-form solution exists. To alleviate these difficulties a new notion of weak solutions has been proposed, the notion of viscosity solutions, see, e.g., Ref. [4] and references therein.

The following example (taken from [2]), for simplicity illustrates this point in a problem similar to the HJB equation, the Hamilton-Jacobi equation, which is valid for the optimal control of deterministic differential equations (rather than Itō processes).

■ EXAMPLE 9.4

Assume that the state of the system is given by the equation

$$dX = u\,dt, \quad u \in \{-1, 1\},$$

such that $X(0) = x$. This can be considered as a "singular" limit of an Itō process with vanishing diffusion coefficient. Assume that we wish to minimize the functional

$$J(x,a) = \int_0^\infty U(X(t), u(t))e^{-rt}dt,$$

where $u(t)$ is the control protocol used (in this case it is a function consisting of alternating 1's and -1's). The function U is assumed to depend only on x and to be smooth and symmetric around 0, and such that $U(x) = x$ and $xU^{'}(x) < 0$ for $|x| > R$. Since the minimization of J is equivalent to the maximization of -J we can formally write the HJB equation [which will now be a first-order and not a second-order equation as the diffusion term is missing and therefore it is called simply a Hamilton-Jacobi (HJ) equation] as

$$rV(x) + |V'(x)| - U(x) = 0.$$

There is no explicit time dependence in the value function and no temporal derivatives in the HJ equation, since the problem is an infinite horizon problem. To make the link with the HJB equation we have studied here, simply assume the temporal dependence of the value function as $V(t,x) = e^{-rt}V(x)$ and substitute that in the temporal derivative to obtain the term rV.

On the other hand, one may easily figure out what the optimal control protocol would look like. It is such that $u(t) = \text{sgn}(x(t))$ as long as $x \neq 0$. At $x = 0$ then $u(t)$ can either be 1 or -1. That means the feedback control function at $x = 0$ is not single valued! This is a familiar phenomenon in optimal control. A quick calculation shows that the value function is

$$V(x) = \begin{cases} \int_0^\infty U(x + \text{sgn}(x)t)e^{-rt}dt, & x \neq 0, \\ \int_0^\infty U(-t)e^{-rt}dt = \int_0^\infty U(t)e^{rt}dt, & x = 0. \end{cases}$$

Clearly, the value function, though continuous, it is not differentiable at $x = 0$. That means that the Hamilton-Jacobi equation we have written formally above makes no sense. How can we interpret the Hamilton-Jacobi equation so that it can be of use in situations such as the one presented in this example?

A more general interpretation of the HJB equation is in terms of viscosity solutions.

Definition 8 *A continuous function ψ is called a viscosity sub(super)-solution of the HJB equation if for all $\phi \in C^\infty$*

$$-\frac{\partial \phi}{\partial t}(t_0, x_0) + \sup_u\{-\mathcal{L}^u\phi(t_0, x_0) - L(t_0, x_0, u)\} \leq (\geq) 0$$

at every (t_0, x_0) *which is a local maximum (minimum) of the difference* $\psi - \phi$.

A continuous function ψ *which is at the same time a viscosity sub- and super-solution of the HJB equation is called a viscosity solution.*

This definition is motivated by the observation that if ψ is a $C^{1,2}$ solution of the HJB equation, then it satisfies this property. In fact it can be shown that the "classical" solution of the HJB equation is also a viscosity solution (of course this does not work the other way around). However, even if the value function does not have the necessary regularity so as to qualify as a classical solution of the HJB equation, then an application of the dynamic programming principle and the use of test functions ϕ, which are as smooth as possible, allow us to derive a "version" of the HJB equation, which is modified in the sense that all derivatives are to be calculated on the smooth test function rather than on the actual value function, which may not have the required smoothness properties. Furthermore, the terminology viscosity solutions is motivated by the approximation properties of such notions of solutions, for example, under certain conditions we add a small artificial viscosity term in the HJB equation and we let the magnitude of this term tend to zero, then the solution of the perturbed HJB equation will tend to its viscosity solution. As for the inequalities involved in the definition of viscosity solutions, these are inspired by the maximum principle that holds for parabolic operators. For more on these, one may consult the specialized literature on viscosity solutions. We simply emphasize here that the numerical treatment of HJB equations makes important use of the concept of viscosity solutions.

9.9 NUMERICAL METHODS

In general the analytical treatment of the PDEs encountered in financial engineering is not possible and it is thus necessary to obtain numerical solutions of the equations. In this section we present some numerical methods which may be used for this purpose. Since the Black-Scholes-Merton equation is reducible to the heat equation, for the purpose of presentation, we will present the numerical methods in terms of the heat equation and leave to the reader the extension to more generall parabolic equations.

9.9.1 The Crank-Nicholson Method

The Crank-Nicholson method is a general method for the numerical treatment of parabolic equations. It is a discretization method, according to which we calculate $U(t, x)$ at a discrete grid in space and time, $\{i\,\delta t\}$, $i = 1, \ldots, N$ and $\{j\delta x\}$, $j = 1, \ldots, M$. We then identify $U(i\delta t, j\delta x)$ by U_j^i. The assumption behind this is that the solution of the heat equation is a smooth enough function U so that it may be well approximated by its value at specific grid points.

Using the Taylor expansion theorem we approximate the derivatives of the function with finite differences. For example, we approximate

$$\frac{\partial U}{\partial t}(i\delta t, j\delta x) \simeq \frac{1}{\delta t}\left(U_j^{i+1} - U_j^i\right),$$
$$\frac{\partial U}{\partial x}(i\delta t, j\delta x) \simeq \frac{1}{\delta x}\left(U_j^{i} - U_{j-1}^i\right),$$
$$\frac{\partial^2 U}{\partial x^2}(i\delta t, j\delta x) \simeq \frac{1}{(\delta x)^2}\left(U_{j+1}^{i} - 2U_j^i + U_{j-1}^i\right).$$

It is a simple exercise in calculus (applying the Taylor expansion theorem with a remainder) to show that if $U \in C^{1,3}$ then the above approximation for the first derivative in time holds to order $O(\delta t^2)$. If $U \in C^{1,3}$ then the above approximation for the first derivative in space holds to order $O(\delta x^2)$. If $U \in C^{1,4}$ the above approximation for the second derivative in space holds to order $O(\delta x^2)$.

If we substitute the above approximations in the heat equation this gives us a difference equation of the form

$$U_j^{i+1} - U_j^i = \alpha\,(U_{j+1}^i - 2U_j^i + U_{j-1}^i),$$

where $\alpha = \frac{\delta t}{(\delta x)^2}$. In this difference equation we know U_j^1 for all j (this is simply our initial condition), therefore iterating in i we may obtain U_j^i for all $i = 2, \ldots, N$. This difference equation is more compactly written in matrix form as

$$\bar{U}^{i+1} = A\bar{U}^i,$$

where $\bar{U}^i = (U_1^i, \ldots, U_N^i)^{tr}$ and A is the tridiagonal matrix

$$A = \begin{pmatrix} 1-2\alpha & \alpha & 0 & \cdots & 0 \\ \alpha & 1-2\alpha & \ddots & \ddots & \vdots \\ 0 & \ddots & \ddots & \ddots & 0 \\ \vdots & \ddots & \ddots & 1-2\alpha & \alpha \\ 0 & \cdots & 0 & \alpha & 1-2\alpha \end{pmatrix}.$$

This is the simplest possible numerical method, called the explicit method, since knowledge of $\bar{U}^1$ alone provides us, upon iteration, the solution at any time

$$\bar{U}^2 = A\bar{U}^1,\ \bar{U}^3 = A^2\bar{U}^1,\ \ldots,\ \bar{U}^N = A^{N-1}\bar{U}^1.$$

This method, even though it is simple, has certain drawbacks. One of the major drawbacks is regarding its stability properties, since α must be chosen rather small.

An alternative method would be to approximate the heat equation as

$$U_j^{i+1} - U_j^i = \alpha \left(U_{j+1}^{i+1} - 2U_j^{i+1} + U_{j-1}^{i+1} \right),$$

i.e., we approximate the second-order derivative in space using the value of the function at time $i+1$ rather than with the value of the function at time i as we did before in the explicit method. The drawback of this method is that knowing $\bar{U}^i$ we may only find $\bar{U}^{i+1}$ by solving a system of linear equations. For this reason this method is called the implicit method.

Writing this equation in compact form we obtain

$$\bar{U}^i = B\,\bar{U}^{i+1},$$

where

$$B = \begin{pmatrix} 1+2\alpha & -\alpha & 0 & \cdots & 0 \\ -\alpha & 1+2\alpha & \ddots & \ddots & \vdots \\ 0 & \ddots & \ddots & \ddots & 0 \\ \vdots & \ddots & \ddots & 1+2\alpha & -\alpha \\ 0 & \cdots & 0 & -\alpha & 1+2\alpha \end{pmatrix}.$$

If the matrix B is invertible (which it is) then the implicit method gives

$$\bar{U}^{i+1} = B^{-1}\,\bar{U}^i.$$

The implicit method is unconditionally stable, in the sense that it displays stability properties for any value of α.

The Crank-Nicholson method is a combination of the explicit and the implicit method (see, e.g., Refs. [11], [12]). We add the implicit and explicit method and divide by 2, so as to approximate the solution by the mean result of the two methods. This yields

$$U_j^{i+1} - U_j^i = \frac{\alpha}{2} \left(U_{j+1}^i - 2U_j^i + U_{j-1}^i + U_{j+1}^{i+1} - 2U_j^{i+1} + U_{j-1}^{i+1} \right).$$

We now write this in a more compact form using matrices. Define the matrices

$$\bar{A} = \begin{pmatrix} 1-\alpha & \frac{\alpha}{2} & 0 & \cdots & 0 \\ \frac{\alpha}{2} & 1-\alpha & \ddots & \ddots & \vdots \\ 0 & \ddots & \ddots & \ddots & 0 \\ \vdots & \ddots & \ddots & 1-\alpha & \frac{\alpha}{2} \\ 0 & \cdots & 0 & \frac{\alpha}{2} & 1-\alpha \end{pmatrix}$$

and

$$\bar{B} = \begin{pmatrix} 1+\alpha & -\frac{\alpha}{2} & 0 & \cdots & 0 \\ -\frac{\alpha}{2} & 1+\alpha & \ddots & \ddots & \vdots \\ 0 & \ddots & \ddots & \ddots & 0 \\ \vdots & \ddots & \ddots & 1+\alpha & -\frac{\alpha}{2} \\ 0 & \cdots & 0 & -\frac{\alpha}{2} & 1+\alpha \end{pmatrix}.$$

In terms of these matrices the Crank-Nicholson method assumes the compact form

$$\bar{B}\,\bar{U}^{i+1} = \bar{A}\,\bar{U}^{i}. \tag{9.35}$$

Iteration of (9.35) yields the solution of the heat equation at all times $i = 2, \cdots, N$. One way of iterating is to invert the matrix $\bar{B}$, form the product $\bar{B}^{-1}\bar{A}$ and then iterate. A better and more efficient way is to use the LU decomposition from numerical linear algebra. To use this method suppose that we known $\bar{U}^i$. Then by a matrix multiplication we calculate the vector $c^i := \bar{A}\,\bar{U}^i$. To obtain $\bar{U}^{i+1}$ we must solve the system of linear equations

$$\bar{B}\,\bar{U}^{i+1} = c^i.$$

This can be done fast and efficiently using the LU decomposition. We decompose $\bar{B}$ as $\bar{B} = LU$, where L and U are lower and upper diagonal matrices, respectively. This requires first solving the lower diagonal system $Ly^i = \bar{U}^i$ for y^i and then the upper diagonal system $U\bar{U}^{i+1} = y^i$ for $\bar{U}^{i+1}$. This completes the Crank-Nicholson method.

A generalization of the Crank-Nicolson method is

$$U_j^{i+1} - U_j^i = \alpha \left[\theta \left(U_{j+1}^i - 2U_j^i + U_{j-1}^i\right) + (1-\theta)\left(U_{j+1}^{i+1} - 2U_j^{i+1} + U_{j-1}^{i+1}\right)\right],$$

where $\theta \in [0,1]$. In the special case where $\theta = 1/2$ we recover the Crank-Nicolson method. In the case where $\theta = 0$ we obtain the implicit method and when $\theta = 1$ we obtain the explicit method. All other cases are in between. The general method may be written in compact matrix form using similar arguments as above, and the resolution of the problem may be provided using the LU decomposition. The details are left as an exercise to the interested reader.

The Crank-Nicolson method may be written directly for the original form of the Black-Scholes equation (exercise).

9.9.2 Numerical Treatment of Variational Inequalities

We now give a short introduction to the numerical treatment of variational inequalities, with special emphasis on the study of the pricing of American options. To facilitate the treatment we assume that we have transformed the Black-Scholes-Merton equation to the heat equation and then study the relevant variational inequality for the heat equation. Of course the analysis may go through for the original version of the Black-Scholes-Merton equation or any parabolic equation, but we will only sketch this in the end of the section.

Applying the transformation (9.7) the variational inequality for the pricing of American options assumes the form

$$\max\left\{\bar{\Phi} - U, \frac{\partial U}{\partial t} + \mathcal{L}_0 U\right\} = 0$$

or equivalently

$$\min\left\{U - \bar{\Phi}, -\frac{\partial U}{\partial t} - \mathcal{L}_0 U\right\} = 0,$$

and $U = \bar{\Phi}$ on ∂D, where $\mathcal{L}_0$ is the operator $\mathcal{L}_0 U = -\frac{\partial^2 U}{\partial x^2}$.

To solve this problem we must work with a discretized version. To this end we choose to use the Crank-Nicolson discretized version of the heat equation (9.35), which in terms of the differential inequality becomes

$$\bar{B}\bar{U}^{i+1} - \bar{A}\bar{U}^i \geq 0, \tag{9.36}$$

where the inequality is to be understood componentwise. Therefore, in terms of the discretized approximation, the linear complementarity problem becomes

Given $c^i := \bar{A}\bar{U}^i$ find the vector $\bar{U}^{i+1}$ such that

$$\bar{B}\bar{U}^{i+1} - c^i \geq 0, \ \ \bar{U}^{i+1} - \bar{\Phi} \geq 0, \ \ \left(\bar{B}\bar{U}^{i+1} - c^i\right)\left(\bar{U}^{i+1} - \bar{\Phi}\right) = 0.$$

This is repeated for all $i = 2, \ldots, N$.

This problem is far from being simple. In fact, it is an interesting problem in linear algebra that may be treated using an iterative algorithm, the projected SOR algorithm.

We consider the auxiliary problem:

Given a vector b, find vectors x, y such that

$$Bx - y = b, \quad x \geq 0, \quad y \geq 0 \ x^{tr} y = 0 \tag{9.37}$$

This problem is equivalent to the original problem if we set $x = \bar{U}^{i+1} - \bar{\Phi}$, $y = B\bar{U}^{i+1} - c^i$, $b = c^i - B\bar{\Phi}$. Then $\bar{U}^{i+1} = x + \bar{\Phi}$ is the solution of the original problem.

The auxiliary problem (9.37) is solved by the following iterative method: We solve the problem $Bx = b$ by the iterative procedure

$$x_\ell^{(k)} = x_\ell^{(k-1)} + \omega \frac{r_\ell^{(k)}}{b_{\ell\ell}},$$

$$r_\ell^{(k)} = b_\ell - \sum_{j=1}^{\ell-1} a_{\ell j} x_j^{(k)} - b_{\ell\ell} x_\ell^{(k-1)} + \sum_{j=\ell+1}^{M} b_{\ell j} x_j^{(k-1)},$$

where ω is a relaxation parameter which is chosen by the user for the best convergence of the method. The iterative method starts with an initial choice $x^{(0)} \geq 0$ and continues trying to ensure that $x^{(k)} \geq 0$ at all levels of the iteration. This is effected by modifying the SOR algorithm so that

$$x_\ell^{(k)} = \max\left\{0, x_\ell^{(k-1)} + \omega \frac{r_\ell^{(k)}}{b_{\ell\ell}}\right\},$$

$$r_\ell^{(k)} = b_\ell - \sum_{j=1}^{\ell-1} a_{\ell j} x_j^{(k)} - b_{\ell\ell} x_\ell^{(k-1)} + \sum_{j=\ell+1}^{M} b_{\ell j} x_j^{(k-1)}.$$

We then need to find the vector y. This is done as follows

$$y_i^{(k)} = -r_i^{(k)} + b_{\ell\ell}(x_i^{(k)} - x_i^{(k-1)}).$$

Undoing the transformations we obtain the solution to the original problem.

9.9.3 Numerical Treatment of HJB Equations

The numerical treatment of the HJB equation is a very interesting but also complicated task, due to the high nonlinearity of the equation. One of the most important methods for the resolution of such equations is through the use of viscosity solutions; we refer the interested reader to the relevant literature (see, e.g., the appendix by Falcone in Ref. [2], or Refs. [10], [7] for some applications and references therein). Other methods are through the discretization of the Itō equation to a Markov chain and the use of dynamic programming techniques for the approximation of the value function and the optimal control (see, e.g., [9] and references therein).

9.10 CONCLUSION

In this chapter we have tried to present a brief introduction to the use of partial differential equations, linear and nonlinear, in financial engineering. We have tried to motivate their use through a simple yet fundamental model in financial mathematics, the Black-Scholes model, and have presented how this model leads in a natural way to a linear parabolic PDE, the Black-Scholes equation. The solution of this equation provides a way of pricing a derivative asset, given the current value of the fundamental asset this derivative is based on. We have then introduced stopping problems related to the pricing of American-type derivative assets and have shown how these lead to free boundary-value problems, introducing in this manner the concept of variational inequality which is very useful in a number of problems of financial engineering. We then moved to fully nonlinear PDEs that were motivated by their occurrence in optimal control problems when treated through dynamic programming. We have shown how the treatment of optimal control problems may lead to the Hamilton-Jacobi-Bellman equation for the value function, which is a fully nonlinear second-order PDE. After providing some examples in which the HJB admits solutions in closed form, we consider the case of weak solutions, called viscosity solutions, which are more widely applicable and furthermore are important to numerical applications. We close the chapter with an introduction to the numerical treatment of PDEs arising in financial mathematics, including variational inequalities.

Needless to say, there are many important aspects of the theory that were not possible to be included in this chapter for lack of space. To mention just a few; the problem of pricing and hedging derivatives in incomplete markets has been and remains a problem of fundamental importance (see Ref. [6] for an excellent treatment; see also Ref. [14] for an alternative approach and references therein). There are also important applications of this basic methodology presented here to other types of assets, e.g., energy derivatives (see, e.g., Ref. [13] and references therein), insurance or reinsurance modeling (see, e.g., Ref. [1] and references therein), etc. However, we hope that this very brief first encounter will motivate some of the readers of the present volume to further explore the exciting fields of stochastic analysis and its connections with applied mathematics, as well as stochastic control theory and mathematical finance.

EXERCISES

9.1 Apply Itō's lemma to show that the solution of the SDE $dS = \mu S dt + \sigma S dW$ with initial condition $S(0) = s$ is the stochastic exponential

$$S(t) = s \exp\Big(\Big(\mu - \frac{\sigma^2}{2}\Big)t + \sigma W(t)\Big).$$

9.2 Complete the integrations involved in the derivation of the pricing formulae for the put and the call options and verify the stated results.

9.3 Verify that the Black-Scholes-Merton equation can be reduced to the heat equation with the stated coordinate change.

9.4 Using the properties of the Wiener process or of the Itō integral calculate the first two statistical moments of $S(t)$.

9.5 Write down the HJB equation for the portfolio optimization example if the utility function $U_1(c) = e^{-\delta t} \ln c$ and try to find a solution.

REFERENCES

1. Baltas, I., Frangos, N. E. and Yannacopoulos, A. N., Optimal reinvestment and reinsurance policies in insurance markets under the effect of inside information, Applied Stochastic Models in Business and Industry, Vol. 27, 203–217 (2011).
2. Bardi, M. and Capuzzo-Dolcetta, I. *Optimal Control and Viscosity Solutions of Hamilton-Jacobi-Bellman Equations*, Birkäuser, Boston (1997).
3. Björk, T., *Arbitrage theory in continuous time*, Oxford University Press, Oxford (1999).
4. Bardi, M., Crandall, M. G., Evans, L. C., Soner, H. M., and Souganidis, P. E., *Viscosity Solutions and Applications*, Lecture Notes in Mathematics, Volume 1660, Springer (1997).
5. Karatzas, I. and Shreve, S., *Brownian motion and Stochastic Calculus*, 2nd Ed., Springer, Berlin (1991).
6. Karatzas, I. and Shreve, S., *Methods of Mathematical Finance*, Springer, Berlin (1998).
7. Kossioris, G., Plexousakis, M., and Yannacopoulos, A. N., A Hamilton-Jacobi-Belman approach to the control of trapping time of a soliton in an external potential, Quarterly of Applied Mathematics 63, no. 2, 309–324 (2005).
8. Kreyszig, E., *Advanced Engineering Mathematics*, 6th Edition, Wiley & Sons, New York (1988).
9. Kushner, H. P., and Dupuis, P. G., *Numerical Methods for Stochastic Control Problems in Continuous Time*, Springer, New York (2000).
10. Nikolopoulos, C. and Yannacopoulos, A. N., A model for optimal stopping in advertisement, Nonlinear Analysis: Real World Applications, Vol. 11, 1129–1242 (2010).
11. Press, W. H., Teukolsky, S. A., Vetterling, W. T., and Flannery, B. P., *Numerical Recipes in C++: The Art of Scientific Computing*, 2nd Edition, Cambridge University Press, Cambridge (2002).
12. Smith, G. D., *Numerical Solution of Partial Differential Equations*, Oxford University Press, Oxford (1974).

13. Tsitakis, D., Xanthopoulos, S., and Yannacopoulos, A. N., A closed form solution for the price of cross commodity electricity derivatives, Physica A: Econophysics Section, Vol. 371, 543–551 (2006).

14. Xanthopoulos, S. and Yannacopoulos, A. N., Scenarios for price determination in incomplete markets, International Journal of Theoretical and Applied Finance, Vol. 11, 415–445 (2008).

CHAPTER 10

DECISION MODELING IN SUPPLY CHAIN MANAGEMENT

HUAJUN TANG

Faculty of Management and Administration, Macau University of Science and Technology, Macau

10.1 INTRODUCTION TO DECISION MODELING

10.1.1 The Origin of Decision Modeling

Decision modeling has been widely used since Fredrick W. Taylor, in the early 1900's, applied the principles of the scientific approach to management. During World War II, numerous novel quantitative and scientific methods were developed to support the military. These new developments became so successful that many companies started to apply similar methods in managerial decision making and planning after World War II. Recently, more and more companies recruit staffs in the field of operations research or management

Mathematical Modeling with Multidisciplinary Applications.
Edited by Xin-She Yang

science to use the principles of scientific management to solve real business problems.

10.1.2 Definition of Decision Modeling

There are varying definitions for *decision modeling.* Here it is defined as a scientific approach to managerial decision making. Decision modeling is also usually referred to as operations research, management science or quantitative analysis. In this chapter, we adopt the term *decision modeling* since we will discuss some modeling techniques in a managerial decision making context.

10.1.3 Data in Decision Modeling

Any decision modeling process begins with data. Like raw materials for a manufacturer, these data are processed into information, which is important to the decision making. The processing of raw data into meaningful information is the heart of decision modeling.

In dealing with a decision-making problem, managers would have to consider both qualitative and quantitative factors. For instance, suppose that we are considering supplier alternatives of CPU products, such as Intel, AMD, and IBM. We can use quantitative factors such as price, capacity, and transportation cost in our decision model to assist our ultimate decision. However, in addition to these factors, we may also have to consider qualitative factors such as quality, lead time, and credit. It could be difficult to quantify these qualitative factors.

Because of the presence of qualitative factors, quantitative decision modeling can play different roles in the decision-making process. When the problem, model and input data remain stable over time without qualitative factors, a decision model can make the decision-making process automatic. For instance, some corporations use quantitative inventory models to determine automatically when to order and how much to order. However, in most cases, decision modeling would be only one of several aids to the decision-making process. The outputs of decision modeling should be combined with qualitative information while making decisions in reality.

10.1.4 Role of Spreadsheets in Decision Modeling

To keep with the fast development of technology in the last three decades, computers have become a fundamental part of the decision modeling process in today's business environments. Until 1990's, many modeling techniques discussed in this chapter required specialized software packages. However, widely available spreadsheet packages such as Microsoft Excel have been increasingly used to set up and solve most of the decision modeling techniques in practical situations. Hence the current trend in many university courses in decision modeling focuses on spreadsheet-based instruction. In keeping with this trend, we will discuss the role and use of spreadsheets (specifically Microsoft Excel's Solver add-in) during the study of the different decision modeling techniques in this chapter.

10.1.5 Types of Decision Models

According to the type and nature of the problem environment under consideration, decision models can be classified into two components: (1) deterministic models and (2) probabilistic models. In the following we define each of these two types of models.

Deterministic models assume that all the relevant input data are known with certainty. That is to say, all the information related to modeling the decision-making problem environment is available, with known values. For example, ABC corporation manufactures several different types of PC products (e.g., desktops, laptops), all of which compete for the same resources (e.g., hard disks, chips, labor). Suppose ABC knows the specific amounts of each resource required to make one unit of each type of PC. In such an environment, if ABC determines a specific production plan, it is easy to compute the quantity required of each resource to satisfy this production plan. For instance, if ABC plans to ship 2,000 units of a specific model and each unit includes two speakers, then ABC will need 4,000 speakers. Perhaps the most common and popular deterministic modeling technique is linear programming (LP). In this chapter, most of the models can be set up and solved by LP.

Compared to deterministic models, *probabilistic models* assume that some input data are not known with certainty. That is to say, the values of some important variables will not be known before decisions are made. Hence it is important to incorporate this uncertainty into the model. An example of this type of model would be the decision of when to order and how many CPU products to order when ABC managers face random demand. Probabilistic modeling techniques take uncertainty into account by using probabilities on these random variables. Because of space limit, we mainly focus on deterministic models in this chapter.

10.1.6 Steps of Decision Modeling

Regardless of the size and complexity of the decision-making problem at hand, the decision modeling process involves three distinct steps: (1) formulation, (2) solution, and (3) interpretation. Figure 10.1 (Balakrishnan *et al.*, 2007) provides a schematic overview of these steps along with the components of each step. In the following we will discuss each of these steps.

It is fundamental to recognize that an iterative process (shown as dotted lines in Figure 10.1) usually occurs between these three steps before the final solution is obtained. For instance, during the solution step it may be realized that the model is incomplete or that some of the input data are erroneous.

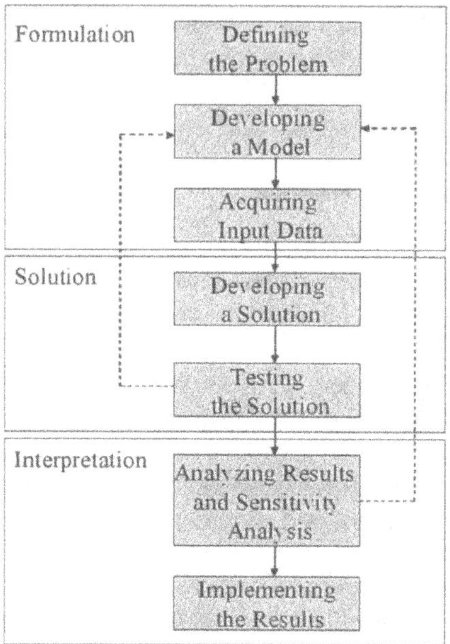

Figure 10.1 The decision modeling process.

This suggests that the formulation needs to be modified, which causes all of the subsequent steps to be changed.

Formulation is the process by which each of the problem scenario is translated and expressed in terms of a mathematical model. This is perhaps the most important and challenging step in decision modeling. Since the results of a posed built problem will certainly be wrong, it is important for the decision maker to analyze the problem rationally. The objective in formulation is to ensure that the mathematical model completely presents all the issues that are relevant with respect to the problem. Formulation can be divided into three components: (a) defining the problem, (b) developing a model, (c) acquiring input data.

Defining the problem, which means the development of a clear, concise description of the problem, can be the most important part of formulation. This description will give direction and meaning to all the parts following it. Once we define the problem to be analyzed, the next part is to *develop a decision model.* The models we develop in this chapter are mathematical. *A mathematical model* is a set of mathematical relationships. In most cases, these relationships are expressed as equations, and inequalities. All models should be built carefully. They must be solvable, realistic, and easy to understand and modify, and the input data should be available. Once we have developed a model, we must obtain the input data to be used in the model. It is necessary to obtain accurate data, since incorrect data may result in misleading decisions, even if the model is an excellent presentation of reality.

The *Solution* step is to identify the optimal solution by solving the mathematical expressions resulting from the formulation step. The solution step could be further classified into two components: (a) developing a solution and (b) testing the solution.

Developing a solution focuses on processing the model to arrive at the best solution to the problem. In some cases, we may find the optimal solution by applying a systematic algorithm to solve a set of mathematical expressions. In other cases, we could use a trial and error method to find the best decision among a set of alternatives. Before a solution can be analyzed and implemented, it must be tested completely. As we know, the solution depends on the input data and the model, both of which require testing. There exist ways to test the data. One is to collect additional data from a different sample source and use statistical tests to contrast these new data with the original data. If there are significant differences, we should make more efforts to obtain accurate input data. If the data are accurate but the model's outputs are not consistent with the problem, we have to modify the model to make sure that it is logical and represents the real situation.

Supposing that the formulation is correct and has been successfully carried out and solved, we have to consider the implications of the results. We will discuss this step in two parts: (a) analyzing the results and their sensitivity analysis and (b) implementing the results. *Analyzing the results* starts with determining the implications of the solution. In most cases, a solution to the problem will lead to some changes in the way a company is operating. The implications of these changes must be determined and analyzed before the results are implemented. Considering that a model is only an approximation of reality, the sensitivity of the solution to changes in the model and input data is an important issue of analyzing the results. This type of analysis is called *sensitivity analysis*. Sensitivity analysis identifies how much the solution will vary if there are changes in the model or the input data. If the optimal solution is very sensitive to changes in the input data and the model, then additional testing must be performed to make sure that the model and input data are valid.

The final part is to *implement* the results. This could be much more difficult than you might imagine. You need to be able to present and explain the results of your study to management. If the manager rejects the new solution, the model is of no value, even if the optimal solution might have results in millions of dollars in additional profits. If and when the new solution is implemented, the supply chain should be closely monitored.

10.2 MATHEMATICAL PROGRAMMING MODELS

10.2.1 Introduction of Linear Programming Models

Since the mid-twentieth century, linear programming (LP) has been applied extensively to typical problems in supply chain management such as manufacturing, transportation, scheduling, assignment, and operational problems. Regardless of the size and the complexity of the decision-making problem in these applications, the development of all LP models can be divided into three distinct steps as defined above: (1) formulation, (2) solution, and (3) interpretation. We now briefly discuss each step with respect to LP models. Formulation is the process by which each aspect of the problem scenario is expressed in terms of mathematical expressions. The aim is to make sure that the set of mathematical expressions present all the issues relevant to the problem. The solution step is to solve the mathematical expressions resulting from the formulation process to identify an optimal solution. In this chapter we mainly focus on solving LP models and other mathematical programming models with spreadsheets. Supposing that the formulation is correct and can be solved with an LP software package, the manager can carry out a sensitivity analysis by using the software to evaluate the impact of several different types of what-if questions.

10.2.2 Properties of a Linear Programming Model

All LP models have some common properties as listed below.

1. All problems aim to maximize or minimize some quantity, usually profit or cost, which is refereed to as the *objective function* of an LP problem. For instance, a typical manufacturer usually seeks to maximize profits. In a trucking distribution system, the objective could be to minimize shipping costs.

2. There are usually some constraints on the allowable values of variables in LP models. For instance, we are restricted by the available raw materials and machinery time when we try to decide how many units of each product in a company's product line should be produced. Furthermore, LP models usually include a set of constraints known as *nonnegativity* constraints, which make sure that the variables in the model take on only nonnegative values. This is feasible and reasonable since negative values of physical quantities are impossible, and we cannot produce a negative number of computers.

3. There must be alternatives among which we can choose. For example, if a factory produces three different products, the manager could use LP to decide how to allocate his limited production resources (e.g., labor and machine hours, raw material) among these products. Should it devote

all manufacturing capacity to only the first product, or equal amounts to each product, or some other ratios. If there are no alternatives to select from, we do not have a decision problem and do not need LP.

4. The objective and constraints in LP models must be expressed in terms of *linear* functions of the variables. This means that all terms in the objective function and in the constraint equations and inequalities are of the first degree. Hence the equation $A + B = 10$ is a valid linear equation, while the equation $A^2 + B + C = 10$ is not linear because the variable A is squared. Examples of linear inequalities are $A + B \leq C$ or $A + B \geq C$.

One example of LP models is presented in the following.

EXAMPLE 10.1

This LP model is to maximize the objective function subject to some constraints.

$$\begin{aligned} \text{Maximize } Z &= 10A + 5B, \qquad (10.1) \\ \text{subject to} & \\ 4A + 3B &\leq 200, \\ 2A + B &\geq 60, \\ A, B &\geq 0. \end{aligned}$$

The constraint $A, B \geq 0$ (the notation means "$A \geq 0$ and $B \geq 0$") is called a *non-negative* constraint.

10.2.3 Assumptions of a Linear Programming Model

Technically, there are four basic assumptions in an LP model. These are listed below.

1. *Certainty.* Numbers used in the objective function and constraints are known with certainty and do not change during the period being studied.

2. There exists *proportionality* in the objective function and constraints. For instance, if the production of 1 unit of a product need 4 labor hours, then making 10 units of that product should need 40 labor hours.

3. *Additivity.* The total of all the activities equals the sum of the individual activities. For instance, when the profit is \$10 per unit of the first product and \$5 per unit of the second product, and 1 unit of each product is manufactured, then the resulting profit should \$15.

4. *Divisibility.* The solutions need not necessarily be in integers. That is, they could take any fractional (i.e., real-number) value.

10.2.4 Other Mathematical Programming Models

The LP models discussed above have three characteristics:
(1) the decision variables are allowed to have fractional values,
(2) there is a single objective function, and
(3) all expressions (objective function and constraints) are linear.

However, there exist other important mathematical programming models that relax these LP conditions: integer programming (IP), goal programming (GP), and nonlinear programming (NLP).

Integer programming is the extension of LP that deals with problems requiring integer solutions, and has two types of variables: *general integer* variables and *binary* variables. General integer variables are those taking on any nonnegative integer values. Binary variables are a special type of integer variables taking values 0 or 1. IP problems can be classified into four types as below.

1. *Pure IP problems* are problems in which all decision variables must have integer values.

2. *Mixed IP problems* are problems in which some, but not all, decision variables must have integer values. The noninteger variable could have fractional values.

3. *Pure binary IP problems* are problems in which all decision variables are binary.

4. *Mixed binary IP problems* are problems in which some decision variables are binary and other decision variables are either general integers or continuous values.

Goal programming considers optimization problems having several objective functions, instead of forcing the decision maker to focus on only a single objective as in LP. *Nonlinear programming* is the extension of LP to problems in which the objective or the constraints are nonlinear. For example, if the objective function in (10.1) is changed to be $Z = 10A + 5B^2$, then the model is a nonlinear programming model.

In the following, we briefly introduce the discipline of supply chain management (SCM), and then apply the above mathematical programming models, especially LP models, to deal with typical problems in SCM.

10.3 INTRODUCTION OF SUPPLY CHAIN MANAGEMENT

A *supply chain* is the flow of products and services from raw material manufacturers to intermediate product manufacturers, end product manufacturers, wholesalers, distributors, and retailers. The supply chain is connected by transportation and storage activities, and integrated through information, planning and integration activities. *Supply chain management* is a set of

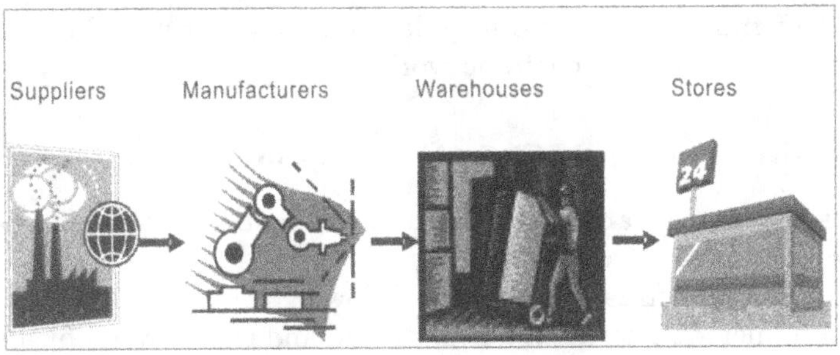

Figure 10.2 One Example of the Supply Chain

approaches to efficiently integrate suppliers, manufacturers, warehouses, and stores, so that commodities are produced and distributed at the right quantities in the right locations at the right time. One example of the supply chain is presented in Figure 10.2 .

Supply chain management was first proposed by a U.S. industry consultant in the early 1980's, and developed through the 1990's with the introduction of electronic data interchange (EDI), and enterprise resource planning (ERP) systems. It developed in the 21st century with the expansion of Internet-based collaborative systems over national boundaries. In the 1990's, industries began to focus on *core competencies* and adopted a specialization model. Companies abandoned vertical integration, sold off non-core operations, and outsourced those functions to third-party companies. This led to the rise of companies specializing in outsourced manufacturing and distribution.

10.3.1 Importance of Supply Chain Management

During the past decades, globalization, outsourcing and information technology have enabled many organizations, such as Dell, Hewlett Packard, and Wall-mark to successfully operate solid collaborative supply networks, in which each specialized business partner focuses on only a few key strategic activities. This inter-organizational supply network can be acknowledged as a new form of organization. Traditionally, companies in a supply network concentrate on the inputs and outputs of the processes, with little concern for the internal management of other individual co-operators. In the 21st century, two changes in the business environment have contributed to the development of supply chain networks. Firstly, as an outcome of globalization and the proliferation of multinational companies, joint ventures, strategic alliances and business partnerships, significant success factors were identified, complementing the earlier "Just-In-Time," "Lean Manufacturing" and "Agile Manufacturing" practices. Secondly, technological changes, particularly the dramatic fall in information communication costs, which are a significant component of transaction costs, have led to changes in coordination among the members of the supply chain network.

10.3.2 Activities in Supply Chain Management

Supply chain activities can be grouped into strategic, tactical, and operational levels. We present them as below.

The strategic level usually covers the following activities: strategic network optimization, including the number, location, and size of warehousing, distribution centers, and facilities; strategic partnerships with suppliers, distributors, and customers, creating communication channels for critical information and operational improvements such as cross docking, direct shipping, and third-party logistics; product life cycle management, so that new and existing products can be optimally integrated into the supply chain and capacity management activities; and where-to-make and make-or-buy decisions.

The tactical level often includes: sourcing contracts and other purchasing decisions; production decisions, including contracting, scheduling, and planning process definition; inventory decisions, including quantity, location, and quality of inventory; and transportation strategy, including frequency, routes, and contracting.

Operational level usually covers: daily production and distribution planning, including all nodes in the supply chain; production scheduling for each manufacturing facility in the supply chain; sourcing planning, including current inventory and forecast demand, in collaboration with all suppliers; inbound operations, including transportation from suppliers and receiving inventory; production operations, including the consumption of materials and flow of finished goods; outbound operations, including all fulfillment activities, warehousing and transportation to customers; and order promising, including all suppliers, manufacturing facilities, distribution centers, and other customers.

In the following, we apply LP and other mathematical programming models with spreadsheet to investigate some typical problems in supply chain managment.

10.4 APPLICATIONS IN SUPPLY CHAIN MANAGEMENT

10.4.1 Manufacturing Applications

Manufacturing issues are the most common and important in supply chain management, because they impact on the later activities, such as transportation, inventory management, and final marketing.

10.4.1.1 Product Mix Problem A popular use of an LP is to solve product mix problems. Most corporations have to meet a number of constraints, ranging from financial concerns to sales demands to material contracts to union labor demands. The corporation's primary goal is to generate the largest profit possible. In the following we consider a simple product mix problem.

Prada corporation, a manufacturer of women's-wear, produces four types of skirts. One is an expensive, all-silk skirt, one is an all-polyester skirt, and two are "blends" made of pieces of polyester and cotton. Table 10.1 illustrates the cost and availability (per monthly production period) of three materials used in the production process. In addition, the labor cost is \$0.80 per skirt. The company has fixed contracts with several major department store chains

Table 10.1 The cost and availability of the three materials.

Material	Cost Per Yard (\$)	Material Available Per Month (Yards)
Silk	18	780
Polyester	9	2,800
Cotton	12	1,600

to supply skirts. The contracts require that Prada supply a minimum and a maximum of monthly quantity of each skirt. Table 10.2 summarizes the contract information for each of the four styles of skirts, the selling price per skirt, and the fabric requirements of each skirt. Prada's goal is to maximize

Table 10.2 Product data for Prada skirt.

Variety of skirt	Selling price per skirt (\$)	Monthly minimum	Monthly maximum	Material per skirt (Yards)	Material Requirements
All silk	7.80	6,000	7,000	0.12	100% silk
All polyester	4.25	11,500	15,000	0.09	100% polyester
Poly-cotton blend 1	5.20	13,000	17,000	0.12	50% polyester-50% cotton
Poly-cotton blend 2	5.80	6,500	8,800	0.10	30% polyester-70% cotton

its monthly profit. It must decide upon a policy for product mix. Let

$$\begin{aligned} S &= \text{number of all-silk skirts produced per month} \\ P &= \text{number of polyester skirts produced per month} \\ B_1 &= \text{number of blend 1 poly-cotton skirts produced per month} \\ B_2 &= \text{number of blend 2 poly-cotton skirts produced per month} \end{aligned}$$

To determine the objective function, the unit profits must be first calculated. We illustrate the net profit calculation for all-silk skirts (S). Each all-silk skirt requires 0.12 yards of silk at a cost of \$18 per yard, resulting in a material cost of \$2.16. The selling price per all-silk skirt is \$7.80, leading to a net profit of \$7.80 − \$2.16 − \$0.80 = \$4.84 per skirt. Similarly, we find that the remain net unit profits are \$2.64, \$3.14, and \$3.89, for all-polyester, poly-cotton blend 1, and poly-cotton blend 2 skirts, respectively.

The objective function could be presented as

$$\begin{aligned} \text{Maximize profit} &= \$4.84S + \$2.64P + \$3.14B_1 + \$3.89B_2 \quad (10.2) \\ \text{subject to} & \\ 0.12S &\leq 780 \text{ (yards of silk)}, \\ 0.09P + 0.06B_1 + 0.03B_2 &\leq 2,800 \text{ (yards of polyester)}, \\ 0.06B_1 + 0.07B_2 &\leq 1,600 \text{ (yards of cotton)}, \\ 6,000 \leq S &\leq 7,000 \text{ (min. and max. of all silk)}, \\ 11,500 \leq P &\leq 15,000 \text{ (min. and max. of all polyester)}, \\ 13,000 \leq B_1 &\leq 17,000 \text{ (min. and max. of blend 1)}, \\ 6,500 \leq B_2 &\leq 8,800 \text{ (min. and max. of blend 2)}, \\ S, P, B_1, B_2 &\geq 0 \text{ (non-negativity)}. \end{aligned}$$

In implementing the model in Excel, we have split the objective function into three components: a revenue components, a labor cost component and a material cost component. The Excel layout and Solver entries for this model are shown in Figure 10.3. Cell F5 defines the revenue component, cells F6 and F7 define the labor cost and material cost components. Cell F8 is the difference between cell F5 and the sum of cells F6 and F7. Figure 10.3 shows that the optimal solution is to produce 6,500 all-silk skirts, 15,000 all-polyester skirts, 16,400 poly-cotton blend 1 skirts, and 8,800 poly-cotton blend 2 skirts. This results in a total revenue of \$250,770 and a net profit of \$156,788.

10.4.1.2 Make-Or-Buy Decision Problem In make-or-buy decision problems, a company manager could satisfy the demand for a product by making some of it in-house and by outsourcing the remainder to another company. For each product, the manager needs to determine how much of the product to make in-house and how much of it to outsource to another firm. Let's continue the product-mix problem in Section 10.4.1.1. Now the Prada manager would like

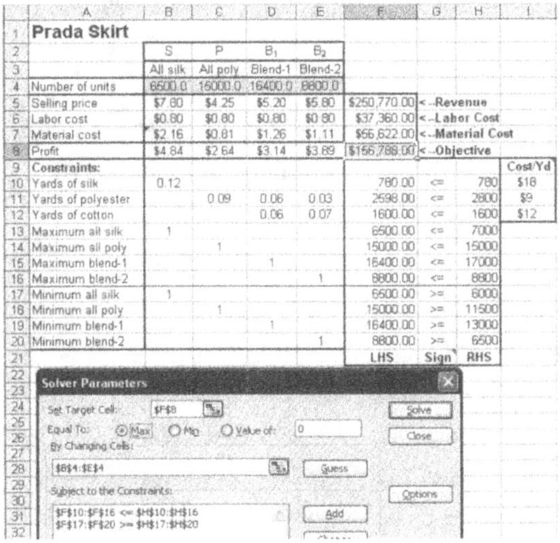

Figure 10.3 Excel layout and solver entries for Prada skirt.

Table 10.3 Outsourcing price from Skirt unlimited

Variety of skirt	All Silk	All Polyester	Blend 1	Blend 2
Outsourcing price	\$4.20	\$1.60	\$2.10	\$2.20

to consider the make-or-buy decision, that is, satisfy the demand by making some of it in house and by outsourcing the remainder to another corporation, named Skirt Unlimited. Skirt Unlimited would sell skirts to Prada with the price information in Table 10.3.

To build a make-or-buy decision model, in addition to the variables defined in Section 10.4.1.1, we should define some additional variables as below.

$$\begin{aligned} S_o &= \text{number of all-silk skirts to buy per month} \\ P_o &= \text{number of polyester skirts to buy per month} \\ B_{1o} &= \text{number of blend 1 poly-cotton skirts to buy per month} \\ B_{2o} &= \text{number of blend 2 poly-cotton skirts to buy per month} \end{aligned}$$

Then the objective function is to maximize profit=revenue-labor cost-material cost-outsourcing cost, in which revenue $= \$7.80(S + S_o) + \$4.25(P + P_o) + \$5.20(B_1 + B_{1o}) + \$5.80(B_2 + B_{2o})$, labor cost $= \$0.80(S + P + B_1 + B_2)$, material cost $= \$2.16S + \$0.81P + \$1.26B_1 + \$1.11B_2$, and Outsourcing cost =

$\$4.20S_o + \$1.60P_o + \$2.10B_{1o} + \$2.20B_{2o}$. Hence the model could be stated as follows.

$$\begin{aligned}
\text{Maximize profit} &= \$4.84S + \$2.64P + \$3.14B_1 + \$3.89B_2 \\
&\quad +\$3.60S_o + \$2.65P_o + \$3.10B_{1o} + \$3.60B_{2o} \\
\text{subject to} & \\
0.12S &\leq 780 \text{ (yards of silk)}, \\
0.09P + 0.06B_1 + 0.03B_2 &\leq 2,800 \text{ (yards of polyester)}, \\
0.06B_1 + 0.07B_2 &\leq 1,600 \text{ (yards of cotton)}, \\
S + S_o &\leq 7,000 \text{ (demand of all silk)}, \\
P + P_o &\leq 15,000 \text{ (demand of all polyester)}, \\
B_1 + B_{1o} &\leq 17,000 \text{ (demand of blend 1)}, \\
B_2 + B_{2o} &\leq 8,800 \text{ (demand of blend 2)}, \\
S, S_o, P, P_o, B_1, B_{1o}, B_2, B_{2o} &\geq 0 \text{ (non-negativity)}.
\end{aligned}$$

The Excel layout and Solver entries for the make-or-buy decision model is shown in Figure 10.4. As we can see, the optimal solution is to produce 6,500 all-silk skirts, 16,400 poly-cotton blend 1 skirts, and 8,800 poly-cotton blend 2 skirts; buy 500 all-silk skirts, 15,000 all-polyester skirts, and 600 poly-cotton blend 1 skirts, respectively. This results in total net profit of $160,598, which is much more than the optimal net profit ($156,788) in the product-mix problem, since the company is now able to satisfy more of the demand.

10.4.2 Transportation Applications

A transportation or shipping problem focuses on determining the amount of goods or items to be transported from a number of origins to a number of destinations. The objective is usually to minimize total transportation costs or maximize the total value of goods loaded. Constraints usually deals with capacities or supplies at each origin and demand at each destination.

10.4.2.1 Vehicle Loading Problem The vehicle loading problem mainly focuses on deciding which items to load onto a vehicle (e.g., truck, ship, plane) so as to maximize the value of a load shipped from the origin to the destination. In the following example, we consider a vehicle loading problem for a shipment by DHL from its warehouse to Asia Airfreight Terminal in Hong Kong. One of its trucks, with a weight capacity of 16,000 pounds and a volume capacity of 1,400 cubic feet, is ready to be loaded. Awaiting transportation are the items shown in Table 10.4. Each of the six items has an associated total dollar value, available weight, and volume per pound that the item takes up. The objective is to maximize the total value of the items loaded on the truck without exceeding the truck's weight and volume capacities.

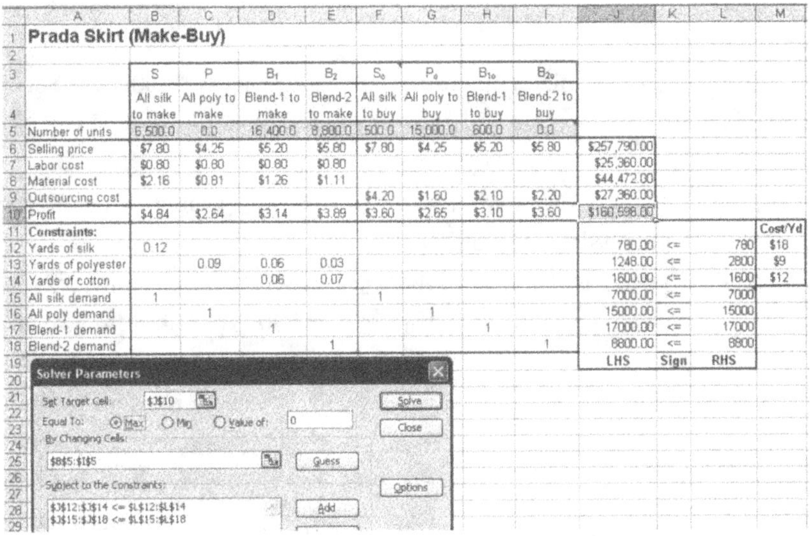

Figure 10.4 Excel layout and solver entries for a make-or-buy decision model.

Table 10.4 Shipments for DHL.

Item	Value	Weight (pounds)	Volume (cubic feet per pound)
1	\$16,000	5,500	0.128
2	\$15,000	5,000	0.068
3	\$11,000	3,500	0.166
4	\$14,825	4,000	0.462
5	\$13,800	4,500	0.046
6	\$9,820	4,000	0.022

The decision variables in this problem are defined as the number of pounds of each item that should be loaded on the truck. There are six decision variables (one for each item) in the model. In this problem, the dollar value of each item should be scaled for use in the objective function. For instance, if the total value of the 5,500 pounds of item 1 is \$16,000, then the value per pound equals \$16,000/5,500=\$2.91. Similarly, the values per pound of items 2 through 6 are \$3.00, \$3.14, \$3.71, \$3.07, and \$2.46, respectively.

Let W_i be the weight in pounds of each item i loaded on the truck. The LP model could be formulated as below.

$$
\begin{aligned}
\text{Maximize load value} &= \$2.91W_1 + \$3.00W_2 + \$3.14W_3 \\
&\quad +\$3.71W_4 + \$3.07W_5 + \$2.46W_6 \\
\text{subject to} & \\
W_1 + W_2 + W_3 + W_4 + W_5 + W_6 &\leq 16,000 \text{ (weight limit of truck)}, \\
0.128W_1 + 0.068W_2 + 0.166W_3 + & \\
0.462W_4 + 0.046W_5 + 0.022W_6 &\leq 1,400 \text{ (volume limit of truck)}, \\
W_1 &\leq 5,500 \text{ (volume item 1 availability)}, \\
W_2 &\leq 5,000 \text{ (volume item 2 availability)}, \\
W_3 &\leq 3,500 \text{ (volume item 3 availability)}, \\
W_4 &\leq 4,000 \text{ (volume item 4 availability)}, \\
W_5 &\leq 4,500 \text{ (volume item 5 availability)}, \\
W_6 &\leq 4,000 \text{ (volume item 6 availability)}, \\
W_1, W_2, W_3, W_4, W_5, W_6 &\geq 0 \text{ (non-negativity)}.
\end{aligned}
$$

According to the Excel layout and solver entries in Figure 10.5, the optimal solution yields a total value of \$48,047.48, and requires DHL to ship 1,943.40 pounds of item 1, 5,000 pounds of item 2, 3,500 pounds of item 3, 4,500 pounds of item 5 and 1,056.60 pounds of item 6. The truck is fully loaded from both weight and volume perspectives. It is interesting to note that, the only item that is not included for loading is item 4, which has the highest dollar value per pound. However, its relative high volume makes it unattractive as cargo.

10.4.2.2 Sensitivity Analysis of The DHL Vehicle Loading Problem As we know, the above LP model of the DHL vehicle loading problem is solved under deterministic assumptions. That is, the load value and weight limit of each item are fixed, and the weight and volume limits of the truck also remain unchanged. DHL managers may be interested in studying the impact of changes in these values. In this case, we can make use of Sensitivity Report from Solver to analyze how sensitive the optimal solution is to changes in the input parameters, and determine a range of values within which the current optimal solution will remain optimal. The Sensitivity Report of the DHL vehicle loading problem is presented in Figure 10.6. As we can see, Sensitivity Report has two distinct parts: a part titled *Adjustable Cells* and a part titled *Constraints.* These parts permit us to answer several what-if questions regarding the problem solution.

Sensitivity analysis focuses on the changes in an objective function coefficient (OFC) and changes in a right-hand side (RHS) value of a constraint. Here are some properties of changes in OFC and RHS.

One Change in OFC. An OFC change has no effect on the feasible region. There is a range for each OFC where the current optimal solution re-

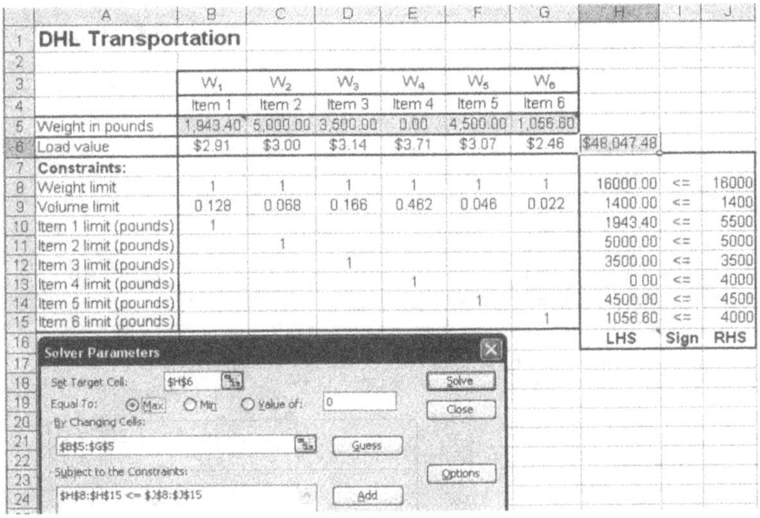

Figure 10.5 Excel layout and solver entries for DHL transportation.

mains optimal. If the OFC change is beyond the allowable range, then a new solution will become optimal.

One Change in RHS. Constraints are classified into binding constraints (those with LHS=RHS at the optimal solution) and nonbinding constraints. For a binding constraint, the impact of one change in RHS is related to the *Shadow Price* (i.e., the change in the objective function value per one-unit increase in the RHS of the constraint). For each nonbinding constraint, there is an allowable range where the corresponding shadow price remains unchanged if its RHS change is within the range. In this case, the change of objective value equals the corresponding shadow price times the RHS change. However, the optimal solution must vary. (2) There is an allowable range for each nonbinding constraint where the optimal solution and objective value remain unchanged if its RHS change is within the range.

Simultaneous Changes in OFCs or RHS. The current information in Sensitivity Report is still valid if $\sum$(change/allowable change) ≤ 1.

In the following we use the above Sensitivity Report of the DHL vehicle loading problem (Figure 10.6) to analyze some possible what-if cases.
Case 1. What is the impact on the optimal solution if the load value of item 6 increases by \$0.15?
A: According to Sensitivity Report, the increase \$0.15 of item 6 is within the allowable increase \$0.20, so the current optimal solution remains optimal.

Microsoft Excel 11.0 Sensitivity Report
Worksheet: [DHL Transportation.xls]3-7A
Report Created: 2011-12-31 15:37:19

Adjustable Cells

Cell	Name	Final Value	Reduced Cost	Objective Coefficient	Allowable Increase	Allowable Decrease
B5	Weight in pounds Item 1	1,943.40	0.00	2.909090909	0.052248377	0.152653409
C5	Weight in pounds Item 2	5,000.00	0.00	3	1E+30	0.347941681
D5	Weight in pounds Item 3	3,500.00	0.00	3.142857143	1E+30	0.070978927
E5	Weight in pounds Item 4	0.00	-0.63	3.70625	0.63365566	1E+30
F5	Weight in pounds Item 5	4,500.00	0.00	3.066666667	1E+30	0.508853631
G5	Weight in pounds Item 6	1,056.60	0.00	2.455	0.201100299	0.197993848

Constraints

Cell	Name	Final Value	Shadow Price	Constraint R.H. Side	Allowable Increase	Allowable Decrease
H8	Weight limit	16000.00	2.36	16000	2437.5	875
H9	Volume limit	1400.00	4.28	1400	112	206
H10	Item 1 limit (pounds)	1943.40	0.00	5500	1E+30	3556.603774
H11	Item 2 limit (pounds)	5000.00	0.35	5000	1866.666667	5000
H12	Item 3 limit (pounds)	3500.00	0.07	3500	1430.555556	2618.055556
H13	Item 4 limit (pounds)	0.00	0.00	4000	1E+30	4000
H14	Item 5 limit (pounds)	4500.00	0.51	4500	1365.853659	3804.878049
H15	Item 6 limit (pounds)	1056.60	0.00	4000	1E+30	2943.396226

Figure 10.6 Sensitivity report of DHL transportation.

Case 2. What is the impact on the objective value if the volume limit of the truck decreases by 180 cubic feet?
A: As we see, the constraint of volume limit is binding. And a decrease of 180 in the truck volume is within the allowable decrease 206. Hence the shadow price \$4.28 is valid, and the objective value will decrease by $\$4.28 \times 180 = \$770.4$.
Case 3. What is the impact on the objective and the optimal solution if the upper limit of item 1 decrease by 3000?
A: According to Sensitivity Report, the upper limit of item 1 is non-binding, and a decrease of 3000 is also within the allowable decrease 3556.6. Thus this change has no impact on either the objective or the optimal solution.
Case 4. What is the impact on the objective if we simultaneously decrease the load value of item 1 and item 6 by \$0.08 and \$0.12, respectively?
A: Since the allowable decreases of item 1 and item 6 are 0.153 and 0.197, respectively, we have $0.08/0.153 + 0.12/0.19 > 1$, which suggests that the optimal objective would vary.

10.4.2.3 Allocation Problem In the example of Section 10.4.2.1, DHL has only one truck and needs to load all the items onto the same truck. However, in the reality, DHL usually has many different trucks in its warehouse. Let us consider the case of 2 trucks where DHL has the option of replacing its single truck (with a weight capacity of 16,000 pounds and a volume capacity of 1,400 cubic feet) with two smaller trucks (each with a weight capacity of 11,000 pounds and a volume capacity of 950 cubic feet). We still use the data in Table 10.4. If DHL uses two trucks, DHL requires that they are loaded with the same total weight. However, total volumes in the two trucks could be different. If the fixed cost of operating the two smaller trucks is $6,000 more than the current cost of operating just a single truck, which alternative should be chosen? In this problem, DHL has to decide how to allocate the six items between the two trucks. Note that it is possible for the total quantity of an item to be split between the two trucks.

To formulate this problem, the decision variables need to specify how much of each item should be loaded on each truck. Let the double-subscripted variable W_{i1} represent the weight of the ith item on the first truck, and W_{i2} the weight of the ith item on the second truck. Then the LP model can be stated in the following.

$$
\begin{aligned}
\text{Maximize load value} \quad &= \quad \$2.91(W_{11}+W_{12})+\$3.00(W_{21}\\
&\quad\; +W_{22})+\$3.14(W_{31}+W_{32})\\
&\quad\; +\$3.71(W_{41}+W_{42})+\$3.07(W_{51}\\
&\quad\; +W_{52})+\$2.46(W_{61}+W_{52})\\
\text{subject to} \quad &\\
W_{11}+W_{21}+W_{31}+W_{41}+W_{51}+W_{61} \quad &\le \quad 11,000 \text{ (weight limit of truck 1)},\\
0.128W_{11}+0.068W_{21}+0.166W_{31}+ \quad &\\
0.462W_{41}+0.046W_{51}+0.022W_{61} \quad &\le \quad 950 \text{ (volume limit of truck 1)},\\
W_{12}+W_{22}+W_{32}+W_{42}+W_{52}+W_{62} \quad &\le \quad 11,000 \text{ (weight limit of truck 1)},\\
0.128W_{12}+0.068W_{22}+0.166W_{32}+ \quad &\\
0.462W_{42}+0.046W_{52}+0.022W_{62} \quad &\le \quad 950 \text{ (volume limit of truck 1)},\\
W_{11}+W_{12} \quad &\le \quad 5,500 \text{ (availability of item 1)},\\
W_{21}+W_{22} \quad &\le \quad 5,000 \text{ (availability of item 2)},\\
W_{31}+W_{32} \quad &\le \quad 3,500 \text{ (availability of item 3)},\\
W_{41}+W_{42} \quad &\le \quad 4,000 \text{ (availability of item 4)},\\
W_{51}+W_{52} \quad &\le \quad 4,500 \text{ (availability of item 5)},\\
W_{61}+W_{62} \quad &\le \quad 4,000 \text{ (availability of item 6)},\\
W_{11}+W_{21}+W_{31}+W_{41}+W_{51}+W_{61} \quad &=\\
W_{12}+W_{22}+W_{32}+W_{42}+W_{52}+W_{62} \quad & \quad\quad \text{(same weight in both trucks)},\\
W_{11},W_{21},W_{31},W_{41},W_{51},W_{61}, \quad &\\
W_{12},W_{22},W_{32},W_{42},W_{52},W_{62}, \quad &\ge \quad 0 \text{ (non-negativity)}.
\end{aligned}
$$

Figure 10.7 shows the Excel layout and solver entries for DHL's allocation problem. For the constraint that makes sure that the same total weight is

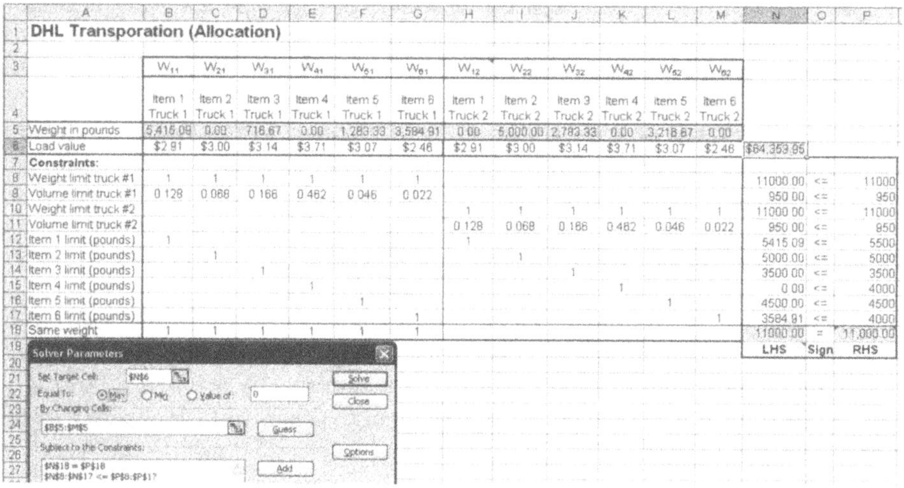

Figure 10.7 Excel layout and solver entries for allocation problem.

loaded on both trucks, the Excel layout includes formulas for both the LHS (cell N18) and RHS (cell P18) entries. Cell N18: =sum(B5:G5), and cell P18: =sum(H5:M5).

According to Figure 10.7, the optimal solution to DHL Transportation's allocation problem yields a total value of $63,353.95. This is $16306.5 more than the load value of $48047.48 that is achievable with just a single truck. Since this increase is more than the increased $6,000 operating cost, DHL should replace its single truck with the two smaller ones. From both weight and volume perspectives, the two trucks are fully loaded. Furthermore, all available quantities of items 2, 3 and 5 are loaded. Most of items 1 and 6 are loaded, while none of item 4 is loaded.

10.4.3 Assignment Applications

Assignment problems focus on determining the most efficient assignment of people to jobs, machines to tasks, police cars to city sectors, salespeople to territories, and so on. The assignments are made on a one-to-one basis. For instance, in a person-to-job assignment problem, each person is assigned to exactly one job, and each job is assigned to exactly one person. Fractional assignments are not allowed. The objective could be to minimize the total cost of the assignments or maximize the total effectiveness or benefit of the assignments.

10.4.3.1 Labor Planning Problem with LP Labor planning problems focus on satisfying staffing needs over a specific planning horizon such as a day, week, month, or year. The ability to solve labor planning problems is particulary useful when staffing needs are different during the different time periods in the planning horizon. In addition, managers have some flexibility in assigning workers to jobs that require overlapping or interchangeable talents.

The main branch of HSBC (Hong Kong and Shanghai Banking Corporation) in Hong Kong is a busy bank that has requirements for between 9 and 19 tellers, depending on the time of day. The afternoon period, from noon to 3 p.m. is usually the busiest. Table 10.5 shows the numbers of workers needed at various time periods. Now HSBC employs 10 full-time tellers and several part-time staffs. A part-time employee must put in exactly 4 hours per day but can start anytime between 9 a.m. and 1 p.m. Part-time tellers are paid relatively little without retirement and lunch benefits. Full-time tellers, work from 9 a.m. to 5 p.m., but are allowed one hour for lunch. Half of the full-time tellers take their lunch break at 11 a.m., and the other at noon. Each full-time teller provides 35 hours per week of productive labor time. According to HSBC policy, the bank limits part-time hours to a maximum of 50% of the day's total requirement. Part-time tellers earn \$35 per day on average, and full-time tellers earn \$100 per day in salary and benefits, on average. Now HSBC would like to determine a schedule that would minimize its total salary costs. It is allowed to release one or more of its full-time employees if it is cost-effective to do so.

In this labor planning problem, we should determine how many employees need to start their work at the different starting times allowed. For instance, in the case of HSBC, we have full-time tellers who all start at 9 a.m., and part-time tellers who can start at anytime between 9 a.m. and 1 p.m.. Define

$$\begin{aligned}
F &= \text{number of full-time tellers to use (all start at 9 a.m.)}\\
P_1 &= \text{number of part-time tellers to start at 9 a.m. and leave at 1 p.m.}\\
P_2 &= \text{number of part-time tellers to start at 10 a.m. and leave at 2 p.m.}\\
P_3 &= \text{number of part-time tellers to start at 11 a.m. and leave at 3 p.m.}\\
P_4 &= \text{number of part-time tellers to start at noon and leave at 4 p.m.}\\
P_5 &= \text{number of part-time tellers to start at 1 p.m. and leave at 5 p.m.}
\end{aligned}$$

Then the objective function could be stated as

$$\text{Minimize total daily personnel cost} = \$100F + \$35(P_1 + P_2 + P_3 + P_4 + P_5).$$

In the following we investigate the constraints. For each hour, the available number of tellers must be no less than the required number. It is simple to count how many different tellers are working during each time period. On the other hand, we should notice that half of the full-time tellers break for lunch between 11a.m. and noon, and the other half break between noon and 1p.m..

Table 10.5 Tellers required for HSBC.

Time Period	Number Required
9 a.m.–10 a.m.	9
10 a.m.–11 a.m.	11
11 a.m.–Noon	15
Noon–1 p.m.	17
1 p.m.–2 p.m.	19
2 p.m.–3 p.m.	18
3 p.m.–4 p.m.	16
4 p.m.–5 p.m.	12

Hence we could present the constraints as below.

$$\begin{aligned}
F + P_1 &\geq 9 \text{ (9 a.m.–10 a.m.)},\\
F + P_1 + P_2 &\geq 11 \text{ (10 a.m.–11 a.m.)},\\
1/2F + P_1 + P_2 + P_3 &\geq 13 \text{ (11 a.m.–noon)},\\
1/2F + P_1 + P_2 + P_3 + P_4 &\geq 17 \text{ (noon–1 p.m.)},\\
F + P_2 + P_3 + P_4 + P_5 &\geq 19 \text{ (1 p.m.–2 p.m.)},\\
F + P_3 + P_4 + P_5 &\geq 18 \text{ (2 p.m.–3 p.m.)},\\
F + P_4 + P_5 &\geq 16 \text{ (3 p.m.–4 p.m.)},\\
F + P_5 &\geq 12 \text{ (4 p.m.-5 p.m.)}.
\end{aligned}$$

Furthermore, at most 10 full-time tellers can be available, so $F \leq 10$. In addition, part-time working hours cannot exceed 50% of the total hours required each day, which is the sum of the tellers needed during each hour. That is, $4(P_1+P_2+P_3+P_4+P_5) \leq 0.5(9+11+13+17+19+18+16+12) = 57.5$. Also, remember to add the constraint of nonnegativity, $F, P_1, P_2, P_3, P_4, P_5 \geq 0$. Excel layout and solver entries for this model are presented in Figure 10.8.

According to Figure 10.8, the optimal solution has fractional values, which is possible because this model is an LP model. Because only integer solutions can be implemented, a possible recourse at this stage is to round off the fractional values to the nearest integers. In this case, the nearest integer solution is to employ 10 full-time tellers, 5 part-time tellers at 9 a.m., 1 part-time teller at 10 a.m., 2 part-time tellers at 11 a.m., 4 part-time tellers at noon, and 2 part-time tellers at 1 p.m. for a total cost of $1,490 per day,

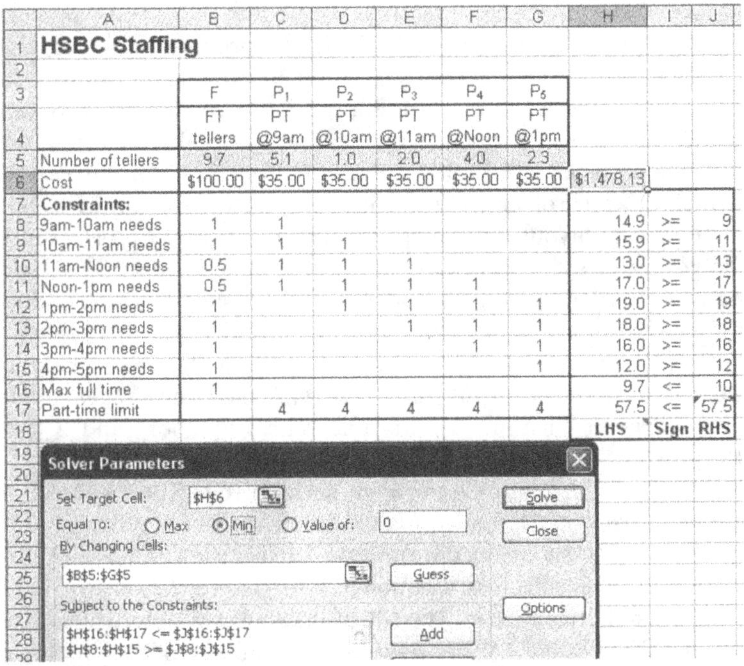

	A	B	C	D	E	F	G	H	I	J
1	**HSBC Staffing**									
2										
3		F	P1	P2	P3	P4	P5			
4		FT tellers	PT @9am	PT @10am	PT @11am	PT @Noon	PT @1pm			
5	Number of tellers	9.7	5.1	1.0	2.0	4.0	2.3			
6	Cost	$100.00	$35.00	$35.00	$35.00	$35.00	$35.00	$1,478.13		
7	**Constraints:**									
8	9am-10am needs	1	1					14.9	>=	9
9	10am-11am needs	1	1	1				15.9	>=	11
10	11am-Noon needs	0.5	1	1	1			13.0	>=	13
11	Noon-1pm needs	0.5	1	1	1	1		17.0	>=	17
12	1pm-2pm needs	1		1	1	1	1	19.0	>=	19
13	2pm-3pm needs	1			1	1	1	18.0	>=	18
14	3pm-4pm needs	1				1	1	16.0	>=	16
15	4pm-5pm needs	1					1	12.0	>=	12
16	Max full time	1						9.7	<=	10
17	Part-time limit		4	4	4	4	4	57.5	<=	57.5
18								**LHS**	**Sign**	**RHS**

Figure 10.8 Excel layout and solver entries for HSBC staffing with LP.

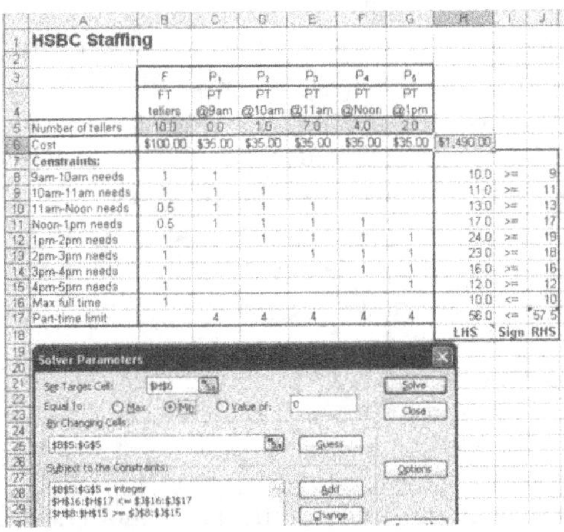

	A	B	C	D	E	F	G	H	I	J
1	**HSBC Staffing**									
2										
3		F	P1	P2	P3	P4	P5			
4		FT tellers	PT @9am	PT @10am	PT @11am	PT @Noon	PT @1pm			
5	Number of tellers	10.0	0.0	1.0	7.0	4.0	2.0			
6	Cost	$100.00	$35.00	$35.00	$35.00	$35.00	$35.00	$1,490.00		
7	**Constraints:**									
8	9am-10am needs	1	1					10.0	>=	9
9	10am-11am needs	1	1	1				11.0	>=	11
10	11am-Noon needs	0.5	1	1	1			13.0	>=	13
11	Noon-1pm needs	0.5	1	1	1	1		17.0	>=	17
12	1pm-2pm needs	1		1	1	1	1	24.0	>=	19
13	2pm-3pm needs	1			1	1	1	23.0	>=	18
14	3pm-4pm needs	1				1	1	16.0	>=	16
15	4pm-5pm needs	1					1	12.0	>=	12
16	Max full time	1						10.0	<=	10
17	Part-time limit		4	4	4	4	4	56.0	<=	57.5
18								**LHS**	**Sign**	**RHS**

Figure 10.9 Excel layout and solver entries for HSBC staffing with IP.

which is slightly more than the optimal cost $1,478.13 that is obtained in the LP model.

10.4.3.2 Labor Planning Problem with IP In the previous example, a natural question will arise: is the round-off solution optimal with respect to integer solutions? To answer this question, we solve an integer programming (IP) problem for the same model, that is, the same objective function and constraints but with the additional constraint that all decision variables be integers. In the Excel/Solver software, it is easy to convert an LP model into an IP model. Excel layout and solver entries for IP are presented in Figure 10.9. In Solver Entries, we use the *Add* option to include an integer constraint. According to Figure 10.9, we find that the optimal cost is also $1,490, which indicates that the round-off solution that was found earlier is optimal. However, Figure 10.9 shows a different optimal solution (having the same same cost), which is to arrange 10 full-time tellers, 1 part-time teller at 10 a.m., 7 part-time tellers at 11 a.m., 4 part-time tellers at noon, and 2 part-time tellers at 1 p.m.

Comparing the labor planning problem with IP to the one with LP, we observe that the integer restriction results in an objective function value ($1,490 of cost) that is no better than the optimal LP solution ($1,478 of cost). It is easy to understand why this inequality is a general feature of IP vs. LP models. The feasible region (i.e., the set of decision variables that satisfy the constraints) of the original LP problem includes all the points in the feasible region of the IP problem, but not vice versa: fractional points in the LP feasible region are not feasible in the IP problem. The LP problem corresponding to the IP problem is called the relaxed problem. At best, the two solutions could be equal when the optimal LP solution is integer-valued.

The HSBC staffing problem focuses on the labor planning during the different time periods of a single day. In fact, in many other labor planning problems, the planning horizon may consist of a week, a month, or even a year, which correspond to operational, tactical, and strategic decisions from the hierarchical perspective. In addition, the time period may include specific shifts. For instance, if there are two shifts per day (day shift and night shift), then there are total 14 time periods in the problem. Worker requirements need to be specified for each of these 14 time periods. In this case, the work schedules need to specify the exact days and shifts. In addition, in many real-world problems, they typically have hundreds or even thousands of decision variables.

10.5 SUMMARY

In this chapter we firstly introduced the decision model, including its origin, definition, data types and steps involved in it. We then briefly described linear programming models and more general programming models, the definition and activities of supply chain management. Finally, we used decision modeling to deal with typical problems in supply chain management, such as product mix problems, make-or-buy problems, vehicle loading problems, allocation problems, and labor planning problems.

EXERCISES

10.1 A furniture manufacturer produces two different types of china cabinets: a European model and a Chinese model. Each cabinet produced must go through three departments: carpentry, painting, and finishing. Table 10.6 contains all relevant information concerning production times (hours per cabinet), production capacities for each operation per day, and revenue ($ per unit). The firm has a contract with a distributor to produce a minimum of 58 cabinets of each type per day. The manufacturer would like to determine the product mix that maximizes the daily revenue. Formulate the problems in an LP model and solve it using Excel. (Answer: $3,428.)

Table 10.6 Information for Exercise 10.1.

Type	Carpentry	Painting	Finishing	Revenue
European	2.80	1.4	0.70	30
Chinese	2.60	1.2	0.70	26
Capacity(hours)	330	220	130	

10.2 A manufacturer company produces three different types of bicycles: B1, B2 and B3. It has a fixed order from another distributor for 2,100 B1, 3,820 B2 and 1,820 B3 bicycles. Between now and when the order is due to delivered, it has 16,800 fabrication hours and 1,800 inspection hours, which are not enough to manufacture the total quantity ordered. The time required in each department by the various bicycles are shown in Table 10.7. Also shown are the costs to manufacture the bicycles in house and the costs to outsource them. For labeling considerations, the company wants to manufacture in-house at least 65% of each type of bicycle that will be shipped to the distributor. How many bicycles of each type should be made in-house and how many should be outsourced? (Hint: fractional solutions are not allowed.) What will be the total cost to fill the distributor's order? (Answer: $153,797.)

10.3 A tramp freighter's cargo officer wants to determine the mix of cargo to be carried on the next trip. The ship's volume limit for cargo is 110,000 cubic meters, and its weight capacity is 2,450 tons. The cargo officer has five different types of cargo from which to select and wishes to maximize the value of the selected shipment. However, to make sure that none of the customers

Table 10.7 Time required and costs for Exercise 10.2.

Type	Fabrication Hours	Inspection Hours	In-house cost	Outsource Cost
B1	2.4	0.26	$16.5	$20.40
B2	3.3	0.32	$19.0	$20.85
B3	3.9	0.48	$22.5	$24.76

Table 10.8 Specifications of five cargoes for Exercise 10.3.

Cargo Type	Tons Available	Value per Ton	Volume per Ton (CU.M.)
A	980	$1,400	28
B	880	$1,780	56
C	1,980	$1,280	32
D	2,300	$920	48
E	3,680	$1,380	38

Table 10.9 Min. number of workers for Exercise 10.4.

Period	Time	Workers Required
1	3 a.m.-7 a.m.	4
2	7 a.m.-11 a.m.	12
3	11 a.m.-3 p.m.	17
4	3 p.m.-7 p.m.	10
5	7 p.m.-11 p.m.	14
6	11 p.m.-3 a.m.	5

are ignored, the officer would like to make sure that at least 22% of each cargo's available weight is selected. The specifications for the five cargoes are shown in Table 10.8. What mix of cargo should the load master carry on the next trip? What is the optimal shipment's value? (Answer: $3,302,272.)

10.4 An Italian restaurant is open 24 hours a day. Waiters and busboys report for duty at 3 a.m., 7 a.m., 11 a.m., 3 p.m., 7 p.m., or 11 p.m., and each works an 8-hour shift. Table 10.9 shows the minimum number of workers needed during the six periods into which the day is divided. How should the restaurant schedule its workers so that the total staff (number of workers) required for one day's operation is minimized? (Answer: 35.)

REFERENCES

1. Balakrishnan, N., Render, B. and Stair, Jr. R. M., *Managerial Decision Modeling with Spreadsheets*, Pearson Education, New Jersey (2007).
2. Dantzig, G.B., *Linear Programming and Extensions*, Princeton University Press, New Jersey (1963).
3. Peterso,n R., and Silver, E., *Decision Systems for Inventory Management and Production Planning*, John Wiley & Sons, New York (1985).
4. Shapiro, J.F., *Modeling the Supply Chain*, Duxbury Press, Duxbury: MA (2007).
5. Simchi-Levi, D., Kaminsky, P., and Simchi-Levi, E., *Designing and Managing the Supply Chain: Concepts, Strategies and Case Studies*, McGraw-Hill, New York (2010).
6. Tayur, S., Ganeshar, R., and Magazine, M., *Quantitative Models for Supply Chain Management*, Kluwer, Boston (1999).
7. Winston, W.L., *Introduction to Mathematical Programming*, Duxbury Press, Duxbury: MA (1995).
8. Winston, W.L., and Albright, S.C., *Practical Modeling and Applications: Spreadsheet Modeling and Applications*, Duxbury Press, Duxbury: MA (1997).
9. Westphal, C., and Blaxton, T., *Data Mining Solution: Methods and Tools for Solving Real-World Problems*, John Wiley & Sons, New York (1998).

CHAPTER 11

MODELING TEMPERATURE FOR PRICING WEATHER DERIVATIVES

FRED ESPEN BENTH

Center of Mathematics for Applications, University of Oslo, Blindern, Oslo, Norway

11.1 INTRODUCTION

In this chapter we will analyze stochastic models for the dynamics of the surface air temperature in a given location. Our motivation comes from the financial market for so-called weather derivatives, where one can buy and sell weather. The Chicago Mercantile Exchange has for some years organized a trade in financial contracts where one can earn (or lose...) money on weather events. The exchange offers contracts on temperature, snowfall and hurricanes. We shall focus on the temperature contracts here.

The most popular temperature derivatives are the futures contracts. A futures contract on temperature can be thought of as fixing the temperature financially, in the sense that one is securing a fixed temperature over a period rather than a varying one as observed in nature. In fact, such a contract

Mathematical Modeling with Multidisciplinary Applications.
Edited by Xin-She Yang

gives the owner money every time the temperature goes above a threshold, while the owner must pay when the observed temperature is below. The contracts are not specified on temperature directly, but on a temperature index. The indices are *cooling degree days* (CDD), *heating degree days* (HDD) and *cumulative average temperature* (CAT). These indices are measured for various cities worldwide, including several US, Canadian, European and Asian cities.

A CDD index is essentially providing a measure for the demand of air-condition cooling. In a pre-defined period of time, the index is computed as the aggregated temperatures above a threshold, defined to be 18° C in the market. Mathematically, one writes

$$\mathrm{CDD}(\tau_1, \tau_2) = \sum_{t=\tau_1}^{\tau_2} \max(T(t) - 18, 0), \tag{11.1}$$

where τ_1 and τ_2 are, resp., the start and end of the *measurement period*, and t ranges over the days in this period. Moreover, $T(t)$ is the daily average temperature, defined as the average of the maximum and minimum temperature of day t. The HDD index measures the aggregated temperatures when heating is required, and is analogously defined as

$$\mathrm{HDD}(\tau_1, \tau_2) = \sum_{t=\tau_1}^{\tau_2} \max(18 - T(t), 0). \tag{11.2}$$

Most futures are written on these two indices. The CAT index is used for temperatures measured in European cities in the summer period, and is defined as

$$\mathrm{CAT}(\tau_1, \tau_2) = \sum_{t=\tau_1}^{\tau_2} T(t)\,. \tag{11.3}$$

For example, we can buy a CAT index futures measured in the city of Stockholm, Sweden, for the period of July. Then we will aggregate the daily average temperatures of July, and get that amount paid in cash at the end of July. In return we have to pay the fixed (futures) price $F(t, \text{July1}, \text{July31})$ agreed at the entry of the contract at time $t \leq \text{July1}$.

The typical measurement periods for temperature futures are the months in the year, but one can also trade in futures on indices measured over a season (consisting of two or more months).

The temperature market offers a financial tool to hedge weather risk. For example, a holiday resort may suffer large losses in case of unfavorable weather in their high season. Thinking about a summer holiday resort in the Mediterranean, it would face large losses if the summer turns out to be too cold. They could use temperature contracts to insure themselves against this risk. For example, buying an HDD futures measured over their high season would result in an income in case of low temperatures, and an expense when temperatures are high. But high temperatures would on the other hand result in profits from tourism, so this expense could be viewed as an insurance premium.

The main actors in this market is naturally the energy sector. Producers and retailers of energy, say, face large volume risk impacted from weather changes. For example, a producer of electricity knows that mild temperatures in the summer, or warm winters lead to low demand for power, and thus reduces the income of the producers. They have a natural need for temperature hedging tools.

The Chicago Mercantile Exchange also offers call and put options written on the various futures. For example, a call option on the July CAT futures with *strike price* K and exercise time $\tau \leq \text{July1}$, will pay the owner

$$\max\left(F(\tau, \text{July1}, \text{July31}) - K, 0\right) . \tag{11.4}$$

In fact, such an option gives the owner the right to enter a CAT contract with a futures price K rather than the market futures price at time τ. This right can be abandoned if it is not favorable to the owner of the option.

In this chapter we will introduce a class of stochastic processes which are suitable for modeling the temperature dynamics in time at a given location. A stochastic process is a family of random variables indexed over time. We will back up our temperature model with an empirical example. We next apply probability theory to derive CAT futures prices and to valuate options. When trading in such instruments in the market, one must have a clear knowledge on the relationship between prices and temperature evolution. We provide a framework for this.

11.2 STOCHASTIC TEMPERATURE MODELING

The surface air temperature at a given geographical location evolves dynamically in time according to complex physical laws for weather. The observed temporal variations in temperature are governed by wind, cloud density, sun, exchange of heat with the sea, topology of the location, etc., and can be modeled by highly complex partial differential equations. In Figure 11.1 we have plotted the daily average temperatures observed in Stockholm, Sweden ranging from May 25, 1996 until May 24, 2006. The daily average temperature is computed as the mean of the maximum and minimum temperature observed over the day. We observe the gray curve varying apparently randomly around a seasonal mean, which is plotted as a black curve in the figure.

We will in this chapter apply a so-called *stochastic reduced-form* approach in modeling the temperature evolution at a specific location, based on stochastic processes. From the temperature series plotted in Figure 11.1, it seems natural to separate the dynamics of temperature into a seasonal *deterministic* component measuring the average temperature, and a stochastic component modeling the “random” fluctuations around this mean. Thus, we assume that

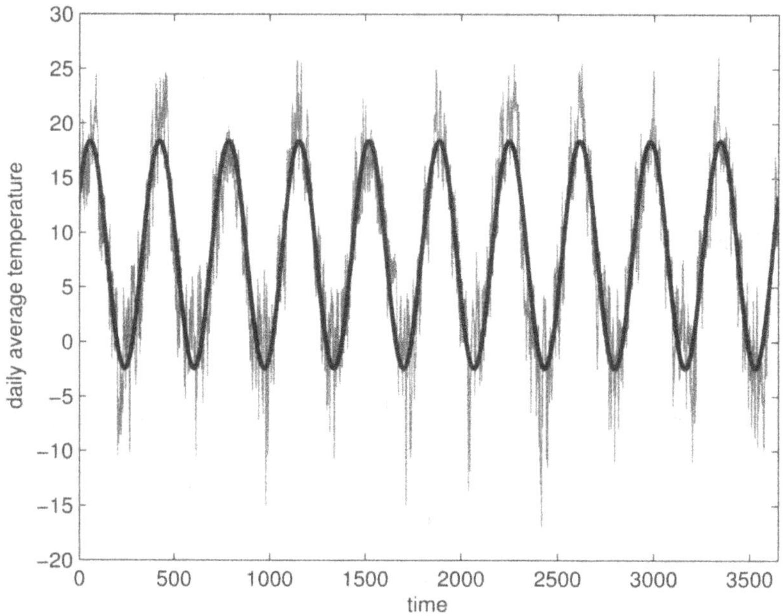

Figure 11.1 Daily average temperatures (in gray) from Stockholm, Sweden, together with the seasonal mean function (in black). Temperatures are ranging from May 25, 1996 until May 24, 2006.

the temperature at time $t \geq 0$ is given by the model

$$T(t) = \Lambda(t) + X(t)\,, \tag{11.5}$$

with Λ being a real-valued continuous and bounded function on $[0, \infty)$. As temperature is evolving continuously in time, it is natural to state its dynamics for all $t \geq 0$, and not only for discrete times $t = 0, 1, 2, 3, \ldots$.

For the specific case of Stockholm data as shown in Figure 11.1, it turned out that a seasonal function of the form

$$\Lambda(t) = a_0 + a_1 t + a_2 \cos\left(\frac{2\pi(t - a_3)}{365}\right)\,, \tag{11.6}$$

was appropriate. Here, a_i for $i = 0, .., 3$ are constants. Such a seasonal function explains, via the trigonometric term, the yearly cycles of cold and warm seasons. The trend $a_0 + a_1 t$ may be interpreted as global warming, or effects from urbanization. One expects an increase in temperature from global warming. Arguably, such an increase is not necessarily linear, but in the short term this provides a good approximation. As buildings and traffic within a city create a local urban climate warmer than outside the city borders, one may experience a temperature increase when looking at a long time series of measurement of temperature at a given location. The measurement station could in the old days be placed outside the city, but over the years the city has grown to include the station as well. The seasonal function in (11.6) seems appropriate for temperatures measured in many locations, but other specifications are of course also possible and relevant. The black curve in Figure 11.1 is the best estimate of (11.6) in the least squares sense on a more than 40 year long data series of daily temperature recordings in Stockholm. We shall come back to this later.

After explaining the seasonality of temperature, we move our attention to the much more difficult task of presenting a model for the stochastic component $X(t)$. From Figure 11.1 it seems that the temperature is randomly fluctuating around its mean, but not drifting too far away. In view of the physical laws of energy conservation, one actually expects the temperature to be pushed back towards its mean value. We present in the next subsection a simple stochastic model with such a feature. This model, frequently referred to as an Ornstein-Uhlenbeck process, could be used for temperature modeling. But as it turns out, a more general class of stochastic processes are more appropriate. We analyze Ornstein-Uhlenbeck processes first in order to set ideas without having to involve too many technicalities.

11.2.1 Simple Stochastic Mean Reverting Processes

A typical model for *mean reversion* is the first-order ordinary differential equation

$$\frac{du(t)}{dt} = -\alpha u(t), \tag{11.7}$$

with an initial value $u(0) = u_0$ and $\alpha > 0$. The solution of this equation can be found to be

$$u(t) = u_0 \exp(-\alpha t). \tag{11.8}$$

Note that by inspecting the differential equation for u, we see that if $u(t) < 0$, then the derivative $du(t)/dt$ becomes positive, and hence the value of $u(t)$ is pushed upwards towards the origin. On the other hand, if $u(t) > 0$, the derivative becomes negative, giving a push down towards the origin. We have a reversion to the "mean value" zero, and any deviation from this will be dampened. From the solution, we get an exponentially decaying function to zero as long as $u_0 > 0$, and the opposite if $u_0 < 0$. Obviously, $u(t)$ has zero as its asymptotic limit.

We find the *half life* of $u(t)$ to be

$$\tau = \frac{\ln 2}{\alpha}, \tag{11.9}$$

being the time it takes before the $u(t)$ is equal to $u_0/2$, that is, the time it takes before a deviation from zero, u_0 is halved in value. This is a measure of how fast the dynamics is reverting to its mean. Note that the ordinary differential equation (11.7) is often used for modeling population growth, then with $\alpha < 0$. In that case, we get a dynamics drifting *away* from zero.

Obviously, from the time series data we have looked at, the dynamics of temperature is far more volatile than the smooth exponential curves implied by the model (11.7). A way to get stochastic fluctuations *and* mean reversion at the same time, is to add noise to the dynamics in (11.7). Heuristically, we write

$$\frac{dX(t)}{dt} = -\alpha X(t)\,dt + \sigma W(t)\,, \tag{11.10}$$

where W is *white noise* and $\sigma > 0$ is a parameter scaling the size of noise, frequently called the *volatility* of temperature.

White noise is often interpreted as the derivative of a *Brownian motion* $B(t)$, that is, $W(t) = dB(t)/dt$. A Brownian motion is a family of random variables parametrized by time, $\{B(t)\}_{t\geq 0}$, with the three properties

1. For each $t > s \geq u > v \geq 0$, $B(t) - B(s)$ is independent of $B(u) - B(v)$.
2. For each $t > s \geq 0$, $B(t) - B(s)$ is normally distributed, with mean zero and variance $t - s$.
3. $B(0) = 0$

By Kolmogorov's extension theorem (see Oksendal [7], Theorem 2.1.5), we are ensured that there exists a probability space $(\Omega, \mathcal{F}, P)$[10] and a family of random variables with the three properties above. Moreover, by Kolmogorov's continuity theorem (see Oksendal [7], Theorem 2.2.3), Brownian motion has continuous paths, that is, for almost every $\omega \in \Omega$, $t \mapsto B(t, \omega)$ is a continuous function. Brownian motion is an example of a *stochastic process*, that is, a parametric family of random variables in time.

Brownian motion can be viewed as a *random walk* in continuous time. Let $Z_i = B(t_i)$ for a uniform partition $\{t_i\}_{i=0,\ldots}$ of the time line. Then, letting $\Delta := t_{i+1} - t_i$, we have from the properties of Brownian motion that $\{Z_i\}_{i=1}^{\infty}$ is a sequence of independent and identically distributed random variables, where $Z_i \sim \mathcal{N}(0, \Delta)$, that is, centered normally distributed with variance Δ.

We note that the random variable

$$\frac{B(t) - B(s)}{t - s}$$

will be normally distributed, with mean zero and variance $1/t - s$, for $t > s \geq 0$. But then, letting $s \to t$, the limit will not exist as the variance explodes.

[10]A probability space is a set Ω equipped with a σ-algebra $\mathcal{F}$ and a probability P, that is, a measure with the property $P(\Omega) = 1$.

This shows that the time derivative of Brownian motion does not exist, and so does not $W(t)$ as we have defined it above. Apparently, the definition of a stochastic dynamics in (11.10) is not meaningful.

This is true in a strict sense, however, we can change our intepretation of (11.10) slightly to get something meaningful. Integrating both sides informally with respect to time, and using $dB(t) = W(t)dt$ gives us

$$X(t) = X(0) - \alpha \int_0^t X(s)\,ds + \sigma B(t)\,. \tag{11.11}$$

Thinking about integration as summation, we have used the natural property that $\int_0^t dB(s) = B(t) - B(0) = B(t)$. As Brownian motion is a meaningful object, we may pose the question whether there exists a stochastic process $X(t)$ solving the *integral equation* (11.11), and if so, if this process is unique or not. Here, $X(0)$ is a given real number being the initial condition of the dynamics.

The integral equation in (11.11) is a special case of a so-called *stochastic differential equation*, and frequently one presents it on a *differential form* as

$$dX(t) = -\alpha X(t)\,dt + \sigma\,dB(t)\,. \tag{11.12}$$

The precise meaning of (11.12) is nothing but (11.11).

Let us investigate the existence and uniqueness of a solution of (11.12). We have the following Lemma:

Proposition 1 *Let $X_i(t)$ for $i = 1, 2$ be two solutions of (11.12) with $X_i(0) = x_i$. If $X_i(t)$ has finite variance, then*

$$\mathbf{E}\left[|X_1(t) - X_2(t)|^2\right]^{1/2} \le |x_1 - x_2| \mathrm{e}^{\alpha t}\,,$$

for every $t \ge 0$.

Proof: It holds from (11.11) that

$$X_1(t) - X_2(t) = x_1 - x_2 - \alpha \int_0^t X_1(s) - X_2(s)\,ds\,.$$

But, by the triangle and Minkowski's inequalities (see Folland [5]), it follows

$$\begin{aligned}\mathbf{E}\left[|X_1(t) - X_2(t)|^2\right]^{1/2} &\le |x_1 - x_2| + \alpha \mathbf{E}\left[\left(\int_0^t X_1(s) - X_2(s)\,ds\right)^2\right]^{1/2} \\ &\le |x_1 - x_2| + \alpha \int_0^t \mathbf{E}\left[|X_1(s) - X_2(s)|^2\right]^{1/2} ds\,.\end{aligned}$$

The Lemma follows by Gronwall's inequality. ∎

This result immediately gives uniqueness of solutions of (11.12) in the class of stochastic processes with finite variance.

Not unexpectedly, there exists an explicit solution to (11.12). We may derive the solution candidate by the means of assuming $W(t)$ in (11.10) being a regular continuous function. If this would be the case, the variation of parameters would give the solution (see Exercise 11.1)

$$X(t) = X(0)\mathrm{e}^{-\alpha t} + \sigma \int_0^t \mathrm{e}^{-\alpha(t-s)} W(s)\, ds\,.$$

As we already know, $W(s)$ does not exist, but by exchanging $W(s)ds$ by $dB(s)$, we may get something which makes sense, namely

$$X(t) = X(0)\mathrm{e}^{-\alpha t} + \sigma \int_0^t \mathrm{e}^{-\alpha(t-s)}\, dB(s)\,. \tag{11.13}$$

In order to prove that this is the solution, we first must understand the precise meaning of the *stochastic integral.*

As integration is summation, consider the sum

$$I_n(t) = \sum_{i=0}^{n-1} \mathrm{e}^{\alpha s_i}\, \Delta B(s_i)\,,$$

where $\{s_i\}_{i=0,\ldots,n}$ is a partition of the interval $[0,t]$, with $s_0 = 0$ and $s_n = t$, and $\Delta B(s_i) = B(s_{i+1}) - B(s_i)$. By the independence of the increments of Brownian motion, and the fact that $\Delta B(s_i) \sim \mathcal{N}(0, s_{i+1} - s_i)$, we find that $I_n(t)$ is normally distributed with mean zero and variance

$$\mathrm{Var}(I_n(t)) = \sum_{i=0}^{n} \mathrm{e}^{2\alpha s_i}(s_{i+1} - s_i)\,.$$

Choosing nested partitions, we see that

$$\lim_{n\to\infty} \mathrm{Var}(I_n(t)) = \int_0^t \mathrm{e}^{2\alpha s}\, ds\,.$$

Indeed, we find that $I_n(t)$ is a Cauchy sequence of random variables with finite variance, hence a Cauchy sequence in the Hilbert space $L^2(\Omega, \mathcal{F}, P)$. Since this is a complete space, there exists a random variable $I(t)$ with finite variance such that $I_n(t) \to I(t)$ as $n \to \infty$. We use this as the definition of the stochastic integral, that is,

$$\int_0^t \mathrm{e}^{\alpha s}\, dB(s) = \lim_{n\to\infty} I_n(t)\,, \tag{11.14}$$

where the limit is taken in $L^2(\Omega, \mathcal{F}, P)$. It is easily seen that the stochastic integral becomes normally distributed, with mean zero and variance

$$\mathrm{Var}(\int_0^t \mathrm{e}^{\alpha s}\, dB(s)) = \int_0^t \mathrm{e}^{2\alpha s}\, ds\,. \tag{11.15}$$

This integral is sometimes called the *Wiener integral*, which is a special case of the more general *Ito integral.* The latter allows for stochastic processes as integrands. The term $\exp(-\alpha t)$ can be moved in or out of the integral by linearity. Hence, this defines the stochastic integral term in (11.13).

Since we have

$$\begin{aligned}\sum_{i=0}^{n-1} \mathrm{e}^{\alpha s_i} \Delta B(s_i) &= \sum_{i=0}^{n-1} \{\mathrm{e}^{\alpha s_{i+1}} B(s_{i+1}) - \mathrm{e}^{\alpha s_i} B(s_i)\} - \sum_{i=0}^{n-1} B(s_{i+1}) \Delta \mathrm{e}^{\alpha s_i} \\ &= \mathrm{e}^{\alpha t} B(t) - \sum_{i=0}^{n-1} B(s_{i+1}) \Delta \mathrm{e}^{\alpha s_i}\,,\end{aligned}$$

it holds that

$$\int_0^t \mathrm{e}^{\alpha s}\, dB(s) = \mathrm{e}^{\alpha t} B(t) - \alpha \int_0^t B(s) \mathrm{e}^{\alpha s}\, ds\,. \tag{11.16}$$

Sometimes, one actually *defines* the Wiener integral by this integration-by-parts formula.

Of course, we can repeat the derivations above for a general measurable function $f(t)$ as integrand rather than $\exp(\alpha t)$. The condition for the stochastic integral $\int_0^t f(s)\, dB(s)$ to exist will in this case be that

$$\int_0^t f^2(s)\, ds < \infty\,.$$

Under this condition, the stochastic integral $\int_0^t f(s)\, dB(s)$ can be defined as a limit of partial sums following the same procedure as above. The integral will be normally distributed, with mean zero and variance given by the $\int_0^t f^2(s)\, ds$. If in addition f is continuously differentiable, we can derive the integration-by-parts formula

$$\int_0^t f(s)\, dB(s) = f(t)B(t) - \int_0^t B(s) f'(s)\, ds\,.$$

These more general considerations will become useful in the next section.

We can now prove that (11.13) is the (unique) solution of (11.12).

Proposition 2 *The stochastic process $X(t)$ defined in (11.13) is the unique solution to the stochastic differential equation (11.12).*

Proof: For simplicity, let $X(0) = 0$. By the integration-by-parts formula in (11.16) we find

$$\begin{aligned}\int_0^t X(s)\,ds &= \int_0^t \sigma \int_0^s \mathrm{e}^{-\alpha(s-u)}\,dB(u)\,ds \\ &= \sigma\int_0^t B(s)\,ds - \alpha\sigma\int_0^t \mathrm{e}^{-\alpha s}\int_0^s B(u)\mathrm{e}^{\alpha u}\,du\,ds\,.\end{aligned}$$

Using the Fubini Theorem (see Folland [5]) on the double integral yields

$$\begin{aligned}\int_0^t X(s)\,ds &= \sigma\int_0^t B(s)\,ds - \alpha\sigma\int_0^t\int_u^t \mathrm{e}^{-\alpha s}\,ds B(u)\mathrm{e}^{\alpha u}\,du \\ &= \sigma\int_0^t B(u)\mathrm{e}^{-\alpha(t-u)}\,du\,.\end{aligned}$$

But then, again using integration-by-parts in (11.16),

$$\begin{aligned}X(t) + \alpha\int_0^t X(s)\,ds &= \sigma\int_0^t \mathrm{e}^{-\alpha(t-s)}\,dB(s) + \alpha\sigma\int_0^t B(s)\mathrm{e}^{-\alpha(t-s)}\,ds \\ &= \sigma B(t) - \sigma\alpha\int_0^t B(s)\mathrm{e}^{-\alpha(t-s)}\,ds \\ &\quad +\alpha\sigma\int_0^t B(s)\mathrm{e}^{-\alpha(t-s)}\,ds \\ &= \sigma B(t)\,.\end{aligned}$$

This shows that $X(t)$ is a solution. Uniqueness follows from the fact that $X(t)$ has finite variance. This proves the proposition. ∎

We end this section with a study of the stationary properties of $X(t)$. For x being an arbitrary real number, let $\psi_{X(t)}(x)$ denote the cumulant function of $X(t)$,

$$\psi_{X(t)}(x) = \ln \mathbf{E}\left[\exp\left(\mathrm{i}xX(t)\right)\right]\,, \tag{11.17}$$

with $\mathrm{i} = \sqrt{-1}$ being the imaginary unit. Inserting $X(t)$ from (11.13) into the defintion of $X(t)$ we find from the normality of the stochastic integral:

$$\begin{aligned}\psi_{X(t)}(x) &= \mathrm{i}xX(0)\mathrm{e}^{-\alpha t} + \ln\mathbf{E}\left[\exp\left(\mathrm{i}x\sigma\int_0^t \mathrm{e}^{-\alpha(t-s)}\,dB(s)\right)\right] \\ &= \mathrm{i}xX(0)\mathrm{e}^{-\alpha t} - \frac{\sigma^2x^2}{4\alpha}\left(1-\mathrm{e}^{-2\alpha t}\right)\,.\end{aligned}$$

See Exercise 11.2 for the computation of the expectation. Letting time go to infinity, we see that the cumulant will converge. In fact, we easily see that

$$\lim_{t\to\infty} \psi_{X(t)}(x) = -\frac{\sigma^2 x^2}{4\alpha}\,.$$

But this is the cumulant function of a normally distributed random variable with mean equal to zero and variance given by $\sigma^2/2\alpha$. We therefore say that $X(t)$ is stationary, since its distribution has a limit. Note that in the deterministic case as we started off with, the limit of the solution was simply zero. In the stochastic case, $X(t)$ is not zero in the limit, but normally distributed. This means in practice that $X(t)$ in stationarity will randomly fluctuate around the mean zero, where the fluctuations are distributed according the normal distribution. The Brownian motion will constantly kick $X(t)$ away from its mean, while the mean reversion α will force the process back to its zero level. As we see, these two forces balance out in the long run.

We remark that the process $X(t)$ is called an Ornstein-Uhlenbeck process driven by Brownian motion. The study of such processes are important for modeling energy markets and not only temperatures. Indeed, by considering jump processes as the stochastic driver rather than Brownian motion, one can develop realistic models for the prices of electricity, gas, oil, etc. We refer the interested reader to the monograph by Benth *et al.* [2].

11.3 CONTINUOUS-TIME AUTOREGRESSIVE STOCHASTIC PROCESSES

The simple stochastic mean reversion dynamics we analyzed in the previous section turns out to be a special case of a much wider class of stationary stochastic processes. Moreover, for the accurate modeling of temperature, the simple mean reversion process is not sufficient for explaining all the probabilistic features of the temperature dynamics.

The basic stochastic dynamics for modeling the time evolution of temperature is the so-called continuous-time autoregressive process. We shall refer to these models as CAR processes and define them following the introduction in Benth *et al.* [2].

For each time $t \geq 0$, let $\mathbf{X}(t)$ be a random variable with values in R^p, for $p \geq 1$ a natural number, defined as the unique solution of the stochastic differential equation

$$d\mathbf{X}(t) = A\mathbf{X}(t)\,dt + \mathbf{e}_p\sigma(t)\,dB(t)\,, \tag{11.18}$$

with $X(0) = X_0$. Here, $\sigma(t)$ is a positive continuous *volatility* function and $\mathbf{e}_k$, for $k = 1, \ldots, p$ is the canonical basis in R^p. Furthermore, A is a $p \times p$-matrix on the specific form

$$A = \begin{bmatrix} 0 & 1 & 0 & \cdots & 0 \\ 0 & 0 & 1 & \cdots & 0 \\ \cdot & \cdot & \cdot & \cdot & \cdot \\ \cdot & \cdot & \cdot & \cdot & \cdot \\ 0 & 0 & 0 & \cdots & 1 \\ -\alpha_p & -\alpha_{p-1} & -\alpha_{p-2} & \cdots & -\alpha_1 \end{bmatrix}. \tag{11.19}$$

The constants α_k, for $k = 1, \ldots, p$ are all positive. We define the continuous autoregressive process $X(t)$ of order p by

$$X(t) = \mathbf{e}_1^* \mathbf{X}(t), \tag{11.20}$$

where we have used $*$ to denote matrix transposition. Such a dynamics $X(t)$ is frequently referred to as a CAR(p) process. Note that tradtionally, one assumes a constant volatility function $\sigma(t)$, but we extend the definition here since temperatures possesses an interesting seasonal pattern in the volatility.

EXAMPLE 11.1

Note that by letting $p = 1$, we recover the simple Ornstein-Uhlenbeck process studied in the previous section as long as $\sigma(t) = \sigma$, a constant. Indeed, for $p = 1$ we have that the matrix A collapses into a real value $A = -\alpha_1$. Then, $\mathbf{X}(t)$ is a stochastic process with values on the real line solving the equation

$$d\mathbf{X}(t) = -\alpha_1 \mathbf{X}(t)\, dt + \sigma\, dB(t).$$

Trivially, $\mathbf{e}_1 = 1$ in the case $p = 1$, and therefore the CAR(1) process $X(t) = \mathbf{e}_1^* \mathbf{X}(t)$ solves the same stochastic differential equation as our Ornstein-Uhlenbeck process in the previous section. There is no big difficulty in extending the analysis of the Ornstein-Uhlenbeck process in the previous section to time-dependent $\sigma(t)$.

As for the simple Ornstein-Uhlenbeck process, our first question is if there exists such a CAR(p) process? We need to have existence and uniqueness of $\mathbf{X}(t)$ solving (11.18), which we now study. For uniqueness, we proceed as in Proposition 1:

Proposition 3 *Let $\mathbf{X}_i(t)$ for $i = 1, 2$ be two solutions of (11.18) with $\mathbf{X}_i(0) = \mathbf{x}_i$. If $\mathbf{X}_i(t)$ has finite variance, then*

$$\mathbf{E}\left[|\mathbf{X}_1(t) - \mathbf{X}_2(t)|_2^2\right]^{1/2} \leq |\mathbf{x}_1 - \mathbf{x}_2|_2 e^{\|A\|_2 t},$$

for every $t \geq 0$. Here, $|\cdot|_2$ is the vector 2-norm on R^p and $\|\cdot\|_2$ is the associated matrix (operator) norm.

Proof: From the definition of the matrix norm and triangle inequality, we find

$$|\mathbf{X}_1(t) - \mathbf{X}_2(t)|_2 \leq |\mathbf{x}_1 - \mathbf{x}_2|_2 + \|A\|_2 \int_0^t |\mathbf{X}_1(s) - \mathbf{X}_2(s)|_2 \, ds .$$

The rest of the argument is similar to the proof of Proposition 1. ∎

Note that

$$\|A\|_2^2 := \sum_{i,j}^{p} a_{ij}^2 = (p-1) + \sum_{i=1}^{p} \alpha_i^2$$

from the definition of A, with a_{ij} being the elements of the matrix.

The above Proposition ensures uniqueness of the solution of (11.18), if there exists such in the space of all processes with finite variance. Motivated from the considerations on the simple Ornstein-Uhlenbeck process, we guess a solution of the form

$$\mathbf{X}(t) = \exp(At)\mathbf{X}_0 + \int_0^t \exp(A(t-s))\mathbf{e}_p \sigma(s) \, dB(s) . \tag{11.21}$$

Here, $\exp(A)$ is the matrix exponential, defined in the usual way as the $p \times p$-matrix

$$\exp(A) = \sum_{n=0}^{\infty} \frac{A^n}{n!} .$$

Note that we have

$$\|\exp(A)\|_2 \leq \sum_{n=0}^{\infty} \frac{\|A\|_2^n}{n!} = \exp(\|A\|_2)$$

which shows that this exponential is well-defined. Moreover, the stochastic integral is defined as the vector in R^p with coordinates given by

$$\int_0^t \{\exp(-A(t-s))\mathbf{e}_p\}_i \sigma(s) \, dB(s)$$

for $i = 1, \ldots, p$, where we use the notation $\{\mathbf{x}\}_i$ for the ith coordinate of a vector $\mathbf{x}$. Each coordinate of the stochastic integral is thus a real-valued stochastic integral of the form we defined in the previous section. We must check that the integrand is square integrable in order for this to be well-defined. To this end, we note that

$$|\exp(A(t-s))\mathbf{e}_p|_2 \leq \exp\left(\|A\|_2(t-s)\right) ,$$

and since $\sigma(t)$ is supposed continuous we find

$$\int_0^t \mathrm{e}^{\|A\|_2(t-s)}\sigma^2(s)\,ds < \infty\,,$$

for every $t < \infty$. Hence, we conclude that the vector-valued stochastic integral in (11.21) is well-defined. Moreover, each coordinate is a normally distributed random variable with mean zero. We next compute its cumulant function in order to show that this becomes a p-variate Gaussian random variable.

We define the cumulant $\psi(x)$ of the stochastic integral in (11.21) as

$$\psi(x) = \ln \mathbf{E}\left[\exp\left(\mathrm{i}\mathbf{x}^* \int_0^t \mathrm{e}^{A(t-s)}\mathbf{e}_p\sigma(s)\,dB(s)\right)\right].$$

Since

$$\mathbf{x}^* \int_0^t \mathrm{e}^{A(t-s)}\mathbf{e}_p\sigma(s)\,dB(s) = \int_0^t \mathbf{x}^*\mathrm{e}^{A(t-s)}\mathbf{e}_p\sigma(s)\,dB(s)\,,$$

we find by normality of the stochastic integral that

$$\begin{aligned}\psi(x) &= \ln \mathbf{E}\left[\exp\left(\mathrm{i}\int_0^t \mathbf{x}^*\mathrm{e}^{A(t-s)}\mathbf{e}_p\sigma(s)\,dB(s)\right)\right] \\ &= -\frac{1}{2}\int_0^t \left(\mathbf{x}^*\mathrm{e}^{A(t-s)}\mathbf{e}_p\right)^2\sigma^2(s)\,ds \\ &= -\frac{1}{2}\mathbf{x}^*\int_0^t \mathrm{e}^{As}\mathbf{e}_p\mathbf{e}_p^*\mathrm{e}^{As}\sigma(s)\,ds\,.\end{aligned}$$

This shows that the stochastic integral is a p-variate Gaussian random variable with a mean zero and variance-covariance matrix defined by

$$\int_0^t \mathrm{e}^{As}\mathbf{e}_p\mathbf{e}_p^*\mathrm{e}^{As}\sigma(s)\,ds$$

In particular, $\mathbf{X}(t)$ has finite variance.

Let us discuss the stationarity of the process $\mathbf{X}(t)$ in (11.21). First, if λ_i, $i = 1, \ldots, p$ are the eigenvalues of A with respective eigenvectors $\mathbf{v}_i$, then we find

$$\exp(At)\mathbf{X}_0 = \sum_{i=1}^p x_i \exp(At)\mathbf{v}_i = \sum_{i=1}^p x_i \exp(\lambda_i t)\mathbf{v}_i$$

with $\mathbf{X}_0 = \sum_{i=1}^p x_i\mathbf{v}_i$. As long as the real part of the eigenvalues is negative, we have that

$$\lim_{t\to\infty} \exp(At)\mathbf{X}_0 = 0\,.$$

In addition, under this condition on the eigenvalues, Ichihara and Kunita [6], Proposition 6.2, prove that the integral

$$\int_0^\infty e^{As}\mathbf{e}_p\mathbf{e}_p^* e^{As}\, ds$$

is finite, showing that $\mathbf{X}(t)$ has a stationary distribution in the case of constant volatility σ. We are interested in stationary models, so from now on we assume that the eigenvalues of the matrix A all have negative real part.

We now prove that the explicit dynamics in (11.21) solves the stochastic differential equation (11.18).

Proposition 4 *Suppose that σ is continuously differentiable. Then the process $\mathbf{X}(t)$ defined in (11.21) is the unique solution to (11.18).*

Proof: Let us assume for simplicity that $\mathbf{X}(0) = 0$. Using the integration-by-parts formula for the stochastic intergral, we find

$$\begin{aligned}
\int_0^t \mathbf{X}(s)\, ds &= \int_0^t \int_0^s (e^{A(s-u)}\mathbf{e}_p)\sigma(u)\, dB(u)\, ds \\
&= \int_0^t \mathbf{e}_p\sigma(s)B(s)\, ds + \int_0^t \int_0^s Ae^{A(s-u)}\mathbf{e}_p\sigma(u)B(u)\, du\, ds \\
&\quad - \int_0^t \int_0^s e^{A(s-u)}\mathbf{e}_p\sigma'(u)B(u)\, du\, ds \\
&= \int_0^t \mathbf{e}_p\sigma(s)B(s)\, ds + \int_0^t \int_u^t Ae^{A(s-u)}\mathbf{e}_p\, ds\sigma(u)B(u)\, du \\
&\quad - \int_0^t \int_u^t e^{A(s-u)}\mathbf{e}_p\, ds\sigma'(u)B(u)\, du\,.
\end{aligned}$$

In the last equality we have applied the Fubini Theorem (see Folland [5]). Hence, after integrating the inner integrals in the last two terms, we find

$$\begin{aligned}
\int_0^t \mathbf{X}(s)\, ds &= \int_0^t e^{A(t-u)}\mathbf{e}_p\sigma(u)B(u)\, du - \int_0^t A^{-1}e^{A(t-u)}\mathbf{e}_p\sigma'(u)B(u)\, du \\
&\quad + \int_0^t (A^{-1}\mathbf{e}_p)\sigma'(u)B(u)\, du\,.
\end{aligned}$$

Again appealing to integration-by-parts for the stochastic integral, we get

$$\begin{aligned}
\mathbf{X}(t) - A\int_0^t \mathbf{X}(s)\, ds &= \mathbf{e}_p\sigma(t)B(t) + \int_0^t Ae^{A(t-u)}\mathbf{e}_p\sigma(u)B(u)\, du \\
&- \int_0^t e^{A(t-u)}\mathbf{e}_p\sigma'(u)B(u)\, du - \int_0^t Ae^{A(t-u)}\mathbf{e}_p\sigma(u)B(u)\, du
\end{aligned}$$

$$+\int_0^t e^{A(t-u)}\mathbf{e}_p\sigma'(u)B(u)\,du - \int_0^t \mathbf{e}_p\sigma'(u)B(u)\,du$$

$$= \mathbf{e}_p\left(\sigma(t)B(t) - \int_0^t \sigma'(u)B(u)\,du\right) = \int_0^t \mathbf{e}_p\sigma(u)\,dB(u) \tag{11.22}$$

This concludes the proof. ■

One may wonder if we need to assume that A is invertible in the above proof. However, one can show that

$$\det(A) = -\alpha_p > 0,$$

so invertibility follows by the definition of the matrix A.

We continue our analysis of CAR(p) processes with the goal to understand the connection with autoregressive time series. Let us spell out the coordinates of the vector stochastic differential equation in (11.18). We find the system

$$\begin{aligned}
dX_1(t) &= X_2(t)\,dt \\
dX_2(t) &= X_3(t)\,dt \\
&\ \ \cdot\quad\cdot\quad\cdot \\
&\ \ \cdot\quad\cdot\quad\cdot \\
dX_{p-1}(t) &= X_p(t)\,dt \\
dX_p(t) &= -\sum_{k=1}^{p}\alpha_k X_k(t)\,dt + \sigma(t)\,dB(t)\,.
\end{aligned}$$

Here we let the kth coordinate of $\mathbf{X}(t)$ be denoted by $X_k(t)$, or in other words, $X_k(t) = \mathbf{e}_k^*\mathbf{X}(t)$ for $k = 1,\ldots,p$. Consider the *ordinary* pth order linear differential equation

$$u^{(p)}(t) = -\sum_{k=1}^{p}\alpha_k u^{(k-1)}(t) + \xi(t) \tag{11.23}$$

where $u(t)$ is a real-valued function and $u^{(k)}$ denotes its kth order derivative. We use the short-hand notation $\xi(t) = \sigma(t)B'(t)$. To have (11.23) well-defined, we need to have $\xi(t)$ reasonably smooth. However, from our definition of $\xi(t)$, it is not even existing as we recall that the paths of a Brownian motion are not differentiable. However, at this stage we bluntly assume this to be the case.

The standard approach to solve such a higher-order linear ordinary differential equation is to associate recursively functions to each of the derivatives and create a linear *system* of a first-order differential equation. To this end, define $v_k(t) = u^{(k-1)}(t)$ for $k = 1,\ldots,p$, where we see that $v_1(t) = u^{(0)}(t) = u(t)$.

Inserting this into (11.23) yields the equation

$$v_p'(t) = -\sum_{k=1}^{p} \alpha_k v_k(t) + \xi(t) .$$

But we also have that $v_k'(t) = v_{k+1}(t)$ for $k = 1, \dots p-1$. Using the vector notation $\mathbf{v}(t) = (v_1(t), \dots, v_p(t))^*$ then yields the first-order linear system of an ordinary differential equation

$$\mathbf{v}'(t) = A\mathbf{v}(t) + \mathbf{e}_p \xi(t) . \tag{11.24}$$

We notice, not unsurprisingly, that this equation in fact is nothing but (11.18), where we informally have differentiated Brownian motion with respect to time in the term $\xi(t)$.

Suppose we want to solve the differential equation (11.23) numerically. A discretization of (11.23) will be based on finite differences approximating the derivatives. Using forward differencing, we have that the approximation of $u^{(k)}(t)$ is

$$u^{(k)}(t) \approx \frac{1}{h^k} \sum_{i=0}^{k} (-1)^i \binom{k}{i} u(t + (k-i)h) \tag{11.25}$$

for a given time step $h > 0$. Applying these differences on the differential equation (11.23) yields a linear sum of $u(t + (p-i)h)$ for $i = 0, \dots, p$ on the left-hand side, and a linear combination of $u(t+kh)$ for $k = 0, \dots, p-1$ on the right-hand side. Additionally, we will have the term $\xi(t)$ on the right-hand side. Solving this linear equation with respect to $u(t + ph)$, we see that we will find

$$u(t + ph) = \sum_{i=0}^{p-1} a_i u(t + ih) + h^p \xi(t) ,$$

where a_i, $i = 0, \dots, p-1$ are expressible in terms of linear combinations of α_k, $k = 1, \dots, p$. Note that we may approximate the term $\xi(t)$ if we interpret this as $\xi(t) = \sigma(t)B'(t)$ again using finite differences. Indeed, for a given h we find

$$\xi(t) \approx \sigma(t) \frac{B(t+h) - B(t)}{h} = \frac{\sigma(t)}{\sqrt{h}} \epsilon(t) ,$$

where $\epsilon(t)$ is standard normally distributed and the equality is in distribution. Here we have used the fact that $B(t+h) - B(t)$ is normally distributed, with mean zero and variance h. Hence, letting time t run on a discrete grid $t = 0, h, 2h, 3h, ...$ and denoting $x(k) := u(kh)$, we find the time series

$$x(k + p) = \sum_{i=0}^{p-1} a_i x(k + i) + h^{p-1/2} \sigma(k) \epsilon(k) . \tag{11.26}$$

We note that $\epsilon(k)$ are independent random variables, since the increments of Brownian motion are independent. In conclusion, we have linked the CAR(p) process $X_1(t) = \mathbf{e}_1^*\mathbf{X}(t)$ to an autoregressive time series of order p with time dependent variance. Knowing the exact relationship between a_i and α_k will enable us to identify the α_k's from a time series estimation of an autoregressive model.

EXAMPLE 11.2

Consider the case $p = 2$. We find that

$$\begin{aligned} u''(t) &\approx \frac{1}{h^2}\sum_{i=0}^{2}(-1)^i \binom{2}{i} u(t+(2-i)h) \\ &= \frac{u(t+2h) - 2u(t+h) + u(t)}{h^2}, \end{aligned}$$

and

$$\begin{aligned} u'(t) &\approx \frac{1}{h}\sum_{i=0}^{1}(-1)^i \binom{1}{i} u(t+(1-i)h) \\ &= \frac{u(t+h) - u(t)}{h}. \end{aligned}$$

Hence, from the differential equation we find the relationship

$$\begin{aligned} u(t+2h) - 2u(t+h) + u(t) &= -\alpha_2 hu(t+h) + \alpha_2 hu(t) - \alpha_1 h^2 u(t) \\ &\quad + h^{3/2}\sigma(t)\epsilon(t), \end{aligned}$$

or, after reorganization,

$$x(k+2) = (2 - \alpha_2 h)x(k+1) + (\alpha_2 h - \alpha_1 h^2 - 1)x(k) + h^{3/2}\sigma(k)\epsilon(k).$$

Therefore, we see that the CAR(2) process $X(t) = \mathbf{e}_2^*\mathbf{X}(t)$ for $p = 2$ is associated with an autoregressive time series of order 2. The regression coefficients are $2 - \alpha_2 h$ for lag 1 and $\alpha_2 h - \alpha_1 h^2 - 1$ for lag 2. These coefficients depend naturally on the spacing of the discretization h.

It is an exercise (see Exercise 11.3) to perform this for the case $p = 3$. As we shall see next, $p = 3$ is the relevant case for modeling temperatures.

11.3.1 An Empirical Study

We report here the empirical analysis on Stockholm temperature data from Chapter 10 in Benth *et al.* [2]. Our aim is to demonstrate that the proposed temperature dynamics fits data very well. For details on the statistical analysis, we refer to Benth *et al.* [2].

We had available daily average temperatures from Stockholm over a period ranging from 1 January 1961 to 25 May 2006, resulting in 16,581 records. The measurements on February 29 were removed from the sample in each leap year, resulting in a time series of 16,570 observations. As already mentioned and shown in Figure 11.1, we first estimate a seasonal function with trend $\Lambda(t)$. After removing the fitted $\Lambda(t)$ from the temperature observations, we have a data set of so-called *de-seasonalized* temperatures that we claim can be modeled accurately by a CAR(p) process $X(t)$.

In Figure 11.2 we have plotted the partial autocorrelation function of the de-seasonalized temperatures. The partial autocorrelation at lag k of a time series $z(t)$ is defined as the correlation between $z(t)$ and $z(t+k)$ not accounted for by the lags 1 up to $k-1$. The partial autocorrelation function was in-

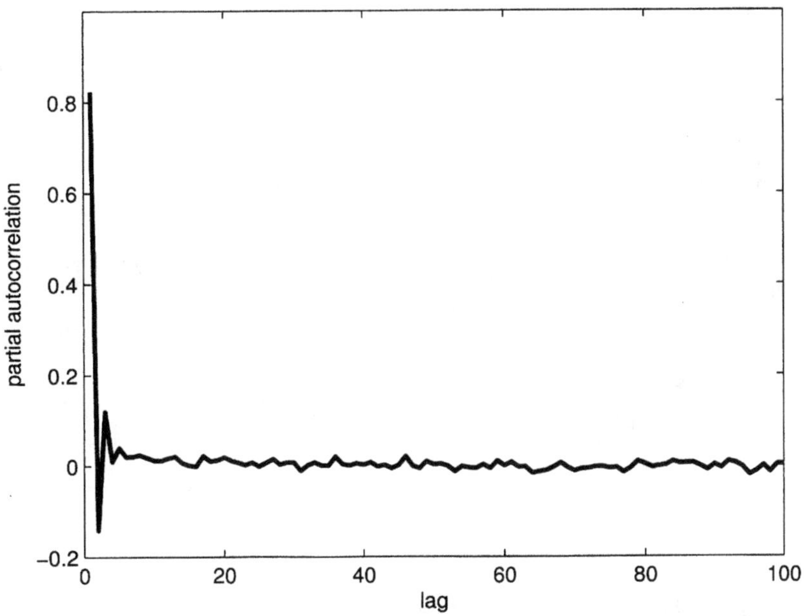

Figure 11.2 The partial autocorrelation function of de-seasonalized temperature data.

troduced by Box and Jenkins (see Box, Jenkins and Reinsel [4] for a recent account) to identify the order of an autoregressive process, and we clearly see from Figure 11.2 that the three first lags seem to be significantly different than zero. From lag four and higher, the partial autocorelation function wiggles around zero, pointing to an autoregressive process of order 3 as appropriate.

After fitting the regression parameters in an autoregressive time series of order 3 for the de-seasonalized data, we obtained the estimates $\widehat{\alpha}_1 = 2.043$,

$\widehat{\alpha}_2 = 1.339$ and $\widehat{\alpha}_3 = 0.177$. In Exercise 11.4 the reader is asked to compute the eigenvalues of the matrix A, and it turns out that these values have negative real parts, thus yielding a stationary model.

Removing the autoregressive part of the time series of de-seasonalized data leaves us with the residuals. Interestingly, the autocorrelation function of these residuals are essentially wiggling around zero, but looking at the autocorrelation function for their squares shows a clear seasonal pattern. Figure 11.3 shows a seasonally varying autocorrelation function for squared residuals. This points towards a seasonality in the volatility, which we model by

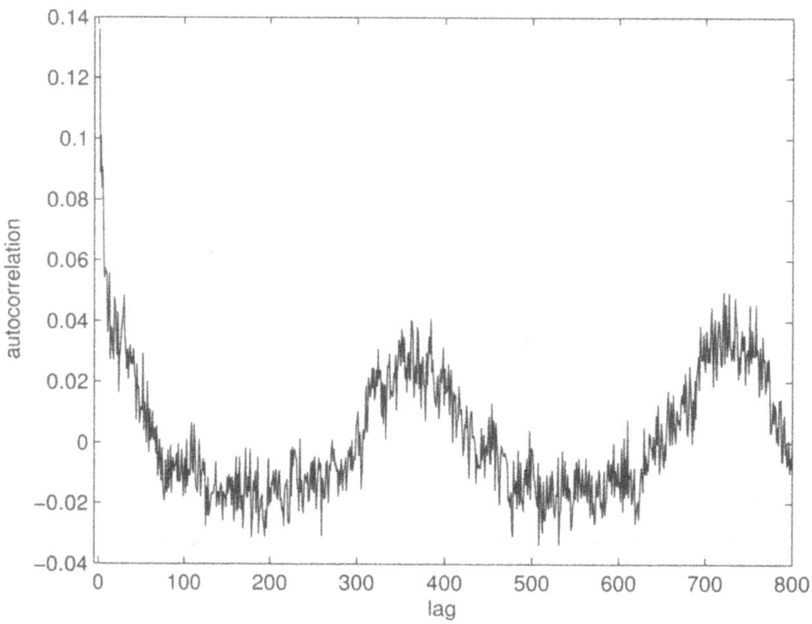

Figure 11.3 The autocorrelation function of squared residuals.

$\sigma(t)$.

Since we have approximately 45 years of daily data, we will have about 45 observations of residuals at each day in the year. Taking the variance of the 45 residual observations on each day gives us the empirical seasonality of the volatility observed over a year. We have depicted the seasonal variance in Figure 11.4 along with the fitted function $\sigma^2(t)$ assumed to be

$$\sigma^2(t) = c_1 + \sum_{k=1}^{4} \{c_{2k}\cos(2k\pi t/365) + c_{2k+1}\sin(2k\pi t/365)\} . \tag{11.27}$$

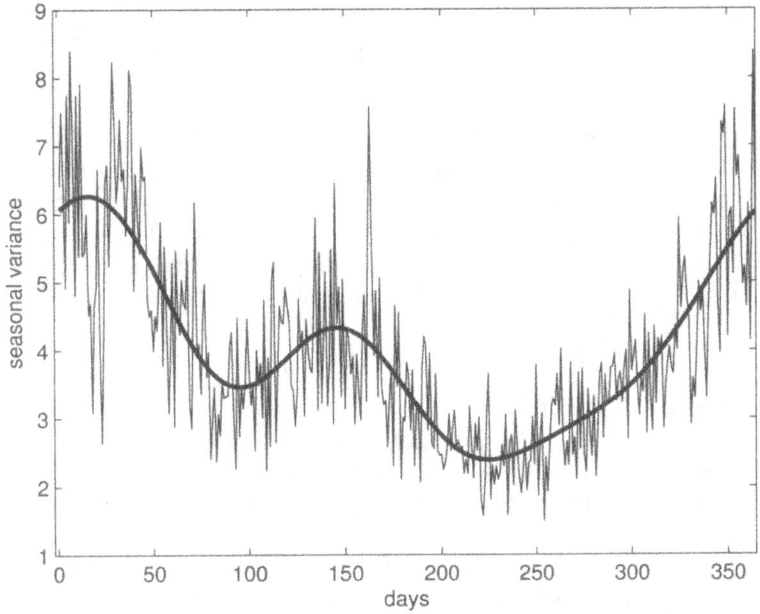

Figure 11.4 The seasonal variance with the fitted $\sigma^2(t)$.

We used nonlinear least squares to fit the parameters of $\sigma^2(t)$ to the daily variances. Note that the function $\sigma^2(t)$ is periodic, with a yearly cycle. From the figure, we note that the variation in temperatures is highest in the winter, but interestingly, it is lowest in the early spring and autumn, while it rises up in the summer again. The most stable temperatures are therefore observed usually in the spring and autumn.

Using the estimated function $\sigma(t)$, we are able to explain the seasonal variance, and the remaining residuals show no pattern. In conclusion, we have shown that the daily average temperature data observed in Stockholm can be fitted using our proposed model.

11.4 PRICING OF TEMPERATURE FUTURES CONTRACTS

We will investigate the pricing of temperature futures contracts in this section. As we discussed in the Introduction, there are three different temperature indices on which there are written futures contracts at the Chicago Mercantile Exchange. We will focus on the simpler one here, namely, the CAT futures.

Recall that a CAT futures contract is settled on the observed cumulative average temperature over a time period $[\tau_1, \tau_2]$, that is, the owner of the contract receives an amount of money proportional to the index

$$I(T_1, T_2) = \sum_{s=\tau_1}^{\tau_2} T(s),$$

where s are running over the days in the measurement period. The amount of money is received at time τ_2, after one has measured the index. The value of the index times a money conversion factor $c > 0$ is the amount paid to the owner. In return, she has to pay the *CAT futures price* $F(t, \tau_1, \tau_2)$, that is, the price agreed at time $t \leq \tau_1$ when entering the futures contract. Hence, the owner of the CAT futures contract experiences at time τ_2 a profit/loss of size

$$cI(\tau_1, \tau_2) - F(t, \tau_1, \tau_2) .$$

It is to be noted that a futures contract is organized so that it is costless to buy, but at time of purchase one agrees on a price to be paid at the end of the measurement period. This agreed price is referred to as the *futures price* of the contract. The question we want to investigate is what this price should be.

In finance theory, one approach to fix the futures price is by the *rational expectation hypothesis*, which states that at time $t \leq \tau_1$, the futures price should be so that the profit/loss function has zero expectation,

$$\mathbf{E}\left[cI(\tau_1, \tau_2) - F(t, \tau_1, \tau_2) \,|\, \mathbf{X}(t)\right] = 0 . \tag{11.28}$$

Note that we have conditioned on $\mathbf{X}(t)$ in the expectation, which is simply a mathematical way to express that we include the information about the temperatures up until today. When the contract is entered, both the seller and the buyer will take into account this information when agreeing on the price, so $F(t, \tau_1, \tau_2)$ naturally becomes a function of $\mathbf{X}(t)$. Thus, by a reorganization of (11.28) we obtain the equation

$$F(t, \tau_1, \tau_2) = c\mathbf{E}\left[I(\tau_1, \tau_2) \,|\, \mathbf{X}(t)\right] \tag{11.29}$$

for the CAT futures price. After commuting expectation and the finite sum in the definition of $I(\tau_1, \tau_2)$, we get

$$F(t, \tau_1, \tau_2) = c \sum_{s=\tau_1}^{\tau_2} \left\{\Lambda(s) + \mathbf{E}\left[\mathbf{e}_1^* \mathbf{X}(s) \,|\, \mathbf{X}(t)\right]\right\} . \tag{11.30}$$

Thus, to find the CAT futures price we need to compute the conditional expectations in the sum above. By normality of our CAR(p) process, this is feasible to do analytically. We derive this in the next Proposition:

Proposition 5 *For $s \geq t \geq 0$, it holds that*

$$\mathbf{E}\left[\mathbf{e}_1^*\mathbf{X}(s) \,|\, \mathbf{X}(t)\right] = \mathbf{e}_1^* \mathrm{e}^{A(s-t)}\mathbf{X}(t).$$

Proof: First, we observe that the solution $\mathbf{X}(s)$ for $s \geq t$ to the stochastic differential equation (11.18), when starting at time t with the initial condition $\mathbf{X}(t)$, is

$$\mathbf{X}(s) = \mathrm{e}^{A(s-t)}\mathbf{X}(t) + \int_t^s \mathrm{e}^{A(s-u)}\mathbf{e}_p\sigma(u)\,dB(u)\,.$$

Hence,

$$\begin{aligned}\mathbf{E}\left[\mathbf{e}_1^*\mathbf{X}(s) \,|\, \mathbf{X}(t)\right] &= \mathbf{e}_1^*\mathrm{e}^{A(s-t)}\mathbf{X}(t) + \mathbf{E}\left[\int_t^s \mathbf{e}_1^*\mathrm{e}^{A(t-u)}\mathbf{e}_p\sigma(u)\,dB(u)\right] \\ &= \mathbf{e}_1^*\mathrm{e}^{A(s-t)}\mathbf{X}(t)\,.\end{aligned}$$

This shows the result. ∎

We may express the CAT futures price as

$$F(t,\tau_1,\tau_2) = c\sum_{s=\tau_1}^{\tau_2}\Lambda(s) + c\mathbf{e}_1^*\left\{\sum_{s=\tau_1}^{\tau_2}\mathrm{e}^{As}\right\}\mathrm{e}^{-At}\mathbf{X}(t)\,. \tag{11.31}$$

Hence, the stochastic dynamics of $t \mapsto F(t,\tau_1,\tau_2)$ is driven by $\mathbf{X}(t)$. Note that the futures price is dependent on the vector process $\mathbf{X}(t)$, and not only on the first component, which would be the CAR(p) process, or, in other words, the temperature less the seasonal function. Therefore, it is not sufficient to observe only today's temperature (today meaning at time t) to find the futures price, but one must recover all the coordinates of the vector $\mathbf{X}(t)$. This is equivalent to saying that the futures price is depending not only on today's temperature [which would mean $\mathbf{e}_1^*\mathbf{X}(t)$], but also the $p-1$ previous day's temperatures. Autoregressive models has a memory, which we clearly see the effect of here.

If $p = 1$, then $\mathbf{X}(t) = \mathbf{e}_1^*\mathbf{X}(t) = X(t) = T(t) - \Lambda(t)$, and the futures price becomes

$$F(t,\tau_1,\tau_2) = c\sum_{s=\tau_1}^{\tau_2}\Lambda(s) + c\left\{\sum_{s=\tau_1}^{\tau_2}\mathrm{e}^{-\alpha_1 s}\right\}\mathrm{e}^{\alpha_1 t}(T(t) - \Lambda(t))\,.$$

In this case we have a dependency on the current temperature, and no memory in the price dynamics in the sense that the temperatures on previous days do not influence the futures price.

Let us return back to the general case, and analyze the dynamics of the futures price. Since

$$\begin{aligned}\mathbf{e}_1^* \mathrm{e}^{A(s-t)}\mathbf{X}(t) &= \mathbf{e}_1^* \mathrm{e}^{A(s-t)} \left\{ \mathrm{e}^{At}\mathbf{X}(0) + \int_0^t \mathrm{e}^{A(t-u)}\mathbf{e}_p \sigma(u)\, dB(u) \right\} \\ &= \mathbf{e}_1^* \mathrm{e}^{As}\mathbf{X}(0) + \int_0^t \mathbf{e}_1^* \mathrm{e}^{A(s-u)}\mathbf{e}_p \sigma(u)\, dB(u)\,.\end{aligned}$$

This implies that

$$F(t,\tau_1,\tau_2) = F(0,\tau_1,\tau_2) + c \int_0^t \mathbf{e}_1^* \sum_{s=\tau_1}^{\tau_2} \mathrm{e}^{A(s-u)}\mathbf{e}_p \sigma(u)\, dB(u)\,,$$

where

$$F(0,\tau_1,\tau_2) = c \sum_{s=\tau_1}^{\tau_2} \Lambda(s) + c\mathbf{e}_1^* \sum_{s=\tau_1}^{\tau_2} \mathrm{e}^{As}\mathbf{X}(0)\,.$$

But this means that we can write the dynamics as

$$dF(t,\tau_1,\tau_2) = c\mathbf{e}_1^* \sum_{s=\tau_1}^{\tau_2} \mathrm{e}^{A(s-t)}\mathbf{e}_p \sigma(t)\, dB(t)\,. \tag{11.32}$$

Intepreting $dF(t,\tau_1,\tau_2)$ as the incremental change in the futures price from time t to $t+dt$, we see that this moves as the change in Brownian motion, scaled by the temperature volatility $\sigma(t)$ and the function

$$g(t) = c\mathbf{e}_1^* \sum_{s=\tau_1}^{\tau_2} \mathrm{e}^{A(s-t)}\mathbf{e}_p,$$

with $t \le \tau_1$.

Define the function

$$f(v) = \mathbf{e}_1^* \mathrm{e}^{Av}\mathbf{e}_p \tag{11.33}$$

and observe that

$$g(t) = c \sum_{s=\tau_1}^{\tau_2} f(s-t)\,.$$

Since $\exp(A0) = I_p$, with I_p being the $p \times p$ identity matrix, we have

$$f(0) = \mathbf{e}_1^* \mathbf{e}_p = \begin{cases} 1, & p = 1, \\ 0, & p > 1. \end{cases}$$

On the other hand, due to the negative real parts of the eigenvalues of A, $f(v) \to 0$ as $v \to \infty$. We have $f(v) = \exp(-\alpha_1 v)$ for $p = 1$, which is an exponentially decaying function. The behavior is completely different for $p > 1$, since then $f(0) = 0$ and not 1. The scaling of the temperature volatility

$\sigma(t)$ will be very different for $p = 1$ and $p > 1$. We refer to Benth *et al.* [2] for more discussions of this feature, which can be related to the so-called Samuelson effect for futures prices.

We next consider the problem of pricing a call option written on the CAT futures. We suppose that the option has strike price K, at the exercise time $\tau \le \tau_1$. The option will then pay the owner at time τ

$$\max\left(F(\tau, \tau_1, \tau_2) - K, 0\right) . \tag{11.34}$$

The price of this option is given as the present expected value of the payoff function in (11.34), that is,

$$C(\tau, K) = \mathrm{e}^{-r\tau} \mathbf{E}\left[\max\left(F(\tau, \tau_1, \tau_2) - K, 0\right)\right] . \tag{11.35}$$

Here, r is the risk-free rate of return on a bank deposit. As the CAT futures price is explicit in terms of $\mathbf{X}(t)$, we can derive a reasonably explicit expression for $C(\tau, K)$. However, we want first to discuss briefly the rationale behind the price (11.35).

From the theory of options (see, for example, Benth [1] for a basic introduction, or Björk [3] for a more advanced treatment), one finds that the price of an option is given as the cost of replication. A replicating portfolio, vaguely spoken, is defined as a dynamical investment strategy in the underlying (here the CAT futures) and a bank account yielding an interest rate r (or, a treasury bond, if one likes), with the value being equal to the option's payoff at the exercise time τ. As it turns out, there exists a probability Q such that the cost or replication can be expressed as a present expected value of the payoff. The expectation is computed with respect to the probability Q. Noteworthy is that this probability turns the underlying instruments into martingales. The expected future value of a martingale is equal to its current value, which is the key defining property of such processes.

As we have derived a price for our CAT futures as an expected value conditional on the current information $\mathbf{X}(t)$, it will be a martingale. This means, in accordance with the option pricing theory, that we can use P as the so-called pricing measure Q above. This validates our pricing rule (11.35) as the cost of hedging the call option.

In order to compute the price $C(\tau, K)$, let us represent the CAT futures price in (11.31) as

$$F(t, \tau_1, \tau_2) = \lambda(\tau_1, \tau_2) + \mathbf{a}(\tau_1, \tau_2)\mathrm{e}^{-At}\mathbf{X}(t) , \tag{11.36}$$

where

$$\lambda(\tau_1, \tau_2) = c \sum_{s=\tau_1}^{\tau_2} \Lambda(s) \tag{11.37}$$

and

$$\mathbf{a}(\tau_1,\tau_2) = \mathbf{c}\mathbf{e}_1^* \sum_{s=\tau_1}^{\tau_2} \mathrm{e}^{As} . \tag{11.38}$$

We have the following result for the option price:

Proposition 6 *The call option price $C(\tau, K)$ is given by*

$$C(\tau,K) = \mathrm{e}^{-r\tau}\left((\mu - K)\Phi\left(\frac{\mu - K}{\Sigma(\tau)}\right) + \frac{\Sigma(\tau)}{\sqrt{2\pi}}\exp\left(-\frac{(\mu-K)^2}{2\Sigma^2(\tau)}\right)\right),$$

where Φ is the cumulative standard normal distribution function and

$$\mu = \lambda(\tau_1,\tau_2) + \mathbf{a}(\tau_1,\tau_2)\mathbf{X}(0),$$

for $\lambda(\tau_1,\tau_2)$ and $\mathbf{a}(\tau_1,\tau_2)$ given in (11.37) and (11.38), resp. Finally,

$$\Sigma^2(\tau) = \int_0^\tau (\mathbf{a}(\tau_1,\tau_2)\mathrm{e}^{-As}\mathbf{e}_p)^2\sigma^2(s)\,ds .$$

Proof: Since

$$\mathbf{X}(\tau) = \mathrm{e}^{A\tau}\mathbf{X}(0) + \int_0^\tau \mathrm{e}^{A(\tau-s)}\mathbf{e}_p\sigma(s)\,dB(s),$$

we find

$$\begin{aligned} F(\tau,\tau_1,\tau_2) &= \lambda(\tau_1,\tau_2) + \mathbf{a}(\tau_1,\tau_2)\mathrm{e}^{-A\tau}\mathbf{X}(\tau) \\ &= \lambda(\tau_1,\tau_2) + \mathbf{a}(\tau_1,\tau_2)\mathbf{X}(0) + \int_0^\tau \mathbf{a}(\tau_1,\tau_2)\mathrm{e}^{-As}\mathbf{e}_p\sigma(s)\,dB(s). \end{aligned}$$

As we have shown, the stochastic integral is normally distributed, with mean zero and variance $\Sigma^2(\tau)$. Hence, we have

$$F(\tau,\tau_1,\tau_2) = \mu + \Sigma(\tau)\epsilon$$

where ϵ is standard normal random variable, and the equality is in distribution. We have

$$\begin{aligned} \mathrm{e}^{r\tau}C(\tau,K) &= \mathbf{E}\left[\max\left(F(\tau,\tau_1,\tau_2) - K, 0\right)\right] \\ &= \mathbf{E}\left[\max\left(\mu - K + \Sigma(\tau)\epsilon, 0\right)\right] \\ &= \frac{1}{\sqrt{2\pi}}\int_{\frac{K-\mu}{\Sigma(\tau)}}^\infty (\mu - K + \Sigma(\tau))\mathrm{e}^{-\frac{x^2}{2}}\,dx \\ &= (\mu - K)\frac{1}{\sqrt{2\pi}}\int_{-\infty}^{\frac{\mu-K}{\Sigma(\tau)}} \mathrm{e}^{-\frac{x^2}{2}}\,dx + \Sigma(\tau)\frac{1}{\sqrt{2\pi}}\int_{\frac{K-\mu}{\Sigma(\tau)}}^\infty x\mathrm{e}^{-\frac{x^2}{2}}\,dx . \end{aligned}$$

A direct calculation using the definition of the cumulative probability distribution yields the result. ■

One may price CDD and HDD futures as well, and options on these. However, they will not give as explicit results as the CAT futures. We refer to Benth *et al.* [2] for a detailed account on the pricing of temperature futures and options.

Acknowledgement: Financial support for the project "Managing Weather Risk in Electricity Markets (MAWREM" funded by the Norwegian Research Council under grant RENERGI 216096 is kindly acknowledged."

EXERCISES

11.1 Solve the ordinary differential equation

$$u'(t) = -\alpha u(t) + g(t)\,, t > 0\,,$$

with $u(0) = u_0$. Here, g is a continuous function and α, u_0 constants.

11.2 Compute the characteristic function of the stochastic integral $\int_0^t e^{\alpha s}\,dB(s$

11.3 Find the associated autoregressive time series to a CAR(3) process $X(t) = \mathbf{e}_1^* \mathbf{X}(t)$.

11.4 Find the eigenvalues of the matrix

$$A = \begin{bmatrix} 0 & 1 & 0 \\ 0 & 0 & 1 \\ -0.177 & -1.339 & -2.043 \end{bmatrix}$$

Conclude that the corresponding CAR(3) model is stationary.

REFERENCES

1. Benth, F. E., *Option Theory with Stochastic Analysis – An Introduction to Mathematical Finance*, Springer Verlag, Berlin (2004).
2. Benth, F. E., Šaltytė Benth, J., and Koekebakker, S. *Stochastic Modelling of Electricity and Related Markets*, World Scientific (2008).
3. Björk, T., *Arbitrage Theory in Continuous Time*, Oxford University Press, Oxford (2004).
4. Box, G. E. P., Jenkins, G. M., and Reinsel, G. C. *Time Series Analysis, Forecasting and Control*, (4th ed.), Wiley, New York (2008).
5. Folland, G. B., *Real Analysis*, Wiley Interscience, New York (1984).
6. Ichihara, K., and Kunita, H., A classification of the second order elliptic operator and its probabilistic characterization. *Z. Wahrsch. Verw. Geb.*, **30**, pp. 235–254, (1974).

7. Oksendal, B., *Stochastic Differential Equations – An Introduction with Applications* (6th ed.), Springer Verlag, Berlin (2005).

CHAPTER 12

DECISION THEORY UNDER RISK AND APPLICATIONS IN SOCIAL SCIENCES: I. INDIVIDUAL DECISION MAKING

E. V. PETRACOU[1] AND A. N. YANNACOPOULOS[2]

[1]Department of Geography, University of the Aegean, Greece
[2] Department of Statistics, Athens University of Economics and Business, Greece

If the facts do not fit the theory change the facts—A. Einstein

12.1 INTRODUCTION

It is the aim of this chapter to provide a very brief and first encounter with mathematical modeling in the social sciences. This is a very exciting field, combining solid mathematics (ranging from pure to the applied) with fundamental concepts from philosophy, political and social theory, in an attempt to provide benchmarks for human behavior and understanding motives and patterns for human actions. A large part of this field has been developed into an independent discipline, that of decision theory in mathematical economics, and is now the dominant tool in understanding the phenomena of the

Edited by Xin-She Yang

economy. On the other hand, there is a lot of interest in developing these techniques to understand social phenomena not directly related to the economy, such as human action and institutions, voting patterns etc. This is now a very active field, blending techniques from mathematics and statistics (e.g., game theory, decision analysis, probability models, optimization techniques, differential equations, etc.), physics and engineering (e.g., particle systems, mean field theory, etc.) and social sciences (e.g., economics, political science and international relations, psychology, sociology, etc.).

This short introduction covers some fundamental aspects of decision theory, starting from decision theory for single agents, moving to noncooperative game theory and finally introducing basic concepts from cooperative game theory. We try to motivate the mathematics with examples from the social sciences, which in some cases have been the driving force for the creation of the mathematical theory. In this effort we introduce important concepts such as the concept of utility, expected utility and its maximization as a tool for individual decision making. These mathematical tools have universal validity and are useful in other branches of applied or pure mathematics as well.

12.2 THE FUNDAMENTAL FRAMEWORK

You cannot expect to start a book on the mathematical theory of the Navier-Stokes equation without an introduction to fluid mechanics. You cannot expect to start a book on the mathematical theory of Maxwell's equations without an introduction to elementary electromagnetic theory. You cannot start a book on symplectic manifolds and Hamiltonian dynamics without mentioning at some point Newton's laws. It is equally impossible to start a chapter on the decision theory in the social sciences without a brief introduction to the fundamental theoretical concepts of choice theory which at the same time introduce the intricacies and specificities of the subject. This should be a word of caution to any student wishing to embark into mathematical modeling of any sort: First approach with respect the science that calls for modeling, understand the basic concepts (including possible problems and critiques involved) and then offer the mathematics. To paraphrase General Jack Ripper's lines from Dr. Strangelove: If in doubt ask questions and read first and shoot your equations later!

In general, decision making is an action based on judgments. Decisions are made in certain social and historical contexts, under certain and specific conditions and have intended as well as unintended outcomes. As a result, decision making both in theory and practice is heavily influenced by the particular social and historical situations and contexts involved and thus it is impossible to separate it from them. It is important to realize that the ways in which theories are constructed in social science are always influenced by the vantage point of the scientist involved in stating them; there is no such thing as the impartial observer in social sciences you are trying to understand and make scientific statements about a system in which you are heavily involved and it is even worse than that: your actions influence the system and may well change its state.

Decision theory, game theory and utility theory constitute mathematical theories and normative[11] disciplines concerning rational behavior of individuals and their interaction. These theories have been developed in connection to economics, mainly in the 20th century. The main assumption of rational theory is that human activity is based on reason and individuals make a choice among possible actions in order to fulfill their own interest. Then, departing from this major assumption, we may formulate rules governing human decision making, "predict" or explain modes or patterns of behavior, etc.

Rationality is a major issue in social science. According to the great sociologist Max Weber, there are different kinds of rationalities; these are ideal types such as purposive or instrumental, value-oriented, emotional and habitual rationality. Individuals are rational in the sense that they can choose the most appropriate or optimal means to fulfill their ends or goals. Logical conduct à la Pareto is: appropriate means for ends, both from the standpoint of an individual—subjective—and from other persons—objective. However, the dominion of rationality as axiomatic foundations for decision theory should neither be taken for granted, nor should be globally acceptable. Rationality reflects a particular type of logic that was developed in Western thought especially after the Enlightenment. It can be argued that it is a theoretical assumption, a hypothesis made by scientists concerning beliefs, preferences, intentions, actions and ends and as an assumption it is subject to criticism. Especially, instrumental rationality presupposes that the individual has preferences and because of her knowledge or sufficient information may choose the best means in order to achieve her end, which is defined by her own (expected) utility. Clearly, one may argue against that and construct an equally interesting theory, refuting the axiom of rationality. In fact, there are very interesting and popular theories starting from this vantage point; the theory of bounded rationality, proposed by the eminent political philosopher and economist Herbert Simon,[12] is a very good example of that.

The science of understanding and modeling human decision making can be divided into two major parts: individual decision making (decision theory) and game theory and ethics. While the division is sometimes not very clearcut, one can generally say that decision theory refers to individual rational behavior while game theory and ethics refers to the theory of rational behavior in a social setting [7]. Individual rational decision theory can be: (i) under certainty—the outcome of an action is uniquely predictable, (ii) under risk—objective probabilities with alternative possible outcomes, and (iii)

[11] A normative discipline or theory is one that provides rules on how things should be and thus forms an idealized view of the world. This in contrast with a descriptive or a positive theory, which tries to provide an explanation of how things are.

[12] Nobel Laureate in Economics, 1978.

under uncertainty—a number of objective probabilities are unknown. Game theory refers to individuals who try to maximize their own interests in a rational way against all the other (rational) individuals while ethics refer to individuals which try to promote a common interest even though they have different interests [7]. In this short presentation we will try to provide an overview of all these aspects, that is, individual decision theory under risk and uncertainty, game theory and cooperative game theory (which is an approximation of what Harsanyi calls ethics).

One can argue that the whole construction of decision and game theory is based upon the philosophical movement of utilitarianism. Utilitarianism is a philosophical school of thought of normative ethics whose main claim is that the morally appropriate action is the action which produces the overall good. The right action is determined by the consequent result (which is maximization of utility, satisfaction and welfare). Each individual's well being is equally important and the overall well being is a result of all individuals' well being. Utilitarianism developed in the 18th and the 19th centuries in Britain and the main classical thinkers of utilitarian ethics were Jeremy Bentham, John Stuart Mill and Henry Sidgwick. The way to evaluate an action in moral terms is Bentham's greatest happiness principle for the greatest number of people.

Utilitarianism has at least two different forms. It is divided into act utilitarianism and rule utilitarianism [10].

- Act utilitarianism: The right act is the act that produces the most well being, i.e., maximizes welfare.

- Rule utilitarianism: It is a code of moral rules, that if each individual follows then she is going to maximize her utility (welfare).

Act utilitarianism has been the subject of critique, and has been accused of promoting injustice and immorality. Rule utilitarianism has also been the subject of critique as an unnecessary concept since it is very closely related to act utilitarianism. An important question is whether politics and morality may be combined in order to contribute to human welfare.

Utilitarianism has had equally ardent followers as well as opponents. It has inspired many philosophical studies, and leading moral philosophers (such John Rawls, John Harsanyi, Amartya Sen, etc.) have spent a lot of effort in unraveling the perplexing issues arising from the above simple ideas. This chapter is by no means the appropriate place to develop this discourse. Since a large part of positive political and social theory follows the basic notions of utilitarianism we will adopt them here, as a working hypothesis, and present their ways of thinking, theorizing and modeling social phenomena.

A very important difficulty of the general theory of rational behavior is how we can turn from individual action to a collective one since collectivity cannot be defined as a mere aggregation of individual acts. This is a result of the development of rational theory which treats individuals outside of social and historical contexts and it is based on the concept of homo economicus, a selfish individual as an abstract and universal man, ignoring homo sociologus (according to Hollis see, e.g., Ref. [2]). The problem of aggregation is very important, has led to the development of a branch of decision theory called social choice theory and is still under very active consideration.

As mentioned above, in social sciences the way you construct a theory concerning phenomena may depend strongly on the school of thought with which you are affiliated. It is thus important to mention that there are different theoretical schools in social sciences. These schools deviate in fundamental issues connected with questions such as "what is it possible to know," "how and under which procedures is it possible to know," etc. One important school of thought is the positivist school, which was developed in the 19th century. Positivism is a philosophical approach that argues that there is a unity of the scientific method in all sciences and the goal of natural and social sciences alike is verification by experiment and prediction. The focus is on systematic observation in a deterministic point of view of the world (through scientific laws). However, this viewpoint is not universally acceptable. Other schools of thought propose different approaches, such as structuralism and humanism [8], and reject this unity and these assumptions for the social sciences. Different approaches mean different perceptions of science and interpretations of the world and also the reasons for adoption of a certain perspective.

We hope that the above thoughts have made clear to the reader that individuals are not alone but they are agents in interactions and groups. Furthermore, they are not self-defined entities but the products of specific conditions. Actors or agents, and that includes scientists and modelers, are shaped by and shape social relations in the processes of social becoming and their relations are constructed by unequal power relations, domination, coercion, enforcement, consensus, cooperation, struggles, plurality and different identities. So, actors internalize norms and rules; furthermore, this happens not in an "one-dimensional" or calculable way. Preferences, information, utility maximization, cost and benefit analysis are concepts that are not neutral, but bear a very specific ideological meaning: that of accepting the notion of agents as selfish maximizers who are free and equal by nature or contract. Clearly this is a major, and for some perhaps a simplistic, assumption.

Based on the above comments and concerns many social scientists have refuted the assumptions of rational theory and have raised a serious critique against rational choice theory and its significance as well as against decision modeling theoretically and empirically [11], [1], [13]. This critique leads us to the important conclusion that in order to improve understanding of social action under mathematical lenses, we need more elaboration on the social theoretical concepts and more empirical research that will play the role of experimental facts upon which we can start building more realistic theories.

However, this word of caution does not diminish the importance of decision theories as a means of modeling and understanding in social sciences. We can, and we should, explore some social phenomena under the prism of decision theories as long as we always keep in mind that what we propose is just a model, which at best may provide benchmarks for understanding and quantifying some aspects of social life. Surely we cannot expect from mathematical modeling in the social sciences the triumphs and global acceptance it has in physics or engineering. We have to come to terms with the fact that there is not a unique and commonly accepted standpoint from which to start our modeling endeavor, and that the results should be treated accordingly. However, this does not mean that mathematical modeling in the social sciences is a pointless exercise. On the contrary it may provide very important insights as well as very challenging mathematical problems to work on. To support the second point we should do no more than mention the extensive list of brilliant mathematicians that devoted their careers to such issues. To support this point, let us just quote the words of the famous philosopher and social scientist Michel Foucault [3]:

> "...like any other domain of knowledge, these sciences (human) may, in certain conditions, make use of mathematics as a tool; some of their procedures and a certain number of their results can be formalized. It is undoubtedly of the greatest importance to know those tools, to be able to practice those formalizations and to define the levels upon which they can be performed; it is no doubt of interest historically to know how Condorcet was able to apply the calculation of probabilities to politics."

12.3 A BRIEF INTRODUCTION TO THEORY OF CHOICE

Suppose we wish to choose between two wallets containing a sum of money, one containing x euros and one containing y euros, with $y > x$. Which one should we choose? Of course the one containing y. In this simple case our choice criterion is easy: count the money in each wallet, see which one contains the biggest sum and choose this. Therefore our decision criterion reduces to checking an inequality.

Can life be that easy? No, since our choice decisions are never so simple as to reduce to the comparison of two real numbers. Consider the next level of difficulty in such a problem: suppose we wish to choose between two baskets of goods (say, apples and pears to fix ideas) and let us assume that the first basket contains (x_1, x_2) kilos of each good and that the second basket contains (y_1, y_2) kilos of each good. Which one should you prefer? The situation is even worse if, e.g., $x_1 > y_1$ and $x_2 < y_2$. In this case we need to know which of the two fruit we prefer, do we prefer apples to pears and if yes "how much" do we prefer one fruit to the other so that we may make our mind up of how much more apples shall we have in a basket so that this excess makes up for the less pears in this basket? Clearly a simple comparison of two numbers is no longer enough, and we need to think of something better!

Suppose that we need to choose between two baskets of n goods each modeled by two vectors $x = (x_1, \cdots, x_n)$ and $y = (y_1, \cdots, y_n)$. If prefer x to y we denote it as $x \succ y$. If we prefer x at least as much as y then we denote it as $x \succeq y$. But which criterion may be used to make this choice up? The set where our choices live is $\mathbb{X} := \mathbb{R}^n_+$ (since the quantities involved are positive quantities). We may not define the operation of inequality in $\mathbb{X}$ as we have done in $\mathbb{R}_+$. One way around that is to define a function $U : \mathbb{X} \to \mathbb{R}_+$ such that

$$(x_1, \cdots, x_n) \succ (y_1, \cdots, y_n) \text{ if and only if } U(x_1, \cdots, x_n) > U(y_1, \cdots, y_n).$$

This function is called a utility function. The use of the utility function is to transfer the preference relation $\succ$ on $\mathbb{X}$ to the operation of inequality $>$ on $\mathbb{R}_+$. This function contains all the necessary information concerning our choice between the two baskets, e.g., it contains the information of how much we prefer one fruit to the other and if yes how much so as to decide between the various baskets.

The existence of a utility function is not always guaranteed. There are certain properties that a preference relation must display so that a utility function may exist. These are the properties of rationality and continuity:

Definition 9 *A preference relation is called rational if*

- *For every $x \in \mathbb{X}$, $x \succeq x$.*
- *For all $x, y \in \mathbb{X}$ either $x \succeq y$ or $y \succeq x$.*
- *For all $x, y, z \in \mathbb{X}$ such that $x \succeq y$ and $y \succeq z$ we have that $x \succeq z$.*

To define continuity we need to define some way of checking when two elements in $\mathbb{X}$ are close enough. This requires either a topology, or not wishing this level of generality at this stage, just a metric. If we consider $\mathbb{X}$ to be a metric space, with metric $d : \mathbb{X} \times \mathbb{X} \to \mathbb{R}_+$ then continuity of the preference relation means that for two elements $x, \bar{x} \in \mathbb{X}$ such that $d(x, \bar{x}) < \epsilon$, if $x \succeq y$ then it also holds that $\bar{x} \succeq y$. In the case where $\mathbb{X} = \mathbb{R}^n_+$ the metric d can be chosen as the Euclidean metric. Not all preference relations are continuous! An example of noncontinuous preference relation is the lexicographic preference relation which chooses out of every bundle the one that contains most in the earliest possible coordinate. E.g., if $n = 3$ we choose $x = (x_1, x_2, x_3)$ from $y = (y_1, y_2, y_3)$ if $x_1 > y_1$ or if $x_1 = y_1$ and $x_2 > y_2$ or if $x_1 = y_1$, $x_2 = y_2$ and $x_3 > y_3$. This reminds us of the way that words are being classified in dictionaries, hence the term lexicographic preference relation. It may be shown that the lexicographic preference relation may not be represented by a utility function! If it did then we would have an one to one correspondence between the rational numbers and the real numbers which of course is not possible!

We thus have the following important theorem:

Theorem 7

(i) A preference relation which is represented by a utility function is rational.

(ii) A rational and continuous preference relation can be represented by a utility function.

The proof of the first claim is more or less trivial. The proof of the second claim is more involved and requires topological tools that may not be presented within the scope of the present book.

12.4 COLLECTIVE CHOICE

In certain situations of interest a group rather than an individual must make a choice. Eventhough the theory of individual choice is well understood, the theory of collective choice is more delicate and complicated.

Consider the following simple example. Assume that we have 3 individuals wishing to make up their minds between 3 choices $x, y, z \in \mathbb{X}$. Each individual i has a rational preference relation $\succeq_i$ that may be represented by a utility function U_i. Suppose that the preference relations are as follows:

$$\begin{array}{ll} \text{individual 1} & x \succeq_1 y \succeq_1 z \\ \text{individual 2} & y \succeq_2 z \succeq_2 x \\ \text{individual 3} & z \succeq_3 x \succeq_3 y \end{array}$$

Let us assume that the 3 individuals decide to use the majority rule in making up their collective preference $\succeq$. Voting so as to decide between x and y we see that there are 2 against 1 which prefer x to y, therefore, the collective preference is $x \succeq y$. Voting between y and z there are 2 votes against 1 which prefer y to z, therefore, the collective preference is $y \succeq z$. Finally, when voting between x and z there are 2 votes against 1 that prefer z to x, therefore $z \succeq x$. But if we look at the collective preference

$$\text{group} \quad x \succeq y, \ y \succeq z, \ z \succeq x$$

The collective preference is no longer rational! Therefore rationality is a fragile property that may not be satisfied by the group decision. Even though the individuals are rational beings, the collective may act irrationally (and therefore cannot have a utility function). This example is a very old one, first written down by Condorcet, who made this observation in the 18th century, and this is a paradox that bears his name.

The transition from individual preferences to collective preferences is not a trivial task. In fact it is a field of very active research since the 1950's when Kenneth Arrow proved an important theorem, the Arrow impossibility theorem. This theorem states that even though individual preferences may be rational, this is not necessarily true for collective preferences. This theorem has started the very important field of social choice theory which is still an active research field.

12.5 PREFERENCES UNDER UNCERTAINTY

In real life we very seldom know the exact content of a bundle x. Depending on contingencies, the value of a bundle will vary. For example a ton of oil will have different value depending on contingencies, it will have higher value under adverse circumstances, e.g., if there is rough political circumstances in the Middle East and lower value if not. We would like to extend our theory of preferences for the case of uncertainty.

Our model will be as follows. We consider, to start with, two time instances $t = 0$ and $t = 1$. At time $t = 1$ only one of S different states of the world will materialize but at $t = 0$ when we wish to make our minds up we do not know exactly which. All we know is the probability of occurrence of each state $P(\text{State} = s) = p_s$, $\sum_{s=1}^{S} p_s = 1$, $p_s \in [0, 1]$. Depending on the state of the world each choice made at $t = 0$ will deliver a different payoff at time $t = 1$. We call x_s the payoff of choice x if the state of the world s materializes. We will write that as the vector $x = (x_1, \cdots, x_S)$. One way of considering x is as a discrete random variable such that $P(x = x_s) = p_s$, $s = 1, \cdots, S$.

Suppose now that we have to make up our minds at $t = 0$ between two choices $x = (x_1, \cdots, x_S)$ and $y = (y_1, \cdots, y_S)$. Which one should we choose? Clearly the probability of occurrence of different payoffs at $t = 1$ should play a rôle in this decision process. For instance, suppose that we want to choose between two lotteries; one paying 30 euros with probability 1/3, 10 euros with probability 1/3 and 0 with probability 1/3 and one paying 100 euros with probability 1/6, 20 euros with probability 1/6 and 0 with probability 2/3. Which one is preferable?

To answer this question we need to define a utility function $U : \mathbb{X} := \mathbb{R}^S \to \mathbb{R}_+$ such that

$$x \succeq y \quad \text{if and only if} \quad U(x_1, \cdots, x_S) \geq U(y_1, \cdots, y_S).$$

The existence of this utility function is covered by Theorem 7. However, we may do better than that.

The simplest idea that crosses our mind when trying to compare two lotteries is to compare their average payoff (the mathematical expectation of the payoff). This is defined as follows:

$$\mathbb{E}[x] := \sum_{s=1}^{S} p_s x_s.$$

We would then pick the lottery which has the highest average payoff,

$$x \succeq y \text{ if and only if } \mathbb{E}[x] \geq \mathbb{E}[y].$$

However, this simple idea has certain drawbacks, coming from the observation that there are well-defined random variables that do not have a well-defined expectation. One classic example is the St. Petersburg paradox, in which a game is defined whose average payoff is infinite so that the above scheme may not go through.

An extension of the above simple and intuitive idea is that of expected utility. This idea was introduced by Bernoulli in the 18th century in an attempt to explain the St. Petersburg paradox. However, the existence of such a concept was only proved mathematically much later, in the 1940's, by Von Neumann and Morgenstern in their study of the theory of games.

Definition 10 *A utility function is said to have the expected utility property if* $U(x) = \sum_{s=1}^{S} p_s U(x_s)$.

Therefore, if we consider x as the random variable X such that $P(X = x_s) = p_s$ the expected utility function is such that $U(x) = \mathbb{E}[U(X)]$.

A very important concept is that of risk aversion. This quantifies how much one is willing to undertake risk. This is shown in the properties of the expected utility function.

Definition 11 *An agent is said to be risk averse if she chooses a certain payoff of value equal to the expected payoff of a lottery rather than the lottery itself, i.e., a risk averse agent has an expected utility function such that*

$$U(\mathbb{E}[X]) \geq U(x) = \mathbb{E}[U(X)].$$

Functions with the above property are called concave functions. Therefore, the expected utility function of a risk averse agent is a concave function. Since more is always better this function must also be an increasing function. Examples of functions with this property are $U_1(w) = \frac{1-e^{-\lambda w}}{\lambda}$, $U_2(w) = \ln(w)$ and $U_3(w) = \frac{w^\gamma}{\gamma}$, $\gamma < 1$. The coefficients λ and γ play an important rôle in these functions. Observe that $-\frac{U_1''(w)}{U_1'(w)} = \lambda$, and $-\frac{wU_3''(w)}{U_3'(w)} = \gamma$, constant for any w. These numbers, called Arrow's measure of absolute and relative risk aversion respectively characterize the risk preferences of the agent.

Expected utility has been the subject of criticism and there are indications that the axioms leading to the existence of expected utility may not hold for certain circumstances (see, e.g., the discussion of the Allais paradox, or the discussion and experimental data leading to alternative theories such as prospect theory, etc.) see, e.g., Ref. [12]. Furthermore, other popular description of choice are random utility models, in which it is considered that the utility function is perturbed by random terms. Such models lead to interesting descriptions that some time may also lead to parametric models which are subject to empirical tests. Concerning random utility theory and discrete choice theory see e.g. [9].

■ EXAMPLE 12.1 Should I Break the Law or Not?

Suppose that your morals being weak your only inhibition in committing a minor misdemeanor is the fear of getting caught and paying a fine. Suppose for instance that you may avoid honoring an obligation that will cost you C but if you are caught doing that then you will have to pay a fine F. Of course there is a probability of getting caught π and a probability $1-\pi$ of getting away with it. What should you do?

Well, suppose you are utility maximizer and your preferences can be modeled by an expected utility U. Suppose your initial wealth is W. In trying to figure out whether you will break the law or not you essentially face the following lotteries. If you do break the law then you face

$$L_{cheat} = \begin{cases} W & \text{with probability}\, 1-\pi \\ W-F & \text{with probability}\, \pi. \end{cases}$$

If you don't you face $L_{honest} = W - C$ with certainty. You will cheat if $L_{cheat} \succ L_{honest}$, i.e., if

$$(1-\pi)U(W) + \pi U(W-F) > U(W-C),$$

and choose to be honest otherwise. Clearly, this will depend on your personal profile (the utility function U), the sum you are about to escape C, the fine you will be subjected to F and of course the probability of being caught π. If for instance $U(x) = ax + b$ (a risk neutral agent) then a quick calculation shows that you will cheat as long as $\pi F < C$. If on the other hand the agent is a risk averse agent with $U(x) = -e^{-\lambda x}$ then a quick calculation shows that the agent will cheat as long as $\pi e^{\lambda F} \leq e^{\lambda C} - (1-\pi)$. Certain observations of a qualitative nature can be made: the more risk averse the agent is the less the fine required, the higher the probability to get caught the less is the required fine, etc.

■ EXAMPLE 12.2 A Utility Approach to the Theory of Voting

Utility theory has been extensively used in the theory of voting. In this example we give an idea of how the concept of utility may be used to model and understand voters' behavior.

A candidate's program can be thought of as a point in a possibly high-dimensional space, called policy space, $\mathcal{P}$. In most models, it is considered that $\mathcal{P} \subset \mathbb{R}^d$ for some $d \in \mathbb{N}$. A policy x then is a point in $\mathcal{P} \subset \mathbb{R}^d$, meaning that a policy x can be represented as a vector $x = (x_1, \cdots, x_d)$. Each component of this vector corresponds to the candidate's position on one of d issues, e.g., x_1 can be public expenditures for health, x_2 for education, x_3 for armaments, etc.

Consider a set of voters. Each voter i has her personal attitudes towards the ideal policy that a candidate should follow. This will again be a point in policy space $\mathcal{P}$, denoted by $x^{(i)} = (x_1^{(i)}, \cdots, x_d^{(i)})$. A voter will be happier with the program of a candidate the closer the candidate's program x is to her ideal view $x^{(i)}$. There are many ways to measure distance. An obvious choice can be the Euclidean distance on $\mathbb{R}^d$, $d(x, x^{(i)}) = \left(\sum_j^d (x_j - x_j^{(i)})^2\right)^{1/2}$. This essentially means that for the voter, all issues are of equal importance. Of course, it may well be that certain issues are of more importance than others. This situation could be modeled by using the distance $d_w(x, x^{(i)}) = \left(\sum_j^d w_j (x_j - x_j^{(i)})^2\right)^{1/2}$, where $w = (w_1, \cdots, w_d)$ is a set of weights such that $0 \leq w_j \leq 1$, $j = 1, \cdots, d$ and $\sum_{j=1}^d w_j = 1$, which measure the relative importance of the various issues for the voter. Clearly, these weights may depend upon the voter, but we avoid the notation $w = w^{(i)}$ so as not to clutter too much the notation.

Having defined the notion of distance in policy space, we may now define the concept of utility that voter i will enjoy if the candidate is elected. We may assume this to be of the form $u_i(x) = U_i(d_w(x, x^{(i)}))$, where $U_i : \mathbb{R}_+ \to \mathbb{R}$ is a decreasing function. This representation means that the utility that voter i will enjoy if the candidate with policy x is elected is decreasing as long as the distance of the candidate's policy from the voters ideal policy is increasing. Such preferences are called Euclidean preferences because they are expressed in terms of a Euclidean type distance in policy space.

Consider now that there are two candidates to choose from: one candidate has policy $x \in \mathcal{P}$ whereas the other candidate has policy $y \in \mathcal{P}$. Which one is voter i likely to support? This decision can be modeled as comparing policy x with policy y. Under the assumption that the preference relation can be expressed in terms of the utility function above, we can say that

$$x \succ y \text{ if and only if } U_i(d_w(x, x^{(i)})) > U_i(d_w(y, x^{(i)})),$$

therefore, since U_i is a decreasing function

$$x \succ y \text{ if and only if } d_w(x, x^{(i)}) < d_w(y, x^{(i)}).$$

That means that voter i will support the candidate that is closer to her perception in policy space. This finding is very intuitive and is sometimes (but not always!) supported by empirical findings.

The next level of complication is to add some uncertainty to this model. A voter does not always know very well what she wants. Furthermore, there is no certainty that a candidate will abide to her proposed program, in fact, experience often points to the opposite direction. Therefore, we may modify our utility model to $u_i(x) = \epsilon_{i1} + U_i(d_w(x, x^{(i)}))$, where ϵ_{i1} is a random variable, related to the credibility of candidate 1 as perceived by voter i, and $u_i(y) = \epsilon_{i2} + U_i(d_w(y, x^{(i)}))$ where ϵ_{i2} is a random variable related to the credibility of candidate 2 as perceived by voter 2. Since a voting decision is reduced to comparing utilities, but now utility itself is a random variable, the decision of choosing between the two is an event that has a probability assigned to it. In particular,

$$\begin{aligned}
p_i(x \succ y) &:= P(\text{voter } i \text{ prefers 1 to 2}) \\
&= P(\epsilon_{i1} + U_i(d_w(x, x^{(i)})) > \epsilon_{i2} + U_i(d_w(y, x^{(i)}))) \\
&= P(\epsilon_{i1} - \epsilon_{i2} > U_i(d_w(y, x^{(i)})) - U_i(d_w(x, x^{(i)})) \\
&= 1 - F(U_i(d_w(y, x^{(i)})) - U_i(d_w(x, x^{(i)}))),
\end{aligned}$$

where F is the cumulative distribution function of the random variable $\epsilon_{i1} - \epsilon_{i2}$. A common assumption is that the distribution of the error is Gaussian. Then this model will provide an expression for the probability that voter i prefers candidate 1 to candidate 2 in terms of the weights w and the opinion of the voter $x^{(i)}$ (these are the personal characteristics of the voter) and the proposed policies of the two candidates x and y, similar to the probit model. Based on that surveys can be, and are, constructed, trying to parameterize the model and fit it to survey data, using techniques from statistics. Of course, other choices for either the distribution of the error or the utility of the voters are possible and are being used.

■ EXAMPLE 12.3 Willingness to Pay

Suppose that you have a common good, say, a park, for example, and you want to evaluate it. Of course there is no reason why you should, but if you must assign a pecuniary value to it, then a reasonable question to ask is "if one was not allowed the use of this good how much would she be willing to pay in order to use it."

To answer this question, assume that an agent has certain characteristics (e.g., age, sex, education level, etc.) which can be summarized in the vector z. This vector has as many coordinates as the characteristics we take into account for the agent. Furthermore, this agent is characterized by an income, which we call y. This agent derives utility from her income, and we may assume that this utility function is of the form

$$u(y) = a \cdot z + \beta(y) + \epsilon,$$

where a is a vector the components of which give some information on how the change in characteristics may change the utility level, b is a known function and ϵ is a random term. This random term is useful in our modeling, since we may not expect all individuals sharing some characteristics (as included in z) having similar behavior. Therefore, ϵ takes care of possible variability of the utilities of individuals with common characteristics z.

It is very reasonable to assume that the utility of an agent depends on the particular state of the world that she is in. If state of the world 1 is the state of the world in which she has full access to the park then her utility level will be of the form

$$u_1(y) = a_1 \cdot z + \beta_1(y) + \epsilon_1,$$

where as above a_1 is a vector modeling the effect of attributes to the utility functions, b_1 is a known deterministic function and ϵ_1 is a random variable. Similarly, if state of the world 2 is the state of the world in which she is denied access to the park then her utility level will be of the form

$$u_2(y) = a_2 \cdot z + \beta_2(y) + \epsilon_2,$$

where a_2, b_2 and ϵ_2 have similar meaning as a_1, b_1 and ϵ_1 but in general $a_1 \neq a_2$, $b_1 \neq b_2$ and $\epsilon_1 \neq \epsilon_2$.

Suppose now that this agent may gain access to the common good by paying an admission fee of W. How much would she be willing to pay in order to access the common good? This sum is called willingness to pay (WTP) and of course depends on the individual's characteristics z as well as on other parameters of the utility function.

The calculation of W is obtained by a simple utility calculation. If she accepts to pay W to access the common good, then her income is lowered by that sum and becomes $y - W$ but she is compensated by the use of the common good. That means that by lowering her income to $y - W$ at state 2 she gets the same utility level that she enjoyed at state 1 when her income was y. This gives us the equality

$$a_2 \cdot z + \beta_2(y - W) + \epsilon_2 = a_1 \cdot z + \beta_1(y) + \epsilon_1,$$

which upon a simple rearrangement gives

$$W = y - \beta_2^{-1}\left(\alpha \cdot z + \beta_1(y) + \epsilon\right),$$

where $\alpha = a_1 - a_2$, $\epsilon = \epsilon_1 - \epsilon_2$. Clearly, willingness to pay is a random variable whose distribution depends on the distribution of ϵ. Certain choices are possible. For instance, if $b_i(y) = \ln(y)$, $i = 1, 2$ then the willingness to pay is given by

$$W = y\left(1 - \exp\left(-(\alpha \cdot z + \epsilon)\right)\right).$$

Then depending on the distribution of ϵ we obtain different models that can be quite useful for statistical investigation. If the error terms are normally distributed, $\epsilon \sim \mathcal{N}(0, \sigma^2)$ where σ^2 is an unknown variance, then we obtain the so-called probit model according to which

$$\mathbb{E}[W \mid \alpha, z, y] = y\left(1 - \exp\left(-\alpha \cdot z + \frac{1}{2}\sigma^2\right)\right).$$

In the above by $\mathbb{E}[W \mid \alpha, z, y]$ we denote the expected willingness to pay given the values of the characteristics α, z and y of the agent. This is a very useful result, because since it is a simple parametric formula it can be used in questionnaires and surveys, and using the data obtained calibrate the model, i.e., find values of the parameters α, such that the willingness to pay obtained from a sample with known z and y matches this model. Another common choice is to assume that ϵ is logistically distributed, in which case we obtain the logit model. Models of this type are frequently used in marketing, environmental science, etc., see, e.g., Ref. [6].

12.6 DECISIONS OVER TIME

In many cases of interest we must make a decision and choose between bundles received at different time instances. For instance, should I get one euro now or should I get it next year? Humans by nature[13] are impatient. They will rather prefer to have the euro now than wait and get it next year. If for some reason an agent B (e.g., the bank) wishes to postpone the payment of the euro to agent A until next year there must be some incentive offered by B to A so as to persuade her to wait. This incentive will be a higher payment than 1 euro if paid on the next year, say, $1(1 + r)$ where $r \geq 0$. This is nothing else but the concept of interest.

How can we model this impatience effect within our utility function approach? Suppose that we have different time instances, t and certain payoffs (a sequence of payoffs) $(x_1, x_2, \cdots, x_t, \cdots)$. To set ideas consider two different

[13]It is extremely dangerous to invoke arguments concerning the human nature, especially in a chapter bearing social science on its title. Human behavior is formed by the society the agents live in, however, we ask the reader to allow us this slip of the tongue (especially since our work is embedded in a volume on mathematics).

payment streams $(1,0)$ and $(0,1)$ The first corresponds to one euro being paid at $t = 0$ and nothing at $t = 1$, whereas the second corresponds to nothing being paid at $t = 0$ and one euro being paid at $t = 1$ (i.e., we postpone the payment of the euro for one time period). A utility function for the first payment can be given as $U_1 := U(1) + \delta U(0)$ whereas for the second payment $U_2 := U(0) + \delta U(1)$. Without loss of generality let us assume that $U(0) = 0$. Then $U_1 \geq U_2$ if $\delta < 1$ and this means that we prefer the payment now to the payment in the next year. The factor $\delta < 1$ is called a discount factor and its presence is very important in choice theory over time. This δ is related to the interest rate r, by $\delta = (1+r)^{-1}$.

This discussion may be generalized to any number of time instances. To evaluate a payment stream $x = (x_0, x_1, x_2, \cdots)$ we may use the intertemporal utility function

$$U(x) = \sum_{t=0}^{\infty} \delta^t U(x_t), \quad \delta < 1. \tag{12.1}$$

This utility function has the property that the relative preference between, e.g., $(1,0,0,\cdots)$, $(0,1,0,\cdots)$, $(0,0,1,\cdots)$ is equal to δ, i.e., at all times in the life of an individual there is a relative preference of δ between payments delayed by one time unit. The intertemporal utility function (12.1) is used very often for decision making over time. A fruitful way to understand this utility function is to understand it as a functional, i.e., as a mapping that maps sequences into the space of real numbers.

An alternative is to assume that time is continuous and that the payment stream is approximated by a continuous function x such that $x(t)dt$ gives the payment in the interval $[t, t+dt)$. Then the intertemporal utility function is given by

$$U(x) = \int_0^{\infty} e^{-rt} U(x(t))dt. \tag{12.2}$$

This is considered again as a functional, i.e., as a mapping from the space of continuous functions to the real numbers.

The exponential form of the intertemporal utility functions (12.1) or its continuous version (12.2) is used extensively in the modeling of preferences over time. In fact, it is a form of intertemporal utility function that leads to temporally consistent decisions. However, the exponential intertemporal utility function is of limited use when decisions over long time periods have to be taken. There is support by experimental evidence that when humans or animals make decisions over large periods of time, there is not a constant ratio of the utility between different time periods. In fact agents tend to be more patient in the beginning of the time period than towards the end of the horizon. This effect is called hyperbolic discounting. An intertemporal utility function displaying this hyperbolic discounting effect is

$$U(\{x_s\}_{s=t}^{\infty}) = u(x_t) + \beta \sum_{\tau=1}^{\infty} \delta^{\tau} u(x_{t+\tau}), \ \ \beta \in (0,1], \ \ \delta \in (0,1). \tag{12.3}$$

When $\beta = 1$ this utility function is the exponential utility function.

■ EXAMPLE 12.4 The Cost of Climate Change

Climate change will lead to long term effects that may affect a number of generations to come. However, if something is to be done to fight it, action should be taken today. Is that worth or not?

One way to answer this question is to estimate the future costs of climate change to the future generations. Let us denote by c_i the cost (losses) incurred to generation i as an effect of climate change. If we assume an exponential discounting scheme, with constant discount factor δ per generation, then the value of the future costs calculated in today's units is

$$R = \sum_{i=1}^{N} \delta^i c_i.$$

How should a politician decide whether to engage in action against climate change or not? Simply by comparing the amount of money she is willing to spend on measures, R_W, with R. A very simple decision rule would be to decide to take action if $R_W \geq R$ and not otherwise.

Cynical as it may sound this is the way political decision is often made. However, this decision rule is not robust since the discount factor δ that enters the calculation is to be interpreted as some sort of social (or aggregate) discount factor and this is quite difficult to measure! What makes things worse is that the value R is very sensitive with respect to the choice of δ. The choice of discount factor has sparked a great debate among economists, moral philosophers, mathematicians, environmental scientists, etc., and has led to the theory of hyperbolic discounting.

12.7 THE PROBLEM OF AGGREGATION

12.7.1 Aggregation of Time Preferences

One of the important problems in going from the individual to the collective is the aggregation of time preferences [4]. The problem is a very natural one (see, e.g., Example 12.4).

The problem has a nice optimization interpretation. Consider a group of agents $\mathcal{I}$, each member of the group having intertemporal utility function

$$U_i(x) = \int_0^t u_i(t, x(t))dt, \quad i \in \mathcal{I},$$

where the discount type is taken into acount in the definition of the function u_i. For instance, if all the agents have an exponential discount function each with discount rate r_i, then $u_i(t, x(t)) = e^{-r_i t}\hat{u}_i(x(t))$. The group is endowed by an intertemporal consumption flow X which is to be divided among the group, each member of the group getting a portion x_i. Obviously the constraint

$$X(t) = \sum_{i \in \mathcal{I}} x_i(t) \tag{12.4}$$

must hold for all $t \in \mathbb{R}_+$. We are interested in Pareto efficient allocations, i.e., in allocations $\{x_i\}$, $i \in \mathcal{I}$ of the group endowment process so that all available resources of the group are divided between the group.[14] In this case, if one member of the group increases her income by some amount, another member of the group must suffer a decrease in her income. Such a Pareto efficient allocation (which may be far from being a fair allocation) can be characterized in terms of the following variational problem,

$$\max_{x_i \in \mathcal{I}} \sum_{i \in \mathcal{I}} \lambda_i U_i(x_i)$$

subject to the constraint (12.4), for a given choice of weights $\lambda = \{\lambda_i\}$, $i \in \mathcal{I}$ such that $\lambda_i \in [0, 1]$ and $\sum_{i \in \mathcal{I}} \lambda_i = 1$. This corresponds to an allocation of the common resource (consumption flow) X to the members of the group so that each member i receives a portion x_i. This would not be the member's choice for consumption if she were alone, i.e., it does not maximize U_i. But it is the member's choice if the other members of the group are taken into account, with a relative weight λ_i. In some sense the weighted sum of utilities is some sort of social welfare function for the whole group.

To define the utility function of the representative agent we need to define the utility function U_g of a single agent, which is consistent with this Pareto efficient allocation. In particular, according to Ref. [4] the group utility function is such that

$$u_g(t, X) := \max_{x_i \in \mathcal{I}} \sum_{i \in \mathcal{I}} \lambda_i u_i(t, x_i),$$

for some choice of $\lambda_i \in [0, 1]$ such that $\sum_{i \in \mathcal{I}} \lambda_i = 1$, subject to the constraint (12.4). The group intertemporal utility function U_g is given in terms of u_g

[14]Though not necessarily in an equal fashion!

as $U_g(X) = \int_0^\infty u_g(t, X(t))dt$ and of course depends on the allocation $\{\lambda_i\}$, $i \in \mathcal{I}$. This is a resource sharing problem. By concavity the solution of the problem is unique. The sharing of the individual consumption is given by u_g as

$$x_i(t, X(t)) = \frac{T_i(t, x_i(t))}{\sum_{i \in \mathcal{I}} T_i(t, x_i(t))},$$

where $T_i(t, x_i) := -\frac{\partial u_i/\partial x}{\partial^2 u_i/\partial x^2}$. The group's rate of impatience is given in terms of the functions T_i according to the rule

$$r_g(t, X) = \frac{\sum_{i \in \mathcal{I}} r_i(t, x_i(t)) T_i(t, x_i(t))}{\sum_{i \in \mathcal{I}} T_i(t, x_i(t))}.$$

Therefore, the group impatience is a weighted average of the individual discount rates, and the weighting factors are related to the optimum consumption of each individual in the chosen allocation. An important consequence of the above result is that even though all individuals may have exponential discounting with constant in time discount functions r_i, the group discounting may be hyperbolic with increasing (decreasing) in t, $r_g(t)$ depending on whether the individual utility functions $\hat{u}_i$ are decreasing (increasing) with respect to x_i [4]. This gives rise to fluctuations in $r_g(t)$ which are driven by fluctuations on the consumption $x_i(t)$.

12.7.2 Aggregation of Beliefs

In certain circumstances of interest a group of agents presents inhomogeneous beliefs. How can we obtain an estimate for the belief of the group? This is a problem which is very intriguing and has been dealt with by a number of authors. We present a recent approach to this problem by Golier [5]. The basic idea is similar to that employed in the problem of aggregation of time preferences, i.e. use of a Pareto optimal allocation through which the collective utility function is defined.

Consider as above a group of agents $\mathcal{I}$, each agent denoted by i. Each agent has her own idea concerning the future states of the world $s \in \mathcal{S} := \{1, \cdots, S\}$, $p_i := \{p_{i,s}\}_{s=1}^S$. The agents derive utility from the consumption of the sharing of a common good $X = \{X(t)\}$, according to a sharing rule $\{x_i(t)\}$, $i \in \mathcal{I}$. We assume that this allocation is a Pareto efficient allocation, so that there exists a vector $\Lambda = \{\lambda_i\}$, $i \in \mathcal{I}$, $\lambda_i \in [0, 1]$, $\sum_{i \in \mathcal{I}} \lambda_i = 1$, such that

$$\max_{\{x_i\}, i \in \mathcal{I}} \sum_{i \in \mathcal{I}} \lambda_i \sum_{s \in \mathcal{S}} p_{i,s} u_i(x_i(t)), \tag{12.5}$$

subject to the sharing rule $\sum_{i \in \mathcal{I}} x_i(t) = X(t)$. Since $X(t)$ and $x_i(t)$ are random variables this equation is considered as an equation of random variables, i.e., holding for every $s \in \mathcal{S}$. This allocation of risk is considered as the socially optimum according to the chosen sharing rule.

This problem may be decomposed into S problems,

$$\max_{\{x_i(t;s)\}, i\in\mathcal{I}} \sum_{i\in\mathcal{I}} \lambda_i p_{i,s} u_i(x_i(t;s)), \quad s \in \mathcal{S}$$

subject to the constraint $\sum_{i\in\mathcal{I}} x_i(t;s) = X(t;s)$ one for each state of the world.[15] Solving the problem for each $s \in \mathcal{S}$ we denote the maximum by $v(X(t;s);\{p\})$, where by $\{p\}$ we denote the set of beliefs of the group of agents. The solution of problem (12.5) is expressed as the weighted sum $W = \sum_{s\in\mathcal{S}} v(X(t;s);\{p\})$. The quantity W measures the welfare of the representative agent.

It is important to characterize the utility function that represents the representative agent. The properties of this utility function depend on the properties of the function v, which encompasses all the individual subjective beliefs. For instance, the utility function of the representative agent has the expected utility property if v is such that $\frac{\partial^2 v}{\partial p \partial X} = 0$. The probability of the group, p_g may be defined via the following relationship:

$$\frac{p_{g,s}}{p_{g,s'}} = \frac{\partial v_g(X(s);P(s))/\partial X}{\partial v_g(X(s');P(s'))/\partial X},$$

where of course the condition $\sum_{s\in\mathcal{S}} p_{g,s} = 1$ must hold, and by P we denote the vector of the beliefs (subjective probabilities) of the individuals in the group. As before, we assume that this Pareto efficient allocation is compatible with a group expected utility function U_g, which is generated by a common belief $p_g = \{p_{g,s}\}$, $s \in \mathcal{S}$.

12.8 CONCLUSION

In this chapter we have introduced some basic concepts of decision theory, focusing on individual decision theory. In particular, after introducing the fundamental theoretical framework of rationality and utilitarianism, we have developed the concept of utility theory as a decision tool. We have further discussed decision theory under uncertainty, and in particular the theory of expected utility and random utility. Their use as a decision making tool in the social sciences was illustrated through a number of examples.

[15] We have introduced s explicitly into the equations to stress the contingency dependence of the state variables.

Acknowledgments

The authors wish to thank Professor Nicholas Yannacopoulos for his constructive comments and enjoyable conversations that maximized their utility (in the Benthamian use of the concept). They also acknowledge the useful comments of the three referees that led to considerable improvement of this chapter. They also wish to thank Dr. Yang for putting this effort together and offering them the tribune from which to present this first introduction to this exciting subject.

EXERCISES

12.1 (St. Petersbourg paradox) Suppose you wish to participate in the following game: You bet 1 euro on a fair coin. If the coin lands tails up in the first time you bet 2 euros. If the coin lands again tails up you bet 4 euros. It the coin land tails up in the nth play you bet 2^n euros. You win the sum of your last bet (and stop) the first time that the coin lands head up. How much should you pay as an entrance fee to the game? Show that the expected winning from this game is infinite (hence you should not pay this sum as an entrance fee) but the expected logarithmic utility of the winnings is finite.

12.2 An agent has intertemporal utility $u(y_1, y_2) = u_0(y_1) + \delta u_0(y_2)$ with $\delta < 1$ and u_0 a strictly increasing function. Suppose that this agent has two options. One to consume 1 unit now and 0 in the second $(y_1, y_2) = (1, 0)$ and one to consume 0 now and consume p in the second $(y_1, y_2) = (0, p)$. How much should p be so that the second option is preferred?

REFERENCES

1. Archer, M. and Tritter J. (eds.), Rational Choice Theory: Resisting Colonisation, Routledge, London (2000).
2. Bermúdez, J. L., Decision Theory and Rationality, Oxford University Press, Oxford (2009).
3. Foucault, M., The Order of Things. An Archaeology of the Human Sciences, Routledge, London (1989).
4. Gollier, C. and Zeckhauser, R., Aggregation of heterogeneous time preferences, Journal of Political Economy, Vol. 113, 878–896 (2005).
5. Gollier, C., Whom should we believe? Aggregation of heterogeneous beliefs, J. Risk Uncertainty, Vol. 35, 107-127 (2007).
6. Haab, T. C. and McConnel, K. E., Valuing Environmental and Natural Resources, Edward Elgar Publishing, (2002).
7. Harsanyi, J., Game and decision theoretic models in ethics, in: Aumann R. and Hart, S. (Eds.), Handbook of Game Theory with Economic Applications, Vol. 1, Elsevier, (1992).

8. Johnston, R. J., Philosophy and human geography: An introduction to contemporary approaches. 2nd Ed. Edward Arnold, (1986).

9. Manski, C. F., The structure of random utility models, Theory and decision, Vol. 8, 229–254 (1977).

10. Mulgan, T., Understanding Utilitarianism, Acumen, (2007).

11. Shapiro, I., The Flight from Reality in the Human Sciences, Princeton University Press, 2005.

12. Starmer, C., Developments in Non-expected Utility Theory: The Hunt for a Descriptive Theory of Choice under Risk, Journal of Economic Literature, Vol. 38(2), 332-382 (2000).

13. Taylor, M., Rationality and the Ideology of Disconnectedness, Cambridge University Press, Cambridge (2006).

CHAPTER 13

FRACTALS, WITH APPLICATIONS TO SIGNAL AND IMAGE MODELING

H. KUNZE[1] AND D. LA TORRE[2]

[1]Department of Mathematics and Statistics, University of Guelph, Canada
[2]Department of Economics, Business and Statistics, University of Milan, Italy

Looking for the definition of "fractals" on the internet, you find fractals are sets whose Hausdorff-Besicovitch dimension exceeds their topological dimension.

That description seems very heavy, so we will write instead that fractals are complicated, often irregular, sets generally produced by iteration of some kind of operation.

In the next section, we'll see how to construct fractal sets through iterated function systems and we will develop an understanding of some odd facts about their dimension or size. In the subsequent section, we will see some modeling uses of fractals.

Mathematical Modeling with Multidisciplinary Applications.
Edited by Xin-She Yang

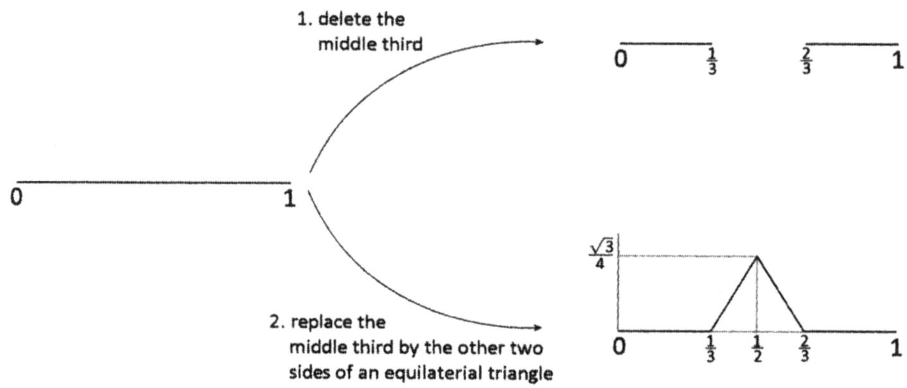

Figure 13.1 Start with the unit line segment and 1. delete the middle third, or 2. replace the middle third by the other two sides of the corresponding equilateral triangle.

13.1 ITERATED FUNCTION SYSTEMS

We develop the ideas through two examples, in parallel. In each case, we consider the line segment of length one lying on the interval $[0, 1]$ along the x-axis. We think of two operations:

1. Delete the middle third of the line segment. To be careful, we mean delete the *open* middle third of the line segment, $\left(\frac{1}{3}, \frac{2}{3}\right)$, leaving behind the *closed* set $\left[0, \frac{1}{3}\right] \cup \left[\frac{2}{3}, 1\right]$ on the x-axis.
2. Replace the middle third of the line segment by the other two sides of the corresponding equilateral triangle. Introducing the y-axis, our object now has corners at the points $\left(\frac{1}{3}, 0\right)$, $\left(\frac{2}{3}, 0\right)$, and, using some geometry, the raised point $\left(\frac{1}{2}, \frac{\sqrt{3}}{6}\right)$.

We illustrate the result in Figure 13.1. A very interesting thing occurs when we *iterate.* That is, repeatedly apply operation 1 to the collection of line segments (or sets) produced by the previous application of operation 1. Keep deleting middle thirds. At first, you might think you delete everything. That is not the case, since the endpoints 0 and 1 will never be in the middle of a segment. Similarly, looking at Figure 13.1, after the first iteration, the points $\frac{1}{3}$ and $\frac{2}{3}$ are endpoints, so they will never be deleted in future steps. In fact, an infinite number of points survive the process, for example, all points of the form $\frac{1}{3^n}$, where n is a nonnegative integer. Similarly, when we repeatedly apply operation 2 to the four line segments produced by the first application of operation 2, we get a "kinkier" curve. Think about iterating forever.

Approximations of the objects that result from iterating forever are presented in Figure 13.2. The object on the left in the figure, produced by

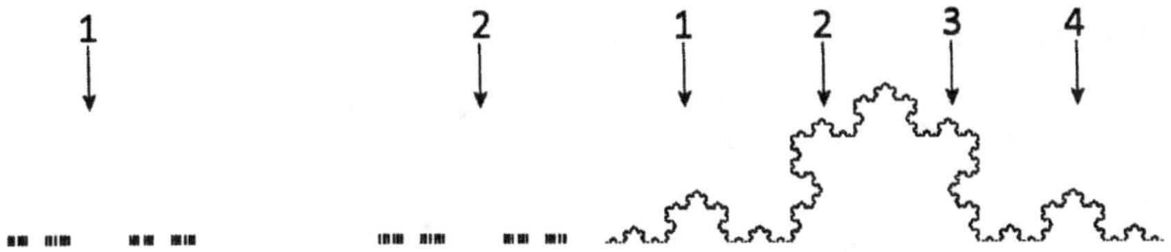

Figure 13.2 The middle-thirds Cantor set and the von Koch curve.

operation 1, is called the (middle-thirds) Cantor set. The object on the right, produced by operation 2, is called the von Koch curve. A little thought about the figure and the two operations will convince you that the Cantor set is precisely the set of points at which the von Koch curve touches the x-axis.

These objects feature the hallmark characteristic of fractal sets: self-similarity. The Cantor set consists of two shrunken copies of itself, as labeled in Figure 13.2. If we shrink the entire set by a factor of $\frac{1}{3}$, keeping $x = 0$ fixed, we get the left "half" of the Cantor set. If we translate that shrunken copy by $\frac{2}{3}$ units along the x axis, we get the right half. The function that shrinks lengths on the x-axis by $\frac{1}{3}$ without moving $x = 0$ is given by $\frac{1}{3}x$. To translate by $\frac{2}{3}$ unit, we just add $\frac{2}{3}$. The resulting system of functions is

$$\begin{aligned} f_1(x) &= \frac{1}{3}x, \\ f_2(x) &= \frac{1}{3}x + \frac{2}{3}. \end{aligned}$$

Letting S denote some set of points on the x-axis, we can consider

$$f_i(S) = \{f_i(x)|x \in S\}, \; i = 1, 2,$$

the set-valued analogues of the two maps. Finally, if we denote the Cantor set by C, then our initial observation that the Cantor set consists of two shrunken copies of itself is expressed as

$$C = f_1(C) \cup f_2(C). \tag{13.1}$$

We can perform similar work for the von Koch curve, which consists of four shrunken copies of itself, as labeled in Figure 13.2. Copy 1 is produced by shrinking *both* x and y by a factor of $\frac{1}{3}$, and copy 4 adds a translation by $\frac{2}{3}$ units in x afterwards. Copies 2 and 3 also shrink by $\frac{1}{3}$, but involve a rotation as well as a translation. For copy 2, for example, we must shrink by first, keeping the origin fixed, then rotate by 60° counterclockwise, and finally translate by $\frac{1}{3}$ in the x direction. Notice the order of the shrinking and the rotation when we write down the function:

$$f_2(x,y) = \begin{pmatrix} \frac{1}{3} & 0 \\ 0 & \frac{1}{3} \end{pmatrix} \begin{pmatrix} \cos\frac{\pi}{3} & \sin\frac{\pi}{3} \\ -\sin\frac{\pi}{3} & \cos\frac{\pi}{3} \end{pmatrix} + \begin{pmatrix} \frac{1}{3} \\ 0 \end{pmatrix}.$$

Similarly, copy 3 involves a counterclockwise rotation by $-60°$ and a translation by $\frac{1}{2}$ unit in x and $\frac{\sqrt{3}}{6}$ units in y. The resulting system of functions is

$$\begin{aligned}
f_1(x,y) &= \begin{pmatrix} \frac{1}{3} & 0 \\ 0 & \frac{1}{3} \end{pmatrix} \begin{pmatrix} x \\ y \end{pmatrix} + \begin{pmatrix} 0 \\ 0 \end{pmatrix}, \\
f_2(x,y) &= \begin{pmatrix} \frac{1}{6} & \frac{\sqrt{3}}{6} \\ -\frac{\sqrt{3}}{6} & \frac{1}{6} \end{pmatrix} \begin{pmatrix} x \\ y \end{pmatrix} + \begin{pmatrix} \frac{1}{3} \\ 0 \end{pmatrix}, \\
f_3(x,y) &= \begin{pmatrix} \frac{1}{6} & -\frac{\sqrt{3}}{6} \\ \frac{\sqrt{3}}{6} & \frac{1}{6} \end{pmatrix} \begin{pmatrix} x \\ y \end{pmatrix} + \begin{pmatrix} \frac{1}{2} \\ \frac{\sqrt{3}}{6} \end{pmatrix}, \\
f_4(x,y) &= \begin{pmatrix} \frac{1}{3} & 0 \\ 0 & \frac{1}{3} \end{pmatrix} \begin{pmatrix} x \\ y \end{pmatrix} + \begin{pmatrix} \frac{2}{3} \\ 0 \end{pmatrix}.
\end{aligned}$$

Let K be all points in the plane that lie on the von Koch curve. Then we conclude that

$$K = \bigcup_{i=1}^{4} f_i(K). \tag{13.2}$$

13.2 FRACTAL DIMENSION

When you need to find the distance between two objects in the same room, you might walk heel to toe and count how many of your feet are needed to span the distance. If your foot has size $\epsilon > 0$ and the number you need is $N(\epsilon)$, then you approximate the distance by $N(\epsilon) \cdot \epsilon$. Typically, you only obtain an approximation because the last little bit of the distance does not exactly equal the length of your foot. In order to be able to measure any distance, you would need to calculate $\lim_{\varepsilon \to 0} N(\epsilon) \cdot \epsilon$. This notion is similar to the calculation of an area via a definite integral, where we approximate the area by the sum of areas of rectangles, needing to take a limit to make sure that the approximation error approaches zero. In fact, in grade school, we are often asked to find the area of a leaf gathered in the school yard. We trace it on graph paper, count squares that contain some of the leave, and then multiply the area of a grid square ϵ^2 by the number of squares $N(\epsilon)$ to get our approximation. We can combine these two observations in one equation, writing

$$\text{Size} = N(\epsilon) \cdot \varepsilon^D,$$

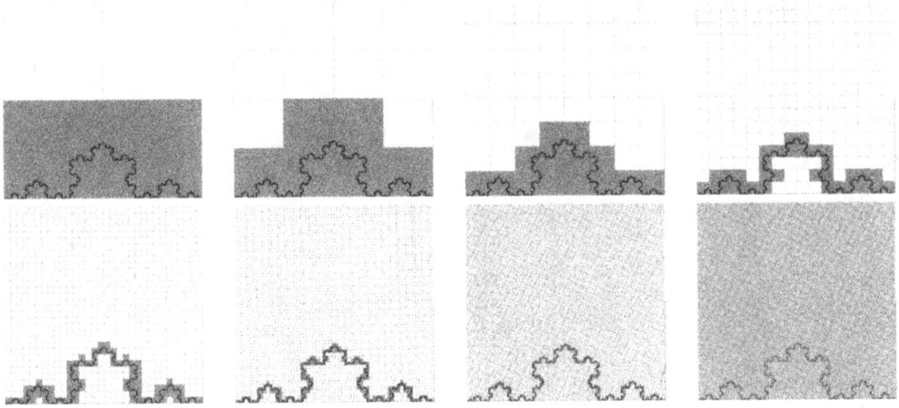

Figure 13.3 Box counting for the von Koch curve.

where $D = 1$ if we are finding length and $D = 2$ if we are finding area, and we need a limit to make the equal sign hold. Rearranging, we have

$$\begin{aligned} N(\epsilon) &= (\text{Size}) \cdot \varepsilon^{-D} \\ \Rightarrow \ln(N(\varepsilon)) &= \ln(\text{Size}) + D \ln \frac{1}{\varepsilon} \\ \Rightarrow \qquad D &= \frac{\ln(N(\varepsilon))}{\ln \frac{1}{\varepsilon}} - \frac{\ln(\text{Size})}{\ln \frac{1}{\varepsilon}}. \end{aligned} \tag{13.3}$$

Since $\ln \frac{1}{\varepsilon} \to \infty$ as $\epsilon \to 0^+$ and $\ln(\text{Size})$ is a constant, we get

$$D = \lim_{\varepsilon \to 0^+} \frac{\ln(N(\varepsilon))}{\ln \frac{1}{\varepsilon}}.$$

This formula for the dimension D is not very helpful to calculate the dimension of C and K, since we can only obtain approximations of these objects. Instead, we use (13.3), which tells us that a plot of $\ln(N(\varepsilon))$ versus $\ln \frac{1}{\varepsilon}$ should be a straight line with slope D. This observation presents a practical method for calculating the value of D, $0 \leq D \leq 2$: just like the leaf exercise in grade school, superimpose grids of with squares of side length ε, count square, plot $\left(\ln \frac{1}{\varepsilon}, \ln(N(\varepsilon))\right)$, and repeat, letting ε get smaller. Fit a line to the plotted points, with the slope approximating D.

Using a computer, we perform the box counting exercise for the von Koch curve. See Figure 13.3, and Table 13.1. When we plot the points and fit a line, we determine that our approximation of the dimension of the von Koch curve is $D \approx 1.306$. It turns out that when all of the copies have the same

Table 13.1 Box counts for the von Koch curve; see Figure 13.3.

ε	$N(\varepsilon)$
$\frac{1}{2}$	2
$\frac{1}{4}$	6
$\frac{1}{8}$	14
$\frac{1}{16}$	32
$\frac{1}{32}$	88
$\frac{1}{64}$	202
$\frac{1}{128}$	512
$\frac{1}{256}$	1204

shrinking or contraction factor, the dimension can be exactly calculated as

$$D = \frac{\ln(\text{number of copies})}{\ln(\text{contraction factor})},$$

which in this case gives that the fractal dimension of the von Koch curve is

$$D = \frac{\ln 4}{\ln 3} \approx 1.26,$$

not far from our box counting result. On the other hand, we can calculate that the dimension of Cantor set is $D = \frac{\ln 2}{\ln 3} \approx 0.63$.

We can summarize, in order to connect to the heavy definition that opened this chapter.

The von Koch curve, even though it lives in the plane, is just a kinky line; we say it has "topological" dimension one. The fractal dimension we calculated, 1.26, is also called the Hausdorff-Besicovitch dimension (see Ref. [1]).

The Cantor set, even though it lives on a line, is just a bunch of "dust"; is has topological dimension zero. The fractal dimension we calculated is 0.63.

In both cases, we see that the fractal dimension exceeds the topological dimension, as stated in that opening definition.

13.3 MORE ON THE DEFINITION OF ITERATED FUNCTION SYSTEM

From the examples at the beginning of Section 13.1 we can now draw a more general definition of Iterated Function System. Suppose that X is a closed and bounded (and then compact) subset of $\mathbb{R}^2$; an *Iterated Function System* (briefly IFS) is a finite collection of maps $f_i : X \to X$, $i = 1, \ldots n$, which satisfy the inequality

$$\|f_i(x) - f_i(y)\| \leq c_i\|x - y\| \tag{13.4}$$

for certain $c_i \in [0, 1)$ (see Refs. [1–3]). When a function f_i holds the property shown in (13.4), we say that f_i is a *contraction* and that c_i is its *contraction factor*. Roughly speaking, when a function f_i is a contraction then it "shrinks" the distance between the vectors x and y and the one between the images $f_i(x)$ and $f_i(y)$. The following theorem shows how it possible to construct fractals and self-similar objects starting from an IFS.

Theorem 8 *Given an Iterated Function System $\{f_1, f_2, \ldots, f_n\}$ there exists a unique compact subset $A \subset X$ which satisfy the equation*

$$A = \bigcup_{i=1}^{n} f_i(A). \tag{13.5}$$

The set A is called the *attractor* and (13.5) states that it is a *self-similar object*, that is it is union of shrunken and distorted copies of itself. The examples showed in Section 13.1, namely, the Cantor set C and the Sierpinski gasket K, are two examples of attractors of an IFS. We see that (13.1) and (13.2) are just particular cases of (13.5).

Of course a crucial question is how to determine, once an IFS is fixed, its attractor (or at least an approximation of it). The algorithm to reconstruct A consists of a list of steps which, as result, generates a sequence of sets approximating A: starting with any set $A_0 \subset X$, let us construct the set A_1 by taking the union of all images $f_i(A_0)$, $i = 1 \ldots n$, that is,

$$A_1 = \bigcup_{i=1}^{n} f_i(A_0). \tag{13.6}$$

It is now possible to repeat the same calculations starting from A_1 and generating the set A_2 in the following manner:

$$A_2 = \bigcup_{i=1}^{n} f_i(A_1). \tag{13.7}$$

In general we can construct the sequence

$$A_{t+1} = \bigcup_{i=1}^{n} f_i(A_t), \tag{13.8}$$

and a theorem says that the goodness of the approximation of A by A_t increases when t tends to $+\infty$.

Figure 13.4 The Sierpinski gasket.

Figure 13.5 A sequence of approximating sets for the Sierpinski gasket.

■ **EXAMPLE 13.1**

A well-known example in real applications is the Sierpinski gasket shown in Figure 13.4. The Sierpinski gasket A is self-similar with respect to the IFS $\{f_1, f_2, f_3\}$ acting on $\mathbb{R}^2$, where $f_i(x, y) = \frac{1}{2}((x, y) - P_i) + P_i$, $i = 1, 2, 3$, and the points P_1, P_2, P_3 are the vertices of the outer triangle. Figure 13.5 shows the first four steps of a sequence of sets approximating the Siepinski gasket A.

Reconstructing the attractor A of an IFS is, in general, time-consuming and inefficient from a computational point of view. At each step a new figure is constructed by applying each f_i to the current figure and then taking the union of all the results. The next section shows how to use the so-called chaos game to determine an approximation of the attractor A. This requires the notion of Iterated Function System with Probabilities.

13.4 THE CHAOS GAME

We are now going to extend the previous definition of Iterated Function Systems to a more general settings. As in the previous section, let us consider a set of contraction mappings $f_i : X \to X$, $i = 1 \dots n$, where c_i is the contraction factor and X is a compact subset of $\mathbb{R}^n$. With respect to the above definition, let us suppose that to each map f_i there is associated a probability $p_i \in [0, 1]$, $i = 1 \dots n$, $\sum_{i=1}^{n} p_i = 1$; in other words, in this context we add "weights" to each function f_i through the set of probabilities p_i. The union of a collection of contractions f_i together with associated probabilities p_i defines an *Iterated Function System with Probabilities* (briefly IFSP). The introduction of a set of probabilities enriches the mathematical complexity, providing then the possibility to represent more complex objects.

Given a vector $x_0 \in X$, consider the random dynamical system generated through the sequence

$$x_{t+1} = f_i(x_t), \tag{13.9}$$

where the map f_i is taken from the set $\{f_1, f_2, \dots f_n\}$ according to the set of corresponding probabilities $\{p_1, p_2, \dots, p_n\}$. At each step the choice of one f_i depends only on the set of probabilities and it is independent from the previous choices; given x_t the next value x_{t+1} of the sequence is obtained by applying only one map f_i which is chosen according to the set of probability p_i, independently from the previous steps.

The process of producing the sequence (13.9) is called the *chaos game* and the use of the word *chaos* is justified by the fact that the set of probabilities affects the behavior of x_t. The union of all x_t, that is $\bigcup_{t=0}^{+\infty}\{x_t\}$ is called the *orbit*; it is worthwhile to notice that different sequences of probabilities generate different element x_t and, therefore, different orbits. Once the sequence of probabilities p_i is fixed, then the elements of the sequence x_t are determined and this allows us to pose the question whether or not this sequence is convergent to a limit l.

Unfortunately, there are infinite orbits associated with the chaos game and this implies, for those sequences which are convergent, the existence of several limit points. It is, in general, misleading to talk about the convergence of an orbit toward a limit—because this is not unique and depends on the probabilities p_i—and it is more correct to take all possible limit points of all possible converging paths. Some of them will occur more frequently than others and this means that, associated with the set of all possible limit points l_i, there is a set of frequencies ξ_i. Each frequency ξ_i describes how many orbits will tend to l_i when $t \to +\infty$. This allows us to conclude that a reasonable definition of limit of a chaos game can not be merely reduced to a number, but involves the construction of the set of all possible limits with associated frequencies.

There is a strong relationship between one orbit and the invariant set A of the Iterated Function System $\{f_1, f_2, \dots f_n\}$; it is possible to prove that each orbit is dense in A, allowing the possibility to reconstruct the attractor A of an IFS by simulating just one orbit of the above random process (13.9).

■ **EXAMPLE 13.2**

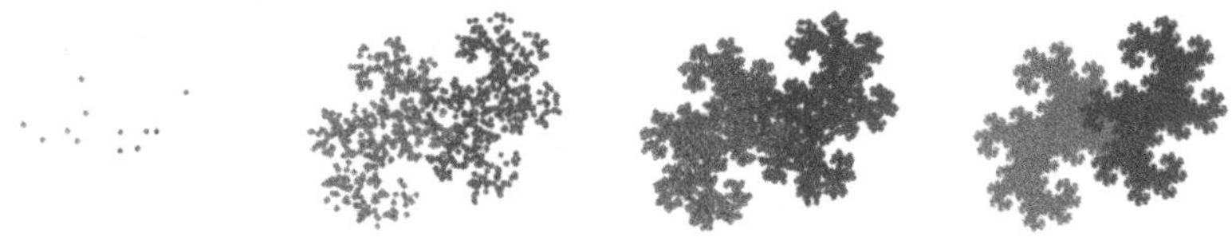

Figure 13.6 Play the chaos game to draw the Twin Dragon: 11 points, 1000 points, 3000 points, and many more points.

We define the IFSP

$$\begin{aligned} f_1(x,y) &= \begin{pmatrix} 0.5 & -0.5 \\ 0.5 & 0.5 \end{pmatrix}\begin{pmatrix} x \\ y \end{pmatrix} + \begin{pmatrix} -1 \\ 0 \end{pmatrix}, \quad p_1 = 0.5, \text{ and} \\ f_2(x,y) &= \begin{pmatrix} 0.5 & -0.5 \\ 0.5 & 0.5 \end{pmatrix}\begin{pmatrix} x \\ y \end{pmatrix} + \begin{pmatrix} 1 \\ 0 \end{pmatrix}, \quad p_2 = 0.5. \end{aligned}$$

To understand the transformation, write the matrix as

$$\begin{pmatrix} 0.5 & -0.5 \\ 0.5 & 0.5 \end{pmatrix} = \underbrace{\begin{pmatrix} \frac{1}{\sqrt{2}} & -\frac{1}{\sqrt{2}} \\ \frac{1}{\sqrt{2}} & \frac{1}{\sqrt{2}} \end{pmatrix}}_{\substack{\text{rotate CCW} \\ \text{by } 45^\circ}} \underbrace{\begin{pmatrix} \frac{1}{\sqrt{2}} & 0 \\ 0 & \frac{1}{\sqrt{2}} \end{pmatrix}}_{\substack{\text{shrink} \\ \text{by } \frac{1}{\sqrt{2}}}}$$

In this case, the probabilities each equal 0.5, so when we play the chaos game, we choose either of the maps with equal likelihood. The IFS acts on $S = [-3,3] \times [-3,3]$. Starting with $x_0 = (0,0)$, we generate the sequence of points x_t via the chaos game. In Figure 13.6, we illustrate the result, drawing small circles centered at the point x_i. The light gray circles correspond to points that were plotted by the function f_1 and the darker gray circles correspond to points that were plotted by the function f_2. As we move from left to right across the first four pictures in the figure, the number of plotted points increases. Some form appears after just 1000 points (the second picture). After 100000 points have been plotted, we essentially see the attractor (the fourth picture). Thanks the coloring, we see that the attractor indeed consists of two shrunken copies of itself. By reading the above description of the action of the matrix in f_1 and f_2, you should be able to verify that the light gray copy is f_1 applied to the entire attractor, and the dark gray copy is f_2 applied to the entire attractor. This attractor is called the Twin Dragon. We can approximate the dimension of the Twin Dragon by using box counting. The process is illustrated in Figure 13.7 with the box counts, obtained from a computer, tabulated in Table 13.2. When we fit a

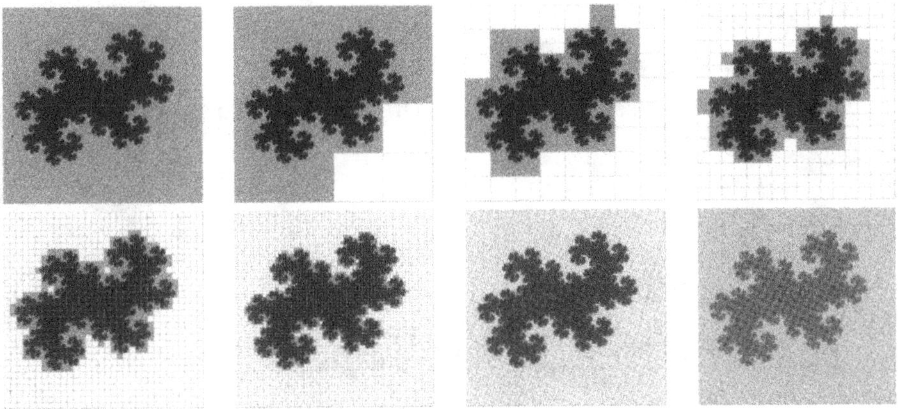

Figure 13.7 Box counting for the the Twin Dragon curve.

Table 13.2 Box counts for the Twin Dragon; see Figure 13.7.

ε	$N(\varepsilon)$
$\frac{1}{2}$	4
$\frac{1}{4}$	13
$\frac{1}{8}$	33
$\frac{1}{16}$	112
$\frac{1}{32}$	388
$\frac{1}{64}$	1380
$\frac{1}{128}$	4987
$\frac{1}{256}$	18427

line to the computer-obtained box counts in Table 13.2, we find that the slope is $1.739 \approx D$. If we discard the points corresponding to larger boxes (since we are more interested in small values of ϵ), we can nudge the approximation higher. Of course, the true fractal dimension of the Twin Dragon is

$$\frac{\ln(\#\text{copies})}{\ln\left(\frac{1}{r}\right)} = \frac{\ln(2)}{\ln(\sqrt{2})} = 2,$$

which should make sense because the Dragon has no holes in it. We say it is a space-filling curve.

EXAMPLE 13.3

We define the IFSP

$$\begin{aligned}
f_1(x,y) &= \begin{pmatrix} 0 & 0 \\ 0 & 0.16 \end{pmatrix}\begin{pmatrix} x \\ y \end{pmatrix} + \begin{pmatrix} 0 \\ 0 \end{pmatrix}, \ p_1 = 0.01, \\
f_2(x,y) &= \begin{pmatrix} 0.2 & -0.26 \\ 0.23 & 0.22 \end{pmatrix}\begin{pmatrix} x \\ y \end{pmatrix} + \begin{pmatrix} 0 \\ 1.6 \end{pmatrix}, \ p_2 = 0.07, \\
f_3(x,y) &= \begin{pmatrix} -0.15 & 0.28 \\ 0.26 & 0.24 \end{pmatrix}\begin{pmatrix} x \\ y \end{pmatrix} + \begin{pmatrix} 0 \\ 0.14 \end{pmatrix}, \ p_3 = 0.07, \text{ and} \\
f_4(x,y) &= \begin{pmatrix} 0.85 & 0.04 \\ -0.04 & 0.85 \end{pmatrix}\begin{pmatrix} x \\ y \end{pmatrix} + \begin{pmatrix} 0 \\ 1.6 \end{pmatrix}, \ p_4 = 0.85,
\end{aligned}$$

acting on the unit square $[0,1] \times [0,1]$. The functions in this IFSP are much stranger looking than those in earlier examples, and we lose all ability to interpret them geometrically. We can readily verify that the entries in the matrices are such that each f_i is contractive. We play the chaos game again and produce the remarkable attractor in Figure 13.8, called (Michael) Barnsley's spleenwort fern. Since the IFSP has consists of four functions, the fern must be the union of four shrunken copies of itself. The first function, with only one non-zero entry, collapses the entire fern to the stem. Notice that this function is chosen with probability 0.01; this choice means that we do not visit the stem so much when we play the chaos game. The fourth function is selected 85% of the time. This choice is made because, as we see from the figure, we need the most points to plot this largest copy. The fern has holes. Its fractal dimension is less than 2. Figure 13.9 and Table 13.3 present the plots and counts. When we fit the box counts to a line, we find that the fractal dimension of the fern is $D \approx 1.715$.

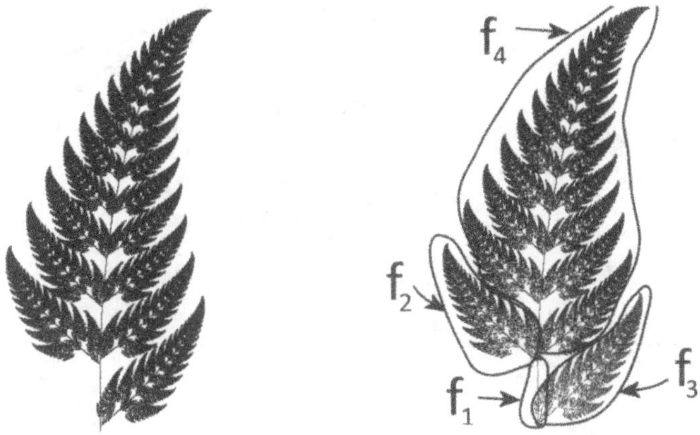

Figure 13.8 Barnsley's spleenwort fern consists of four shrunken copies of itself.

Figure 13.9 Box counting for the Barnsley's spleenwort fern.

ε	$N(\varepsilon)$
$\frac{1}{2}$	4
$\frac{1}{4}$	9
$\frac{1}{8}$	28
$\frac{1}{16}$	94
$\frac{1}{32}$	316
$\frac{1}{64}$	1113
$\frac{1}{128}$	3975
$\frac{1}{256}$	13989

Table 13.3 Box counts for the spleenwort fern; see Figure 13.9.

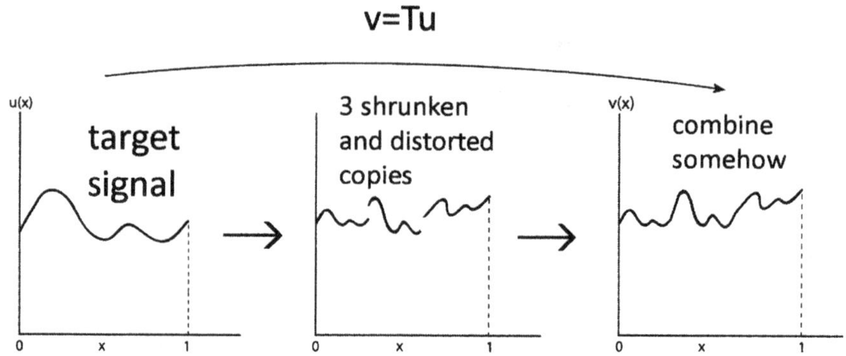

Figure 13.10 The signal approximation process.

13.5 AN APPLICATION TO IMAGE ANALYSIS

Remembering (13.5), suppose that a target set S is very close to the attractor A. Then we expect that each $f_i(S) \approx f_i(A)$, which means

$$S \approx A = \bigcup_{i=1}^{n} f_i(A) \approx \bigcup_{i=1}^{n} f_i(S) \;\Rightarrow\; S \approx \bigcup_{i=1}^{n} f_i(S).$$

We interpret the final equation as saying that S is approximated by shrunken and distorted copies of itself.

We first think about a signal $u(x)$. We could write "function" instead of "signal," but we want to start thinking in terms of images, and a signal is a one-dimensional image. We are motivated to pursue the following idea: create some shrunken copies of the signal, adjust the copies somehow, and try to recover an approximation of the original signal, as in Figure 13.10. The full process is described by $v = Tu$, where T is called the *fractal transform*. We flesh things out with some numbers by considering a particular situation. We pick the two-function IFS $\{f_1, f_2\} = \left\{\frac{6}{10}x, \frac{6}{10}x + \frac{4}{10}\right\}$ on $X = [0,1]$. We see that

$$w_1 : [0,1] \to [0,0.6] \quad \text{and} \quad w_2 : [0,1] \to [0.4,1],$$

so both functions are contractive. Shrink a signal $u(x)$ to produce two copies, one sitting on $f_1(X) = \left[0, \frac{6}{10}\right]$ and the other on $f_2(X) = \left[\frac{4}{10}, 1\right]$, as depicted in Figure 13.11. The equations for the shrunken copies in the picture are

$$\begin{aligned} a_1(x) &= u(f_1^{-1}(x)), \quad x \in f_1(X), \\ a_2(x) &= u(f_2^{-1}(x)), \quad x \in f_2(X). \end{aligned}$$

Now, we modify the "gray values of the signals" (just think "function values") by using gray level maps. For example, we might multiply all values $a_1(x)$

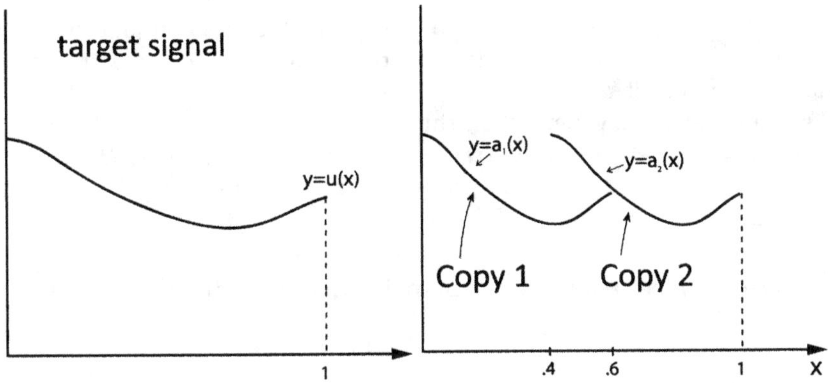

Figure 13.11 Make two shrunken copies on $f_1(X)$ and $f_2(X)$.

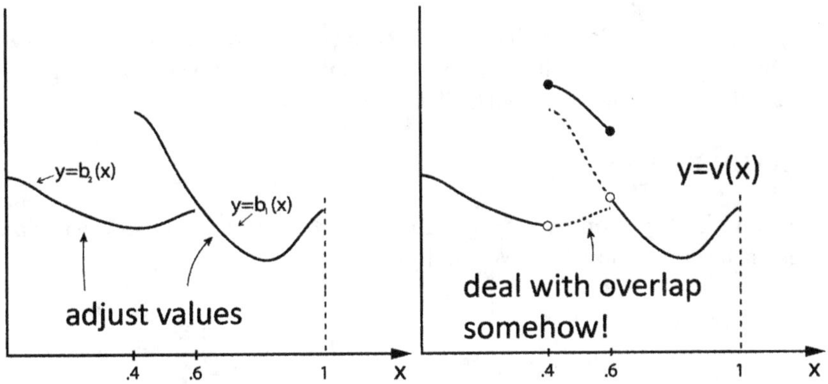

Figure 13.12 Adjust and combine the shrunken copies to get $y = v(x)$.

by $\frac{1}{2}$ and add $\frac{1}{2}$, and multiply all values $a_2(x)$ by $\frac{3}{4}$. Then we combine the shrunken copies, dealing with overlaps. In Figure 13.12, we just add values on the overlap region. The adjusted copies of the signal have equations

$$\begin{aligned} b_1(x) &= \varphi_1(a_1(x)) = \varphi_1(u(f_1^{-1}(x))), \quad x \in f_1(X), \\ b_2(x) &= \varphi_2(a_2(x)) = \varphi_2(u(f_2^{-1}(x))), \quad x \in f_2(X), \end{aligned}$$

where

$$\varphi_1(t) = \frac{1}{2}t + \frac{1}{2} \quad \text{and} \quad \varphi_2(t) = \frac{3}{4}t.$$

The function $v(x) = (Tu)(x)$ is produced by the two-function IFS with gray level Maps = IFSM. T is called a *fractal transform*.

When T is iterated on some initial signal $u_0(x)$ defined on $[0, 1]$ we obtain a sequence of signals $\{u_n(x)\}$ that converges to a signal $\overline{u}$, the attractor and unique fixed point of T.

We can construct a general T in this fashion, using functions $f_i(x) = s_i x + a_1$, with $c_i = |s_i| < 1$ so f_i is contractive, and gray level maps $\phi_i(t) = \alpha_i t + \beta_i$, $i = 1, \ldots, n$. In this case, we can prove that T is contractive (with respect to the "$\mathcal{L}^2$ metric") when $\sum_{i=1}^{n} \sqrt{c_i}|\alpha_i < 1$.

The inverse problem for function approximation using IFSM is stated as follows.

Inverse Problem: Given a target function (or image) u and $\varepsilon > 0$, find an N-map IFSM $(\underline{w}, \underline{\Phi})$ $(N = N(\varepsilon) < \infty)$ with associated fractal transform T_ε such that

$$\int_X (u(x) - (T_\varepsilon u)(x))^2 \, dx < \varepsilon.$$

The expression on the left of the inequality is the squared $\mathcal{L}^2$ distance between $u(x)$ and $(T_\varepsilon u)(x)$, so the inequality says that we want these two signals to be arbitrarily close together (with distance measured in this way).

■ **EXAMPLE 13.4**

Consider the function $u(x) = \sqrt{x}$ for x in $X = [0, 1]$. Suppose we plan to use a four-function IFS. We define

$$X_1 = \left[0, \frac{1}{4}\right], \; X_2 = \left[\frac{1}{4}, \frac{2}{4}\right], \; X_3 = \left[\frac{2}{4}, \frac{3}{4}\right], \; X_2 = \left[\frac{3}{4}, 1\right].$$

Then we want to define our maps so that

$$f_i : X \mapsto X_i, \; i = 1, 2, 3, 4,$$

and

$$\bigcup_{i=1}^{4} f_i(X) = X.$$

If we stick to affine maps, there are two choices for each map $f_i(x)$: it either flips X around or not. The formulas are given in Table 13.4 The left picture in Figure 13.13 illustrates $u(x) = \sqrt{x}$ and the four subintervals X_i, $i = 1, 2, 3, 4$. The right picture illustrates $u(f_i^{-1}(x))$, $i = 1, 2, 3, 4$, the shrunken copies of $u(x)$ on each of the subintervals X_i. The green curves are the "no flip" choices, and the brown curves are the "flip" choices. Next, we add in the affine gray level maps

$$\phi_i(t) = \alpha_i t + \beta_i, \; i = 1, 2, 3, 4,$$

Table 13.4 The candidates for the functions in our IFS.

No Flip		Flip	Action
$f_1(x) = \frac{1}{4}x$	or	$f_1(x) = \frac{1}{4} - \frac{1}{4}x$	$f_1 : X \mapsto X_1$
$f_2(x) = \frac{1}{4}x + \frac{1}{4}$	or	$f_1(x) = \frac{2}{4} - \frac{1}{4}x$	$f_2 : X \mapsto X_2$
$f_2(x) = \frac{1}{4}x + \frac{2}{4}$	or	$f_1(x) = \frac{3}{4} - \frac{1}{4}x$	$f_3 : X \mapsto X_3$
$f_2(x) = \frac{1}{4}x + \frac{3}{4}$	or	$f_1(x) = 1 - \frac{1}{4}x$	$f_4 : X \mapsto X_4$

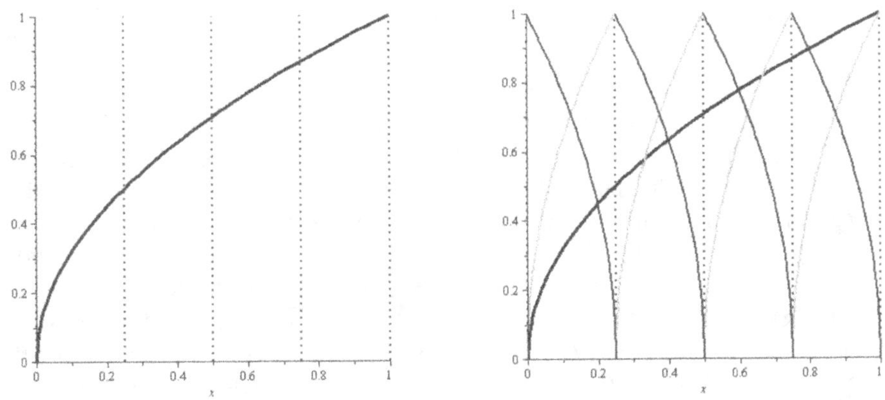

Figure 13.13 (Left) The target image $y = u(x) = \sqrt{x}$ and (right) the shrunken copies on X_i.

Table 13.5 Minimizing values of the gray level map parameters α_i and β_i.

Subinterval	Minimal Collage Distance	Best α_i	Best β_i	Flip?
X_1	0.000000000	0.500000000	0.000000000	No
X_1	0.000459942	-0.465708264	0.643805510	Yes
X_2	0.000022730	0.249142477	0.443380723	No
X_2	0.000050309	-0.245125030	0.772892391	Yes
X_3	0.000016374	0.191245540	0.661744743	No
X_3	0.000026039	-0.189417620	0.915520170	Yes
X_4	0.000012571	0.161160270	0.827175681	No
X_4	0.000017450	-0.160063020	1.041324530	Yes

to get an IFSM. Although we wrote it as a sum last time, on each subinterval X_i the fractal transform of $u(x)$ is

$$(Tu)(x) = \phi_i(u(f_i^{-1}(x))) = \alpha_i u(f_i^{-1}(x)) + \beta_i,\ x \in X_i.$$

The distances of interest are

$$\Delta_i = \int_{X_i} ((Tu)(x) - u(x))^2\, dx = \int_{X_i} \left(\alpha_i u(f_i^{-1}(x)) + \beta_i - u(x)\right)^2 dx.$$

Notice that Δ_i is a *quadratic* function of the two unknown gray level map parameters α_i and β_i. So, when we use calculus to minimize Δ_i, the equations

$$\frac{\partial \Delta_i}{\partial \alpha_i} = 0 \text{ and } \frac{\partial \Delta_i}{\partial \beta_i} = 0$$

give a *linear* system of two equations in the two unknowns α_i and β_i. Using a computer, we get the results in Table 13.5. For each subinterval, the best choice is the "unflipped" f_i; why does this make sense for this example? In Figure 13.14, we present the graphs of $y = u(x)$ and $y = (Tu)(x)$. Notice that the four pieces comprising $Tu(x)$ are shrunken and distorted copies of $u(x)$. The whole point of all of this is we can pick any function $u_0(x)$ on $[0, 1]$ and iterate T: $u_n(x) = (T^{\circ n} u_0)(x)$. The iterates should get closer to $u(x)$. Figure 13.15 shows what happens when we start with $u_0(x) = 0$. If we repeat the entire process, increasing the number of IFS maps to 8, say, then all errors decrease, as illustrated in Figure 13.16.

Note that this process works well because the the original signal is a monotone function. If you repeat the same exercise with the nice function $u(x) = \sin(\pi x)$ on $[0, 1]$, the results are terrible. This fact leads to the modified algorithm:

- Divide X into *parent* intervals I_i.

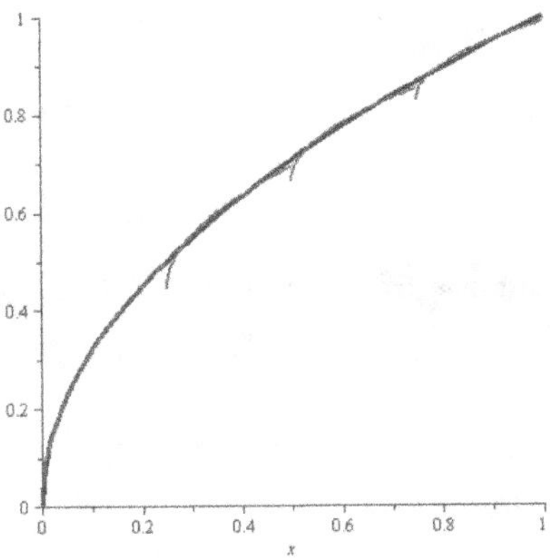

Figure 13.14 The target signal $y = u(x) = \sqrt{x}$ and the fractal transform $y = (Tu)(x)$ consisting of four shrunken and distorted copies of u.

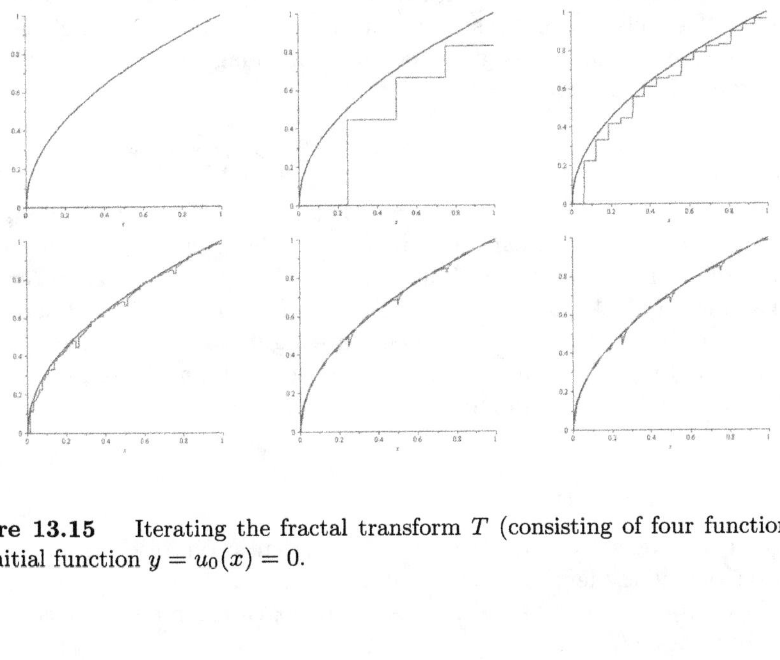

Figure 13.15 Iterating the fractal transform T (consisting of four functions) on the initial function $y = u_0(x) = 0$.

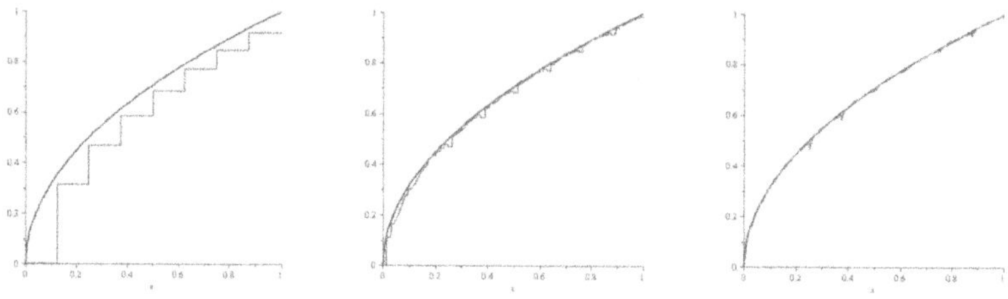

Figure 13.16 Iterating the fractal transform T (consisting of eight functions) on the initial function $y = u_0(x) = 0$.

- Divide X into *child* intervals J_j. The child intervals are smaller than the parent intervals.

- Each parent I_i is mapped to each child J_j by two *known* contraction maps, w_{ij}^{flip} and $w_{ij}^{no\ flip}$.

- For each child, consider each parent with both the "flip" and "no flip" case, finding the gray level map ϕ_i that minimizes each collage distance.

- For each child, the minimum of all collage distances considered determines the "best" parent, α, β, and whether it is flipped or not. The collection of these four parameters for all children defines a fractal transform that gives the minimal collage distance for this parent/child partitioning.

The method is call Local Iterated Function Systems with gray level Maps: LIFSM.

The LIFSM algorithm can do remarkable things with two-dimensional pictures. The entire process can be explained in three pictures, a few words, and one equation. Pick a target image $u(x, y)$. Divide the image into parent blocks and child blocks, and let f_{ij} take parent P_j to child C_i, as described in the three pictures of Figure 13.17. There are eight ways for $f_{ij} : P_j \to C_i$, 4 rotations $\times$ 2 flips. For each child C, we consider each parent and each choice of f, finding α and β that minimize

$$\Delta = \int_C \left(\alpha u(f^{-1}(x,y)) + \beta - u(x,y)\right)^2 \, dA.$$

For each child, we store the parent number, orientation for f, α and β; this collection of values defines the fractal transform T. When we repeatedly apply T to *any* initial image, the iterates approach an image that resembles the target image $u(x, y)$.

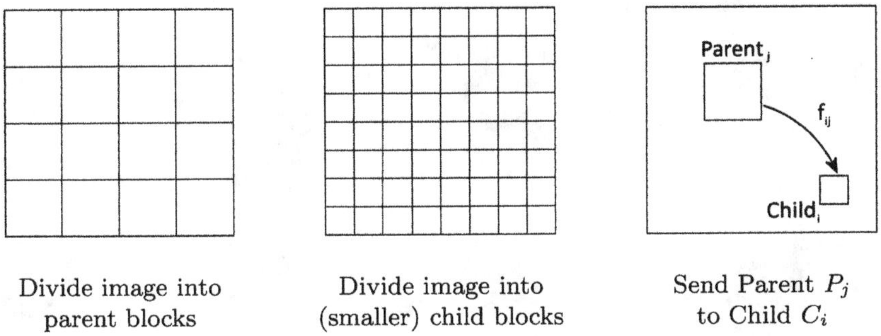

Figure 13.17 The LIFSM algorithm for images.

■ EXAMPLE 13.5

We perform the LISFM algorithm on a 256×256 grayscale image, using parent blocks that are 8×8 pixel^2 in size and child blocks that are 4×4 pixel^2 in size. To repeat, for each child, we store the best parent number, the orientation change, and the α and β values in the gray level map $\phi(t) = \alpha t + \beta$. On the left in Figure 13.18, we present the input image of an assortment of peppers. On the right, we show some parent-child pairs that are identified in the process. Each row on the right shows the parent shrunk to child size (4×4 pixels^2), the orientation change and gray level map applied to the shrunken parent, and the child. The transformed parent and the child look very close, which is the essence of our discussion. Finally, in Figure 13.19, we start with a different image, this time of a frog, and iterate the fractal transform T on it. We show the first four iterations.

REFERENCES

1. Barnsley, M.F., *Fractals Everywhere*, Academic Press, New York (1989).
2. Hutchinson, J., Fractals and self-similarity, *Indiana Univ. J. Math.*, Vol. 30, 713–747 (1981).
3. Kunze, H., La Torre, D., Mendivil, F., and Vrscay, E.R., *Fractal-Based Methods in Analysis*, Springer, New York (2012).

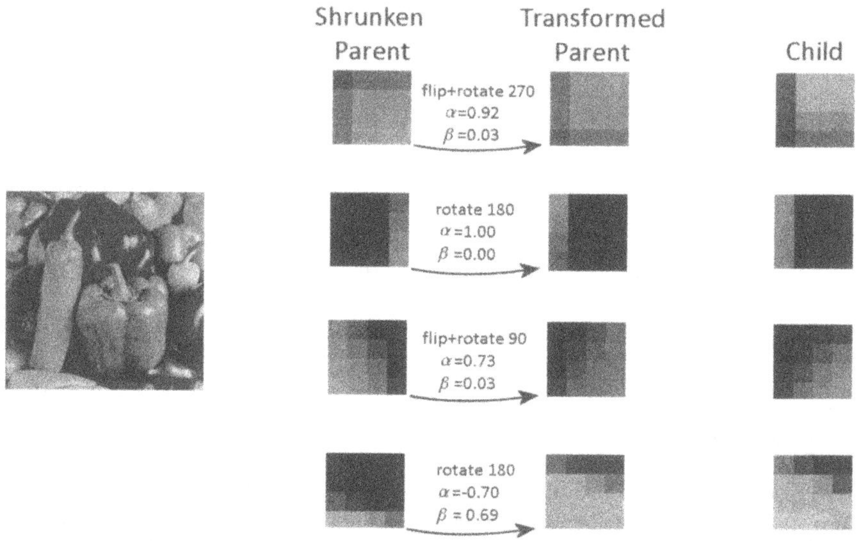

Figure 13.18 The peppers input image and some parent-child pairs identified by the algorithm.

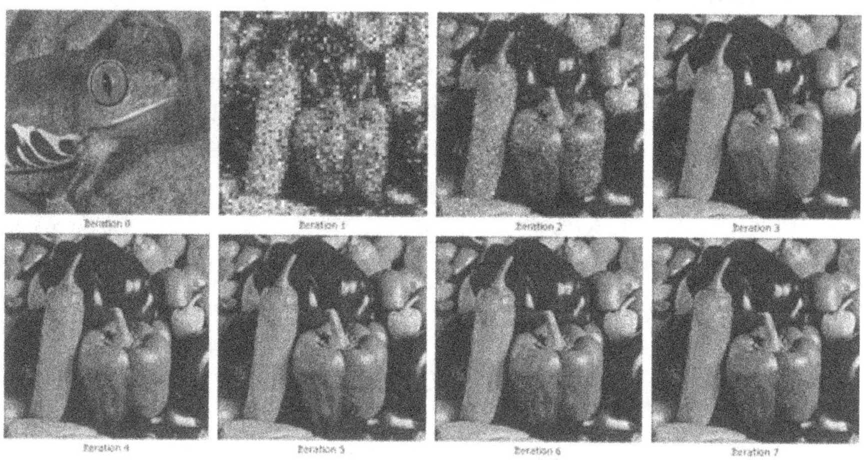

Figure 13.19 Iterating the peppers fractal transform on the initial image of a frog.

CHAPTER 14

EFFICIENT NUMERICAL METHODS FOR SINGULARLY PERTURBED DIFFERENTIAL EQUATIONS

S. NATESAN

Department of Mathematics, Indian Institute of Technology Guwahati, India

14.1 INTRODUCTION

Singular perturbation problems (SPPs) arise in several branches of applied mathematics which include fluid dynamics, quantum mechanics, elasticity, chemical reactor theory, gas porous electrodes theory, etc. The presence of small parameter(s) in these problems prevents us from obtaining satisfactory numerical solutions. It is a well known fact that the solutions of SPPs have a multiscale character. That is, there are thin layer(s) where the solution varies very rapidly, while away from the layer(s) the solution behaves regularly and varies slowly. Even in the case where only the approximate solution of the singularly perturbed boundary-value problem is required, classical numerical methods, such as finite difference schemes and finite element methods exhibit unsatisfactory behavior. This arises because the accuracy of the approximate

Mathematical Modeling with Multidisciplinary Applications.
Edited by Xin-She Yang

solution depends inversely on the perturbation parameter value and thus it deteriorates as the parameter decreases. Therefore, the numerical treatment of SPPs gives major computational difficulties.

Various finite difference schemes have been proposed in the literature to guarantee stability of the schemes for all values of the perturbation parameter. Careful examination of numerical results from such schemes on uniform grids shows that, for fixed (small) values of the perturbation parameter, the maximum pointwise error usually increases as the mesh is refined, because of the presence of the boundary or interior layer, until the mesh diameter is comparable in size to the parameter. This behavior is clearly unsatisfactory. Therefore, a separate treatment is necessary to deal with such problems.

It is well known that most of the incompressible fluid flow problems in fluid dynamics are modeled by the Navier-Stokes equations. Consider the 2D Navier-Stokes equations:

$$\begin{cases} \dfrac{\partial u}{\partial t} - \dfrac{1}{R}\Delta u + (u \cdot \nabla)u + \nabla p = 0, \quad \Omega = (0,1)^2 \times (0,T], \ T > 0, \\ \nabla \cdot u = 0, \\ u = 0, \quad \text{on the boundary } \partial\Omega, \end{cases} \tag{14.1}$$

where R is the Reynolds' number. The velocity $u(x, y, t)$ of the fluid is nonzero in the interior of the domain Ω as well as close to the boundary, whereas it will become zero on the boundary. That is, the nonzero velocity will become zero within the small region closer to the boundary, which is referred as the *boundary layer*. More precisely, the boundary layer is defined as the region of the independent variable, in which the dependent variable changes rapidly in order to satisfy the prescribed boundary conditions. The linearized version of the stationary boundary-value problem (BVP) corresponding to (14.1) is given by

$$\begin{cases} -\varepsilon\Delta u + a \cdot \nabla u + bu = f, \quad \Omega = (0,1)^2, \\ u = 0, \quad \text{on the boundary } \partial\Omega, \end{cases} \tag{14.2}$$

where $0 < \varepsilon \ll 1$ denotes the viscosity of the fluid. The sign of the convection coefficient a determines the location of the boundary layers. In general, the boundary layers locate at the outflow boundaries. Here, the convection coefficient a dominates the viscosity coefficient (diffusion parameter) ε, therefore, these problems are classified as *convection-dominated BVP*. In general, in differential equations the highest-order derivative is more significant than the lower-order derivatives, whereas this phenomenon does not hold for SPPs. If one assumes that the diffusion parameter $\varepsilon = 0$, then the second-order elliptic BVP (14.2) becomes a first-order hyperbolic PDE and it does not satisfy all the boundary conditions. This causes difficulties in finding asymptotic as well as numerical solutions.

The rest of the chapter is organized in the following manner: Section 14.2 studies special characteristics of SPPs, and the asymptotic approximate solution. The difficulties in obtaining uniformly convergent numerical approximate solutions and the exponentially fitted difference scheme are studied in Section 14.3. Four efficient numerical techniques, namely the initial-value technique, boundary-value technique, shooting method and booster method are presented in Section 14.4 for a special type of SPPs arising in chemical reactor theory; also semilinear SPPs are studied in this section. Numerical experiments are carried to show the efficiency and accuracy of these numerical methods. Boundary layer adapted nonuniform meshes for SPPs of the form (14.3) are discussed in Section 14.5.

14.2 CHARACTERIZATION OF SPPs

To understand the concepts of various solution techniques clearly, we deal with the one-dimensional SPP of the following form:

$$\begin{cases} \varepsilon u''(x) + a(x)u'(x) - b(x)u(x) = f(x), & x \in \Omega = (0,1), \\ u(0) = A, \quad u(1) = B, \end{cases} \tag{14.3}$$

where $0 < \varepsilon \ll 1$ is a small parameter, $a(x)$, $b(x)$ and $f(x)$ are sufficiently smooth functions (infinitely differentiable functions) such that $a(x) \geq \alpha > 0$, and $b(x) \geq 0$, $x \in \overline{\Omega} = [0,1]$. The SPP (14.3) admits a unique solution $u(x)$, which exhibits a boundary layer of width $O(\varepsilon)$ at $x = 0$.

The main interest of the SPP (14.3) is to study the analytical and numerical behavior of the solution as the diffusion parameter $\varepsilon \to 0$.

The existence, uniqueness and asymptotic approximate solution of the SPP (14.3) are studied by various authors, for example, one can refer the books of [7], [13], [18] and [19].

■ EXAMPLE 14.1

Consider the following constant coefficient two-point BVP:

$$\begin{cases} \varepsilon u''(x) + u'(x) = 0, & x \in \Omega, \\ u(0) = 1, \quad u(1) = 0. \end{cases} \tag{14.4}$$

The exact solution is given by

$$u(x) = A + B\exp(-x/\varepsilon), \tag{14.5}$$

where

$$A = \frac{\exp(-1/\varepsilon)}{\exp(-1/\varepsilon) - 1}, \quad B = \frac{-1}{\exp(-1/\varepsilon) - 1}.$$

In the BVP (14.4) the diffusion coefficient ε is too small in comparison with the convection coefficient. These problems are also called as convection-dominated BVPs.

The derivative(s) of the solution (14.5) contains negative power of the diffusion parameter ε, therefore it has a steep gradient in the neighborhood of $x = 0$, which is known as the *boundary layer region* and the width of the boundary layer is of $O(\varepsilon)$. In fact, $u(x)$ behaves non-uniformly near $x = 0$, which can be seen from the following result:

$$0 = \lim_{x \to 0} \lim_{\varepsilon \to 0} \exp(-x/\varepsilon) \neq \lim_{\varepsilon \to 0} \lim_{x \to 0} \exp(-x/\varepsilon) = 1.$$

The region of nonuniformity causes various difficulties in solving the BVP (14.4) analytically as well as numerically.

Generally, the diffusion coefficient models the viscosity of the fluid which is too small in comparison with other quantities. If suppose one assumes that of neglecting this small viscosity term, that is, $\varepsilon = 0$, then the second-order ODE (14.3) get reduced to a first-order ODE. This is called the *order-reduction* of the ODE. Since, the second-order ODE becomes a first-order ODE, it is not possible to impose boundary conditions at both the ends. Therefore, one has to drop one of the boundary conditions, this is known as the *loss of boundary conditions.*

Another important characteristics of SPPs is that a straightforward asymptotic expansion (also known as the outer expansion) does not satisfy the DEs along with both the boundary conditions.

Consider the straightforward asymptotic expansion

$$u(x, \varepsilon) = u_0(x) + \varepsilon u_1(x) + \varepsilon^2 u_2(x) + \dots, \tag{14.6}$$

where $u_0(x)$ is the solution of the reduced problem (first-order ODE) obtained by setting $\varepsilon = 0$ in the DE (14.4):

$$\begin{cases} u_0'(x) = 0, \\ u_0(1) = 0, \end{cases}$$

and $u_i(x)$ for $i = 2, \dots$ satisfy the following problems:

$$\begin{cases} u_i'(x) = 0, \\ u_i(1) = 0. \end{cases}$$

The left boundary condition of the BVP (14.4) is not taken into account with the straightforward asymptotic expansion (14.6). Therefore, one more asymptotic expansion is required to deal with the boundary layer region, and it is known as the inner expansion or boundary layer correctors. The inner expansion is given in a suitable stretching variable $\tau = x/\varepsilon$:

$$v(\tau, \varepsilon) = v_0(\tau) + \varepsilon v_1(\tau) + \varepsilon^2 v_2(\tau) + \ldots, \tag{14.7}$$

where $v_0(\tau)$ is the solution of the second-order ODE:

$$\begin{cases} \dfrac{d^2 v_0}{d\tau^2} + \dfrac{dv_0}{d\tau} = 0, & \tau \in (0, \infty), \\ v_0(0) = 1, \end{cases}$$

and $v_i(\tau)$ for $i = 1, 2, \ldots$ satisfy the following second-order ODEs with one-sided boundary condition:

$$\begin{cases} \dfrac{d^2 v_i}{d\tau^2} + \dfrac{dv_i}{d\tau} = 0, & \tau \in (0, \infty), \\ v_i(0) = 0. \end{cases}$$

The second arbitrary constant of integration for $v_i(\tau)$ for $i = 0, 1, \ldots$ will be determined from the *Prandtl's matching condition* (refer, for example, the book by Bush [3]).

14.3 NUMERICAL APPROXIMATE SOLUTION

14.3.1 Failure of Classical Finite Difference Schemes on Uniform Meshes

Assume that the domain $\Omega = (0, 1)$ is divided by N number of intervals with uniform step-size $h = 1/N$, and the mesh points are given by $x_i = ih$, for $i = 0, 1, \ldots, N$. We define the first-order finite difference operators on the uniform mesh by

$$D^+ U_i = \frac{U_{i+1} - U_i}{h}, \quad D^- U_i = \frac{U_i - U_{i-1}}{h}, \quad D^0 U_i = \frac{U_{i+1} - U_{i-1}}{2h}.$$

Assume that we are replacing the derivatives in the ODE (14.4) by the following central difference scheme:

$$\frac{\varepsilon}{h^2}(U_{i+1} - 2U_i + U_{i-1}) + \frac{1}{2h}(U_{i+1} - U_{i-1}) = 0, \tag{14.8}$$

where $U_i \approx u(x_i)$. If we solve the difference equation (14.8), we obtain that

$$U_i = A_1 + B_1 \left(\frac{2\varepsilon - h}{2\varepsilon + h} \right)^i. \tag{14.9}$$

Since the solution of the ODE (14.4) is of monotonically decreasing, we expect the same behavior in the numerical approximate solution (14.9) as well. In order to have a stable (nonoscillatory) numerical solution, one has to restrict the step-size $h < 2\varepsilon$ (refer [21]). This condition is too stringent because the diffusion parameter is too small, for example, $\varepsilon \approx 10^{-4}$, in that case, one has to solve a very large system of linear algebraic equations, even in the one-dimensional case. If one goes for higher dimensions it is almost impossible to meet this stringent condition on the step-size. Figures 14.1(a) and 14.1(b) show the exact and numerical solution obtained by the central difference scheme (14.8) for the SPP (14.4) with $h = 0.05$. The stability condition is satisfied for Figure 14.1(a), and therefore there is no non-physical oscillations, whereas this condition is violated in Figure 14.1(b), and the numerical solution is having spurious oscillations.

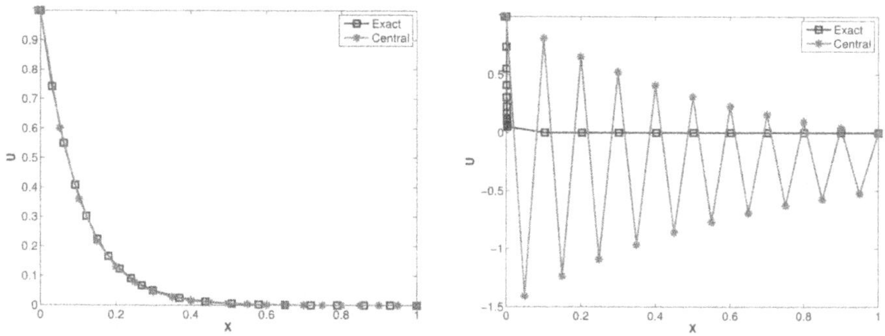

(a) $\varepsilon = 1e - 01$. (b) $\varepsilon = 1e - 03$.

Figure 14.1 Exact solution and the approximate solution obtained by the central difference scheme for Example 14.1 for $h = 0.05$.

To obtain a nonoscillatory numerical approximate solution, one can replace the first-order derivative in the ODE (14.4) by the forward difference (upwind) scheme, which gives the second-order difference equation

$$\frac{\varepsilon}{h^2}(U_{i+1} - 2U_i + U_{i-1}) + \frac{1}{h}(U_{i+1} - U_i) = 0. \tag{14.10}$$

By solving the difference equation (14.10), one can obtain the solution

$$U_i = A_1 + B_1 \left(\frac{\varepsilon}{\varepsilon + h}\right)^i. \tag{14.11}$$

Although the solution of the upwind difference is oscillation-free [21], it won't provide any information about the solution inside the boundary layer region, which is of width $O(\varepsilon)$. In order to study the behavior of the solution inside the boundary layer, one has to take smaller step-size in relation with the diffusion parameter ε, otherwise, the first mesh point itself will go outside the boundary layer region. These observations can be depicted from Figures 14.2(a) and 14.2(b). In fact, Figure 14.2(b) clearly indicates the need of a smaller step-size in relation with ε to have some mesh points inside the boundary layer region.

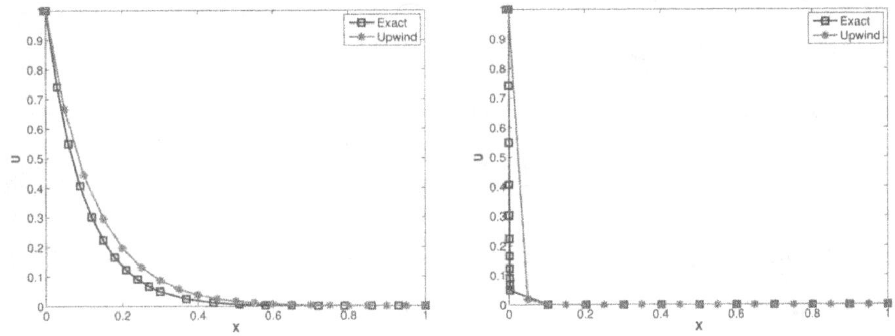

(a) $\varepsilon = 1e - 01$. (b) $\varepsilon = 1e - 03$.

Figure 14.2 Exact solution and the approximate solution obtained by the upwind difference scheme for Example 14.1 for $h = 0.05$.

Let $u(x)$ be the solution of the BVP (14.4) and U_i be the numerical solution of either (14.10) or (14.8). Then, the consistency constant C of the following error estimate:

$$\sup_{0<\varepsilon\ll 1} ||u(x_i) - U_i|| \leq Ch^p, \quad p = 1 \text{ or } 2$$

(where $||\cdot||$ is any suitable norm, throughout this chapter, we consider the maximum norm) depends on the inverse powers of the diffusion parameter ε. Therefore, one has to keep on reducing the step-size h with respect to ε in order to have a meaningful error bound. But in practice it is not possible to have smaller step-sizes than ε.

Definition 12 *Let $u(x)$ be the exact solution of a SPP, and let U_i be the numerical solution of the corresponding discrete problem. We say U_i converges uniformly to $u(x_i)$, if for some fixed $N_0 > 0$ the following condition holds:*

$$\sup_{0<\varepsilon\ll 1} ||u(x_i) - U_i|| \leq CN^{-p}, \quad \textit{for } N \geq N_0,$$

where N_0, C are independent of ε.

14.3.2 Exponentially Fitted Difference Scheme

Applied mathematicians and engineers are interested in finding uniformly convergent numerical approximate solution for SPPs. As we have seen that the classical finite difference schemes fail to provide uniformly convergent numerical solution on uniform grids, one has to look for alternate ways to overcome this difficulty. Allen-Southwell [1] and Il'in [9] proposed the following exponentially fitted difference (EFD) scheme for the SPP (14.3)

$$\begin{cases} \varepsilon\sigma_i D^+D^-U_i + a_i D^0 U_i - b_i U_i = f_i, & 0 \le i \le N, \\ u_0 = A, \quad u_N = B, \end{cases} \tag{14.12}$$

where the fitting factor σ_i is given by

$$\sigma_i = \left(\frac{a_i h}{2\varepsilon}\right) \coth\left(\frac{a_i h}{2\varepsilon}\right). \tag{14.13}$$

The following theorem provides the uniform convergence of the EFD scheme (14.12).

Theorem 9 *[6] Let $u(x)$ and U_i be respectively the solution of the continuous problem (14.3) and (14.12). Then, the error satisfies the following bound*

$$\|u(x_i) - U_i\| \ \le\ Ch,$$

where the constant C is independent of x_i, h and ε.

To see the efficiency of the EFD scheme (14.12), we applied it to the SPP (14.4) given in Example 14.1. The approximate solution along with the exact solution and the corresponding error for $\varepsilon = 1e-02$, $h = 0.05$ are plotted in Figures 14.3(a) and 14.3(b), respectively. These plots reveal the accuracy of the EFD scheme.

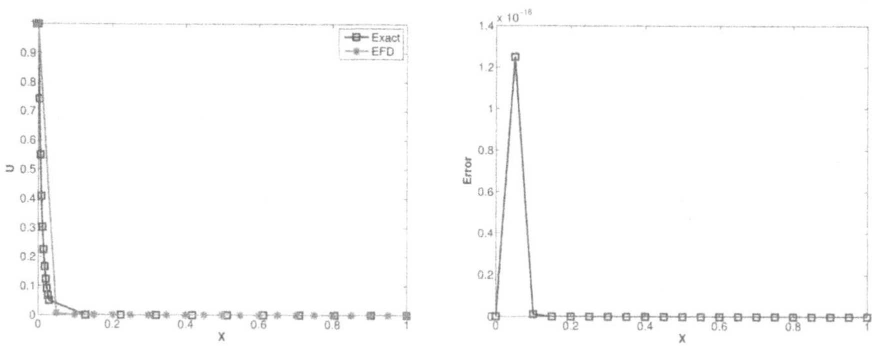

(a) Exact and Approximate Solutions. (b) Error.

Figure 14.3 Numerical solution and error plots of EFD scheme for Example 14.1, for $\varepsilon = 1e-02$, $h = 0.05$.

14.4 SPPs ARISING IN CHEMICAL REACTOR THEORY

In this section, we consider a class of SPP for second-order ODEs with a special type of mixed boundary conditions which arises in the theory of chemical reactions [5]. Consider the following two-point BVP:

$$Lu(x) \equiv F \Leftrightarrow \begin{cases} Pu(x) \equiv \varepsilon u''(x) + a(x)u'(x) - b(x)u(x) = f(x), & x \in \Omega, \\ B_0 u(0) \equiv -u'(0) = A, \quad B_1 u(1) \equiv u(1) + \varepsilon u'(1) = B, \end{cases} \tag{14.14}$$

where $0 < \varepsilon \ll 1$ is a small parameter, $a(x)$, $b(x)$ and $f(x)$ are sufficiently smooth functions such that $a(x) \geq \alpha > 0$, and $b(x) \geq 0$, $x \in \overline{\Omega}$, $L = (P, B_0, B_1)^T$, $F = (f(x), A, B)^T$. Under these assumptions the BVP (14.14) admits a unique solution. For these problems the asymptotic solution converges to the solution of the reduced problem throughout the domain $[0, 1]$ while the derivatives generally converge nonuniformly as $\varepsilon \to 0$ at the left boundary $x = 0$. These problems are classified as weak layer problems. This nonuniformity in the convergence of the derivatives prevents us from obtaining satisfactory numerical approximate solution by classical finite difference or finite element methods on uniform meshes.

The following lemma states that the solution of the SPP (14.14) satisfies the maximum principle. Further, it shows that the solution is uniformly stable.

Lemma 14.4.1 *Let $u(x)$ be any smooth function satisfying $Pu(x) \leq 0$, $x \in \Omega$, $B_0 u(0) \geq 0$ and $B_1 u(1) \geq 0$, then, $u(x) \geq 0$ for all $x \in \overline{\Omega}$. Further, the following stability estimate holds:*

$$|u(x)| \leq C[|B_0 u(0)| + |B_1 u(1)| + \max_{y \in \overline{\Omega}} |Pu(y)|].$$

Proof. The proof of this lemma can be found in Doolan *et al.* [6].

The derivatives of the solution of the SPP (14.14) will satisfy the following bound (for the detailed proof refer the article [17]):

$$|u^{(i)}(x)| \leq C[1 + \varepsilon^{-i+1} \exp(-\alpha x/\varepsilon)], \quad \text{for } i = 1(1)j + 1, \text{ for some fixed } j.$$

In this section, we provide four efficient numerical methods to solve the SPP (14.14), namely initial-value technique (IVT), boundary-value technique (BVT), shooting method and booster method. The methods are described in the following subsections, and a numerical example is solved by all these methods to show the efficiency and accuracy of these methods. The numerical results are presented in the form of tables or figures for the approximate solution or for the errors.

14.4.1 Initial-Value Technique

In initial-value technique, the numerical solution of the SPP (14.14) is obtained by solving some suitable initial-value problems (IVPs) of first-order ODEs obtained through the asymptotic approximate solution. The asymptotic approximate solution is given by

$$\widehat{u}(x) = u_0(x) + \varepsilon[u_1(x) + p(x)v_0(x)], \tag{14.15}$$

where

$$p(x) = \frac{a^2(0)(A + u_0'(0))}{a(x)(a^2(0) - \varepsilon[b(0) - a'(0)])},$$

and $u_0(x)$, $u_1(x)$ and $v_0(x)$ are respectively the solutions of the following first-order IVPs:

$$\begin{cases} a(x)u_0'(x) - b(x)u_0(x) = f(x), \quad x \in \overline{\Omega}, \\ u_0(1) = B, \end{cases} \tag{14.16}$$

$$\begin{cases} a(x)u_1'(x) - b(x)u_1(x) = -u_0''(x), \quad x \in \overline{\Omega}, \\ u_1(1) = -u_0'(1), \end{cases} \tag{14.17}$$

and

$$\begin{cases} \varepsilon v_0'(x) - [a(x) - \varepsilon b(x)/a(x)]v_0(x) = 0, \quad x \in \overline{\Omega}, \\ v_0(0) = 1. \end{cases} \tag{14.18}$$

The asymptotic expansion given in (14.15) provides a second-order approximation to the solution $u(x)$ of the SPP (14.14). That is,

$$|u(x) - \widehat{u}(x)| \ \leq \ C\varepsilon^2,$$

provided that $(a^2(0) - \varepsilon_0[b(0) + |a'(0)|]) \geq \nu > 0$, for $0 < \varepsilon \leq \varepsilon_0 < 1$. By using the barrier function technique one can prove this result; for details refer to Ref. [16].

The first-order derivatives in the IVPs (14.16) and (14.17) are without multiplied by the small parameter ε, therefore, one can apply any classical numerical scheme to solve these IVPs. Here, we apply the fourth-order explicit Runge-Kutta method to solve these problems.

Consider the most general first-order IVP

$$\begin{cases} u'(x) = f(x, u), \quad x \in (0, 1), \\ u(0) = A. \end{cases} \tag{14.19}$$

The fourth-order Runge-Kutta method for the above IVP is given as

$$\begin{cases} U_{i+1} = U_i + \frac{1}{6}[K_1 + 2K_2 + 2K_3 + K_4], \quad i \geq 0, \\ K_1 = hf(x_i, U_i), \quad K_2 = hf\left(x_i + \frac{h}{2}, U_i + \frac{1}{2}K_1\right), \\ K_3 = hf\left(x_i + \frac{h}{2}, U_i + \frac{1}{2}K_2\right), \quad K_4 = hf(x_i + h, U_i + K_3). \end{cases} \tag{14.20}$$

The error of the Runge-Kutta method satisfies the following bound

$$|u(x_i) - U_i| \leq Ch^4,$$

where $u(x)$ and U_i are respectively the solutions of (14.19) and (14.20).

In order to solve the singularly perturbed IVP (14.18), we apply the EFD schemes of Doolan *et al.* [6], [16]. The explicit EFD scheme is given by

$$\begin{cases} \varepsilon\sigma_i D^+ V_i - (a_i - \varepsilon b_i/a_i)V_i = 0, \quad i \geq 0 \\ V_0 = 1, \end{cases} \tag{14.21}$$

where the fitting factor is $\sigma_i = (ha_i/\varepsilon)/(1 - \exp(-a_i h/\varepsilon))$.

Let $v_0(x_i)$ be the solution of the singularly perturbed IVP (14.18), and V_i be the numerical solution of (14.21), then the error satisfies

$$|v_0(x_i) - V_i| \leq Ch.$$

To solve the singularly perturbed IVP (14.18), one can also apply the following implicit EFD scheme ([6]):

$$\begin{cases} \varepsilon\sigma_{i+1} D^+ V_i - (a_{i+1} - \varepsilon b_{i+1}/a_{i+1})V_{i+1} = 0, \quad i \geq 0, \\ V_0 = 1, \end{cases} \tag{14.22}$$

where the fitting factor is $\sigma_i = (ha_i/\varepsilon)/(\exp(a_i h/\varepsilon) - 1)$.

The error of the implicit EFD scheme (14.22) satisfies the following bound

$$|v_0(x_i) - V_i| \leq C\min(h, \varepsilon).$$

The following theorem provides the error bounds for the numerical approximate solution of the SPP (14.14), here the singularly perturbed IVP (14.18) is solved by using the explicit EFD scheme (14.21).

Theorem 10 *[16] Let $u(x)$ be the solution of the SPP (14.14). Let U_{0i} and U_{1i} be respectively the numerical solutions of the IVPs (14.16) and (14.17) obtained by the classical Runge-Kutta method, and V_{0i} be the solution of the singularly perturbed IVP (14.18) obtained by the explicit EFD scheme (14.21),* *then the error satisfies*

$$|u(x_i) - [U_{0i} + \varepsilon(U_{1i} + p_i V_{0i})]| \leq C(\varepsilon^2 + h^4 + \varepsilon h^4 + \varepsilon h).$$

When the implicit EFD (14.22) is used to solve the singularly perturbed IVP (14.18), one can have the following error estimate.

Theorem 11 *[16] Let $u(x)$ be the solution of the SPP (14.14). Let U_{0i} and U_{1i} be respectively the numerical solutions of the IVPs (14.16) and (14.17) obtained by the classical Runge-Kutta method, and V_{0i} be the solution of the singularly perturbed IVP (14.18) obtained by the implicit EFD scheme (14.22), then the error satisfies*

$$|u(x_i) - [U_{0i} + \varepsilon(U_{1i} + p_i V_{0i})]| \leq C(\varepsilon^2 + h^4 + \varepsilon h^4 + \varepsilon \min(h, \varepsilon)).$$

■ EXAMPLE 14.2

Consider the following linear two-point BVP:

$$\begin{cases} \varepsilon u''(x) + u'(x) - u(x) = 0, & x \in \Omega, \\ -u'(0) = 0, \quad u(1) + \varepsilon u'(1) = 1. \end{cases} \tag{14.23}$$

Since it is a constant coefficient linear BVP, one can easily obtain the exact solution as

$$u(x) = \frac{\exp(-m_1)[m_2 \exp(m_1 x) - m_1 \exp(m_2 x)]}{m_2(1 + \varepsilon m_1) - m_1(1 + \varepsilon m_2)\exp(m_2 - m_1)},$$

where $m_{1,2} = (-1 \pm \sqrt{1 + 4\varepsilon})/2\varepsilon$.

The asymptotic approximate solution of the BVP (14.23) is given by

$$\widetilde{u}(x) = u_0 + \varepsilon v_0 = \exp(x - 1) - \varepsilon \exp(-1)\exp(-x/\varepsilon).$$

To validate the theoretical error estimates of the initial-value technique, we apply it to the SPP (14.23). Basically, we have to solve two first-order terminal-value problems without ε (by the fourth-order Runge-Kutta method) and one singularly perturbed initial-value problem [by the explicit EFD scheme (14.21)]. The numerical results are presented in Table 14.1 for Example 14.2.

14.4.2 Boundary-Value Technique

Roberts [20] proposed a non-overlapping domain decomposition method to solve SPP of the form (14.3). The BVPs of the subdomains are solved by the classical finite difference schemes. In [11], the authors solved SPPs arising in chemical reactor theory by the boundary-value technique. In this method,

Table 14.1 Exact solution and error of the IVT for Example 14.2.

Mesh points	$\varepsilon = 10^{-3}$,	$h = 10^{-3}$	$\varepsilon = 10^{-5}$,	$h = 10^{-5}$
	Exact solution	Error	Exact solution	Error
$0 * \varepsilon$	0.367883	$3.7e-04$	0.367883	$3.7e-06$
$3 * \varepsilon$	0.369002	$3.7e-04$	0.367990	$3.7e-06$
$5 * \varepsilon$	0.369724	$3.6e-04$	0.368027	$3.7e-06$
$7 * \varepsilon$	0.370461	$3.6e-04$	0.368137	$3.7e-06$
$9 * \varepsilon$	0.371202	$3.6e-04$	0.368211	$3.7e-06$
$10 * \varepsilon$	0.371573	$3.6e-04$	0.368248	$3.7e-06$
0.20	0.448790	$2.4e-04$	0.449283	$4.3e-05$
0.40	0.548045	$5.5e-04$	0.548755	$5.5e-05$
0.60	0.669250	$9.2e-04$	0.670249	$7.0e-05$
0.80	0.817260	$1.4e-03$	0.818642	$8.8e-05$
1.00	0.999890	$2.0e-03$	0.999890	$1.1e-04$

the inner region (boundary layer region) problem is solved by an EFD scheme and the outer region problem is solved by the classical upwind finite difference scheme. The domain $\overline{\Omega} = [0, 1]$ is divided into two non-overlapping subdomains as $[0, k\varepsilon]$ and $[k\varepsilon, 1]$, for $k > 0$ and $k\varepsilon \ll 1$ is the width of the boundary layer (inner) region near $x = 0$. Two boundary-value problems are obtained for these subdomains, and the boundary condition at $x = k\varepsilon$ is obtained from the asymptotic approximate solution. More precisely,

1. The *inner region (boundary layer) problem* is given by

$$\begin{cases} \varepsilon u''(x) + a(x)u'(x) - b(x)u(x) = f(x), & x \in (0, k\varepsilon), \\ -u'(0) = A, \quad u(k\varepsilon) = \overline{A}, \end{cases} \tag{14.24}$$

and

2. The *outer region problem* is given as

$$\begin{cases} \varepsilon u''(x) + a(x)u'(x) - b(x)u(x) = f(x), & x \in (k\varepsilon, 1), \\ u(k\varepsilon) = \overline{A}, \quad u(1) + \varepsilon u'(1) = B, \end{cases} \tag{14.25}$$

where $\overline{A}$ is solution of the reduced problem calculated at $k\varepsilon$, that is, $\overline{A} = u_0(k\varepsilon)$.

The reduced problem solution is obtained by solving the following first-order ODE:

$$\begin{cases} a(x)u_0'(x) - b(x)u_0(x) = f(x), & x \in \Omega, \\ u_0(1) = B. \end{cases} \tag{14.26}$$

If one wants to have a better approximation at the transition point $k\varepsilon$, then the following asymptotic approximate solution can be used:

$$\widetilde{u}(x) = u_0(x) + \varepsilon(u_1(x) + v_0(\tau)), \quad \text{where } \tau = x/\varepsilon, \tag{14.27}$$

and $u_1(x)$ and $v_0(\tau)$ are respectively the solution of the following problems:

$$\begin{cases} a(x)u_1'(x) - b(x)u_1(x) = -u_0''(x), & x \in \Omega, \\ u_1(1) = 0, \end{cases} \tag{14.28}$$

and

$$\begin{cases} \dfrac{d^2 v_0}{d\tau^2} + \dfrac{dv_0}{d\tau} = 0, \quad x \in (0, \infty), \\ -\dfrac{dv_0}{d\tau}(0) = A + \dfrac{du_0}{dx}(0), \quad u(\infty) = 0. \end{cases} \tag{14.29}$$

It is worthwhile to note that the asymptotic approximation given in (14.27) is of second-order approximation for $u(x)$, that is, $|u(x) - \widetilde{u}(x)| \leq C\varepsilon^2$.

Here, we use only the reduced problem solution to determine the boundary value at the transition point $k\varepsilon$, that is,

$$\overline{A} = u_0(k\varepsilon). \tag{14.30}$$

Now, the boundary layer problem and the outer region problem can be solved numerically. To solve these problems we apply the following numerical schemes, the boundary layer problem is solved by an exponentially fitted difference scheme and the classical finite difference scheme to solve the outer region problem.

To solve the boundary layer problem (14.24) we apply the following exponentially fitted difference scheme as given in Doolan *et al.* [6]:

$$\begin{cases} \varepsilon\sigma_i D^+D^-U_i + a_i D^0 U_i - b_i U_i = f_i, \quad 0 \leq i \leq N, \\ -\dfrac{U_1 - U_0}{h} = A, \quad U_N = \overline{A}, \end{cases} \tag{14.31}$$

where σ_i is as given in (14.13). Also one can obtain the following error estimate for the boundary layer region problem.

If u is the solution of the original SPP (14.14) and U_i is the numerical solution of (14.31). Then the error satisfies the bound

$$\|u(x_i) - U_i\| \leq C(h + \varepsilon), \tag{14.32}$$

where the constant C is independent of ε, x_i and h.

Since the outer region problem is away from the boundary layer, the classical upwind finite difference scheme performs well, the BVP (14.25) is solved by the following upwind scheme:

$$\begin{cases} \varepsilon D^+D^-U_i + a_iD^+U_i - b_iU_i = f_i, \quad 0 \le i \le N, \\ U_0 = \overline{A}, \quad U_N + \varepsilon\left(\dfrac{U_N - U_{N-1}}{h}\right) = B. \end{cases} \tag{14.33}$$

Following the proof error estimate given in Ref. [12], one can have the following error estimate for the outer region problem (14.25). Let $u(x)$ be the solution of the SPP (14.14), and U_i be the solution of (14.33), then

$$\begin{cases} \|u(x_i) - U_i\| \le C\left(\varepsilon + h + h\varepsilon^{-1}\exp(-\gamma x_i/\varepsilon)\right), & h \le \varepsilon, \\ \|u(x_i) - U_i\| \le C\left(\varepsilon + h + \exp(-\alpha x_i/(\varepsilon + \alpha h))\right), & h \ge \varepsilon, \end{cases} \tag{14.34}$$

where $\gamma \in (0, \alpha)$.

By increasing the value of k (thus widening the boundary layer region), we keep on solving the inner region BVP (14.24) by the EFD scheme (14.31) until the solution profiles do not differ materially from iteration to iteration. For computational purposes, we use the following absolute error criteria for any suitable tolerance bound:

$$\|U_i^{m+1} - U_i^m\| \le \text{Tol},$$

where U_i^m is the mth-iteration of the inner region solution. Once, the solution stabilizes in the inner region, keeping that particular value of k, we solve the outer region problem (14.33) and combine both the solutions to obtain the numerical approximation for the whole domain $\overline{\Omega}$.

14.4.3 Shooting Method

In this method, the domain of computation $\overline{\Omega} = [0, 1]$ is divided into two non-overlapping subdomains as $[0, k_1]$ (inner region with k_1 as the width of the boundary layer) and $[k_2, 1]$ (outer region) with $0 < k_1 \le k_2 < 1$. In the inner region $[0, k_1]$, the second-order SPP (14.14) is converted into a system of two first-order ODEs as

$$\begin{cases} u_1' - u_2 = 0, \quad x \in [0, k_1], \\ \varepsilon u_2' + a(x)u_2 - b(x)u_1 = f(x), \\ u_1(0) = \overline{A} = u_0(k_1), \quad u_2(0) = -A. \end{cases} \tag{14.35}$$

To solve the system of first-order ODEs (14.35), we apply the following difference scheme which uses the classical finite difference scheme and an EFD

Table 14.2 Exact solution and error of the BVT for Example 14.2.

Mesh Points	Numerical solution $k = 1$	$k = 10$	$k = 20$	Exact solution	Error
0.000	0.36798861	0.36826893	0.36827264	0.36824640	$2.6241e-05$
0.001	0.36811217	0.36839259	0.36839630	0.36838188	$1.4415e-05$
0.002	0.36896646	0.36867096	0.36867468	0.36866461	$1.0074e-05$
0.003	0.36982076	0.36900644	0.36901016	0.36900167	$8.4828e-06$
0.004	0.37067506	0.36936312	0.36936685	0.36935895	$7.9027e-06$
0.005	0.37152935	0.36972784	0.36973157	0.36972388	$7.6940e-06$
0.006	0.37238365	0.37009575	0.37009948	0.37009186	$7.6220e-06$
0.007	0.37323794	0.37046506	0.37046879	0.37046119	$7.6001e-06$
0.008	0.37409224	0.37083512	0.37083885	0.37083126	$7.5966e-06$
0.009	0.37494653	0.37120568	0.37120943	0.37120183	$7.6000e-06$
0.010	0.37580083	0.37157667	0.37158042	0.37157281	$7.6058e-06$
0.200	0.45508026	0.45102504	0.45102504	0.44923906	$1.7860e-03$
0.300	0.50264254	0.49816350	0.49816350	0.49643640	$1.7271e-03$
0.400	0.55517575	0.55022859	0.55022859	0.54859232	$1.6363e-03$
0.500	0.61319942	0.60773521	0.60773521	0.60622777	$1.5074e-03$
0.600	0.67728738	0.67125209	0.67125209	0.66991844	$1.3336e-03$
0.700	0.74807344	0.74140736	0.74140736	0.74030049	$1.1069e-03$
0.800	0.82625763	0.81889486	0.81889486	0.81807693	$8.1793e-04$
0.900	0.91261316	0.90448088	0.90448088	0.90402460	$4.5628e-04$
1.000	0.99901000	0.99901000	0.99901000	0.99901000	$9.8521e-06$

scheme

$$\begin{cases} D^+U_{1,i} - U_{2,i} = 0, \quad 0 \le i \le N \\ \varepsilon\sigma D^+U_{2,i} + a_i U_{2,i+1} - b_i U_{1,i} = f_i, \\ U_{1,0} = \overline{A}, \quad U_{2,0} = A, \end{cases} \tag{14.36}$$

where $\sigma = (ha(0)/\varepsilon)\exp(-a(0)h/\varepsilon)/[1 - \exp(-a(0)h/\varepsilon)]$.

One can obtain the following error estimate for the numerical solution of the inner region problem (14.36)

$$\|\mathbf{u}(x_i) - \mathbf{U}_i\| \le Ch, \tag{14.37}$$

where $\mathbf{u} = (u_1, u_2)^T$ and $\mathbf{U} = (U_1, U_2)^T$ are respectively the solutions of the continuous system (14.35) and the discrete system (14.36). Here the error constant C is independent of ε, h, x_i, the resultant scheme is of ε-uniform convergent method.

By using the uniqueness of the solution of the BVP (14.14), and the system of IVPs (14.35) one can have the error estimate. Let $u(x)$ be the solution of the original SPP (14.14), further, $\mathbf{U} = (U_1, U_2)^T$ be the solution of (14.36), then

$$|u(x_i) - U_{1,i}| \le C(\varepsilon + h), \quad x_i \in [0, k_1].$$

The outer region problem is given by the BVP:

$$\begin{cases} \varepsilon u''(x) + a(x)u'(x) - b(x)u(x) = f(x), & x \in (k_2, 1), \\ u(k_2) = \overline{B} = u_0(k_2), \quad u(1) + \varepsilon u'(1) = B. \end{cases} \tag{14.38}$$

The above BVP (14.38) is solved by the classical upwind finite difference scheme as given in (14.33), and error estimate given in (14.34) holds true for this BVP also.

Since the inner region problem (14.36) defined in the subdomain $[0, k_1]$ and the outer region problem (14.38) defined in the interval $(k_2, 1)$ [by the difference scheme (14.33)] are independent of each other, this opens the door for parallel computing. By this way, one can reduce the computation time.

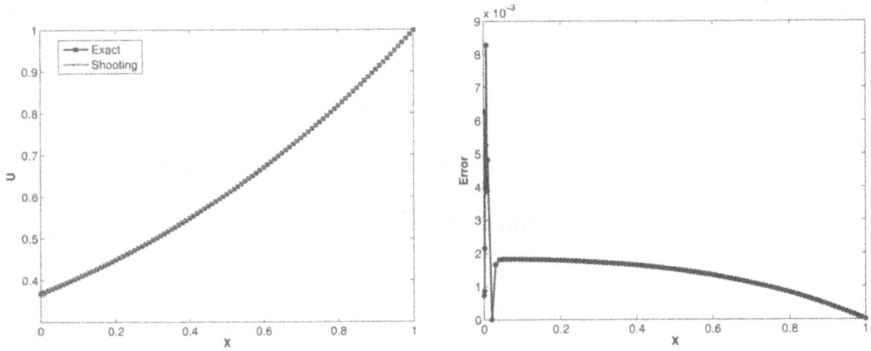

(a) Exact and Approximate solutions. *(b) Error.*

Figure 14.4 Plots of the exact and approximate solution obtained by shooting method for Example 14.2 for $\varepsilon = 10^{-3}$, $k_1 = 10\varepsilon$, $k_2 = 0.02$, $h_1 = \varepsilon$, and $h_2 = 0.01$.

14.4.4 Booster Method

To obtain better approximate numerical solutions to SPPs of the form (14.3), Israeli and Ungarish [10] proposed a numerical method, in which an asymptotic approximate solution is incorporated into a suitable numerical scheme to improve the accuracy of the numerical approximate solution. The basic properties of the numerical scheme will remain as it is, and the accuracy will be improved by this method.

Let U be the numerical solution of (14.14) obtained through the following difference scheme:

$$L_h(U) = F_h, \tag{14.39}$$

where L_h is a suitable difference operator obtained by replacing the derivatives in L by finite difference quotients. Then, the truncation error $T(u)$ of the difference operator L_h is defined as

$$T(u) = L_h(u) - L(u) = L_h(u) - L_h(U).$$

Since the differential operator L and the difference operator L_h are linear, one can rewrite the truncation error as

$$T(u) = L_h(u - U), \tag{14.40}$$

and it can be taken as

$$\|u - U\| \ = \ \|L_h^{-1}T(u)\|.$$

In the booster method, we use an asymptotic approximate solution $\widetilde{u}$ to the original solution $u(x)$ in order to reduce the truncation error. The *improved* numerical solution, denoted by U_B, is obtained by applying the *Booster* operator L_{hB}, defined by

$$L_{hB}(U_B) \equiv L_h(U_B) - L_h(\widetilde{u}) + L(\widetilde{u}) = F_h. \tag{14.41}$$

Then, the truncation error of the booster method $T_B(u)$ can be given as

$$\begin{aligned} T_B(u) &= [L_h(u) - L_h(\widetilde{u}) + L(\widetilde{u})] - L(u) \\ &= [L_h(u) - L_h(\widetilde{u})] - [L(u) - L(\widetilde{u})] \\ &= [L_h(u) - L(u)] - [L_h(\widetilde{u}) - L(\widetilde{u})] = T(u) - T(\widetilde{u}). \end{aligned} \tag{14.42}$$

Thus, if the approximation is such that

$$\|T_B(u)\| \ = \ \|T(u) - T(\widetilde{u})\| \ \leq \ \varepsilon_B\|T(u)\|,$$

where ε_B is small, then U_B is expected to be a better approximation to u than U. From the truncation errors given in (14.40) and (14.42), we can obtain that

$$L_h(u - U_B) \ = \ [L_h(u) - L(u)] - [L_h(\widetilde{u}) - L(\widetilde{u})] \ = \ T_B(u),$$

and therefore,

$$\|u - U_B\| \ = \ \|L_h^{-1}T_B(u)\| \ \leq \ K\varepsilon_B\|L^{-1}T_B(u)\| \ = \ K\varepsilon_B\|u - U\|. \tag{14.43}$$

From the above inequality one can notice that, if ε_B is small, then the error estimate of the booster scheme is much smaller than the original error. This is possible because of the incorporation of the asymptotic approximation inside the numerical scheme.

Further, let us note that $T_B(u) = T(u - \widetilde{u})$. The essence of the Booster method is to make use of the equation (14.41) instead of the regular numerical scheme (14.39), and therefore, one can obtain the better numerical solution U_B, than the original numerical approximation U. For example, if one applies the EFD scheme (14.12) to solve the SPP (14.14), then from Theorem 9 the following error estimate holds:

$$\|u(x_i) - U_i\| \leq Ch.$$

The corresponding error of the booster method (after incorporating an $O(\varepsilon)$-asymptotic approximate solution) satisfy the following bound

$$\|u(x_i) - U_{Bi}\| \leq C\varepsilon h.$$

Whenever ε is too small, one can obtain better results. One can see the article by Natesan and Ramanujam [17], for further details about the booster method, and for numerical results.

To show the efficiency of the booster method, we apply it to the SPP (14.44) given in Example 14.3.

■ EXAMPLE 14.3

Consider the following nonhomogenous BVP:

$$\begin{cases} \varepsilon u''(x) + u'(x) = -(1+2x), & x \in \Omega, \\ -u'(0) = 1, \quad u(1) + \varepsilon u'(1) = 0. \end{cases} \tag{14.44}$$

the exact solution of the above BVP can be calculated as

$$u(x) = 2 - x(1+x) + \varepsilon[1 - 2(\varepsilon[1 - \exp(-x/\varepsilon)] - x)].$$

Table 14.3 presents the error of the EFD scheme and the corresponding booster method for Example 14.3. From this table one can notice the accuracy of the booster method. The errors of the booster method reveals the effect of the incorporation of an $O(\varepsilon)$-asymptotic approximation into the EFD scheme. The booster method can be applied to any type of scheme, and here we applied it only to the EFD scheme.

14.4.5 Semilinear Problems

This section deals with the numerical solution of semilinear singularly perturbed BVPs. These problems arise in various applications, including chemical reactions, gas porous electrodes theory, etc. Generally, semilinear problems admit multiple solutions, and it is difficult to obtain ε-uniformly convergent numerical solutions.

Table 14.3 Error for the EFD scheme and the corresponding Booster method for Example 14.3.

Nodes	Error			
$H = 0.05$	$\varepsilon = 10^{-3}$		$\varepsilon = 10^{-5}$	
	EFD	Booster	EFD	Booster
0.00	$4.8100e-02$	$9.8000e-05$	$4.9981e-02$	$9.9980e-07$
0.10	$4.3298e-02$	$1.1102e-15$	$4.4983e-02$	$6.6613e-16$
0.20	$3.8498e-02$	$4.4409e-16$	$3.9985e-02$	$6.6613e-16$
0.30	$3.3698e-02$	$2.2204e-16$	$3.4987e-02$	0
0.40	$2.8898e-02$	0	$2.9989e-02$	$2.2204e-16$
0.50	$2.4098e-02$	$2.2204e-16$	$2.4991e-02$	$4.4409e-16$
0.60	$1.9298e-02$	$6.6613e-16$	$1.9993e-02$	$6.6613e-16$
0.70	$1.4498e-02$	$3.3307e-16$	$1.4995e-02$	$4.4409e-16$
0.80	$9.6980e-03$	$1.1102e-16$	$9.9970e-03$	$1.1102e-16$
0.90	$4.8980e-03$	$5.5511e-17$	$4.9990e-03$	$5.5511e-17$
1.00	$9.8000e-05$	$4.3368e-19$	$9.9980e-07$	$3.3881e-21$

Consider the semilinear two-point BVP:

$$\begin{cases} \varepsilon u_{xx} + a(x)u_x - b(x,u) = 0, & x \in \Omega, \\ -u_x(0) = A, \quad u(1) + \varepsilon u_x(1) = B, \end{cases} \tag{14.45}$$

where $a(x)$ and $b(x,u)$ are sufficiently smooth functions such that $a(x) \geq \alpha > 0$, $b_u(x,u) \geq 0$, $(x,u) \in \overline{\Omega} \times \mathbb{R}$. Here, the notation u_{xx} is used instead of $u''(x)$ mainly for the sake of convenience to denote the sequence of linear problems. Analytic results such as existence, uniqueness and the asymptotic approximate solutions can be seen in the classical books of Chang and Howes [4] and O'Malley [19].

To solve the semilinear BVP (14.45) numerically, the Newton's method of quasilinearization is applied to obtain a sequence $\{u^m\}_0^\infty$ of approximations with a proper choice of initial approximation $u^0(x)$. For each non-negative integer m, we define u^{m+1} to be the solution of the following linear problem:

$$\begin{cases} \varepsilon u_{xx}^{m+1} + a(x)u_x^{m+1} - b^m(x)u^{m+1} = f^m(x), & x \in \Omega, \\ -u_x^{m+1}(0) = A, \quad u^{m+1}(1) + \varepsilon u_x^{m+1}(1) = B, \end{cases} \tag{14.46}$$

where $b^m(x) = b_u(x,u^m)$ and $f^m(x) = b(x,u^m) - b_u(x,u^m)u^m$.

If the initial approximation $u^0(x)$ is sufficiently close to the solution $u(x)$ of (14.45), then the sequence of approximate solutions $\{u^m\}_0^\infty$ converge to the solution $u(x)$ of the given problem. For each fixed m, the BVP (14.46) is a linear BVP of the form (14.14), and hence this problem can be solved by all the methods discussed above.

The solution of the reduced problem of (14.45) will be taken as the initial approximation $u^0(x)$. For the numerical computations, we use the following stopping criteria:

$$|U_i^{m+1} - U_i^m| \leq \text{Tol}, \quad 1 \leq i \leq N, \ m \geq 0,$$

where U_i^m is the mth-iteration solution at the ith mesh point for any prescribed tolerance bound.

14.5 LAYER-ADAPTED NONUNIFORM MESHES

The extension of the EFD scheme (14.12) to higher-dimensional problems are computationally expensive and in several cases it is not possible. One of the main alternate ways to obtain an ε-uniformly convergent numerical solution is to use classical finite difference/element schemes on nonuniform meshes, which are condensed inside the boundary layer and coarse in the rest of the domain. Let the mesh points in an arbitrary nonuniform mesh with N subintervals be denoted by $\overline{\Omega}^N = \{x_i\}_0^N$, then the mesh width be denoted by $h_i = x_i - x_{i-1}$ for $1 \leq i \leq N$. Before proceeding further, we define the finite difference operators on the nonuniform meshes as

$$D^+U_i = \frac{U_{i+1} - U_i}{h_{i+1}}, \quad D^-U_i = \frac{U_i - U_{i-1}}{h_i},$$
$$D^0U_i = \left(\frac{h_{i+1}D^+ + h_iD^-}{h_i + h_{i+1}}\right)U_i, \quad \delta^2U_i = 2\left(\frac{D^+ - D^-}{h_i + h_{i+1}}\right)U_i.$$

14.5.1 Bakhvalov Meshes

Bakhvalov [2] was the first person to introduce the nonuniform meshes for solving SPPs numerically. Assume that the solution of an SPP is having a boundary layer at the left end $x = 0$, and therefore, the boundary layer function is $u = \exp(-\alpha x/\varepsilon)$, for some $\alpha > 0$. His idea is to use an equidistant u-mesh near $u = 1$ (which corresponds to $x = 0$), then to map this mesh back to the x-axis by means of the boundary layer function. This defines the mesh points near to $x = 0$ by

$$\exp\left(-\frac{\alpha x}{\varepsilon}\right) = 1 - \frac{i}{N},$$

which is equivalent to define the mesh points as

$$x_i = -\frac{\varepsilon}{\alpha}\ln\left(1 - \frac{i}{N}\right).$$

The *Bakhvalov* meshes are obtained from the following mesh generating function:

$$\lambda(t) = \begin{cases} \phi(t) := -\gamma\varepsilon \ln(1 - t/\eta), & \text{for } t \in [0, \tau], \\ \psi(t) := \phi(t) + \phi'(t)(t - \tau), & \text{for } t = [\tau, 1], \end{cases} \tag{14.47}$$

where γ and η are positive constants, and the transition point τ is obtained by solving the nonlinear equation

$$\phi(\tau) + \phi'(\tau)(1 - \tau) = 1.$$

Here, one has to solve a nonlinear problem to obtain the nonuniform Bakhvalov meshes even to solve a linear SPP of the form (14.3). Further, the mathematical proof of the uniform convergence is too difficult.

In order to obtain an ε-uniformly convergent numerical solution to SPP (14.3), we use the classical upwind finite difference scheme on the layer-adapted nonuniform meshes. Define the following upwind finite difference scheme on nonuniform meshes as

$$\begin{cases} \varepsilon\delta^2 U_i + a_i D^+ U_i - b_i U_i = f_i, & 0 \leq i \leq N, \\ u_0 = A, \quad u_N = B. \end{cases} \tag{14.48}$$

Theorem 12 *Let $u(x)$ be the solution of the SPP (14.3), let U_i be the numerical solution of the scheme (14.48) applied on the Bakhvalov meshes defined in (14.47). Then, the error satisfies the following bound*

$$\max_{0 \leq i \leq N} \|u(x_i) - U_i\| \leq CN^{-1},$$

where the constant C is independent of x_i and ε.

14.5.2 Shishkin Meshes

Shishkin [22] proposed piecewise-uniform meshes to obtain the uniform convergent numerical solutions for SPPs of the form (14.3). Assume that the boundary layer is on the left boundary $x = 0$, then the domain $[0, 1]$ is divided into two subdomains as $[0, \tau]$ and $[\tau, 1]$, and $N/2$ mesh intervals are placed in each of the subdomains, and the transition parameter τ is defined by

$$\tau = \min\left\{\frac{1}{2}, \frac{\varepsilon}{\alpha} \ln N\right\}.$$

The piecewise-uniform meshes are given by

$$x_i = \begin{cases} \dfrac{2i\tau}{N}, & i \le N/2, \\ x_{i-1} + \dfrac{2(1-\tau)}{N}, & N/2 < i. \end{cases} \tag{14.49}$$

There is no need to solve any nonlinear equation to obtain the piecewise-uniform Shishkin meshes. Extension of Shishkin meshes to higher-dimensional rectangular domain problems is easy. Also, one can obtain the error estimates by decomposing the solution into regular and singular components of the solution. More details about the Shishkin meshes for various DEs and the error estimates can be found in the books of Farrell *et al.* [8], Miller *et al.* [15] and Roos *et al.* [21].

Theorem 13 *[15] Let $u(x)$ be the solution of the SPP (14.3), let U_i be the numerical solution of the scheme (14.48) applied on the piecewise-uniform Shishkin meshes defined in (14.49). Then, the error satisfies the following bound*

$$\max_{0\le i\le N} \|u(x_i) - U_i\| \;\le\; CN^{-1}\ln N,$$

where the constant C is independent of x_i and ε.

It is worthwhile to note that the error estimate obtained in the above theorem is first-order up to a logarithmic factor, and it is not an optimal bound.

14.5.3 Equidistribution Meshes

This is the most general way of generating layer-adapted nonuniform meshes, by equidistributing a positive monitor function which depends on the derivatives of the solution over the domain of differential equation. The underlying idea of the equidistribution meshes is given by the following identity:

$$\int_{x_{i-1}}^{x_i} M\big(u(x), x\big)dx = \frac{1}{N}\int_0^1 M\big(u(x), x\big)dx, \quad i = 1, \cdots, N, \tag{14.50}$$

where $M(u(x), x) > 0$ is called the monitor function. Equidistribution can also be thought of giving rise to a mapping $x = x(\eta)$, relating a computational coordinate $\eta \in [0,1]$ to the physical coordinate $x \in [0,1]$, defined by

$$\int_0^{x(\eta)} M\big(u(s), s\big)ds = \eta \int_0^1 M\big(u(s), s\big)ds. \tag{14.51}$$

The identity given in (14.50) can be written in the following form as well:

$$\int_{x_{i-1}}^{x_i} M\big(u(x), x\big)dx = \int_{x_i}^{x_{i+1}} M\big(u(x), x\big)dx, \quad i = 1, \cdots, N. \tag{14.52}$$

Since the solution $u(x)$ of the SPP (14.3) has steep gradients, the monitor function contains either the first-order or the second-order derivatives or any suitable combination of both. Some examples of monitor functions are:

(i) $M\big(u(x), x\big) = |u'(x)|$.

(ii) $M\big(u(x), x\big) = \sqrt{\alpha + |u'(x)|^2}$, where $\alpha > 0$.

(iii) $M\big(u(x), x\big) = 1 + \alpha|u'(x)|^p$, where $\alpha > 0$ and $p \in (0, 1)$.

(iv) $M\big(u(x), x\big) = \alpha + |u''(x)|^m$, where $\alpha > 0$ and $m \in (0, 1)$ are user chosen parameter.

To obtain the equidistribution meshes, one has to solve the following nonlinear system of equations, obtained by discreting the identity given in (14.52):

$$M_{i-\frac{1}{2}}(x_i - x_{i-1}) = M_{i+\frac{1}{2}}(x_{i+1} - x_i), \tag{14.53}$$

where $M_{i-\frac{1}{2}} \approx M(u(x_{i-\frac{1}{2}}), x_{i-\frac{1}{2}})$.

This is an iterative process: to start the iteration, one has to use the uniform meshes, and solve the discretized BVP (14.48) to obtain numerical approximate solution U_i. By using this U_i in (14.53), one has to solve the nonlinear equations (14.53). This process has to be repeated until some suitable stopping criteria holds. The following theorem provides an error estimate for the numerical solution.

Theorem 14 *[14] Let $u(x)$ be the solution of the SPP (14.3), let U_i be the numerical solution of the scheme (14.48) applied on the equidistributed meshes defined in (14.53). Then, the error satisfies the following bound*

$$\max_{0 \le i \le N} \|u(x_i) - U_i\| \;\le\; CN^{-1},$$

where the constant C is independent of x_i and ε.

As like in the Bakhvalov meshes case, here also one has to solve a system of nonlinear equations to obtain the nonuniform meshes for the solution of a linear SPP. Further, it is an iterative process, therefore, it is computationally expensive. This method does not require any *a priori* information about the location and the width of the boundary layers. Therefore, several real application problems in higher-dimensions can be solved on these equidistribution meshes.

REFERENCES

1. Allen, D.N. and Southwell, R. V., Relaxation methods applied to determine the motion in 2D of a viscous fluid past a fixed cylinder. *Quater. J. Mech. and Appl. Math.*, **VIII**(2):129–145 (1955).
2. Bakhvalov, A. S., On the optimization of methods for solving boundary value problems with boundary layers. *Zh. Vychisl. Mat. i Mat. Fis.*, **9**:841-859 (1969). (In Russian).
3. Bush, A. W., *Perturbation Methods for Engineers and Scientists.* CRC Press, Boca Raton (1992).
4. Chang, K.W. and Howes, F.A., *Nonlinear Singular Perturbation Phenomena: Theory and Applications.* Springer, New York (1984).
5. Cohen D.S., Multiple stable solutions of nonlinear boundary-value problems arising in chemical reactor theory. SIAM *J. Math. Anal.*, **20**:1–13 (1973).
6. Doolan, E. P., Miller J. J. H., and Schildres, W. H. A., *Uniform Numerical Methods for Problems with Initial and Boundary Layers.* Boole Press, Dublin (1980).
7. Eckhaus, W., *Asymptotic Analysis of Singular Perturbations.* Noth-Holland, Amsterdam (1979).
8. Farrell, P. A., Hegarty, A. F., Miller, J. J. H., O'Riordan, E. and Shishkin, G.I., *Robust Computational Techniques for Boundary Layers.* Chapman & Hall/CRC Press (2000).
9. Il'in A.M., Differencing schemes for a differential equation with a small parameter affecting the highest derivative. *Math. Notes*, **6**:596–602 (1969).
10. Israeli, M. and Ungarish, M., Improvement of numerical solution of boundary layer problems by incorporation of asymptotic approximations. *Numer. Math.*, **39**:309–324 (1982).
11. Jayakumar, J. and Ramanujam, N., A numerical method for singular perturbation problems arising in chemical reactor theory. *Comput. Math. Applic.*, **27**(5):83–99 (1994).
12. Kellogg, R. B. and Tsan, A., Analysis of some difference approximations for a singular perturbation problem without turning points. *Math. Comput.*, **32**(144):1025–1039 (1978).
13. Lagerstrom, P. A., *Matched Asymptotic Expansions.* Springer, New York (1988).
14. Mackenzie, J., Uniform convergence analysis of an upwind finite-difference approximation of a convection-diffusion boundary value problem on an adaptive grid. IMA *J. Numer. Anal.*, **19**:233–249 (1999).
15. Miller, J. J. H., O'Riordan, E. and Shishkin, G.I., *Fitted Numerical Methods for Singular Perturbation Problems.* World Scientific, Singapore (1996).
16. Natesan, S. and Ramanujam, N., Initial-value technique for singularly perturbed boundary-value problems for second-order ordinary differential equations arising in chemical reactor theory. *J. Optim. Theory Appl.*, **97**(2):455–470 (1998).

17. Natesan, S. and Ramanujam, N., A "booster method" for singular perturbation problems arising in chemical reactor theory. *Appl. Math. Comput.*, **100**:27–48 (1999).

18. Nayfeh, A. H., *Perturbation Methods.* John Wiley & Sons, New York (1973).

19. O'Malley, R.E., *Introduction to Singular Perturbations.* Academic Press, New York (1974).

20. Roberts, S.M., A boundary value technique for singular perturbation problems. *J. Math. Anal. Appl.*, **87**:489–508 (1982).

21. Roos, H.-G. , Stynes, M. and Tobiska, L., *Numerical Methods for Singularly Perturbed Differential Equations.* Springer-Verlag, Berlin, (2008), Second edition.

22. Shishkin, G. I., A difference scheme on a nonuniform mesh for a differential equation with a small parameter in the highest derivative. *U.S.S.R. Comput. Maths. Math. Phys.*, **23**:59–66 (1983).

PART III

ADVANCED MODELING TOPICS

CHAPTER 15

FRACTIONAL CALCULUS AND ITS APPLICATIONS

IVO PETRÁŠ

Technical University of Košice, Slovakia

15.1 INTRODUCTION

It is well-known that an important part of mathematical modeling of objects and processes is a description of their dynamics. In this manner we obtain a dynamical mathematical model, usually in the form of differential equations. In such equations we are able to use a mathematical phenomenon, so-called "fractional calculus."

The term fractional calculus is more than 300 years old. It is a generalization of the ordinary differentiation and integration to noninteger (arbitrary) order. The subject is as old as the calculus of differentiation and goes back to times when Leibniz, Gauss, and Newton invented this kind of calculation. In a letter to L'Hopital in 1695 Leibniz raised the following question: "Can the meaning of derivatives with integer order be generalized to derivatives

Mathematical Modeling with Multidisciplinary Applications.
Edited by Xin-She Yang

with noninteger orders?" The story goes that L'Hopital was somewhat curious about that question and replied with another question to Leibniz. "What if the order will be 1/2?" Leibniz in a letter dated September 30, 1695 replied: "It will lead to a paradox, from which one day useful consequences will be drawn." The question raised by Leibniz for a fractional derivative was an ongoing topic for the last 300 years. Several mathematicians contributed to this subject over the years. People like Liouville, Riemann, and Weyl made major contributions to the theory of fractional calculus. The story of the fractional calculus continued with contributions from Fourier, Leibniz, Grünwald, and Letnikov. Nowadays, fractional calculus attracts many scientists and engineers. There are several applications of this mathematical phenomenon in mechanics, physics, chemistry, electrical circuits, control theory, chaos, and so on [3,5,14,16,21,22,25,27,35,38,43,50]. Those applications prove that the fractional calculus is a calculus of the 21st century and in present days should be consider as a basic mathematical tool.

In the past, the main reason for using integer-order models was the absence of solution methods for fractional differential equations. At present there are many methods for the approximation of the fractional derivative and integral and fractional calculus can be easily used in wide areas of applications.

Currently, the number of applications of fractional calculus rapidly grows. These mathematical phenomena allow us to describe and model a real object more accurately than the classical "integer" methods. The real objects are generally fractional [28,38,51], however, for many of them, the fractionality is very low. A typical example of a noninteger (fractional) order system is the voltage-current relation of a semi-infinite lossy transmission line or diffusion of heat through a semi-infinite solid, where the heat flow is equal to the half-derivative of the temperature [38].

It is important to note, what is the main reason for using fractional calculus in mathematical modeling. It is not correct, if we just replace integer-order derivative with a fractional one without any good reason. There are several reasons which lead to the fractional-order models, not necessarily constant but also variable and of distributed order. We can summarize them as follows:

- Memory of the modeled processes or systems, e.g., heat transfer, or inductor hysteresis, where for a general current in the inductor the voltage is $V(t) = L\frac{d^\alpha I(t)}{dt^\alpha}$, and where L is inductance of the inductor and constant α (order) is related to the "proximity effect."

- Hereditary behavior of the process or systems, e.g., viscoelasticity.

- Porous or rough materials, e.g., capacitor electrode. For a general input voltage $V(t)$ the current is $I(t) = C\frac{d^\alpha V(t)}{dt^\alpha}$, where C is capacitance of the capacitor. It is related to the kind of dielectric. Another constant α (order) is related to the losses of the capacitor.

- Recursivity and selfsimilarity (fractality), e.g., RC ladder network, connection of n series (parallel) RC branches, with recursive parameters $R_{k+1} = aR_k$, $C_{k+1} = bC_k$, $k = 1, \ldots, n$, where $0 < a < 1$ and $0 < b < 1$.
- Chaotic behavior of the system, e.g., Brownian motion, diffusion, etc.

In this chapter we bring basic information on fractional calculus, fractional-order systems, and examples of the applications used fractional-order models.

15.2 FRACTIONAL CALCULUS FUNDAMENTALS

15.2.1 Special Functions

Here, we should mention the most important function used in fractional calculus, Euler's *Gamma* function, which is defined as

$$\Gamma(x) = \int_0^\infty t^{x-1}\, e^{-t}\, dt. \tag{15.1}$$

This function is generalization of a factorial in the following form:

$$\Gamma(x) = (x-1)! \tag{15.2}$$

Another function which plays a very important role in the fractional calculus, was in fact introduced by Humbert and Agarwal in 1953. It is a two-parameter function of the *Mittag-Leffler* type defined as [38]

$$E_{\alpha,\beta}(z) = \sum_{k=0}^{\infty} \frac{z^k}{\Gamma(\alpha k + \beta)}, \qquad (\alpha > 0, \quad \beta > 0) \tag{15.3}$$

There are some relationships (given, e.g., in Ref. [38])

$$E_{1,1}(z) = e^z, \;\; E_{1/2,1}(\sqrt{z}) = \frac{2}{\sqrt{\pi}} e^{-z}\mathrm{erfc}(-\sqrt{z}), \;\; E_{2,1}(-z^2) = \cos(z). \tag{15.4}$$

For the numerical evaluation of the Mittag-Leffler function the Matlab routine `mlf()` written by Podlubny and Kacenak can be used [42].

15.2.2 Definitions of Fractional Operator

Fractional calculus is a generalization of integration and differentiation to noninteger-order fundamental operator ${}_aD_t^\alpha$, where a and t are the bounds of the operation and $\alpha \in \mathrm{R}$. The continuous integro-differential operator is defined as

$${}_aD_t^\alpha = \begin{cases} \frac{d^\alpha}{dt^\alpha} & : \alpha > 0, \\ 1 & : \alpha = 0, \\ \int_a^t (d\tau)^{-\alpha} & : \alpha < 0. \end{cases}$$

The three most frequently used definitions for the general fractional differ-integral are: the Grünwald-Letnikov (GL) definition, the Riemann-Liouville (RL) and the Caputo definition [24, 26, 38]. Other definitions are connected with well-known names as, for instance, Weyl, Fourier, Cauchy, Nishimoto, etc.

In this chapter we will consider mainly the GL, the RL, and the Caputo's definitions. This consideration is based on the fact that, for a wide class of functions, the three best known definitions—GL, RL, and Caputo—are equivalent under some conditions [38].

15.2.3 Grünwald-Letnikov Fractional Derivatives

If we consider $n = \frac{t-a}{h}$, where a is a real constant, which expresses a limit value, we can write

$$ {}_aD_t^\alpha f(t) = \lim_{h \to 0} \frac{1}{h^\alpha} \sum_{j=0}^{[\frac{t-a}{h}]} (-1)^j \binom{\alpha}{j} f(t - jh), \tag{15.5}$$

where $[x]$ means the integer part of x, a and t are the bounds of operation for ${}_aD_t^\alpha f(t)$.

For binomial coefficients calculation we can use the relation between Euler's *Gamma* function and factorial, defined as

$$ \binom{\alpha}{j} = \frac{\alpha!}{j!(\alpha-j)!} = \frac{\Gamma(\alpha+1)}{\Gamma(j+1)\,\Gamma(\alpha-j+1)} \quad \text{for} \quad \binom{\alpha}{0} = 1. \tag{15.6}$$

15.2.4 Riemann-Liouville Fractional Derivatives

Formula for the Riemann-Liouville definition of fractional derivative of the order α has the following form:

$$ {}_aD_t^\alpha f(t) = \frac{1}{\Gamma(n-\alpha)} \frac{d^n}{dt^n} \int_a^t \frac{f(\tau)}{(t-\tau)^{\alpha-n+1}} d\tau, \tag{15.7}$$

for $(n - 1 < \alpha < n)$, where a and t are the limits of operation ${}_aD_t^\alpha f(t)$.

15.2.5 Caputo Fractional Derivatives

The Caputo definition of fractional derivatives can be written as

$$ {}_aD_t^\alpha f(t) = \frac{1}{\Gamma(n-\alpha)} \int_a^t \frac{f^{(n)}(\tau)}{(t-\tau)^{\alpha-n+1}} d\tau, \quad \text{for} \quad (n-1 < \alpha < n). \tag{15.8}$$

Let us denote the Riemann-Liouville fractional derivative as ${}_a^{RL}D_t^{\alpha} f(t)$ and the Caputo definition as ${}_a^{C}D_t^{\alpha} f(t)$, then the relationship between them is

$$
{}_a^{RL}D_t^{\alpha} f(t) = {}_a^{C}D_t^{\alpha} f(t) + \sum_{k=0}^{n-1} \frac{(t-a)^{k-\alpha}}{\Gamma(k-\alpha+1)} f^{(k)}(a),
$$

for $f^{(k)}(a) = 0, \ (k = 0, 1, \ldots, n-1)$.

The initial conditions for the fractional-order differential equations with the Caputo derivatives are in the same form as for the integer-order differential equations. It is an advantage because applied problems require definitions of fractional derivatives, where there are clear interpretations of initial conditions, which contain $f(a)$, $f^{'}(a)$, $f^{''}(a)$, ..., $f^{(n-1)}(a)$.

15.2.6 Laplace Transform Method

The Laplace transform method is a very frequently used tool for solving engineering problems.

For zero initial conditions, the Laplace transform of fractional derivatives of order α (Grünwald-Letnikov, Riemann-Liouville, and Caputo's) reduces to

$$
\mathscr{L}\{{}_0D_t^{\alpha} f(t)\} = s^{\alpha} F(s). \tag{15.9}
$$

Moreover, the Laplace transform of the Riemann-Liouville fractional derivative is well-known. However, its practical applicability is limited by the absence of the physical interpretation of the limit values of fractional derivatives at the lower bound $t = 0$. So far, such an interpretation was partially solved only in Ref. [13].

15.2.7 Some Properties of Fractional Calculus

The main properties of fractional derivatives/integrals are as follows [26]:

1. If $f(t)$ is an analytical function of t, then its fractional derivative ${}_0D_t^{\alpha} f(t)$ is an analytical function of t, α.

2. For $\alpha = n$, where n is integer, the operation ${}_0D_t^{\alpha} f(t)$ gives the same result as classical differentiation of integer order n.

3. For $\alpha = 0$ the operation ${}_0D_t^{\alpha} f(t)$ is the identity operator:

$$
{}_0D_t^{0} f(t) = f(t).
$$

4. Fractional differentiation and fractional integration are linear operations:

$$
{}_aD_t^{\alpha} \left(\lambda f(t) + \mu g(t)\right) = \lambda \, {}_aD_t^{\alpha} f(t) + \mu \, {}_aD_t^{\alpha} g(t). \tag{15.10}
$$

5. The additive index law (semigroup property)

$$ {}_0D_t^{\alpha}\,{}_0D_t^{\beta} f(t) = {}_0D_t^{\beta}\,{}_0D_t^{\alpha} f(t) = {}_0D_t^{\alpha+\beta} f(t) $$

holds under some reasonable constraints on the function $f(t)$.

The fractional-order derivative commutes with integer-order derivative

$$ \frac{d^n}{dt^n}\left({}_aD_t^r f(t)\right) = {}_aD_t^r\left(\frac{d^n f(t)}{dt^n}\right) = {}_aD_t^{r+n} f(t), \tag{15.11} $$

under the condition $t = a$ we have $f^{(k)}(a) = 0$, $(k = 0, 1, 2, \ldots, n-1)$. The relationship above says the operators $\frac{d^n}{dt^n}$ and ${}_aD_t^r$ commute.

The geometric and physical interpretation of fractional integration and fractional differentiation was clearly explained in Podlubny's work [40].

Some other important properties of fractional derivatives and integrals as for example Leibniz's rule, translation, chain rule, behavior and dependence on limit and so on, can be found in several other works (e.g., [24,26,38], etc.).

15.2.8 Numerical Methods for Fractional Calculus

For practical implementation of the fractional calculus in engineering applications we need a good approximation techniques.

A description and overview of the various approximation methods and techniques for continuous and discrete fractional-order models in form of IIR and FIR filters can be found in [48]. Besides the mentioned methods, some other approaches were described in [33]. Last but not least, we should mention the matrix approach proposed by Podlubny [39,41].

The frequency domain approximation methods are not always reliable, especially in detecting chaos behavior in nonlinear systems [44,46]. As has been shown, due to an error of approximation, numerical simulation may result in wrong conclusions, e.g., fake chaos is produced due to the implementation of the frequency domain approximation methods [45]. Simulation of the fractional-order system using the time-domain methods is complicated and due to long memory characteristics of these systems requires a very long simulation time but on the other hand it is more accurate. Applying some ideas as, for instance, the short memory principle [38], we can reduce the computational cost of time-domain methods. Results obtained by these methods are more reliable than those determined using the frequency-based approximation [46].

15.2.8.1 Grünwald-Letnikov Method For numerical calculation of fractional-order derivatives we can use the relation (15.12) derived from the GL definition (15.5). This approach is based on the fact that for a wide class of functions, three definitions—GL (15.5), RL (15.7), and Caputo's (15.8)—are equivalent.

The relation for the explicit numerical approximation of qth derivative at the points kh, $(k = 1, 2, \dots)$ has the following form [10, 38, 49]:

$$ {}_{(k-L_m/h)}D^q_{t_k} f(t) \approx h^{-q} \sum_{j=0}^{k} (-1)^j \binom{q}{j} f(t_{k-j}), \tag{15.12} $$

where L_m is the "memory length," $t_k = kh$, h is the time step of calculation and $(-1)^j \binom{q}{j}$ are binomial coefficients $c_j^{(q)}$ $(j = 0, 1, \dots)$. For their calculation we can use the following expression [10]:

$$ c_0^{(q)} = 1, \qquad c_j^{(q)} = \left(1 - \frac{1+q}{j}\right) c_{j-1}^{(q)}. \tag{15.13} $$

Then, the general numerical solution of the fractional differential equation

$$ {}_aD^q_t y(t) = f(y(t), t), $$

can be expressed as

$$ y(t_k) = f\left(y(t_k), t_k\right) h^q - \sum_{j=v}^{k} c_j^{(q)} y(t_{k-j}). \tag{15.14} $$

For the *memory term* expressed by the sum, a "short memory" principle can be used. Then the lower index of the sums in relations (15.14) will be $v = 1$ for $k < (L_m/h)$ and $v = k - (L_m/h)$ for $k > (L_m/h)$, or without using the "short memory" principle, we put $v = 1$ for all k.

Obviously, for this simplification we pay a penalty in the form of some inaccuracy. If $f(t) \le M$, we can easily establish the following estimate for determining the memory length L_m, providing the required accuracy ϵ:

$$ L_m \ge \left(\frac{M}{\epsilon |\Gamma(1-q)|}\right)^{1/q}. \tag{15.15} $$

An evaluation of the short memory effect and convergence relation of the error between short and long memory were clearly described and also proved in [38].

The described numerical method is the so-called Power Series Expansion (PSE) of a generating function. It is important to note that PSE leads to an approximation in the form of polynomials, that is, the discretized fractional operator is in the form of a FIR filter, which has only zeros.

The resulting discrete transfer function, approximating fractional-order operators, can be expressed in the z-domain as follows:

$$ {}_0D^{\pm r}_{kT} G(z) = \frac{Y(z)}{F(z)} = \left(\frac{1}{T}\right)^{\pm r} \mathrm{PSE}\left\{\left(1 - z^{-1}\right)^{\pm r}\right\}_n \approx T^{\mp r} R_n(z^{-1}), \tag{15.16} $$

where T is the sample period, PSE$\{u\}$ denotes the function resulting from applying the power series expansion to the function u, $Y(z)$ is the Z transform of the output sequence $y(kT)$, $F(z)$ is the Z transform of the input sequence $f(kT)$, n is the order of the approximation, and R is the polynomial of degree n, respectively, in the variable z^{-1}, $z = \exp(sT)$, and $k = 1, 2, \ldots$. Matlab routine `dfod2()` of this method can be downloaded from the MathWorks, Inc. website [30].

15.2.8.2 Continuous- and Discrete-Time Approximation Techniques Another approach can be realized by Continued Fraction Expansion (CFE) of the generating function and then the approximated fractional operator is in the form of an IIR filter, which has poles and zeros [48].

Taking into account that our aim is to obtain equivalents to the fractional integro-differential operators in the Laplace domain, $s^{\pm r}$, the result of such an approximation for an irrational function, $G(s)$, can be expressed in the form

$$\begin{aligned} G(s) &\simeq a_0(s) + \cfrac{b_1(s)}{a_1(s) + \cfrac{b_2(s)}{a_2(s) + \cfrac{b_3(s)}{a_3(s)+\ldots}}} \\ &= a_0(s) + \frac{b_1(s)}{a_1(s)+}\,\frac{b_2(s)}{a_2(s)+}\,\frac{b_3(s)}{a_3(s)+}\,\ldots, \end{aligned} \tag{15.17}$$

where $a_i's$ and $b_i's$ are rational functions of the variable s, or are constants. The application of the method yields a rational function, which is an approximation of the irrational function $G(s)$.

In other words, for evaluation purposes, the rational approximations obtained by CFE frequently converge much more rapidly than the PSE and have a wider domain of convergence in the complex plane. On the other hand, the approximation by PSE and the short memory principle is convenient for the dynamical properties consideration.

For interpolation purposes, rational functions are sometimes superior to polynomials. This is, roughly speaking, due to their ability to model functions with poles. These techniques are based on the approximations of an irrational function, $G(s)$, by a rational function defined by the quotient of two polynomials in the variable s in frequency s-domain

$$G(s) \simeq R_{i(i+1)\ldots(i+m)} = \frac{P_\mu(s)}{Q_\nu(s)} = \frac{p_0 + p_1 s + \ldots + p_\mu s^\mu}{q_0 + q_1 s + \ldots + q_\nu s^\nu}, \quad (m+1 = \mu+\nu+1) \tag{15.18}$$

passing through the points $(s_i, G(s_i)), \ldots, (s_{i+m}, G(s_{i+m}))$.

The resulting discrete transfer function, approximating fractional-order operators, can be expressed as [49]:

$$ {}_0D_{kT}^{\pm r}\,G(z) = \frac{Y(z)}{F(z)} = \left(\frac{2}{T}\right)^{\pm r} \text{CFE}\left\{\left(\frac{1-z^{-1}}{1+z^{-1}}\right)^{\pm r}\right\}_{p,n} \approx \left(\frac{2}{T}\right)^{\pm r}\frac{P_p(z^{-1})}{Q_n(z^{-1})}, \tag{15.19} $$

where T is the sample period, CFE$\{u\}$ denotes the function resulting from applying the continued fraction expansion to the function u, $Y(z)$ is the Z transform of the output sequence $y(kT)$, $F(z)$ is the Z transform of the input sequence $f(kT)$, p and n are the orders of the approximation, and P and Q are polynomials of degrees p and n, respectively, in the variable z^{-1}, and $k = 1, 2, \ldots$. Matlab routine `dfod1()` can be downloaded from MathWorks, Inc. web site [31].

In general, the discretization of fractional-order differentiator/integrator $s^{\pm r}$ $(r \in \mathrm{R})$ can be expressed by the *generating function* $s \approx \omega(z^{-1})$. This generating function and its expansion determine both the form of the approximation and the coefficients [20].

In this section, for directly discretizing s^r, $(0 < r < 1)$, we shall concentrate on the IIR form of discretization where as a generating function we will adopt an Al-Alaoui idea on a mixed scheme of Euler and Tustin operators, but we will use a different ratio between both operators. The mentioned new operator, raised to power $\pm r$, has the form [31]

$$ (\omega(z^{-1}))^{\pm r} = \left(\frac{1+a}{T}\,\frac{1-z^{-1}}{1+az^{-1}}\right)^{\pm r}, \tag{15.20} $$

where a is the ratio term and r is the fractional order. The ratio term a is the amount of phase shift and this tuning knob is sufficient for most engineering problems being solved.

In expanding the above in rational functions, we will use the CFE. It should be pointed out that, for control applications, the obtained approximate discrete-time rational transfer function should be stable. Furthermore, for a better fit to the continuous frequency response, it would be of high interest to obtain discrete approximations with poles and zeros interlaced along the line $z \in (-1, 1)$ of the z plane. The direct discretization approximations proposed in this chapter enjoy the desired properties.

The result of such approximation for an irrational function, $\widehat{G}(z^{-1})$, can be expressed by $G(z^{-1})$ in the CFE form [48]

$$ \begin{aligned} G(z^{-1}) &\simeq a_0(z^{-1}) + \cfrac{b_1(z^{-1})}{a_1(z^{-1}) + \cfrac{b_2(z^{-1})}{a_2(z^{-1}) + \cfrac{b_3(z^{-1})}{a_3(z^{-1}) + \ldots}}} \\ &= a_0(z^{-1}) + \frac{b_1(z^{-1})}{a_1(z^{-1})+}\,\frac{b_2(z^{-1})}{a_2(z^{-1})+}\cdots\frac{b_3(z^{-1})}{a_3(z^{-1})+}\cdots, \end{aligned} $$

where a_i and b_i are either rational functions of the variable z^{-1} or constants. The application of the method yields a rational function, $G(z^{-1})$, which is an approximation of the irrational function $\widehat{G}(z^{-1})$.

The resulting discrete transfer function, approximating fractional-order operators, can be expressed as

$$\begin{aligned}(\omega(z^{-1}))^{\pm r} &\approx \left(\frac{1+a}{T}\right)^{\pm r} \mathrm{CFE}\left\{\left(\frac{1-z^{-1}}{1+az^{-1}}\right)^{\pm r}\right\}_{p,q} \\ &= \left(\frac{1+a}{T}\right)^{\pm r} \frac{P_p(z^{-1})}{Q_q(z^{-1})} \\ &= \left(\frac{1+a}{T}\right)^{\pm r} \frac{p_0 + p_1 z^{-1} + \cdots + p_m z^{-p}}{q_0 + q_1 z^{-1} + \cdots + q_n z^{-q}},\end{aligned} \tag{15.21}$$

where CFE$\{u\}$ denotes the continued fraction expansion of u; p and q are the orders of the approximation and P and Q are polynomials of degrees p and q. Normally, we can set the order of approximation $p = q = n$.

■ EXAMPLE 15.1

Here we present some results for fractional order $r = 0.5$ (half-order derivative). The value of approximation order n is truncated to $n = 5$ and weighting factor a was chosen $a = 1/3$. Assume sampling period $T = 0.001\,s$. The approximation of the fractional half-order derivative obtained by routine `dfod1()` is [37]

$$G(z^{-1}) = \frac{985.9 - 1315z^{-1} + 328.6z^{-2} + 36.51z^{-3}}{27 - 18z^{-1} - 3z^{-2} + z^{-3}} \tag{15.22}$$

The Bode plots and unit step response of the digital fractional-order differentiator (15.22) and the analytical continuous solution of a fractional semiderivative are depicted in Figure 15.1. Poles and zeros of the transfer function (15.22) lie in a unit circle.

For simulation purpose, here we also present the Oustaloup's Recursive Approximation (ORA) algorithm [28, 29]. The method is based on the approximation of a function of the form

$$H(s) = s^r, \qquad r \in \mathrm{R}, \qquad r \in [-1;1] \tag{15.23}$$

for the frequency range selected as (ω_b, ω_h) by a rational function:

$$\widehat{H}(s) = C_o \prod_{k=-N}^{N} \frac{s + \omega_k'}{s + \omega_k}, \tag{15.24}$$

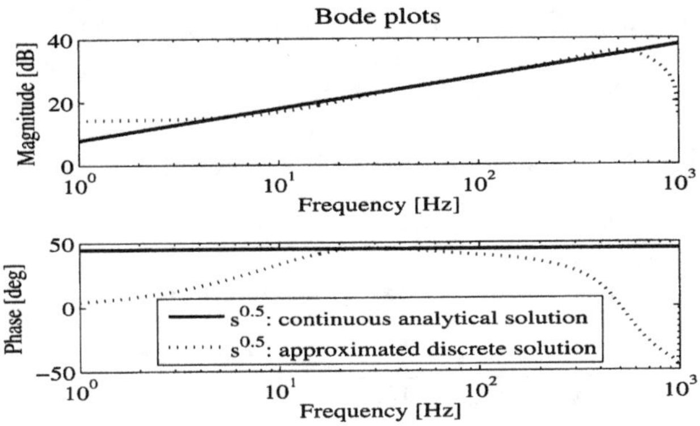

(a) Bode plots for $r = 0.5$, $n = 5$, $a = 1/3$, and $T = 0.001\,s$ in (15.21).

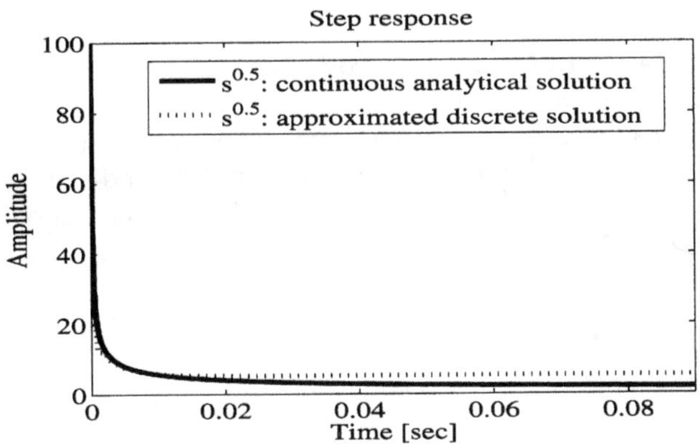

(b) Unit step responses for $r = 0.5$, $n = 5$, $a = 1/3$, and $T = 0.001\,s$ in (15.21).

Figure 15.1 Characteristics of approximated fractional-order differentiator (15.22).

using the following set of synthesis formulas for zeros, poles and the gain:

$$\omega_k' = \omega_b \left(\frac{\omega_h}{\omega_b}\right)^{\frac{k+N+0.5(1-r)}{2N+1}}, \quad \omega_k = \omega_b \left(\frac{\omega_h}{\omega_b}\right)^{\frac{k+N+0.5(1-r)}{2N+1}},$$

$$C_o = \left(\frac{\omega_h}{\omega_b}\right)^{-\frac{r}{2}} \prod_{k=-N}^{N} \frac{\omega_k}{\omega_k'}, \tag{15.25}$$

where ω_h, ω_b are the high and low transitional frequencies. An implementation of this algorithm in Matlab as a function `ora_foc()` is given in [6].

EXAMPLE 15.2

Using the described Oustaloup's Recursive Approximation (ORA) method with

$$\omega_h = 10^3, \quad \omega_b = 10^{-3}, \tag{15.26}$$

the obtained approximation for fractional function $H(s) = s^{-0.5}$ is [37]

$$\widehat{H_5}(s) = \frac{s^5 + 74.97s^4 + 768.5s^3 + 1218s^2 + 298.5s + 10}{10s^5 + 298.5s^4 + 1218s^3 + 768.5s^2 + 74.97s + 1}. \tag{15.27}$$

The Bode plots and the unit step response of the approximated fractional-order integrator (15.27) are depicted in Figure 15.2.

15.2.8.3 Adams-Bashforth-Moulton Method For numerical simulation of the fractional-order system a method on the basis of the Adams-Bashforth-Moulton type predictor-corrector scheme has also been proposed [9]. It is suitable for Caputo's derivative because it just requires the initial conditions and for an unknown function it has clear physical meaning. The method is based on the fact that the fractional differential equation

$$D_t^q y(t) = f(y(t), t), \quad y^{(k)}(0) = y_0^{(k)}, \; k = 0, 1, \ldots, m-1$$

is equivalent to the Volterra integral equation

$$y(t) = \sum_{k=0}^{\lceil q \rceil - 1} y_0^{(k)} \frac{t^k}{k!} + \frac{1}{\Gamma(q)} \int_0^t (t-\tau)^{q-1} f(\tau, y(\tau)) d\tau. \tag{15.28}$$

Discretizing the Volterra equation (15.28) for a uniform grid $t_n = nh$ ($n = 0, 1, \ldots, N$), $h = T_{sim}/N$ and using the short memory principle (fixed or logarithmic) we obtain a good numerical approximation of the true solution $y(t_n)$ of fractional differential equation while preserving the order of accuracy. Assume that we have calculated approximations $y_h(t_j)$, $j = 1, 2, \ldots, n$ and we want to obtain $y_h(t_{n+1})$ by means of the equations

$$\begin{aligned} y_h(t_{n+1}) &= \sum_{k=0}^{m-1} \frac{t_{n+1}^k}{k!} y_0^{(k)} + \frac{h^q}{\Gamma(\alpha+2)} f(t_{n+1}, y_h^p(t_{n+1})) \\ &+ \frac{h^q}{\Gamma(\alpha+2)} \sum_{j=0}^{n} a_{j,n+1} f(t_j, y_n(t_j)), \end{aligned} \tag{15.29}$$

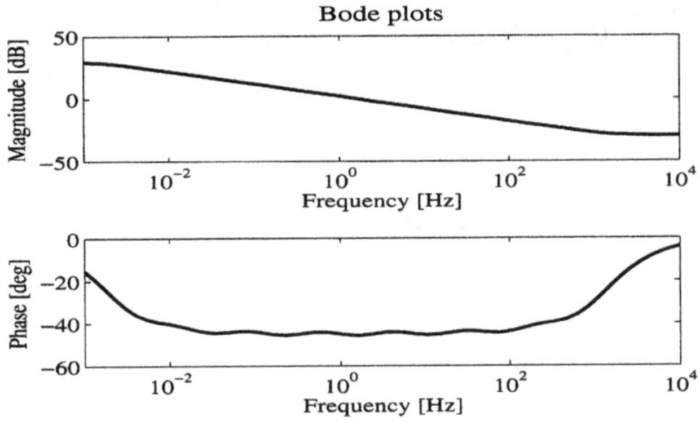

(a) Bode plots for $r = -0.5$ and $N = 5$.

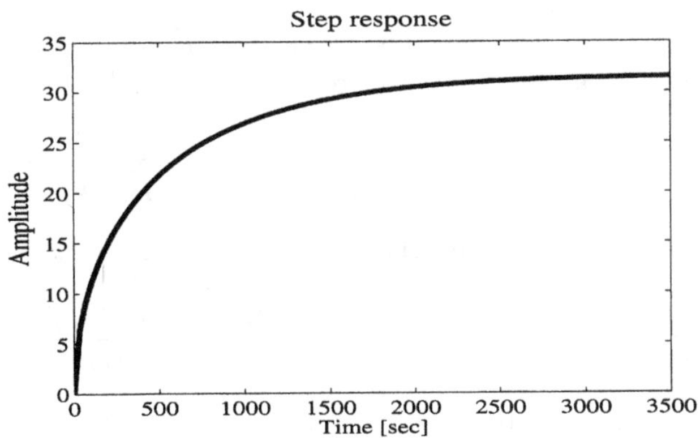

(b) Unit step response for $r = -0.5$ and $N = 5$.

Figure 15.2 Characteristics of approximated fractional-order integrator (15.27).

where

$$a_{j,n+1} = \begin{cases} n^{q+1} - (n-q)(n+1)^q, & : \text{if } j = 0, \\ (n-j+2)^{q+1} + (n-j)^{q+1} + 2(n-j+1)^{q+1} & : \text{if } 1 \le j \le n, \\ 1, & : \text{if } j = n+1. \end{cases}$$

The preliminary approximation $y_h^p(t_{n+1})$ is called a predictor and it is given by

$$y_h^p(t_{n+1}) = \sum_{k=0}^{m-1} \frac{t_{n+1}^k}{k!} y_0^{(k)} + \frac{1}{\Gamma(q)} \sum_{j=0}^{n} b_{j,n+1} f(t_j, y_n(t_j)), \tag{15.30}$$

where

$$b_{j,n+1} = \frac{h^q}{q}((n+1-j)^q - (n-j)^q). \tag{15.31}$$

15.3 FRACTIONAL-ORDER SYSTEMS AND CONTROLLERS

15.3.1 Fractional LTI Systems

A general fractional-order system can be described by a fractional differential equation of the form

$$\begin{aligned} a_n D^{\alpha_n} y(t) + a_{n-1} D^{\alpha_{n-1}} y(t) + \ldots + a_0 D^{\alpha_0} y(t) \\ = b_m D^{\beta_m} u(t) + b_{m-1} D^{\beta_{m-1}} u(t) + \ldots + b_0 D^{\beta_0} u(t), \end{aligned} \tag{15.32}$$

where $D^\gamma \equiv {}_0D_t^\gamma$ denotes the Grünwald-Letnikov, the Riemann-Liouville or the Caputo's fractional derivative [38]. The corresponding transfer function of *incommensurate* real orders has the following form [38]:

$$G(s) = \frac{b_m s^{\beta_m} + \ldots + b_1 s^{\beta_1} + b_0 s^{\beta_0}}{a_n s^{\alpha_n} + \ldots + a_1 s^{\alpha_1} + a_0 s^{\alpha_0}} = \frac{Q(s^{\beta_k})}{P(s^{\alpha_k})}, \tag{15.33}$$

or in the frequency domain it has the form

$$G(j\omega) = \frac{b_m (j\omega)^{\beta_m} + \ldots + b_1 (j\omega)^{\beta_1} + b_0 (j\omega)^{\beta_0}}{a_n (j\omega)^{\alpha_n} + \ldots + a_1 (j\omega)^{\alpha_1} + a_0 (j\omega)^{\alpha_0}} = \frac{Q((j\omega)^{\beta_k})}{P((j\omega)^{\alpha_k})}, \tag{15.34}$$

where a_k $(k = 0, \ldots n)$, b_k $(k = 0, \ldots m)$ are constant, and α_k $(k = 0, \ldots n)$, β_k $(k = 0, \ldots m)$ are arbitrary real or rational numbers and without loss of generality they can be arranged as $\alpha_n > \alpha_{n-1} > \ldots > \alpha_0$, and $\beta_m > \beta_{m-1} > \ldots > \beta_0$.

The incommensurate order system (15.33) can also be expressed in commensurate form by the multivalued transfer function [2]

$$H(s) = \frac{b_m s^{m/v} + \cdots + b_1 s^{1/v} + b_0}{a_n s^{n/v} + \cdots + a_1 s^{1/v} + a_0}, \quad (v > 1). \tag{15.35}$$

Note that every fractional-order system can be expressed in the form (15.35) and the domain of the $H(s)$ definition is a Riemann surface with v Riemann sheets [18].

In the particular case of *commensurate*-order systems, it holds that, $\alpha_k = \alpha k, \beta_k = \alpha k, (0 < \alpha < 1), \forall k \in Z$, and the transfer function has the following form:

$$G(s) = K_0 \frac{\sum_{k=0}^{M} b_k (s^\alpha)^k}{\sum_{k=0}^{N} a_k (s^\alpha)^k} = K_0 \frac{Q(s^\alpha)}{P(s^\alpha)}. \tag{15.36}$$

With $N > M$, the function $G(s)$ becomes a proper rational function in the complex variable s^α which can be expanded in partial fractions of the following form:

$$G(s) = K_0 \left[\sum_{i=1}^{N} \frac{A_i}{s^\alpha + \lambda_i} \right], \tag{15.37}$$

where λ_i ($i = 1, 2, .., N$) are the roots of the pseudo-polynomial $P(s^\alpha)$ or the system poles which are assumed to be simple without loss of generality. The analytical solution of the system (15.37) can be expressed as

$$y(t) = \mathscr{L}^{-1} \left\{ K_0 \left[\sum_{i=1}^{N} \frac{A_i}{s^\alpha + \lambda_i} \right] \right\} = K_0 \sum_{i=1}^{N} A_i t^\alpha E_{\alpha,\alpha}(-\lambda_i t^\alpha), \tag{15.38}$$

where $E_{\mu,\nu}(z)$ is the Mittag-Leffler function defined as (15.3).

A fractional-order plant to be controlled can be described by a typical n-term linear homogeneous fractional-order differential equation (FODE) in time domain

$$a_n\, D_t^{\alpha_n} y(t) + \cdots + a_1\, D_t^{\alpha_1} y(t) + a_0\, D_t^{\alpha_0} y(t) = 0 \tag{15.39}$$

where a_k ($k = 0, 1, \cdots, n$) are constant coefficients of the FODE; α_k ($k = 0, 1, 2, \cdots, n$) are real numbers. Without loss of generality, assume that $\alpha_n > \alpha_{n-1} > \ldots > \alpha_0 \geq 0$.

The analytical solution of the FODE (15.39) is given by a general formula in the form [38]

$$\begin{aligned} y(t) &= \frac{1}{a_n} \sum_{m=0}^{\infty} \frac{(-1)^m}{m!} \sum_{\substack{k_0+k_1+\ldots+k_{n-2}=m \\ k_0 \geq 0; \ldots, k_{n-2} \geq 0}} (m; k_0, k_1, \ldots, k_{n-2}) \\ &\times \prod_{i=0}^{n-2} \left(\frac{a_i}{a_n} \right)^{k_i} \mathcal{E}_m(t, -\frac{a_{n-1}}{a_n}; \alpha_n - \alpha_{n-1}, \alpha_n \\ &+ \sum_{j=0}^{n-2} (\alpha_{n-1} - \alpha_j) k_j + 1), \end{aligned} \tag{15.40}$$

where $(m; k_0, k_1, \ldots, k_{n-2})$ are the multinomial coefficients and $\mathcal{E}_k(t, \lambda; \mu, \nu)$ is the function of Mittag-Leffler type introduced by Podlubny [38]. The function is defined by

$$\mathcal{E}_k(t, \lambda; \mu, \nu) = t^{\mu k + \nu - 1} E_{\mu,\nu}^{(k)}(\lambda t^\mu), \quad (k = 0, 1, 2, \ldots), \tag{15.41}$$

where $E_{\mu,\nu}^{(k)}(z)$ is kth derivative of the Mittag-Leffler function of two parameters given by

$$E_{\mu,\nu}^{(k)}(z) = \sum_{i=0}^{\infty} \frac{(i+k)!\ z^i}{i!\ \Gamma(\mu i + \mu k + \nu)}, \qquad (k = 0, 1, 2, ...). \tag{15.42}$$

The Laplace transform of the function $\mathcal{E}_k(t, \pm\lambda; \alpha, \beta)$ is [38]

$$\mathscr{L}\{\mathcal{E}_k(t, \pm\lambda; \alpha, \beta)\} = \frac{k!s^{\alpha-\beta}}{(s^\alpha \mp \lambda)^{k+1}}$$

for $s > |\lambda|^{1/\alpha}$.

The Laplace transforms for several other Mittag-Leffler type functions are summarized as follows [21]:

$$\begin{aligned} \mathscr{L}\{E_\alpha(-\lambda t^\alpha)\} &= \frac{s^{\alpha-1}}{s^\alpha + \lambda}, \\ \mathscr{L}\{t^{\alpha-1}E_{\alpha,\alpha}(-\lambda t^\alpha)\} &= \frac{1}{s^\alpha + \lambda}, \\ \mathscr{L}\{t^{\beta-1}E_{\alpha,\beta}(-\lambda t^\alpha)\} &= \frac{s^{\alpha-\beta}}{s^\alpha + \lambda}. \end{aligned} \tag{15.43}$$

Consider a control function which acts on the FODE system (15.39) as follows:

$$a_n\, D_t^{\alpha_n} y(t) + \cdots + a_1\, D_t^{\alpha_1} y(t) + a_0\, D_t^{\alpha_0} y(t) = u(t). \tag{15.44}$$

By Laplace transform, we can get a fractional transfer function:

$$G(s) = \frac{Y(s)}{U(s)} = \frac{1}{a_n s^{\alpha_n} + \cdots + a_1 s^{\alpha_1} + a_0 s^{\alpha_0}}. \tag{15.45}$$

The fractional-order linear time-invariant (LTI) system can also be represented by the following state-space model (see, e.g., Ref. [23])

$$\begin{aligned} {}_0D_t^{\mathbf{q}} x(t) &= \mathbf{A}x(t) + \mathbf{B}u(t) \\ y(t) &= \mathbf{C}x(t), \end{aligned} \tag{15.46}$$

where $x \in \mathrm{R}^n$, $u \in \mathrm{R}^r$ and $y \in \mathrm{R}^p$ are the state, input and output vectors of the system and $\mathbf{A} \in \mathrm{R}^{n\times n}$, $\mathbf{B} \in \mathrm{R}^{n\times r}$, $\mathbf{C} \in \mathrm{R}^{p\times n}$, and $\mathbf{q} = [q_1, q_2, \ldots, q_n]^T$ are the fractional orders. If $q_1 = q_2 = \ldots q_n \equiv \alpha$, system (15.46) is called a commensurate-order system, otherwise it is an incommensurate-order system.

15.3.2 Fractional Nonlinear Systems

In this chapter, we will consider the general incommensurate fractional-order nonlinear system represented as follows:

$$\begin{aligned} {}_0D_t^{q_i}x_i(t) &= f_i(x_1(t), x_2(t), \ldots, x_n(t), t), \\ x_i(0) &= c_i, \quad i = 1, 2, \ldots, n, \end{aligned} \tag{15.47}$$

where c_i are initial conditions. The vector representation of (15.47) is

$$D^{\mathbf{q}}\mathbf{x} = \mathbf{f}(\mathbf{x}), \tag{15.48}$$

where $\mathbf{q} = [q_1, q_2, \ldots, q_n]^T$ for $0 < q_i < 2$, $(i = 1, 2, \ldots, n)$ and $\mathbf{x} \in \mathrm{R}^n$.

The equilibrium points of system (15.48) are calculated via solving the following equation:

$$\mathbf{f}(\mathbf{x}) = 0 \tag{15.49}$$

and we suppose that $E^* = (x_1^*, x_2^*, \ldots, x_n^*)$ is an equilibrium point of system (15.48).

15.3.3 Fractional-Order Controllers

15.3.3.1 Definition of Fractional-Order Controllers The fractional-order $PI^\lambda D^\delta$ (a.k.a. $PI^\lambda D^\mu$ controller) controller (FOC) was proposed in [38] as a generalization of the PID controller with integrator of real order λ and differentiator of real order δ. The transfer function of such controller in the Laplace domain has this form:

$$C(s) = \frac{U(s)}{E(s)} = K_p + T_i\, s^{-\lambda} + T_d\, s^{\delta}, \qquad (\lambda, \delta > 0), \tag{15.50}$$

where K_p is the proportional constant, T_i is the integration constant and T_d is the differentiation constant.

The internal structure of the fractional-order controller consists of the parallel connection, the proportional, integration, and derivative part [11]. The transfer function (15.50) corresponds in time domain to a fractional differential equation

$$u(t) = K_p\, e(t) + T_i\, {}_0D_t^{-\lambda}e(t) + T_d\, {}_0D_t^{\delta}e(t), \tag{15.51}$$

or a discrete transfer function given in the following expression:

$$C(z) = \frac{U(z)}{E(z)} = K_p + \frac{T_i}{(\omega(z^{-1}))^{\lambda}} + T_d(\omega(z^{-1}))^{\delta}, \tag{15.52}$$

where $\omega(z^{-1})$ denotes the discrete operator, expressed as a function of the complex variable z or the shift operator z^{-1}.

Taking $\lambda = 1$ and $\delta = 1$, we obtain a classical PID controller. If $\lambda = 0$ and $T_i = 0$, we obtain a PD^δ controller, etc. All these types of controllers are particular cases of the fractional-order controller, which is more flexible and gives an opportunity to better adjust the dynamical properties of the fractional-order control system.

It can also be mentioned that there are many another considerations of the fractional-order controller [7, 25, 35]. For example we can mention several of them: three generations of CRONE controller, TID compensator, fractional lead-lag compensator, etc.

It can be expected that $PI^\lambda D^\delta$ controller (15.50) may enhance the systems control performance due to more tuning knobs introduced. For a wide class of controlled objects we recommend the fractional $PI^n D^\delta$ controller, which is a particular case of $PI^\lambda D^\delta$ controller, where $\lambda = n$, $n \in \mathrm{N}$ and $\delta \in \mathrm{R}$. The integer-order integrator is important for steady-state error cancellation but on the other hand the fractional integral is also important for obtaining a *Bode's ideal loop transfer function* response with a constant phase margin for a desired frequency range [4].

15.4 STABILITY OF FRACTIONAL-ORDER SYSTEMS

Stability as an extremely important property of the dynamical systems can be investigated in various domains [11]. The usual concept of bounded input–bounded output (BIBO) or external stability in *time domain* can be defined via the following general stability conditions [23]:

A causal LTI system with impulse response $h(t)$ will be BIBO stable if the necessary and sufficient condition is satisfied

$$\int_0^\infty ||h(\tau)||d\tau < \infty,$$

where the output of the system is defined by convolution

$$y(t) = h(t) * u(t) = \int_0^\infty h(\tau)u(t-\tau)d\tau,$$

where $u, y \in L_\infty$ and $h \in L_1$.

Another very important domain is *frequency domain.* In the case of a frequency method for evaluating the stability we transform the s-plane into the complex plane $G_o(j\omega)$ and the transformation is realized according to the transfer function of the open loop system $G_o(j\omega)$. During the transformation, all roots of the characteristic polynomial are mapped from the s-plane into the critical point $(-1, j0)$ in the plane $G_o(j\omega)$. The mapping of the s-plane into the $G_o(j\omega)$ plane is conformal, that is, the direction and location of points in the s-plane is preserved in the $G_o(j\omega)$ plane. A frequency investigation method and utilization of the Nyquist frequency characteristics based on argument principle were described in Ref. [25].

However, we cannot directly use algebraic tools as, for example, Routh-Hurwitz criteria for the fractional-order system, because we do not have a characteristic polynomial but pseudo-polynomial with a rational power–*multivalued function.*

When dealing with incommensurate fractional-order systems (or, in general, with fractional-order systems) it is important to bear in mind that $P(s^\alpha)$, $\alpha \in \mathrm{R}$ is a multivalued function of s^α, $\alpha = \frac{u}{v}$, the domain of which can be viewed as a Riemann surface with a finite number of Riemann sheets v, where the origin is a branch point and the branch cut is assumed at R^- (see Figure 15.3). Function s^α becomes holomorphic in the complement of the branch cut line. It is a fact that in multivalued functions only the first Riemann sheet has its physical significance [18]. Note that each Riemann sheet has only one edge at the branch cut and not only poles and singularities originated from the characteristic equation, but branch points and branch cuts of given multivalued functions are also important for the stability analysis [35].

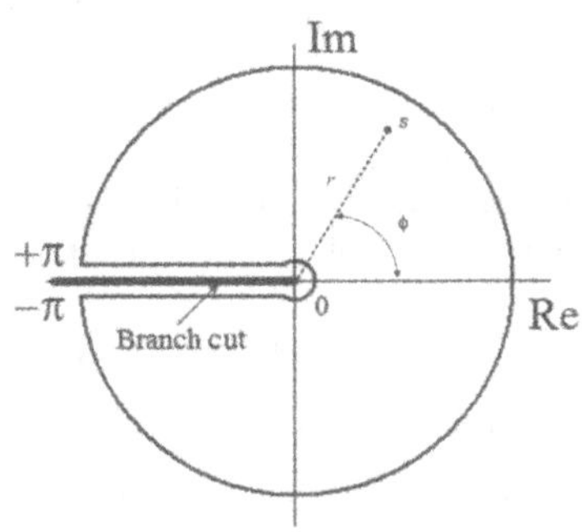

Figure 15.3 Branch cut $(0, -\infty)$ for branch points in the complex plane.

In this chapter the branch cut is assumed at R^- and the first Riemann sheet is denoted by Ω and defined as (see also Figure 15.3)

$$\Omega := \{re^{j\phi} \,|\, r > 0, -\pi < \phi < \pi\}. \tag{15.53}$$

It is well-known that an integer-order LTI system is stable if all the roots of the characteristic polynomial $P(s)$ are negative or have negative real parts if they are complex conjugate (e.g., Ref. [11]). This means that they are located on the left of the imaginary axis of the complex s-plane. System $G(s) = Q(s)/P(s)$ is BIBO stable if

$$\exists, \quad ||G(s)|| \leq M < \infty, \quad M > 0, \quad \forall s, \Re(s) \geq 0.$$

A necessary and sufficient condition for the asymptotic stability is

$$\lim_{t\to\infty}||g(t)|| = 0.$$

EXAMPLE 15.3

Let us investigate the transfer function of fractional-order system (multivalued function) defined as

$$G(s) = \frac{1}{s^{\alpha} + b}, \tag{15.54}$$

where $\alpha \in \mathrm{R}$ $(0 < \alpha \leq 2)$ and $b \in \mathrm{R}$ $(b > 0)$.

The analytical solution of the fractional-order system (15.54) obtained according to relation (15.40) has the following form:

$$g(t) = \mathcal{E}_0(t, -b; \alpha, \alpha). \tag{15.55}$$

The Riemann surface of the function (15.54) contains an infinite number of sheets and infinitely many poles in positions

$$s = b^{\frac{1}{\alpha}} e^{\frac{j(\pi+2\pi n)}{\alpha}}, \quad n = 0, \pm 1, \pm 2, \dots \quad , \text{for } (\alpha > 0) \text{ and } (b > 0).$$

The sheets of the Riemann surface are all different if α is irrational.

For $1 < \alpha < 2$ we have two poles corresponding to $n = 0$ and $n = -1$, and the poles are

$$s = b^{\frac{1}{\alpha}} e^{\pm \frac{i\pi}{\alpha}}.$$

However, for $0 < \alpha < 1$ in (15.54) the denominator is a multivalued function and the singularity of the system can not be defined unless it is made single valued. Therefore we will use the Riemann surface. Let us investigate transfer function (15.54) for $\alpha = 0.5$ (half-order system), then we get

$$G(s) = \frac{1}{s^{\frac{1}{2}} + b}, \tag{15.56}$$

and by equating the denominator to zero we have

$$s^{\frac{1}{2}} + b = 0.$$

Rewriting the complex operator $s^{\frac{1}{2}}$ in exponential form and using the well-known relation $e^{j\pi} + 1 = 0$ (or $e^{j(\pm\pi+2k\pi)} + 1 = 0$) we get the following formula:

$$r^{\frac{1}{2}} e^{j(\phi/2+k\pi)} = a e^{j(\pm\pi+2k\pi)}. \tag{15.57}$$

From relationship (15.57) it can be deduced that the modulus and phase (arg) of the pole are

$$r = b^2 \text{ and } \phi = \pm 2\pi(1+k) \text{ for } k = 0, 1, 2, \ldots.$$

However, the first sheet of the Riemann surface is defined for range of $-\pi < \phi < +\pi$, the pole with the angle $\phi = \pm 2\pi$ does not fall within this range but the pole with the angle $\phi = 2\pi$ falls to the range of the second sheet defined for $\pi < \phi < 2\pi$. Therefore this half-order pole with magnitude b^2 is located on the second sheet of the Riemann surface that consequently maps to the left side of the w-plane (see Figure 15.4). On this plane the magnitude and phase of the single valued pole are b^2 and π, respectively [18].

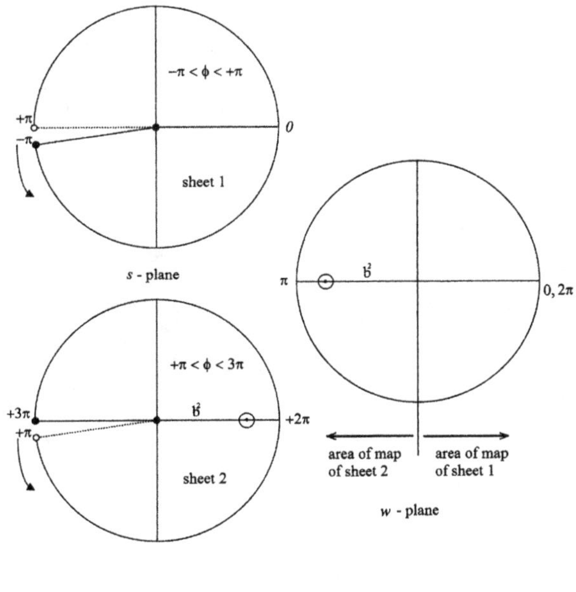

Figure 15.4 Correspondence between the s-plane and the w-plane.

Generally, for the multivalued function defined as follows

$$w = s^{\frac{1}{v}}, \tag{15.58}$$

where $v \in \mathrm{N}$ $(v = 1, 2, 3, \ldots)$ we get the v sheets in the Riemann surface. In Figure 15.5 is shown the relationship between the w-plane and the v sheets of the Riemann surface where sector $-\pi/v < \arg(w) \leq \pi/v$ corresponds to Ω (first Riemann sheet).

Mapping the poles from the s^q-plane into the w-plane, where $q \in \mathrm{Q}$ is such that $q = \frac{k}{m}$ for $k, m \in \mathrm{N}$ and $|\arg(w)| = |\phi|$, can be done by the following rule:

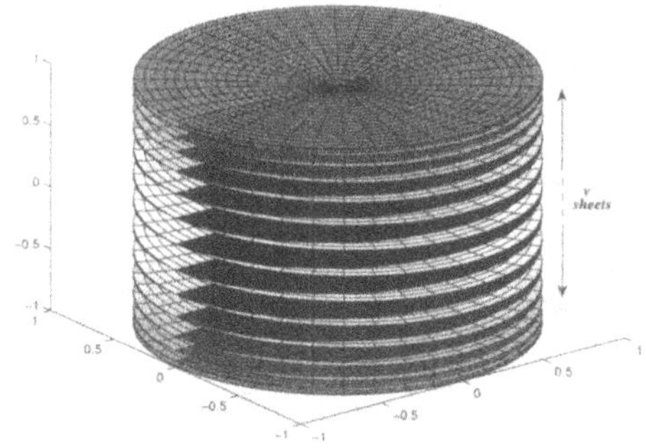

(a) Riemann surface

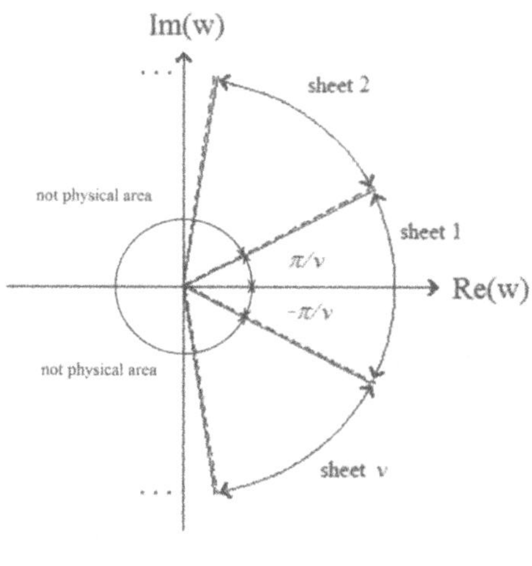

(b) Complex w-plane

Figure 15.5 Correspondence between the w-plane and the Riemann sheets.

If we assume $k = 1$, then the mapping from s-plane to w-plane is independent of k. Unstable region from the s-plane transforms to sector $|\phi| < \frac{\pi}{2m}$ and the stable region transforms to sector $\frac{\pi}{2m} < |\phi| < \frac{\pi}{m}$. The region where $|\phi| > \frac{\pi}{m}$ is not physical. Therefore, the system will be stable if all roots in the w-plane lie in the region $|\phi| > \frac{\pi}{2m}$. Stability regions depicted in Figure 15.6 correspond to the following propositions:

1. For $k < m$ $(q < 1)$ the stability region is depicted in Figure 15.6(a).

2. For $k = m$ $(q = 1)$ the stability region corresponds to the s-plane.

3. For $k > m$ $(q > 1)$ the stability region is depicted in Figure 15.6(b).

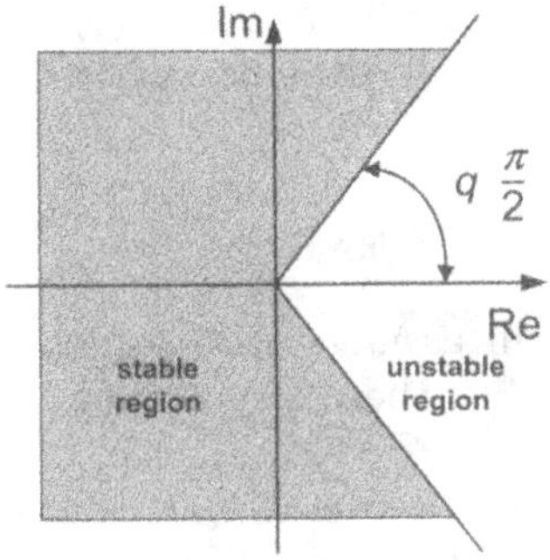

(a) $0 < q < 1$

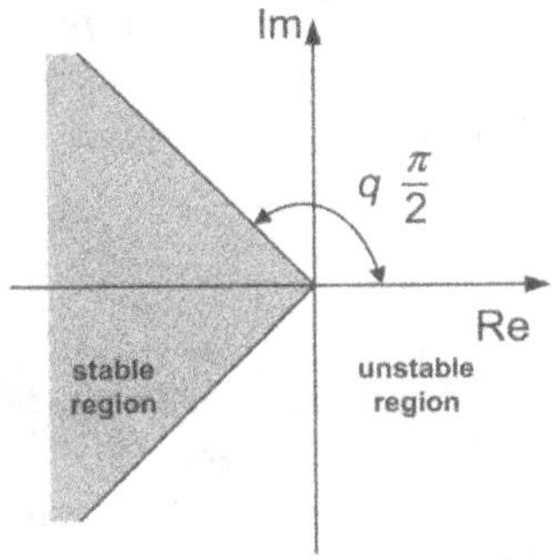

(b) $1 < q < 2$

Figure 15.6 Stability regions of the fractional-order system.

15.4.1 Stability of Fractional LTI Systems

As we can see in previous subsection, in the fractional case, the stability is different from the integer case. It is interesting that a stable fractional system may have roots in the right half of the complex w-plane (see Figure 15.6). Since the principal sheet of the Riemann surface is defined for $-\pi < \arg(s) < \pi$, by using the mapping $w = s^q$, the corresponding w domain is defined by $-q\pi < \arg(w) < q\pi$, and the w-plane region corresponding to the right half plane of this sheet is defined by $-q\pi/2 < \arg(w) < q\pi/2$.

Consider the fractional-order pseudo-polynomial

$$Q(s) = a_1 s^{q_1} + a_2 s^{q_2} + \ldots + a_n s^{q_n} = a_1 s^{c_1/d_1} + a_2 s^{c_2/d_2} + \ldots + a_n s^{c_n/d_n},$$

where q_i are rational numbers expressed as c_i/d_i and a_i are real numbers for $i = 1, 2, \ldots, n$. If for some i, $c_i = 0$ then $d_i = 1$. Let v be the least common multiple (LCM) of $d_1, d_2, \ldots d_n$ denoted as $v = \mathrm{LCM}\{d_1, d_2, \ldots d_n\}$, then [12]

$$\begin{aligned} Q(s) &= a_1 s^{\frac{v_1}{v}} + a_2 s^{\frac{v2}{v}} + \ldots + a_n s^{\frac{v_n}{v}} \\ &= a_1 (s^{\frac{1}{v}})^{v_1} + a_2 (s^{\frac{1}{v}})^{v_2} + \ldots + a_n (s^{\frac{1}{v}})^{v_n}. \end{aligned} \qquad (15.59)$$

The fractional degree (FDEG) of the polynomial $Q(s)$ is defined as [12]

$$\mathrm{FDEG}\{Q(s)\} = \max\{v_1, v_2, \ldots, v_n\}.$$

The domain of definition for (15.59) is the Riemann surface with v Riemann sheets where the origin is a branch point of order $v - 1$ and the branch cut is assumed at R^-. The number of roots for fractional algebraic equation (15.59) is given by the following proposition [1]:

Let $Q(s)$ be a fractional-order polynomial with FDEG$\{Q(s)\} = n$. Then the equation $Q(s)=0$ has exactly n roots on the Riemann surface.

The fractional-order polynomial

$$Q(s) = a_1 s^{\frac{n}{v}} + a_2 s^{\frac{n-1}{v}} + \ldots + a_n s^{\frac{1}{v}} + a_{n+1}$$

is *minimal* if FDEG$\{Q(s)\} = n$. We will assume that all fractional-order polynomials are minimal. This ensures that there is no redundancy in the number of the Riemann sheets [12].

On the other hand, it has been shown, by several authors and by using several methods, that for the case of fractional-order LTI system of commensurate order, a geometrical method of complex analysis based on the argument principle of the roots of the characteristic equation (a polynomial in this particular case) can be used for the stability check in the BIBO sense (see, e.g., Refs. [23, 35]). The stability condition can then be stated as follows [23]:
A commensurate-order system described by a rational transfer function (15.36) is stable if and only if

$$|\arg(\lambda_i)| > \alpha\frac{\pi}{2}, \text{ for all } i \tag{15.60}$$

with λ_i being the ith root of $P(s^\alpha)$.

For the fractional-order LTI system with commensurate order where the system poles are in general complex conjugate, the stability condition can also be expressed as follows [23]:
A commensurate-order system described by a rational transfer function

$$G(w) = \frac{Q(w)}{P(w)}, \tag{15.61}$$

where $w = s^q$, $q \in R^+$, $(0 < q < 2)$, is stable if and only if

$$|\arg(w_i)| > q\frac{\pi}{2},$$

with $\forall w_i \in C$ being the ith root of $P(w) = 0$.

When $w = 0$ is a single root (singularity at the origin) of P, the system cannot be stable. For $q = 1$, this is the classical theorem of pole location in the complex plane: P has no pole in the closed right half plane of the first Riemann sheet. The stability region suggested by this theorem tends to the whole s-plane when q tends to 0, corresponds to the Routh-Hurwitz stability when $q = 1$, and tends to the negative real axis when q tends to 2.

It also has been shown that commensurate system (15.46) is stable if the following condition is satisfied (also if the triplet $\mathbf{A}$, $\mathbf{B}$, $\mathbf{C}$ *is minimal)*

$$|arg(eig(\mathbf{A}))| > q\frac{\pi}{2}, \tag{15.62}$$

where $0 < q < 2$ *and* $eig(\mathbf{A})$ *represents the eigenvalues of matrix* $\mathbf{A}$.

Consider the following autonomous system for internal stability definition

$${}_0D_t^{\mathbf{q}}x(t) = \mathbf{A}x(t), \quad x(0) = x_0, \tag{15.63}$$

with $\mathbf{q} = [q_1, q_2, \ldots, q_n]^T$ and its n-dimensional representation:

$$\begin{aligned}
{}_0D_t^{q_1}x_1(t) &= a_{11}x_1(t) + a_{12}x_2(t) + \cdots + a_{1n}x_n(t),\\
{}_0D_t^{q_2}x_2(t) &= a_{21}x_1(t) + a_{22}x_2(t) + \cdots + a_{2n}x_n(t),\\
&\ldots\\
{}_0D_t^{q_n}x_n(t) &= a_{n1}x_1(t) + a_{n2}x_2(t) + \cdots + a_{nn}x_n(t),
\end{aligned} \tag{15.64}$$

where all q_i's are rational numbers between 0 and 2. Assume m to be the LCM of the denominators u_i's of q_i's , where $q_i = v_i/u_i$, $v_i, u_i \in \mathrm{Z}^+$ for $i = 1, 2, \ldots, n$ and we set $\gamma = 1/m$. Define

$$\det\begin{pmatrix} \lambda^{mq_1} - a_{11} & -a_{12} & \cdots & -a_{1n}\\ -a_{21} & \lambda^{mq_2} - a_{22} & \cdots & -a_{2n}\\ \cdots & & & \\ -a_{n1} & -a_{n2} & \cdots & \lambda^{mq_n} - a_{nn}\end{pmatrix} = 0. \tag{15.65}$$

The characteristic equation (15.65) can be transformed to an integer-order polynomial equation if all q_i's are rational number. Then the zero solution of system (15.64) is globally asymptotically stable if all roots λ_i's of the characteristic (polynomial) equation (15.65) satisfy

$$|\arg(\lambda_i)| > \gamma\frac{\pi}{2} \text{ for all } i.$$

Denoting λ by s^γ in Eq. (15.65), we get the characteristic equation in the form $\det(s^\gamma I - A) = 0$.

Suppose $q_1 = q_2 = \ldots, q_n \equiv q$, $q \in (0, 2)$, all eigenvalues λ of matrix A in (15.64) satisfy $|\arg(\lambda)| > q\pi/2$, the characteristic equation becomes $\det(s^q I - A) = 0$ and all characteristic roots of the system (15.64) have negative real parts.

■ EXAMPLE 15.4

Let us consider two examples of certain class of the fractional-order systems [17]. The first of them has the following form:

$$\begin{aligned} {}_0D_t^{1.1}\begin{bmatrix} x_1(t) \\ x_2(t) \\ x_3(t) \end{bmatrix} &= \begin{bmatrix} -8 & 2 & -3 \\ -1 & -2 & 0.2 \\ 0.5 & -1 & -2 \end{bmatrix}\begin{bmatrix} x_1(t) \\ x_2(t) \\ x_3(t) \end{bmatrix} + \begin{bmatrix} 1 \\ 1 \\ 1 \end{bmatrix} u(t), \\ y(t) &= \begin{bmatrix} 0 & 0 & 1 \end{bmatrix}\begin{bmatrix} x_1(t) \\ x_2(t) \\ x_3(t) \end{bmatrix}, \end{aligned} \tag{15.66}$$

where $x \in \mathrm{R}^3$. For the system matrix $\mathbf{A}$ of the system (15.66) we get the following eigenvalues $\lambda_{1,2} = -2.2719 \pm 0.8119j, \lambda_3 = -7.4562$. All eigenvalues satisfy the stability conditions (15.60), where $|\arg(\lambda_{1,2})| = 2.7984$ and $|\arg(\lambda_3)| = \pi$, thus $|\arg(eig(\mathbf{A}))| > 1.1\frac{\pi}{2}$ and therefore system (15.66) is *stable*.

The second of them has the following form:

$$\begin{aligned} {}_0D_t^{1.2}\begin{bmatrix} x_1(t) \\ x_2(t) \\ x_3(t) \end{bmatrix} &= \begin{bmatrix} -4 & 1 & 1 \\ 0 & -3 & 1 \\ 1 & 3 & 1 \end{bmatrix}\begin{bmatrix} x_1(t) \\ x_2(t) \\ x_3(t) \end{bmatrix} + \begin{bmatrix} 0 \\ 0 \\ 1 \end{bmatrix} u(t), \\ y(t) &= \begin{bmatrix} 0 & 0 & 1 \end{bmatrix}\begin{bmatrix} x_1(t) \\ x_2(t) \\ x_3(t) \end{bmatrix}. \end{aligned} \tag{15.67}$$

For the system matrix $\mathbf{A}$ of the system (15.67) we get the following eigenvalues $\lambda_1 = 1.8284, \lambda_2 = -4$, and $\lambda_3 = -3.8284$. The eigenvalue λ_1 does not satisfy the stability condition (15.60), where $|\arg(\lambda_1)| = 0$ and $|\arg(\lambda_{2,3})| = \pi$, thus $|\arg(\lambda_1)| < 1.2\frac{\pi}{2}$ and therefore system (15.67) is *unstable*.

15.4.2 Stability of Fractional Nonlinear Systems

Stability of the fractional-order nonlinear system is very complex and is different from the fractional-order linear system. The main difference is that for a nonlinear system it is necessary to investigate steady states and there are two types of them: equilibrium point and limit cycle. Nonlinear systems may have several equilibrium points. For nonlinear systems, there are many definitions of stability (asymptotic, global, local, orbital, etc.). The basic idea was formulated by A. M. Lyapunov.

As mentioned in Ref. [23], exponential stability cannot be used to characterize asymptotic stability of fractional-order systems.

Trajectory $x(t) = 0$ of the system (15.47) is t^{-q} asymptotically stable if there is a positive real q such that

$$\forall ||x(t)|| \text{ with } t \leq t_0, \ \exists N(x(t)), \text{ such that } \forall t \geq t_0, ||x(t)|| \leq Nt^{-q}.$$

The fact that the components of $x(t)$ slowly decay towards 0 following t^{-q} leads to fractional systems sometimes being called long memory systems. Power law stability t^{-q} is a special case of the Mittag-Leffler stability [19].

According to stability theorem defined in [47], the equilibrium points are asymptotically stable for $q_1 = q_2 = \cdots = q_n \equiv q$ if all the eigenvalues λ_i, $(i = 1, 2, \ldots, n)$ of the Jacobian matrix $\mathbf{J} = \partial\mathbf{f}/\partial\mathbf{x}$, where $\mathbf{f} = [f_1, f_2, \ldots, f_n]^T$, evaluated at the equilibrium E^*, satisfy the condition [44, 45]

$$|\arg(\text{eig}(\mathbf{J}))| = |\arg(\lambda_i)| > q\frac{\pi}{2}, \quad i = 1, 2, \ldots, n. \tag{15.68}$$

Figure 15.6 shows stable and unstable regions of the complex plane for such a case.

When we consider the incommensurate fractional-order system $q_1 \neq q_2 \neq \cdots \neq q_n$ and suppose that m is the LCM of the denominators u_i's of q_i's, where $q_i = v_i/u_i$, $v_i, u_i \in Z^+$ for $i = 1, 2, \ldots, n$ and we set $\gamma = 1/m$. System (15.48) is asymptotically stable if

$$|\arg(\lambda)| > \gamma\frac{\pi}{2} \tag{15.69}$$

for all roots λ of the following equation:

$$\det(\text{diag}([\lambda^{mq_1} \ \lambda^{mq_2} \ \ldots \ \lambda^{mq_n}]) - \mathbf{J}) = 0. \tag{15.70}$$

A necessary stability condition for fractional-order systems (15.48) to remain chaotic is keeping at least one eigenvalue λ in the unstable region [45]. The number of equilibrium points and eigenvalues for one-scroll, double-scroll and multi-scroll attractors was exactly described in Ref. [46]. Assume that a 3D chaotic system has only three equilibria. Therefore, if the system has a double-scroll attractor, it has two saddle-focus points surrounded by scrolls and one additional saddle point.

Suppose that the unstable eigenvalues of scroll focus points are $\lambda_{1,2} = \alpha_{1,2} \pm j\beta_{1,2}$. The necessary condition to exhibit a double-scroll attractor of system (15.48) is the eigenvalues $\lambda_{1,2}$ remaining in the unstable region [46]. The condition for commensurate derivatives order is

$$q > \frac{2}{\pi}\text{atan}\left(\frac{|\beta_i|}{\alpha_i}\right), \quad i = 1, 2. \tag{15.71}$$

This condition can be used to determine the minimum order for which a nonlinear system can generate chaos [45]. In other words, when the instability measure $\pi/2m - \min(|\arg(\lambda)|)$ is negative, the system cannot be chaotic.

■ EXAMPLE 15.5

Let us investigate Chen's system with a double-scroll attractor in 3D state space. The fractional-order form of such system can be described as [47]

$$\begin{aligned} {}_0D_t^{0.8}x_1(t) &= 35[x_2(t) - x_1(t)], \\ {}_0D_t^{1.0}x_2(t) &= -7x_1(t) - x_1(t)x_3(t) + 28x_2(t), \\ {}_0D_t^{0.9}x_3(t) &= x_1(t)x_2(t) - 3x_3(t). \end{aligned} \tag{15.72}$$

The system has three equilibria at $(0,0,0)$, $(7.94, 7.94, 21)$, and $(-7.94, -7.94, 21)$. The Jacobian matrix of the system evaluated at equilibrium $E^* = (x_1^*, x_2^*, x_3^*)$ is

$$\mathbf{J} = \begin{bmatrix} -35 & 35 & 0 \\ -7 - x_3^* & 28 & -x_1^* \\ x_2^* & x_1^* & -3 \end{bmatrix}. \tag{15.73}$$

The two last equilibrium points are saddle points and are surrounded by a chaotic double-scroll attractor. For these two points, equation (15.70) becomes as follows:

$$\lambda^{27} + 35\lambda^{19} + 3\lambda^{18} - 28\lambda^{17} + 105\lambda^{10} - 21\lambda^8 + 4410 = 0 \tag{15.74}$$

The characteristic equation (15.74) has unstable roots $\lambda_{1,2} = 1.2928 \pm 0.2032j$, $|\arg(\lambda_{1,2})| = 0.1560$ and therefore system (15.72) satisfies the necessary condition for exhibiting a double scroll attractor. The instability measure is 0.0012.

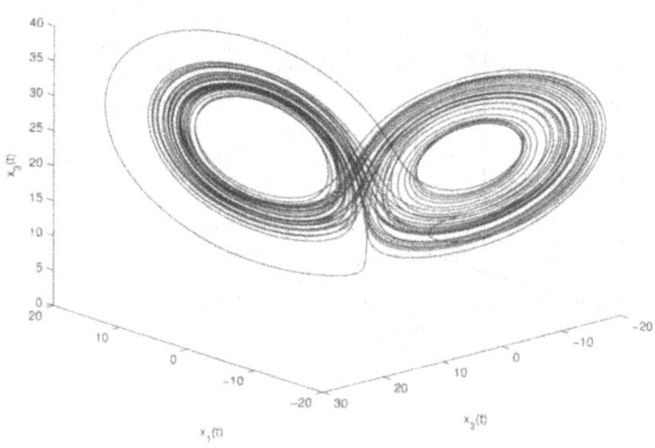

Figure 15.7 Double-scroll attractor of Chen's system (15.72) projected into 3D state space for simulation time 30 s.

Numerical simulation of the system (15.72) for initial conditions $(-9, -5, 14)$ is depicted in Figure 15.7.

15.5 APPLICATIONS OF FRACTIONAL CALCULUS

15.5.1 Control of Electrical Heater

The mathematical model used for the system to be controlled is a two-term differential equation of the fractional-order of the form

$$b_1 \, D_t^{\beta} y(t) + b_0 \, y(t) = u(t), \tag{15.75}$$

for which the parameters b_1, b_0 and β were obtained by an identification method based on the measured step response of the system (see Figure 15.8) and the minimization of the quadratics criteria

$$J = \frac{1}{M+1} \sum_{i=0}^{M} |y_i^* - y_i|^2 , \tag{15.76}$$

being y_i^* the measured values, y_i the model values and $M+1$ the number of measurements.

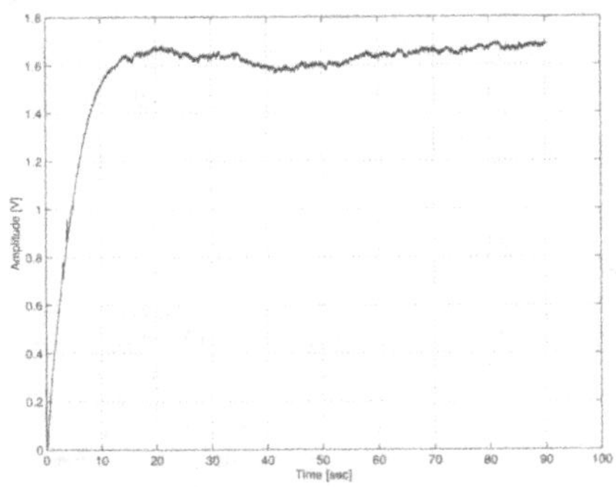

Figure 15.8 Unit-step response of controlled object.

In this case, the obtained parameters are [5, 35]

$$b_1 = 39.69; \qquad b_0 = 0.598; \qquad \beta = 1.25.$$

So, the continuous transfer function used for controller design is

$$G(s) = \frac{1}{39.69 s^{1.25} + 0.598}. \tag{15.77}$$

This mathematical model was used in Refs. [5, 35] for fractional controller design, and an alternative integer-order model was used for traditional PD controller design with comparison purposes.

The alternative integer-order model has the form of first-order system represented by the following transfer function:

$$\widehat{G}(s) = \frac{1}{20.14s + 0.598}. \tag{15.78}$$

This integer-order system was used for comparison of control performance between classical PD controller and fractional PD^δ controller with transfer function:

$$C(s) = K + T_d s^\delta, \tag{15.79}$$

where K, T_d and δ are controller parameters.

The controller design was done in Refs. [5, 35], according to the method (poles placement [11]) described in Ref. [32], for obtaining a stability measure $St \approx 2.0$. The obtained fractional-order PD^δ controller designed for the fractional-order model (15.77) has the continuous transfer function:

$$C(s) = 64.47 + 48.99s^{0.5}. \tag{15.80}$$

The parameters of the integer-order PD controller were designed by the same method and the controller has the following transfer function:

$$\widehat{C}(s) = 64.47 + 12.46s. \tag{15.81}$$

Let us consider the single input—single output (SISO) feedback control system shown in Fig. 15.9, where W is required value, E is control error, U is control value and Y is actual value.

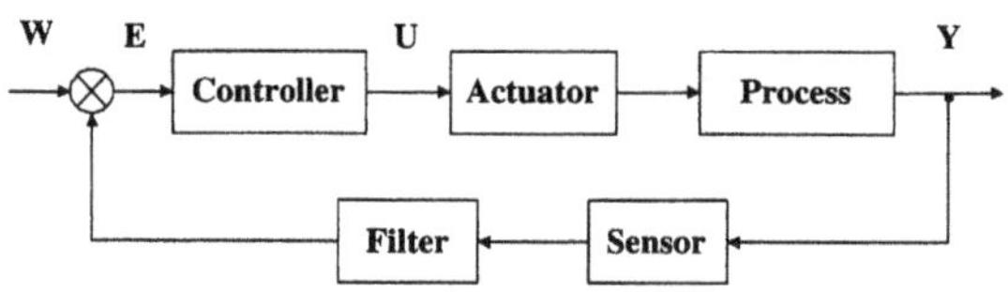

Figure 15.9 General SISO feedback loop system.

The fractional differential equation of a closed control loop, depicted in Figure15.9, with a fractional model of a controlled system and a fractional PD^δ controller, has the following form:

$$b_1\,{}_0D_t^\beta y(t) + T_d\,{}_0D_t^\delta y(t) + (b_0 + K)y(t) = K\ w(t) + T_d\,{}_0D_t^\delta w(t). \tag{15.82}$$

15.5.2 Memristor-Based Chua's Circuit

The fractional-order Chua's system was described and investigated in many works. Similar to the classical one, it contains a capacitor C, an inductor L, a resistor R and a nonlinear resistor, known as the Chua's diode. Since the memristor was postulated by professor L. O. Chua in 1971 and discovered by R. Williams *et al.* (HP laboratory) in 2008, it becomes the fourth circuit element. This fact allow us use a memristor as a nonlinear element in a circuit which exhibits chaos. In the case of Chua's circuit, the nonlinear resistor is replaced by a memristor (M).

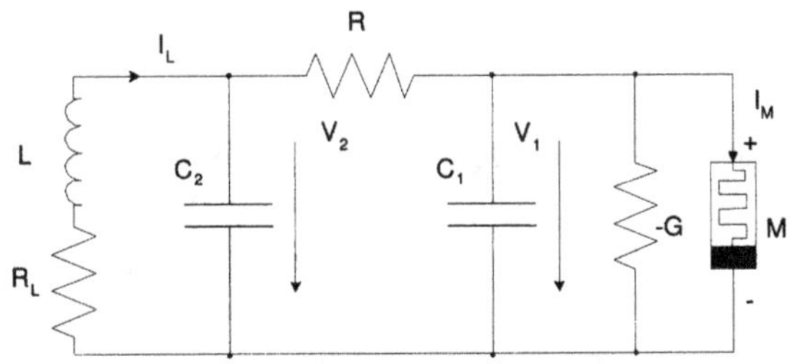

Figure 15.10 Chua's circuit with memristor and negative conductance.

The memristor in Figure 15.10 is a flux-controlled memristor whose characteristic is given by [8]:

$$I_M(t) = W(\phi(t))V_1(t), \tag{15.83}$$

where $W(\phi(t))$ is called the incremental memductance. For the flux-controlled memristor it was assumed to have a monotone-increasing piecewise-linear characteristic [15]. The memristor constitutive relation is expressed as

$$q(\phi) = b\phi + 0.5(a-b) \times (|\phi+1| - |\phi-1|), \tag{15.84}$$

where a, $b > 0$. The memductance function that is obtained from the $q(\phi)$ function is

$$W(\phi) = \frac{dq(\phi)}{d\phi} \begin{cases} a & : \quad |\phi| < 1, \\ b & : \quad |\phi| > 1. \end{cases} \tag{15.85}$$

The dynamic of the Chua's circuit with a passive memristor (flux-controlled memristor and negative conductance) depicted in Figure 15.10 is given by the following set of differential equations:

$$\begin{aligned}
\frac{dV_1(t)}{dt} &= \frac{1}{C_1}\left[\frac{(V_2(t)-V_1(t))}{R} + V_1(t)(G - W(\phi(t)))\right], \\
\frac{dV_2(t)}{dt} &= \frac{1}{C_2}\left[\frac{(V_1(t)-V_2(t))}{R} + I_L(t)\right], \\
\frac{dI_L(t)}{dt} &= \frac{1}{L}\left[-V_2(t) - R_L I_L(t)\right], \\
\frac{d\phi(t)}{dt} &= V_1(t),
\end{aligned} \tag{15.86}$$

where functions $q(\phi)$ and $W(\phi)$ are given by (15.84) and (15.85), respectively.
When we set

$$\begin{aligned}
x &= V_1, \quad y = V_2, \quad z = I_L, \quad w = \phi, \quad C_2 = 1, \\
\alpha &= 1/C_1, \quad \beta = 1/L, \quad \gamma = R_L/L, \quad \zeta = G, \quad R = 1,
\end{aligned} \tag{15.87}$$

then Eq.(15.86) can be transformed into the dimensionless form [15]:

$$\begin{aligned}
\frac{dx(t)}{dt} &= \alpha\left(y(t) - x(t) + \zeta x(t) - W(w)x(t)\right), \\
\frac{dy(t)}{dt} &= x(t) - y(t) + z(t), \\
\frac{dz(t)}{dt} &= -\beta y(t) - \gamma z(t), \\
\frac{dw(t)}{dt} &= x(t),
\end{aligned} \tag{15.88}$$

where piecewise-linear function $W(w)$ is given as

$$W(w) = \begin{cases} a & : \quad |w| < 1, \\ b & : \quad |w| > 1. \end{cases} \tag{15.89}$$

The equilibrium points of the system (15.88) are given by setting the left side of equations to 0 except last one. We set w=constant, which corresponds to the w-axis [15]. The Jacobian matrix at this equilibrium state E^* is

$$\mathbf{J}_W = \begin{bmatrix} \alpha(-1+\zeta - W(w)) & \alpha & 0 & 0 \\ 1 & -1 & 1 & 0 \\ 0 & -\beta & -\gamma & 0 \\ 1 & 0 & 0 & 0 \end{bmatrix}. \tag{15.90}$$

If we consider a fractional-order model for each electrical element in the circuit depicted in Figure 15.10, we can write a more general mathematical model for this circuit. As it was already mentioned in introduction, the real capacitor and real inductor are "fractional" and for the real memristor we postulated a fractional-order model as well ($d^\alpha \phi(t)/dt^\alpha = V(t)$). By using a technique of fractional calculus we obtain the following equations [34]:

$$\begin{aligned} {}_0D_t^{q_1} x(t) &= \alpha \left(y(t) - x(t) + \zeta x(t) - W(w)x(t) \right), \\ {}_0D_t^{q_2} y(t) &= x(t) - y(t) + z(t), \\ {}_0D_t^{q_3} z(t) &= -\beta y(t) - \gamma z(t), \\ {}_0D_t^{q_4} w(t) &= x(t), \end{aligned} \tag{15.91}$$

where function $W(w)$ is given by (15.89) and where q_1, q_2, q_3, and q_4 are the fractional orders of real electrical elements (memristive systems), namely: capacitor C_1, capacitor C_2, inductor L, and memristor M, respectively.

The stability of the new fractional-order memristor-based Chua's system can be investigated by using a condition (15.69).

In the case of piecewise-nonlinearity (15.84), we should investigate a characteristic equation for a linear part with slope a and for a linear part with slope b, respectively.

A necessary stability condition for fractional-order systems (15.91) to remain chaotic is keeping at least one eigenvalue λ in the unstable region. According to condition (15.71), we can also determine a minimal order q for which a nonlinear system has chaotic behavior.

Because the frequency approximation techniques are unreliable in recognizing chaos in fractional-order nonlinear systems [44], for simulation purposes we use a numerical solution of the memristor-based Chua's equations (15.91) obtained by the method described in [35]. That is a time domain method derived by using the relationship (15.12), which leads to equations in the form

$$\begin{aligned} x(t_k) &= \left(\alpha(y(t_{k-1}) - x(t_{k-1}) + \zeta x(t_{k-1}) - W(w(t_{k-1}))x(t_{k-1}))\right)h^{q_1} \\ &\quad - \sum_{i=v}^{k} c_i^{(q_1)} x(t_{k-i}), \\ y(t_k) &= \left(x(t_k) - y(t_{k-1}) + z(t_{k-1})\right) h^{q_2} - \sum_{i=v}^{k} c_i^{(q_2)} y(t_{k-i}), \\ z(t_k) &= \left(-\beta y(t_k) - \gamma z(t_{k-1})\right) h^{q_3} - \sum_{i=v}^{k} c_i^{(q_3)} z(t_{k-i}), \\ w(t_k) &= x(t_k) h^{q_4} - \sum_{i=v}^{k} c_i^{(q_4)} w(t_{k-i}), \end{aligned} \tag{15.92}$$

where

$$\begin{aligned} W(w(t_{k-1})) &= a \quad \text{for} \quad |w(t_{k-1})| < 1, \\ W(w(t_{k-1})) &= b \quad \text{for} \quad |w(t_{k-1})| > 1, \end{aligned} \tag{15.93}$$

and where T_{sim} is the simulation time, $k = 1, 2, 3 \ldots, N$, for $N = [T_{sim}/h]$, and $(x(0), y(0), z(0), w(0))$ is the start point (initial conditions). The binomial coefficients $c_i^{(q)}$ are calculated according to relation (15.13).

■ EXAMPLE 15.6

Let us consider the following parameter set:

$$\alpha = 10,\ \beta = 13,\ \gamma = 0.1,\ \zeta = 1.5,\ a = 0.3,\ b = 0.8. \tag{15.94}$$

When we consider real orders of capacitor models [51]: $q_1 = q_2 = 0.98$, a real order of an inductor model [51]: $q_3 = 0.99$, and we assume a real order of the memristor model: $q_4 = 0.97$; for the parameters (15.94) the initial conditions: $x(0) = 0.8$, $y(0) = 0.05$, $z(0) = 0.007$, $w(0) = 0.6$, simulation time $T_{sim} = 100\,s$, and time step $h = 0.005$, we get the chaotic double-scroll attractor for the total system order 3.92.

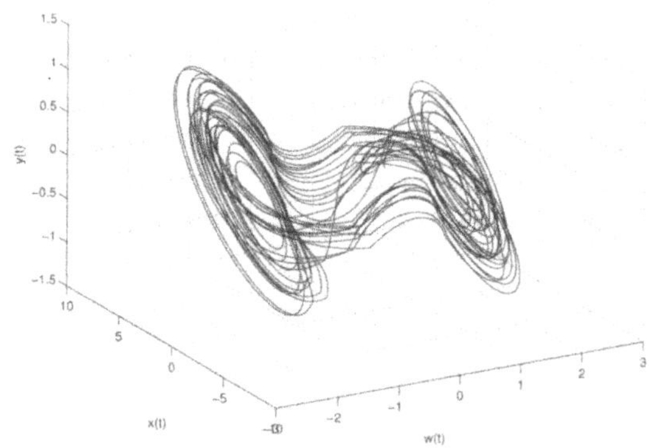

Figure 15.11 Strange attractor of the memristor-based Chua's system (15.91) in $w - x - y$ state space, for parameters $\alpha = 10$, $\beta = 13$, $\gamma = 0.1$, $\zeta = 1.5$, $a = 0.3$, $b = 0.8$, and orders $q_1 = q_2 = 0.98$, $q_3 = 0.99$, $q_4 = 0.97$.

In Figure 15.11 and Figure 15.12 are depicted chaotic attractors in 3D state space for $T_{sim} = 100\,s$. The simulations were performed without using the short memory principle ($v = 1$) for time step $h = 0.005$. Simulations show the double-scroll attractors and we can observe a chaotic behavior.

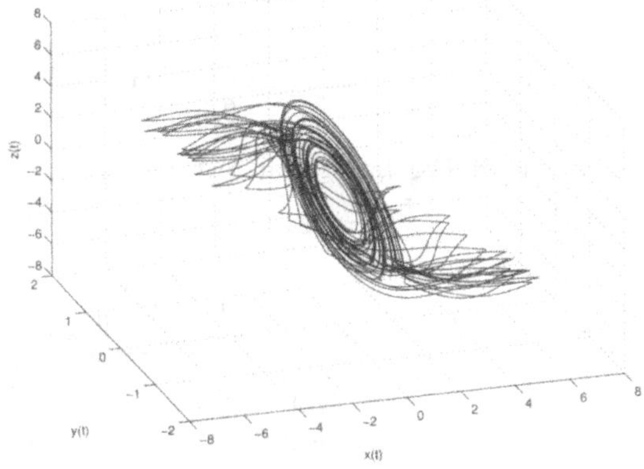

Figure 15.12 Strange attractor of the memristor-based Chua's system (15.91) in $x-y-z$ state space, for parameters $\alpha=10,\ \ \beta=13,\ \ \gamma=0.1,\ \ \zeta=1.5,\ \ a=0.3,$ $b=0.8$, and orders $q_1=q_2=0.98,\ q_3=0.99,\ q_4=0.97.$

15.5.3 Viscoelastic Models of Cells

Cells have essential biological roles and often change shape, attach and detach from a surface, and sometimes divide. Such activities require the deformation in response to local stress. The rheological behavior of these cells can be modeled with the following fractional differential equation [21]:

$$\sigma(t)=G_s\theta(t)+\lambda\,{}_0D_t^{\alpha}\theta(t)+\mu\frac{d\theta(t)}{dt}, \qquad (15.95)$$

where σ is stress, θ is strain, G_s is the static elastic modulus, λ is fractional relaxation time constant, and μ is the viscosity.

If we apply the Laplace transform to system (15.95), assuming that the initial conditions are all zeros, we obtain

$$G(s)=\frac{\Sigma(s)}{\Theta(s)}=G_s+\lambda s^{\alpha}+\mu s. \qquad (15.96)$$

As it was mentioned in Ref. [21], the parameter G_s can be neglected. For a step function $u(t)$ in applied stress, $\sigma(t)=\sigma_0 u(t)$, the creep response can be written as

$$\Theta(s)=\frac{\Sigma_0}{s(\mu s+\lambda s^{\alpha})}=\frac{\Sigma_0 s^{(1-\alpha)-2}}{\mu(s^{(1-\alpha)}+\lambda/\mu)}. \qquad (15.97)$$

The inverse Laplace transform of this expression can be written by using a Laplace transform of the Mittag-Leffler function [38]:

$$\mathscr{L}\{t^{\beta-1}E_{\gamma,\beta}(-zt^{\gamma})\}=\frac{s^{\gamma-\beta}}{s^{\gamma}+z},$$

and we obtain an analytical solution in the form

$$\theta(t) = \frac{\Sigma_0}{\mu} t E_{1-\alpha,2}\left(-\frac{\lambda}{\mu} t^{1-\alpha}\right). \tag{15.98}$$

For numerical solution of the fractional differential equation (15.95) for $G_s = 0$ we can use relations (15.12) and (15.13). The resulting difference equation has the form [36]

$$\theta(t_k) = \frac{\Sigma_0 + \mu h^{-1}\theta(t_{k-1}) - \lambda h^{-\alpha}\sum_{j=v}^{k} c_j^{(\alpha)}\theta(t_{k-j})}{\lambda h^{-\alpha} + \mu h^{-1}}, \tag{15.99}$$

where $t_k = kh$ for $k = 1, 2, 3, \ldots N$, where $N = [T_{sim}/h]$ and h is a time step of calculation, and $\theta(t_0)$ is obtained from an initial condition, e.g., $\theta(t_0) = 0$ for a zero initial condition.

■ EXAMPLE 15.7

Let us assume the following model parameters: $\lambda = \mu = \Sigma_0 = 1$, zero initial condition, $T_{sim} = 5\,s$, $h = 0.001$, and $v = 1$.

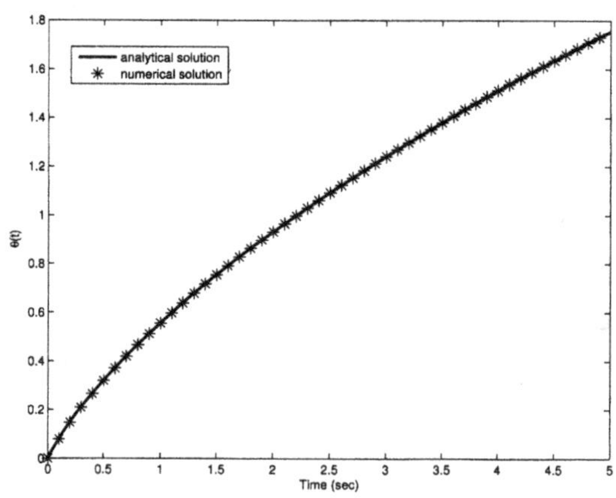

Figure 15.13 Comparison of analytical and numerical solutions of fractional-order viscoelastic models of cell (15.97) for simulation time $5\,s$, step $h = 0.001$, and $v = 1$ in (15.99).

Comparison of the analytical solution (15.98) and the numerical solution (15.99) of the fractional differential equation (15.95) for the parameters $G_s = 0$, $\lambda = \mu = \Sigma_0 = 1$, zero initial condition, $T_{sim} = 5\,s$, $h = 0.001$, and $v = 1$ is depicted in Figure 15.13. As we can observe, the numerical solution fits the analytical solution and we can say that both solutions are consistent.

EXERCISES

15.1 Find the analytical solution (impulse response) for zero initial conditions of a closed loop system consisting of an electrical heater and integer PD controller as described in Section 15.5.1.

15.2 Investigate the stability of a closed loop system consisting of an electrical heater and integer PD controller as described in Section 15.5.1.

15.3 Investigate the stability of the fractional memristor-based Chua's system described in Section 15.5.2.

15.4 For the fractional-order Chen system described by Eq.(15.72) find a numerical solution.

REFERENCES

1. Bayat F. M., Afshar, M. and Ghartemani, M. K., Extension of the root-locus method to a certain class of fractional-order systems, *ISA Transactions*, Vol. 48, pp. 48–53 (2009).
2. Bayat, F. M. and Afshar, M., Extending the root-locus method to fractional-order systems, *Journal of Applied Mathematics*, Article ID 528934 (2008).
3. Baleanu, D., Guvenc, Z. B. and Tenreiro, Machado, J. A. (Eds.), *New Trends in Nanotechnology and Fractional Calculus Applications*, Springer, London (2010).
4. Bode, H. W., *Network Analysis and Feedback Amplifier Design*, Tung Hwa Book Company (1949).
5. Caponetto, R., Dongola, G., Fortuna, L. and Petráš, I., *Fractional Order Systems: Modeling and Control Applications*, World Scientific, Singapore (2010).
6. Chen, Y. Q., Oustaloup-Recursive-Approximation for Fractional Order Differentiators, Matlab Central File Exchange, MathWorks, Inc. (2003), url: *http://www.mathworks.com/matlabcentral/fileexchange/3802.*
7. Chen, Y. Q., Petráš, I. and Xue, D., Fractional Order Control - A Tutorial, *Proc. of the American Control Conference*, Hyatt Regency Riverfront, St. Louis, MO, USA, June 10–12 (2009).
8. Chua, L. O., Memristor - the missing circuit element, *IEEE Transaction on Circuit Theory*, vol. CT-18, pp. 507–519 (1971).
9. Deng, W., Short memory principle and a predictorcorrector approach for fractional differential equations, *Journal of Computational and Applied Mathematics*, Vol. 206, pp. 174–188 (2007).
10. Dorčák, Ľ., Numerical Models for the Simulation of the Fractional-Order Control Systems, *UEF-04-94, The Academy of Sciences, Inst. of Experimental Physic*, Košice, Slovakia (1994).

11. Dorf, R. C. and Bishop, R. H., *Modern Control Systems*, Addison-Wesley, New York (1990).

12. Ghartemani, M. K. and Bayat, F. M., Necessary and sufficient conditions for perfect command following and disturbance rejection in fractional order systems, *Proc. of the 17th World Congress IFAC*, Soul, Korea, July 6-11, pp. 364–369 (2008).

13. Heymans, N. and Podlubny, I., Physical interpretation of initial conditions for fractional differential equations with Riemann-Liouville fractional derivatives. *Rheologica Acta*, vol. 45, no. 5, pp. 765–772 (2006).

14. Hilfer, R., *Applications of Fractional Calculus in Physics*, World Scientific Publishers, Singapore (2000).

15. Itoh, M. and Chua, L. O., Memristor oscillation, *International Journal of Bifurcation and Chaos*, vol. 18, pp. 3183-3206 (2008).

16. Kaczorek, T., *Selected Problems of Fractional Systems Theory*, Springer, Berlin (2011).

17. Kheirizad, I., Tavazoei, M. S. and Jalali, A. A., Stability criteria for a class of fractional order systems, *Nonlinear Dyn.*, Vol. 61, no. 1-2 (2009).

18. LePage, W. R., *Complex variables and the Laplace transform for engineers*, McGraw-Hill (1961).

19. Li, Y., Chen, Y. Q., and Podlubny, I., Mittag-Leffler stability of fractional order nonlinear dynamic system, *Automatica*, Vol. 45, no. 8, pp. 1965–1969 (2009).

20. Lubich, Ch., Discretized fractional calculus, *SIAM J. Math. Anal.*, Vol. 17, no. 3, pp. 704–719 (1986).

21. Magin, R. L., *Fractional Calculus in Bioengineering*, Begell House Publishers (2006).

22. Mainardi, F., *Fractional Calculus and Waves in Linear Viscoelasticity: An Introduction to Mathematical Models*, Imperial College Press, Singapore (2010).

23. Matignon, D., Stability properties for generalized fractional differential systems, *Proc. of Fractional Differential Systems: Models, Methods and App.*, Vol. 5, pp. 145–158 (1998).

24. Miller, K. S. and Ross, B., *An Introduction to the Fractional Calculus and Fractional Differential Equations*, John Wiley & Sons. Inc., New York (1993).

25. Monje, C. A., Chen, Y. Q., Vinagre, B. M., Xue, D. and Feliu, V., *Fractional Order Systems and Control - Fundamentals and Applications*, Advanced Industrial Control Series, Springer, London (2010).

26. Oldham, K. B. and Spanier, J., *The Fractional Calculus*, Academic Press, New York (1974).

27. Ortigueira, M. D., *Fractional Calculus for Scientists and Engineers*, Lecture Notes in Electrical Engineering, Springer, London (2011).

28. Oustaloup, A., *La Derivation Non Entiere: Theorie, Synthese et Applications*, Hermes, Paris (1995).

29. Oustaloup, A., Levron, F., Mathieu, B., and Nanot, F. M., Frequency-band complex noninteger differentiator: characterization and synthesis, *IEEE Trans.*

on Circuits and Systems I: Fundamental Theory and Applications I, vol. 47, no. 1, pp. 25–39 (2000).

30. Petráš, I., Digital Fractional Order Differentiator/integrator - FIR type, Matlab Central File Exchange, MathWorks, Inc. (2003), url: *http://www.mathworks.com/matlabcentral/fileexchange/3673.*

31. Petráš, I., Digital Fractional Order Differentiator/integrator - IIR type, Matlab Central File Exchange, MathWorks, Inc., (2003) url: *http://www.mathworks.com/matlabcentral/fileexchange/3672.*

32. Petráš, I., The fractional-order controllers: methods for their synthesis and application, *Journal of Electrical Engineering*, Vol. 50, pp. 284–288 (1999).

33. Petráš, I., Podlubny, I., O'Leary, P., Dorčák, Ľ., and Vinagre, B. M., *Analogue Realization of Fractional Order Controllers*, FBERG, Technical University of Košice (2002).

34. Petráš, I., Fractional-order memristor-based Chua's circuit, *IEEE Transactions on Circuits and Systems II-Express Briefs*, Vol. 57, no. 12, pp. 975–979 (2010).

35. Petráš, I., *Fractional-Order Nonlinear Systems: Modeling, Analysis and Simulation*, Springer, Berlin (2011).

36. Petráš, I., An effective numerical method and its utilization to solution of fractional models used in bioengineering applications, *Advances in Difference Equations*, Vol. 2011, pp. 1–14 (2011).

37. Petráš, I., Fractional derivatives, fractional integrals, and fractional differential equations in Matlab, In: A. Assi (Eds.) *Engineering Education and Research Using MATLAB*, InTech, Rijeka, chapter 10 (2011).

38. Podlubny, I., *Fractional Differential Equations*, Academic Press, San Diego (1999).

39. Podlubny, I., Matrix approach to discrete fractional calculus, *Fractional Calculus and Applied Analysis*, Vol. 3, no. 4, pp. 359–386 (2000).

40. Podlubny, I., Geometric and physical interpretation of fractional integration and fractional differentiation, *Fractional Calculus and Applied Analysis*, Vol. 5, no. 4, pp. 367–386 (2002).

41. Podlubny, I., Chechkin, A., Skovranek, T., Chen, Y. Q., and Vinagre, B. M., Matrix approach to discrete fractional calculus II: Partial fractional differential equations, *Journal of Computational Physics*, Vol. 228, no. 8, pp. 3137–3153 (2009).

42. Podlubny, I. and Kacenak, M., Mittag-Leffler function, Matlab Central File Exchange, MathWorks, Inc. (2005), url: *http://www.mathworks.com/matlabcentral/fileexchange/8738.*

43. Sabatier, J., Agrawal, O. P. and Tenreiro Machado, J. A. (Eds.), *Advances in Fractional Calculus: Theoretical Developments and Applications in Physics and Engineering*, Springer, Berlin (2007).

44. Tavazoei, M. S. and Haeri, M., Unreliability of frequency-domain approximation in recognizing chaos in fractional-order systems, *IET Signal Proc.*, Vol. 1, no. 4, pp. 171–181 (2007).

45. Tavazoei, M. S. and Haeri, M., A necessary condition for double scroll attractor existence in fractional - order systems, *Physics Letters A*, Vol. 367, pp. 102–113 (2007).

46. Tavazoei, M. S. and Haeri, M., Limitations of frequency domain approximation for detecting chaos in fractional order systems, *Nonlinear Analysis*, Vol. 69, pp. 1299–1320 (2008).

47. Tavazoei, M. S. and Haeri, M., Chaotic attractors in incommensurate fractional order systems, *Physica D*, Vol. 237, pp. 2628–2637 (2008).

48. Vinagre, B. M., Podlubny, I., Hernández, A., and Feliu, V., Some approximations of fractional order operators used in control theory and applications, *Fractional Calculus and Applied Analysis*, Vol. 3, no. 3, pp. 231–248 (2000).

49. Vinagre, B. M., Chen, Y. Q., and Petráš, I., Two direct Tustin discretization methods for fractional-order differentiator/integrator, *Journal of Franklin Institute*, Vol. 340, pp. 349–362 (2003).

50. West, B., Bologna, M., and Grigolini, P., *Physics of Fractal Operators*, Springer, New York (2003).

51. Westerlund, S., *Dead Matter Has Memory!*, Causal Consulting, Kalmar, Sweden (2002).

CHAPTER 16

THE GOAL PROGRAMMING MODEL: THEORY AND APPLICATIONS

BELAID AOUNI[1], CINZIA COLAPINTO,[2] AND DAVIDE LA TORRE[3]

[1]School of Commerce and Administration, Laurentian University, Canada
[2]Department of Management, Ca' Foscari University of Venice, Italy
[3]Department of Economics, Management and Quantitative Methods, University of Milan, Italy

16.1 MULTI-CRITERIA DECISION AID

In any kind of organization, managers are frequently challenged with complex decision making situations which involve several, and often conflicting, objectives and priorities. In fact, the decision setting is unstable because of corporate politics, market conditions or regulations' changes. The decision-making is central within organizations, and managers have to make the best choices, substituting objective issues for casual judgments. The classical and easiest formulation of a decision-making model usually involves an objective function f, which has to be optimized, depending on a set of decision variables and subject to some constraints. The model can be mathematically stated as

$$\text{Optimize} f(x), \tag{16.1}$$

Mathematical Modeling with Multidisciplinary Applications.
Edited by Xin-She Yang

subject to

$$x \in D, \tag{16.2}$$

where the set D describes in compact form the set of all possible constraints. D is called the *feasible set* and, in general, it is a subset of R^m (that is the set of all m-tuples of real numbers). For instance, in many real applications f represents the profit or the cost while the set D is the budget constraint. However there are several decision-making situations in which different and conflicting aspects and objectives have to be considered simultaneously and they cannot merely reduced to a single criterion. Keeney and Howard [15] state that "in complex value problems consequences cannot be adequately described objectively by a single attribute."

In general an agent's utility is a vector, having components such as: cost, life expectancy, profit and quality of life. For instance, in environmental economics and management the Decision Maker (DM) has to plan the use of natural resources (fisheries, forestry, water and land) in a complex process which inevitably involves several incommensurable and conflicting objectives (for instance the level of emissions of a power plant to prevent deaths and disabilities against the benefits of the power); in public economics when a new airport has to be sited the DM has to consider conflicting criteria as the noise of flight operation, safety issues, cost of land, and distance from communities [21].

The Multi-Criteria Decision Aid (MCDA) considers several conflicting and incommensurable objectives or attributes which are optimized simultaneously. If $D \subset R^m$ is the set of feasible solutions, the general formulation of MCDA is as follows [24]:

$$\text{Optimize} f(x) := (f_1(x), f_2(x), \ldots f_n(x)), \tag{16.3}$$

subject to

$$x \in D, \tag{16.4}$$

where f_i represents the ith objective function. The function f is a vector-valued function defined on R^m and it assigns values in R^n. We suppose R^n being ordered by the usual Pareto cone R^n_+ which means $a \geq b$ if and only if $a - b \in R^n_+$ for all $a, b \in R^n$.

Definition 13 *In the case that f is to be maximized, a vector $\hat{x} \in D$ is said to be a Pareto optimal solution or an efficient solution to (16.3) if it is not dominated, that is there is not a $y \in D$ such that $f_i(\hat{x}) \leq f_i(y)$ for all $i = 1 \ldots n$ and $f_j(\hat{x}) < f_j(y)$ for at least one j.*

For necessary and sufficient conditions which characterize Pareto optimal solutions to (16.3) one can see Refs. [24, 26].

Stochastic or Scenario-based Multi-Criteria Decision Aid (SMCDA) represents the natural extension of deterministic multi-criteria programming to stochastic context. There are several decision-making situations where the

DM wishes to optimize objectives which depend on some random parameters (see, for instance, Refs. [2, 3]). In literature many approaches have been proposed to deal with such situations (see, for instance, Refs. [7, 9, 10, 25]). The fact that the objectives depend on random parameters makes the objectives random variables too; so a point in the domain could be a Pareto optimal solution of the problem only when some realizations of the random parameters occur. Several definitions of Pareto optimality have been introduced; the solution of such problems usually involves the transformation into its deterministic equivalent.

16.2 THE GOAL PROGRAMMING MODEL

The Goal Programming (GP) model is a well-known and the most popular model within the field of Multi-Criteria Decision Aid; this takes into account simultaneously many objectives which can be conflicting and provides an aggregating procedure to simplify the model and reduce it to a single-criterion program. Thus, the obtained solution through the GP procedure represents the best compromise that can be made by the DM. The GP model can be considered as a special case of the "Distance Function Model" where the deviations between the achievement and aspiration levels have to be minimized. In fact, both positive and negative deviations are unwanted. The deviations will be positive if the goal is surpassed, otherwise negative.

The first GP formulation was developed by Charnes *et al.* [12] and Charnes and Cooper [13] and then used by Lee [17] and Lee and Clayton [18]. The GP models have received a lot of attention and they are so widespread because of their applications in practical decision-making situations such as accounting and financial aspects of stock management, marketing, quality control, human resources, and production (see Refs. [1, 22]). In recent years, numerous variants of the GP model have been studied including, for instance, Weighted GP, Lexicographical GP, Integer GP, Imprecise GP, Fuzzy GP, and so on. According to Aouni and Kettani [5] the GP is still alive and supported by a well-established network of researchers and practitioners. The popularity of the GP is due to the fact that it is a model that is simple and easy to understand and to apply. Moreover, the GP formulation can be solved through some powerful mathematical programming software such as LINDO, LINGO and CPLEX.

However, it is worth mentioning that the GP model is based on a satisfying philosophy. This means that the obtained solution is the best compromise. When efficiency is a required property by the DM, a GP model has to be integrated with optimality tests in order to check whether the obtained solution is nondominated (see Ref. [16]). In any case, the GP approach provides a solution to a MCDA program with a given level of satisfaction.

The standard mathematical formulation of the GP model (see Ref. [12]) is as follows:

$$\min \sum_{i=1}^{n} (\delta_i^+ + \delta_i^-), \tag{16.5}$$

subject to

$$\begin{cases} f_i(x) + \delta_i^- - \delta_i^+ = g_i, \quad i = 1, 2, \dots n, \\ x \in D, \\ \delta_i^-, \delta_i^+ \geq 0, \quad i = 1, 2, \dots n. \end{cases} \tag{16.6}$$

where δ_i^+ and δ_i^- are, respectively, the positive and the negative deviations with respect to the aspiration levels (goals) g_i.

EXAMPLE 16.1

Solve the following GP model:

$$\min Z = \delta_1^+ + \delta_1^- + \delta_2^+ + \delta_2^-,$$

subject to

$$\begin{cases} 2x_1 + x_2 + 2x_3 + \delta_1^- - \delta_1^+ = 50, \\ 3x_1 + 6x_2 + 3x_3 + \delta_2^- - \delta_2^+ = 150, \\ x_1, x_2, x_3 \geq 0, \\ \delta_1^+, \delta_1^-, \delta_2^+, \delta_2^- \geq 0. \end{cases}$$

The solution can be computed by using LINGO which provides the following solutions: $x_1 = 0$, $x_2 = 16.66667$, $x_3 = 16.66667$, $\delta_1^- = 0$, $\delta_1^+ = 0$, $\delta_2^- = 0$, and $\delta_2^+ = 0$.

In the Weighted Goal Programming (WGP) model each deviation is multiplied by a weight or scaling factor w_i as follows:

$$\min \sum_{i=1}^{n} (w_i^+ \delta_i^+ + w_i^- \delta_i^-), \tag{16.7}$$

subject to

$$\begin{cases} f_i(x) + \delta_i^- - \delta_i^+ = g_i, \quad i = 1, 2, \dots n, \\ x \in D, \\ \delta_i^-, \delta_i^+ \geq 0, \quad i = 1, 2, \dots n. \end{cases} \tag{16.8}$$

EXAMPLE 16.2

Solve the following WGP model:

$$\min Z = 0.2\delta_1^+ + 0.2\delta_1^- + 0.3\delta_2^+ + 0.3\delta_2^-,$$

subject to

$$\begin{cases} 2x_1 - x_2 + 2x_3 + \delta_1^- - \delta_1^+ = 50, \\ 3x_1 - 6x_2 - 3x_3 + \delta_2^- - \delta_2^+ = 150, \\ x_1, x_2, x_3 \geq 0, \\ \delta_1^+, \delta_1^-, \delta_2^+, \delta_2^- \geq 0. \end{cases}$$

The solution can be computed by using LINGO which provides the following solutions: $x_1 = 50$, $x_2 = 0$, $x_3 = 0$, $\delta_1^- = 0$, $\delta_1^+ = 50$, $\delta_2^- = 0$, and $\delta_2^+ = 0$.

The above GP formulation does not include the DM's preferences. Martel and Aouni [20] introduced the concept of satisfaction functions in the goal programming model where the DM can explicitly express his/her preferences for any deviation between the achievement and the aspiration level of each objective. In general, given three positive numbers ξ_i, ξ_d and ξ_v which will be called, respectively, the *indifference threshold*, the *dissatisfaction threshold* and the *veto threshold* in the sequel, a satisfaction function $F : [0, \xi_v] \to [0, 1]$ satisfies the following properties:

- $F(x) = 1$, for all $x \in [0, \xi_i]$,
- $F(x) = 0$ for all $x \in [\xi_d, \xi_v]$,
- F is continuous and descreasing.

Depending on the thresholds' values, positive and negative deviations might be penalized in a different manner and thus affect the probability of reaching the goals. In fact, the threshold values depend on the DM preferences regarding the dispersion of the deviations. The GP model with satisfaction function is formulated as follows:

$$\max \sum_{i=1}^{n} \left(w_i^+ F(\delta_i^+) + w_i^- F(\delta_i^-) \right), \tag{16.9}$$

subject to

$$\begin{cases} f_i(x) + \delta_i^- - \delta_i^+ = g_i, \; i = 1, 2, \ldots n, \\ x \in D, \\ \delta_i^+, \delta_i^- \in [0, \xi_v], \quad i = 1, 2, \ldots n. \end{cases} \tag{16.10}$$

Let us notice that the GP model (16.9) admits a solution because of the continuity of f_i and F, and the compactness of D.

■ EXAMPLE 16.3

Solve the following GP model with a satisfaction function:

$$\max Z = \frac{0.3}{1 + (0.01\delta_1^+)^2} + \frac{0.3}{1 + (0.01\delta_1^-)^2} + \frac{0.2}{1 + (10\delta_2^+)^2} + \frac{0.2}{1 + (10\delta_2^-)^2},$$

subject to

$$\left\{\begin{array}{l} 0.9x_1+0.99x_2+0.21x_3+0.178x_4+0.5724x_5+0.102x_6+ \\ 1.527x_7+0.996x_8+0.3x_9+0.0711x_{10}+0.45x_{11}+0.9306x_{12}+ \\ 0.942x_{13}+0.504x_{14}+0.5888x_{15}+\delta_1^+-\delta_1^+=2.82 \\ 0.84x_1+0.95x_2+0.93x_3+0.94x_4+0.93x_5+0.94x_6+ \\ 0.95x_7+0.9x_8+0.94x_9+0.93x_{10}+0.94x_{11}+0.94x_{12}+ \\ 0.94x_{13}+0.93x_{14}+0.9x_{15}+\delta_2^+-\delta_2^+=5.63, \\ x_1+x_2+x_3+x_4+x_5+x_6+x_7+x_8+x_9+ \\ x_{10}+x_{11}+x_{12}+x_{13}+x_{14}+x_{15}<=7, \\ x_1,x_2,x_3,x_4,x_5,x_6,x_7,x_8,x_9, \\ x_{10},x_{11},x_{12},x_{13},x_{14},x_{15}\geq 0, \\ \delta_1^+,\delta_1^-,\delta_2^+,\delta_2^-\geq 0, \\ \delta_1^+\leq 300, \\ \delta_1^-\leq 300, \\ \delta_2^+\leq 0.3, \\ \delta_2^-\leq 0.3. \end{array}\right.$$

The solution can be computed by using LINGO which provides the following solutions: $x_1 = 0.2855723$, $x_2 = 0.2522141$, $x_3 = 0.5054681$, $x_4 = 0.5155977, x_5 = 0.3880378, x_6 = 0.5402304, x_7 = 0.3964897E-01$, $x_8 = 0.2517976$, $x_9 = 0.4760572$, $x_{10} = 0.5504814$, $x_{11} = 0.4274446$, $x_{12} = 0.2717116$, $x_{13} = 0.2680179$, $x_{14} = 1.284094$, $x_{15} = 0$, $\delta_1^- = 0$, $\delta_1^+ = 0$ $\delta_2^- = 0$, and $\delta_2^+ = 0$.

16.3 SCENARIO-BASED GOAL PROGRAMMING

Let $\Omega = \{\omega_1, \omega_2, \ldots, \omega_l\}$ be a space of events with associated probabilities $p_1, p_2, \ldots, p_l$, $\sum_i p_i = 1$. Consider $f : R^m \times \Omega \to R^n$ to be a function such that $f(x, \cdot)$ is a discrete random variable for each fixed $x \in R^m$ and $f(\cdot, \omega)$ is continuous for $\omega_i \in \Omega$. Let $D \subset R^m$ be a compact set and consider now the following scenario-based multi-criteria problem

$$\min_{x\in D} f(x, \omega_i), \tag{16.11}$$

where $\omega_i \in \Omega$. Using the assumption we set before, we get that program (16.11) has at least a solution over D for all $\omega_i \in \Omega$. According to Refs. [9, 10], we can introduce the following deterministic multi-criteria equivalent problems associated with program (16.11).

Definition 14 *A point $x \in D$ is an expected-value Pareto optimal solution of the scenario-based multi-criteria problem if it is a Pareto optimal solution* *of the following problem:*

$$\min_{x \in D} \mathrm{E}(f(x,\cdot)) := (\mathrm{E}(f_1(x,\cdot)),\dots,\mathrm{E}(f_n(x,\cdot))), \tag{16.12}$$

where $\mathrm{E}(f_i(x,\cdot))$ *is the expectation value of the random variable*[16] $f_i(x,\cdot)$ *for each fixed* $x \in D$.

Definition 15 *A point* $x \in D$ *is a minimum-variance Pareto optimal solution of the scenario-based multi-criteria problem if it is a Pareto optimal solution of the following problem:*

$$\min_{x \in D} \sigma^2(f(x,\cdot)) := \left(\sigma^2(f_1(x,\cdot)),\dots,\sigma^2(f_n(x,\cdot))\right), \tag{16.13}$$

where $\sigma^2(f_i(x,\cdot))$ *is the variance of the random variable* $f_i(x,\cdot)$ *for each fixed* $x \in D$.

Definition 16 *A point* $x \in D$ *is an expected-valued standard deviation Pareto optimal solution of the scenario-based multi-criteria problem if it is a Pareto optimal solution of the following program:*

$$\min_{x \in D}(\mathrm{E}(f(x,\cdot),\sigma(f(x,\cdot)) := (\mathrm{E}(f_1(x,\cdot)),\dots,\mathrm{E}(f_n(x,\cdot),\sigma(f_1(x,\cdot),\dots,\sigma(f_n(x,\cdot)), \tag{16.14}$$

where $\sigma(f(x,\cdot))$ *is the standard deviation of the random variable* $f(x,\cdot)$ *for each fixed* $\omega \in \Omega$.

Consider now the aspiration levels g_i $(i = 1\dots n)$, and suppose they are random variables $g_i : \Omega \to R$. Let $\mathrm{E}(g_i)$ and $\sigma^2(g_i)$ be the expectation values and the variances of g_i, respectively. Consider the following goal programming models associated with formulations (16.12), (16.13) and (16.14), respectively.

GP Model 1:

$$\min \sum_{i=1}^{n} (\delta_i^+ + \delta_i^-), \tag{16.15}$$

subject to

$$\begin{cases} \mathrm{E}(f_i(x,\cdot)) + \delta_i^- - \delta_i^+ = \mathrm{E}(g_i), \; i = 1,2,\dots n, \\ x \in D, \\ \delta_i^+, \delta_i^- \geq 0 \quad i = 1,2,\dots n. \end{cases} \tag{16.16}$$

[16]If $Y : \Omega \to R^n$ is a discrete random variable defined on the space $\Omega = \{\omega_1,\omega_2,\dots,\omega_l\}$ with associated probabilities $p_1, p_2, \dots, p_l$, $\sum_i p_i = 1$, the first moment is the expected value defined as $\mathrm{E}(Y) = \sum_i Y(\omega_i)p_i$, while the second moment is the variance defined as $\sigma^2(Y) = \mathrm{E}(Y - \mathrm{E}(Y))^2$.

GP Model 2:

$$\min \sum_{i=1}^{n} (\delta_i^+ + \delta_i^-) \tag{16.17}$$

subject to

$$\begin{cases} \sigma^2(f_i(x,\cdot)) + \delta_i^- - \delta_i^+ = \sigma^2(g_i) & i = 1,2,\ldots n, \\ x \in D, \\ \delta_i^+, \delta_i^- \geq 0, \quad i = 1 \ldots n. \end{cases} \tag{16.18}$$

GP Model 3:

$$\min \sum_{i=1}^{n} (\delta_i^+ + \delta_i^- + \vartheta_i^+ + \vartheta_i^-), \tag{16.19}$$

subject to

$$\begin{cases} \mathrm{E}(f_i(x,\cdot)) + \delta_i^- - \delta_i^+ = \mathrm{E}(g_i) & i = 1,2,\ldots n, \\ \sigma(f_i(x,\cdot)) + \vartheta_i^- - \vartheta_i^+ = \sigma(g_i) & i = 1,2,\ldots n, \\ x \in D, \\ \delta_i^+, \delta_i^-, \vartheta_i^-, \vartheta_i^+ \geq 0 \quad i = 1,2,\ldots n. \end{cases} \tag{16.20}$$

Suppose we take a sample of observations of the random vector $f(x,\omega)$, say, $(f(x,\omega_1), f(x,\omega_2), \ldots f(x,\omega_s)) \in R^{s\times m}$. If the observations are independent and identically distributed (i.i.d.) we can get an estimation of the mean and the variance of $\mathrm{E}(f(x,\cdot))$, $\sigma^2(f(x,\cdot))$ and $\mathrm{E}(D(\cdot))$ by using the classical statistical formulas

$$\mathrm{E}(f(x,\cdot)) \approx \frac{\sum_{k=1}^{s} f(x,\omega_k)}{s}, \tag{16.21}$$

$$\sigma^2(f(x,\cdot)) \approx \frac{\sum_{k=1}^{s} (f(x,\omega_k) - \mathrm{E}(f(x,\cdot)))^2}{s-1}. \tag{16.22}$$

16.4 APPLICATIONS

In this section we present a set of applications—namely to finance, media management, public economics and software engineering—of the different Goal Programming formulations presented in the previous sections.

16.4.1 A Goal Programming Model for Portfolio Selection

Portfolio managers have to acquire and interpret information related to the movements in security prices. Indeed, portfolio management concerns making decisions about investment mix and policy, balancing risk, liquidity and performance of the chosen assets. The history of returns on different asset classes provides compelling evidence of a risk-return trade-off and the classical finan-

Table 16.1 Return per unit in percentage/relative risk in percentage.

Company/Year	2007
AT & T	0.15/11.42
Walmart	0.01/3.49
Exxon Mobil	0.11/8.44
General Electric	0.02/3.58
Bank of America	0.05/6.93
Ford Motor Company	0.01/9.36
Hewlett-Packard	0.16/9.32
McKesson Corporation	-0.01/4.57
J. P. Morgan Chase	0.08/6.58
Proctor & Gamble	0.04/5.15

cial theory demonstrated that portfolio diversification can reduce variability and investment risk because the assets prices do not move in exact lockstep. How do asset managers decide where and how to allocate their funds?

As a numerical example, let us consider the following financial decision-making situation based on real data coming from the NYSE (New York Stock Exchange). We selected the largest public and private companies by gross revenues in 2007. Data contains stock unit daily closing price and the percentage change in the price of a stock from the previous day's closing price. Table 16.1 represents price of expected return per unit and risk in percentage. We evaluate investment into different companies based on two objectives: maximizing return and minimizing risk. We consider two different GP formulations, namely the weighted GP model and the GP model with satisfaction function.

To simplify the equations we assign to each company a variable. The decision variables will be defined as follows:

$$X_j = \text{the amount of money invested in the security of company } j \quad (16.23)$$

where $j = 1$ for AT & T, $j = 2$ for Walmart,..., $j = 10$ for Proctor & Gamble. The Financial Decision Maker (FDM) prefers to invest by default into three different market sectors: at least $30,000.00\$$ to the financial sector, at least $20,000.00\$$ to the oil/gas sector and at least $10,000.00\$$ to the telecommunication sector. As said before, the multi-criteria problem for the year 2007 consists of maximizing the return and minimizing the risk and can be formulated as follows:

$$\begin{aligned}\max NA_1 := &1.0015X_1 + 1.0001X_2 + 1.0011X_3 + \\ &1.0002X_4 + 1.0005X_5 + 1.0001X_6 + 1.0016X_7 + \\ &0.9999X_8 + 1.0008X_9 + 1.0004X_{10},\end{aligned} \quad (16.24)$$

$$\begin{aligned}\min NA_2 :=&\ 1.1142X_1 + 1.0349X_2 + 1.0844X_3 +\\ &1.0358X_4 + 1.0693X_5 + 1.0936X_6 + 1.0932X_7 +\\ &1.0475X_8 + 1.0658X_9 + 1.0515X_{10},\end{aligned} \tag{16.25}$$

subject to

$$\begin{cases} X_1 + X_2 + X_3 + X_4 + X_5 + X_6 + X_7 + X_8 + X_9 + X_{10} = 100,000.00,\\ X_5 + X_9 \geq 30,000.00,\\ X_3 \geq 20,000.00,\\ X_1 \geq 10,000.00,\\ X_1, X_2, X_3, X_4, X_5, X_6, X_7, X_8, X_9, X_{10} \geq 0. \end{cases} \tag{16.26}$$

Let $g_1 = 100,125.00\$$ (return) and $g_2 = 106,200.00\$$ (risk) be the two aspiration levels for the objective functions NA_1 and NA_2. We propose a WGP model where each weight is a trade-off parameter between risk and return. We consider a different financial situation in which we assume $w_1^+ = w_1^- = 0.75$ and $w_2^+ = w_2^- = 0.25$. This characterizes a FDM with low risk aversion. The FDM solves the following WGP model:

$$\min Z = 0.75\delta_1^+ + 0.75\delta_1^- + 0.25\delta_2^+ + 0.25\delta_2^-, \tag{16.27}$$

subject to

$$\begin{cases} 1.0015X_1 + 1.0001X_2 + 1.0011X_3 + 1.0002X_4 + 1.0005X_5 +\\ 1.0001X_6 + 1.0016X_7 + 0.9999X_8 + 1.0008X_9 + 1.0004X_{10}\\ -\delta_1^+ + \delta_1^- = 100,125.00\\ 1.1142X_1 + 1.0349X_2 + 1.0844X_3 + 1.0358X_4 + 1.0693X_5 +\\ 1.0936X_6 + 1.0932X_7 + 1.0475X_8 + 1.0658X_9 + 1.0515X_{10} +\\ \delta_2^+ + \delta_2^- = 106,200.00\\ X_1 + X_2 + X_3 + X_4 + X_5 + X_6 + X_7 + X_8 + X_9 + X_{10} = 100,000.00,\\ X_5 + X_9 \geq 30,000.00,\\ X_3 \geq 20,000.00,\\ X_1 \geq 10,000.00,\\ X_1, X_2, X_3, X_4, X_5, X_6, X_7, X_8, X_9, X_{10} \geq 0,\\ \delta_1^+, \delta_1^-, \delta_2^+, \delta_2^- \geq 0. \end{cases} \tag{16.28}$$

The results are presented in Table 16.2.

Let us introduce the concept of satisfaction function. It will be utilized to integrate explicitly the FDM's preferences according to the deviations between the achievement and the aspiration levels of each objective. For satisfaction function, let us consider

$$F(\delta_i) = \frac{1}{1 + \alpha^2\delta_i^2}, \tag{16.29}$$

where α is a parameter. This function exhibits the behavior to be considered a satisfaction function and it is trivial to verify that $F(0) = 1$, and $F(+\infty) = 0$, $F''\left(\frac{1}{2\alpha}\right) = 0$ and that if $0 \leq \delta_i \leq \frac{1}{3\alpha}$ it holds $0 \leq F(\delta_i) \leq 0.1$, if $\delta_i \geq \frac{3}{\alpha}$ it holds $0 \leq F(\delta_i) \leq 0.01$. In other words, this function shows a level of satisfaction between 90 and 100 when $0 \leq \delta_i \leq \frac{1}{3\alpha}$ and a level of satisfaction between 0 and 10 when $\delta_i \geq \frac{3}{\alpha}$. Natural candidates for the indifference threshold and the dissatisfaction threshold are, respectively, $\xi_i = \frac{1}{3\alpha}$ and $\xi_d = \frac{3}{\alpha}$. Let us assume the veto threshold is $\xi_v = \frac{6}{\alpha}$. In the following let us choose $\alpha = \frac{1}{10}$, which implies that $\xi_i = \frac{10}{3}$, $\xi_d = 30$ and $\xi_v = 60$. The GP Model with satisfaction function and with weights $w_1^+ = w_1^- = 0.75$ and $w_2^+ = w_2^- = 0.25$ is the following:

$$\max Z = 0.75F(\delta_1^+) + 0.75F(\delta_1^-) + 0.25F(\delta_2^+) + 0.25F(\delta_2^-), \tag{16.30}$$

subject to

$$\begin{cases} 1.0015X_1 + 1.0001X_2 + 1.0011X_3 + 1.0002X_4 + 1.0005X_5 + \\ 1.0001X_6 + 1.0016X_7 + 0.9999X_8 + 1.0008X_9 + 1.0004X_{10} \\ -\delta_1^+ + \delta_1^- = 100,125.00 \\ 1.1142X_1 + 1.0349X_2 + 1.0844X_3 + 1.0358X_4 + 1.0693X_5 + \\ 1.0936X_6 + 1.0932X_7 + 1.0475X_8 + 1.0658X_9 + 1.0515X_{10} + \\ \delta_2^+ + \delta_2^- = 106,200.00 \\ X_1 + X_2 + X_3 + X_4 + X_5 + X_6 + X_7 + X_8 + X_9 + X_{10} = 100,000.00, \\ X_5 + X_9 \geq 30,000.00, \\ X_3 \geq 20,000.00, \\ X_1 \geq 10,000.00, \\ X_1, X_2, X_3, X_4, X_5, X_6, X_7, X_8, X_9, X_{10} \geq 0, \\ 0 \leq \delta_i^+, \delta_i^- \leq 60, \ i = 1, 2. \end{cases} \tag{16.31}$$

We have solved the mathematical program (16.31) by using the software LINDO [19] and we have obtained the following solutions (Table 16.3).

16.4.2 A Goal Programming Model for Media Management and Planning

Media markets are characterized by the presence of two distinct sides whose benefits from interacting through a common platform. The media firms sell the attention of viewers/readers/listeners to advertisers who need media firms to make their products known to potential consumers; more than the past advertisers face a typical multi-criteria problem and they have to make the best choice in terms of audience and costs by taking into account the audiences' preferences [2]. Moreover, in the last decade audience fragmentation and digitalization have changed the media scenario and media diet is getting more and more differentiated. Consequently, it is strategic to build a media plan able to reach the planned target group. A media planner has limited financial

Table 16.2 Results of the above WGP model

Variable	Value
δ_1^+	0.00
δ_1^-	60.00
δ_2^+	0.00
δ_2^-	0.00
AT & T	10,000.00
Walmart	40,000.00
Exxon Mobil	20,000.00
General Electric	0.00
Bank of America	0.00
Ford Motor Company	0.00
Hewlett-Packard	0.00
McKesson Corporation	0.00
J. P. Morgan Chase	30,000.00
Proctor & Gamble	0.00

Table 16.3 Results of the above GP model with satisfaction function.

Variable	Value
δ_1^+	0.00
δ_1^-	59.99
δ_2^+	0.01
δ_2^-	0.00
AT & T	10,000.00
Walmart	39,983.77
Exxon Mobil	20,000.00
General Electric	16.23
Bank of America	0.00
Ford Motor Company	0.00
Hewlett-Packard	0.00
McKesson Corporation	0.00
J. P. Morgan Chase	30,000.00
Proctor & Gamble	0.00

resources and aims to get the best return on investment in terms of attention and engagement with potential customers, and at the same time to minimize total costs of advertising and communication. Several approaches and models have been developed in order to sustain the decision-making process in this area and create tools able to increase advertising productivity: each model has merits and drawbacks (see Ref. [2] and the references therein). Generally, the used objectives in the media selection and planning problem are conflicting and incommensurable such as the consumer exposures, consumer attention levels to a particular advertisement, attention and engagement of customers and the costs of advertising. Usually the available information related to these objectives is stochastic. Moreover, the Decision Maker appreciates differently the deviations between the achievement and the aspiration levels. Let us consider an approach based on the GP with satisfaction function to integrate explicitly the DM's preferences for solving the media selection and planning problem. Our example is based on real data coming from the Italian media market. Table 16.4 shows the evolution of Italians' media diet [11].

Table 16.4 The evolution of Italians' media consumption.

Media vehicle	Media consumption
Tv	94 %
Radio	67.8 %
Newspaper	43 %

Table 16.5 shows an average of official prices' list for an advertising slot for different vehicles (TV, radio, newspaper, and internet).

Let us choose $g_1 = 2,000$ and $900,000$ to be the goals. The integer decision variables X_j are defined as follows:

$$X_j = \text{number of slots to be bought in the vehicle } j, \tag{16.32}$$

The proposed model with satisfaction function is the following:

$$\max Z = w_1^+ F(\delta_1^+) + w_1^- F(\delta_1^-) + w_2^+ F(\delta_2^+) + w_2^- F(\delta_2^-), \tag{16.33}$$

Table 16.5 List of prices for an advertising slot.

Media vehicles	Prices (Euro)
Tv	70,000
Radio	10,000
Newspaper	60,000

subject to

$$\begin{cases} I(v_1)X_1 + I(v_2)X_2 + I(v_3)X_3 - \delta_1^+ + \delta_1^- = g_1, \\ c(v_1)X_1 + c(v_2)X_2 + c(v_3)X_3 - \delta_2^+ + \delta_2^- = g_2, \\ X_j \in D, \; j = 1, 2, 3, \\ 0 \leq \delta_i^+, \delta_i^- \leq \xi_v, \; i = 1, 2. \end{cases} \tag{16.34}$$

where D is the set of integer numbers. The coefficients $I(v_j)$ associated with a vehicle v_j can be interpreted as an expected value that an individual will be exposed to an advertisement placed in media vehicle v_j, while the coefficient $c(v_j)$ is the cost of a slot in the media vehicle v_j. The model is solved by LINDO and provides the following results: $X_1 = 6$, $X_2 = 18$, $X_3 = 5$.

16.4.3 A Goal Programming Model for Site Selection

Site location optimization usually involves the analysis of several conflicting criteria. In this context the DM faces the problem of determining the best site location by considering different objectives such as costs and budget, performance, maintenance, population density, and spatial coverage.

In this illustrative example we utilize the site selection model presented by Brans *et al.* [8] and we extend it to a stochastic context by assuming that the underlying probability space consists of three different scenarios $\Omega = \{\omega_1, \omega_2, \omega_3\}$ with associated probabilities $p_1 = \frac{1}{3}$, $p_2 = \frac{1}{3}$ and $p_3 = \frac{1}{3}$. Six criteria are considered by the DM in order to select one site among six potential locations to build a hydroelectric power-station. The set of potential criteria is as follows: X_1 = Italy, X_2 = Belgium, X_3 = Germany, X_4 = Sweden, X_5 = Austria, and X_6 = France. These countries are evaluated through the following list of criteria: f_1 = manpower, f_2 = power (MW), f_3 = construction costs in dollar (10^9), f_4 = maintenance costs in dollar (10^6), f_5 = number of villages to evacuate, and f_6 = security level. The decision variables will be defined as follows:

$$X_j = \begin{cases} 1, & \text{if the country } j \text{ is selected}, \\ 0, & \text{otherwise}, \end{cases} \tag{16.35}$$

where $j = 1$ for Italy, $j = 2$ for Belgium,..., $j = 6$ for France. We will consider three evaluations (scenarios) of each location according to each stochastic criterion as indicated in Tables 16.6 and 16.7. Table 16.8 describes the stochastic goals, Table 16.9 presents the deterministic equivalent formulation and the expected goals are provided in Table 16.10.

The model we propose to be solved is the following:

$$\min Z = \sum_{i=1}^{6} (w_i^+ \delta_i^+ + w_1^- \delta_i^-), \tag{16.36}$$

Table 16.6 Objective functions

Objectives	X_1	X_2	X_3
f_1 (min)	(78,80,82)	(62,65,66)	(80,83,87)
f_2 (max)	(84,90,92)	(50,58,60)	(54,60,66)
f_3 (min)	(56,60,61)	(18,20,21)	(38,40,45)
f_4 (min)	(5,5.4,5.8)	(9.5,9.7,9.8)	(6.8,7.2,7.6)
f_5 (min)	(6,8,9)	(0.5,1,3)	(3,4,7)
f_6 (max)	(1,5,7)	(0.3,1,1.7)	(3,7,10)

Table 16.7 Objective functions.

Objectives	X_4	X_5	X_6
f_1 (min)	(35,40,42)	(50,52,56)	(90,94,98)
f_2 (max)	(70,80,90)	(70,72,73)	(90,96,99)
f_3 (min)	(96,100,110)	(50,60,67)	(60,70,75)
f_4 (min)	(7,7.5,7.7)	(1.5,2,2.6)	(3,3.6,4)
f_5 (min)	(6,7,9)	(1,3,6)	(3,5,8)
f_6 (max)	(8,10,13)	(6,8,9)	(5,6,9)

Table 16.8 Goals.

Goals	**Samples**
g_1	(38,40,44)
g_2	(90,96,100)
g_3	(18,20,26)
g_4	(1.4,2,5)
g_5	(0.5,1,4)
g_6	(7,10,14)

subject to

$$\begin{cases} 80X_1 + 60.66X_2 + 83.33X_3 + 39X_4 + 52.66X_5 + 94X_6 - \delta_1^+ + \delta_1^- = 40.66, \\ 88.66X_1 + 56X_2 + 60X_3 + 80X_4 + 71.66X_5 + 95X_6 - \delta_2^+ + \delta_2^- = 95.33, \\ 59X_1 + 19.66X_2 + 41X_3 + 102X_4 + 59X_5 + 68.33X6 - \delta_3^+ + \delta_3^- = 21.33, \\ 5.4X_1 + 9.66X_2 + 7.2X_3 + 7.4X_4 + 2.03X_5 + 3.53X_6 - \delta_4^+ + \delta_4^- = 2.38, \\ 7.66X_1 + 1.5X_2 + 4.66X_3 + 7.33X_4 + 3.33X_5 + 5.33X_6 - \delta_5^+ + \delta_5^- = 1.83, \\ 4.33X_1 + X_2 + 6.66X_3 + 10.33X_4 + 7.66X_5 + 6.66X_6 - \delta_6^+ + \delta_6^- = 10.33, \\ X_1 + X_2 + X_3 + X_4 + X_5 + X_6 = 1, \\ X_1, X_2, X_3, X_4, X_5, X_6 \in \{0, 1\}, \\ \delta_i^+, \delta_i^- \geq 0, \; i = 1, 2, \ldots, 6. \end{cases} \tag{16.37}$$

Table 16.9 Deterministic equivalent.

Objectives	X_1	X_2	X_3	X_4	X_5	X_6
E(f_1)	80	60.66	83.33	39	52.66	94
E(f_2)	88.66	56	60	80	71.66	95
E(f_3)	59	19.66	41	102	59	68.33
E(f_4)	5.4	9.66	7.2	7.4	2.03	3.53
E(f_5)	7.66	1.5	4.66	7.33	3.33	5.33
E(f_6)	4.33	1	6.66	10.33	7.66	6.66

Table 16.10 Expected goals.

Goals	**Samples**
E(g_1)	40.66
E(g_2)	95.33
E(g_3)	21.33
E(g_4)	2.8
E(g_5)	1.83
E(g_6)	10.33

The solution of the deterministic equivalent, obtained by LINGO [19], is X_2 (Belgium) and this means that the hydroelectric power-station will be built in Belgium.

16.4.4 A Goal Programming Model for the Next Release Problem

In software engineering, the next release problem (NRP) consists in finding an ideal set of requirements to be developed in a next release by balancing the customers' priorities and the resource constraints of the developing company [6]. Several solutions have been proposed for this problem, e.g. of Bagnall *et al.* [6], Sagrado *et al.* [23] and Zhang *et al.* [27]. Here we propose a new approach based on a GP model for solving a stochastic extension of the multi-objective next release problem (MONRP) proposed in Ref. [27].

The NRP assumes a set of independent requirements $\{r_1, r_2, \ldots, r_n\}$ which are candidates to be considered and then developed in the next release. Each requirement r_k, $k = 1, 2, \ldots, n$, has an associated cost $cost_k$ which is determined from the amount of resources and effort that each requirement needs in order to be implemented. The model also supposes that all requirements are suggested by different customers c_j, $j = 1, 2, \ldots, m$, and the company assigns a different level of importance to each customer. This can be denoted by the following weight values $\{\lambda_1, \lambda_2, \ldots, \lambda_m\}$ where λ_j is the weight associated with customer c_j. We denote by $V(r_k, c_j)$ the value of the requirement

r_k for the customer c_j. Therefore, the overall score or importance of a given requirement r_k to the company is calculated by

$$S_k = \sum_{j=1}^{m} \lambda_j V(r_k, c_j). \tag{16.38}$$

The decision variables will be defined as follows:

$$X_k = \begin{cases} 1, & \text{if the requirement } k \text{ is selected for the next release,} \\ 0, & \text{otherwise.} \end{cases} \tag{16.39}$$

Following Ref. [27], the multiobjective NRP considers two objectives, namely the overall customer satisfaction or score provided by Eq. (16.40) - to be maximized - and the overall cost for the next release given by Eq. (16.41) — to be minimized. The minimization of the cost is treated as an objective instead of a constraint. The objective functions considered in Ref. [27] are

$$\max \sum_{k=1}^{n} S_k X_k \tag{16.40}$$

$$\min \sum_{k=1}^{n} cost_k X_k \tag{16.41}$$

We now consider a stochastic extension of the model (16.40)-(16.41) which suppose that the parameters r_k, $cost_k$ and λ_j are random variables defined on the probability space Ω.

As a numerical example, let us suppose that $\Omega = \{\omega_1, \omega_2, \omega_3\}$ with probabilities $p(\omega_1) = p(\omega_2) = p(\omega_3) = \frac{1}{3}$. In other words we suppose to have three different scenarios with uniform probability distribution. The following examples use the data presented in Tables 16.11 to 16.14 which have been simulated starting from the data presented by Sagrado *et al.* [23]. Table 16.11 presents the cost of each requirement (we drop the currency for simplicity). It is assumed that different scenarios can lead to different costs on the development of each requirement. Table 16.12 presents the priority assigned by each customer to each requirement. In this case it is assumed that the priority is not affected by the different scenarios. Table 16.13 presents the different level of importance of each customer for the company and finally Table 16.14 presents the score of each requirement.

In order to solve the developed model, let us consider the deterministic equivalent formulation of the stochastic model and consider the following GP model with the following set of weights: $w_1^+ = 0.1, w_1^- = 0.1, w_2^+ = 0.4, w_2^- = 0.4$.

$$\min Z = 0.1\delta_1^+ + 0.1\delta_1^- + 0.4\delta_2^+ + 0.4\delta_2^-, \tag{16.42}$$

Table 16.11 Requirements' development costs.

Requirements	r_1	r_2	r_3	r_4	r_5	r_6	r_7	r_8	r_9	r_{10}
ω_1	1	3	2	2	1	6	8	1	1	2
ω_2	1	4	2	3	4	7	10	2	1	3
ω_3	5	5	6	4	6	7	10	4	3	6
Mean	2.33	4	3.33	3	3.67	6.67	9.33	2.33	1.67	3.67

Requirements	r_{11}	r_{12}	r_{13}	r_{14}	r_{15}	r_{16}	r_{17}	r_{18}	r_{19}	r_{20}
ω_1	2	4	6	1	1	4	8	1	8	2
ω_2	2	5	8	2	1	4	10	4	8	4
ω_3	5	8	10	4	5	4	10	6	8	4
Mean	3	5.67	8	2.33	2.33	4	9.33	3.67	8	3.33

Table 16.12 Priority levels assigned by each customer to each requirement.

Priorities	r_1	r_2	r_3	r_4	r_5	r_6	r_7	r_8	r_9	r_{10}
c_1	4	2	1	2	5	5	2	4	4	4
c_2	4	4	2	2	4	5	1	4	4	5
c_3	5	3	3	3	4	5	2	4	4	4
c_4	4	5	2	3	3	4	2	4	2	3
c_5	5	4	2	4	5	4	2	4	5	2

Priorities	r_{11}	r_{12}	r_{13}	r_{14}	r_{15}	r_{16}	r_{17}	r_{18}	r_{19}	r_{20}
c_1	2	3	4	2	4	4	4	1	3	2
c_2	2	3	2	4	4	2	3	2	3	1
c_3	2	4	1	5	4	1	2	3	3	2
c_4	5	2	3	2	4	3	5	4	3	2
c_5	4	5	3	4	4	1	1	2	4	1

Table 16.13 Customer weights.

Weights	c_1	c_2	c_3	c_4	c_5
ω_1	5	1	1	5	3
ω_2	4	4	3	5	5
ω_3	1	4	2	5	4

Table 16.14 Requirements' scores.

Scores	r_1	r_2	r_3	r_4	r_5	r_6	r_7	r_8	r_9	r_{10}
ω_1	64	54	26	42	63	67	29	60	53	50
ω_2	92	78	41	60	88	95	38	84	79	73
ω_3	70	69	33	47	64	71	28	64	58	55
Mean	75.33	67	33.33	49.67	71.67	77.67	31.67	69.33	63.33	59.33

Scores	r_{11}	r_{12}	r_{13}	r_{14}	r_{15}	r_{16}	r_{17}	r_{18}	r_{19}	r_{20}
ω_1	51	47	47	41	60	41	53	36	48	26
ω_2	67	71	57	69	84	47	64	51	68	33
ω_3	55	53	41	54	64	33	49	43	52	24
Mean	57.67	57	48.33	54.67	69.33	40.33	55.33	43.33	56	27.67

subject to

$$\begin{cases} 75.33X_1 + 67X_2 + 33.33X_3 + 49.67X_4 + 71.67X_5 \\ +77.67X_6 + 31.67X_7 + 69.33X_8 + 63.33X_9 + 59.33X_{10} + 57.67X_{11} \\ +57X_{12} + 48.33X_{13} + 54.67X_{14} + 69.33X_{15} \\ +40.33X_{16} + 55.33X_{17} + 43.33X_{18} + 56X_{19} + 27.67X_{20} + \delta_1^- - \delta_1^+ = 500, \\ 2.33X_1 + 4X_2 + 3.3333X_3 + 3X_4 + 3.6667X_5 + 6.67X_6 + 9.33X_7 \\ +2.33X_8 + 1.67X_9 + 3.67X_{10} + 3X_{11} \\ +5.67X_{12} + 8X_{13} + 2.33X_{14} + 2.33X_{15} + 4X_{16} \\ +9.33X_{17} + 3.67X_{18} + 8X_{19} + 3.33X_{20} + \delta_2^- - \delta_2^+ = 25, \\ X_j \in \{0,1\}, \quad j = 1, 2, \ldots 20, \\ \delta_1^+, \delta_1^-, \delta_2^+, \delta_2^- \geq 0. \end{cases} \tag{16.43}$$

The model is solved by LINGO [19] which provides the following values for the deviations: $\delta_1^+ = 0$, $\delta_1^- = 12.67$, $\delta_2^+ = 3.33$, $\delta_2^- = 0$. Furthermore, the chosen requirements are $r_1, r_2, r_4, r_5, r_8, r_9, r_{10}, r_{11}, r_{14}$ and r_{15}, the cost of this solution is 28.33$ while the customer satisfaction is 637.33.

Following the approach by Aouni *et al.* [3], this second example introduces the concept of the satisfaction function. As above we are going to use the following definition of satisfaction function: $F(\delta_i) = \frac{1}{1+\alpha^2\delta_i^2}$. The value chosen for α is $\alpha = 1$ which implies $\xi_i = \frac{1}{3}$, $\xi_d = 3$ and $\xi_v = 6$. Using the same goals and weights of the previous example, the model is rewritten into the following:

$$\max Z = 0.1F(\delta_1^+) + 0.1F(\delta_1^-) + 0.4F(\delta_2^+) + 0.4F(\delta_2^-), \tag{16.44}$$

subject to

$$\begin{cases} 75.33X_1 + 67X_2 + 33.33X_3 + 49.67X_4 + 71.67X_5 + 77.67X_6 + 31.67X_7 \\ +69.33X_8 + 63.33X_9 + 59.33X_{10} + 57.67X_{11} + 57X_{12} \\ +48.33X_{13} + 54.67X_{14} + 69.33X_{15} + 40.33X_{16} + 55.33X_{17} + 43.33X_{18} \\ +56X_{19} + 27.67X_{20} + \delta_1^- - \delta_1^+ = 500, \\ 2.33X_1 + 4X_2 + 3.3333X_3 + 3X_4 + 3.6667X_5 + 6.67X_6 + 9.33X_7 + 2.33X_8 \\ +1.67X_9 + 3.67X_{10} + 3X_{11} + 5.67X_{12} + 8X_{13} + 2.33X_{14} + 2.33X_{15} + 4X_{16} \\ +9.33X_{17} + 3.67X_{18} + 8X_{19} + 3.33X_{20} + \delta_2^- - \delta_2^+ = 25, \\ X_j \in \{0,1\}, \quad j = 1,2,\dots 20, \\ \delta_1^+, \delta_1^-, \delta_2^+, \delta_2^- \geq 0, \\ 0 \leq \delta_1^+ \leq 6, \\ 0 \leq \delta_1^- \leq 6, \\ 0 \leq \delta_2^+ \leq 6, \\ 0 \leq \delta_2^- \leq 6. \end{cases}$$

The solution provided by LINGO is the following: $\delta_1^+ = 4$, $\delta_1^- = 6$, $\delta_2^+ = 6$, $\delta_2^- = 0$. The chosen requirements are $r_1, r_4, r_5, r_6, r_8, r_9, r_{10}, r_{11}, r_{14}$ and r_{15}, the cost of this solution is 31$ while the customer satisfaction is 648.

EXERCISES

This section presents some exercises to help the reader to become more familiar with the material above.

16.1 Solve the following GP model:

$$\min Z = 0.1\delta_1^+ + 0.1\delta_1^- + 0.4\delta_2^+ + 0.4\delta_2^-, \tag{16.45}$$

subject to

$$\begin{cases} x_1 + x_2 + 3x_3 + \delta_1^- - \delta_1^+ = 500, \\ 2x_1 - 4x_2 - 3x_3 + \delta_2^- - \delta_2^+ = 25, \\ x_1, x_2, x_3 \geq 0, \\ \delta_1^+, \delta_1^-, \delta_2^+, \delta_2^- \geq 0. \end{cases} \tag{16.46}$$

16.2 Solve the following GP model:

$$\min Z = 0.1\delta_1^+ + 0.1\delta_1^- + 0.4\delta_2^+ + 0.4\delta_2^-, \tag{16.47}$$

subject to

$$\begin{cases} x_1 + x_2 + 3x_3 + \delta_1^- - \delta_1^+ = 500, \\ 2x_1 - 4x_2 - 3x_3 + \delta_2^- - \delta_2^+ = 1500, \\ x_1, x_2, x_3 \geq 0, \\ \delta_1^+, \delta_1^-, \delta_2^+, \delta_2^- \geq 0. \end{cases} \tag{16.48}$$

16.3 Solve the model presented in Exercise 16.2 by including a different system of preferences through the satisfaction function $F(\delta_i) = \frac{1}{1+0.025^2\delta_i^2}$

(choose the indifference, the dissatisfaction and the veto thresholds being, respectively, equal to $\xi_i = 40$, $\xi_d = 120$ and $\xi_v = 240$):

$$\max Z = 0.1F(\delta_1^+) + 0.1F(\delta_1^-) + 0.4F(\delta_2^+) + 0.4F(\delta_2^-), \tag{16.49}$$

subject to

$$\begin{cases} x_1 + x_2 + 3x_3 + \delta_1^- - \delta_1^+ = 500, \\ 2x_1 - 4x_2 - 3x_3 + \delta_2^- - \delta_2^+ = 25, \\ x_1, x_2, x_3 \geq 0, \\ \delta_1^+, \delta_1^-, \delta_2^+, \delta_2^- \geq 0. \end{cases} \tag{16.50}$$

16.4 Let $\Omega = \{\omega_1 = 1, \omega_2 = 2, \omega_3 = 3\}$ be the underlying space of events with associated probabilities $p_1 = 0.1$, $p_2 = 0.2$ and $p_3 = 0.7$. Let $Y : \Omega \to R$ be a random variable defined as $Y(\omega_i) = w_i^2$.

a) Calculate the expected value and the variance of Y.

b) Consider the following scenario-based GP model:

$$\min Z = 0.1\delta_1^+ + 0.1\delta_1^- + 0.4\delta_2^+ + 0.4\delta_2^-$$

subject to

$$\begin{cases} Y(\omega_i)x_1 + x_2 + 3x_3 + \delta_1^- - \delta_1^+ = 500 \\ 2x_1 - 4x_2 - Y(\omega_i)x_3 + \delta_2^- - \delta_2^+ = 25 \\ x_1, x_2, x_3 \geq 0 \\ \delta_1^+, \delta_1^-, \delta_2^+, \delta_2^- \geq 0 \end{cases}$$

Write the deterministic equivalent formulation of program 16.4 and solve it.

REFERENCES

1. Aouni, B., Linéarisation des expressions quadratiques en programmation mathématique: des bornes plus efficaces. Administrative Sciences Association of Canada, *Management Science*, 17, 38–46 (1996).
2. Aouni, B., Colapinto, C., La Torre, D., Stochastic goal programming model and satisfaction functions for media selection and planning problem, *International Journal of Multicriteria Decision Making*, in press (2012).
3. Aouni, B., Colapinto, C., and La Torre, D., Solving Stochastic Multi-Objective Programming in Multi-Attribute Portfolio Selection through the Goal Programming Model, *Journal of Financial Decision Making*, 6, 17–30 (2010).
4. Aouni, B. and La Torre, D., A generalized stochastic goal programming model, *Applied Mathematics and Computation*, 215, 4347–4357 (2010).
5. Aouni, B. and Kettani, O., Goal programming model: a glorious history and a promising future, *European Journal of Operational Research*, 133(2), 1–7 (2001).

6. Bagnall, A., Rayward-Smith, V., and Whittley, I., The next release problem, *Information and Software Technology*, 43, 883–890 (2001).

7. Ben Abdelaziz, F., Lang, P., and Nadeau, R., Dominance and efficiency in multicriteria decision under uncertainty, *Theory and decisions*, 47(3), 191–211 (1999).

8. Brans, J. P., Vincke, Ph., and Marechal, B., How to select and how to rank projects: the Promethee method, *European Journal of Operations Research*, 24, 228–238 (1986).

9. Caballero, R., Cerda E., Munoz M.M., Rey L., Stancu-Minasian I.M., Efficient solution concepts and their relations in stochastic multiobjective programming, *Journal of Optimization Theory and Applications*, 110(1), 53–74 (2001).

10. Caballero, R., Cerda, E., Munoz, M. M., and Rey, L., Stochastic approach versus multiobjective approach for obtaining efficient solutions in stochastic multiobjective programming problems, *European Journal of Operations Research*, 158, 633–648 (2004).

11. Censis, Centro Studi Investimenti Sociali (2010), available at http://www.censis.it.

12. Charnes, A. and Cooper, W. W., Chance constraints and normal deviates, *Journal of the American Statistical Association*, 57, 134–148 (1952).

13. Charnes, A. and Cooper, W. W., Chance-constrained programming, *Management Science*, 6, 73–80 (1959).

14. Hannan, E., Non-dominance in goal programming, *INFOR Information Systems and Operational Research* 18, 300–309 (1980).

15. Keeney, R. and Howard, R., *Decisions with Multiple Objectives*, Wiley, New York (1976).

16. Larbani, M. and Aouni, B., A new approach for generating efficient solutions within the goal programming model, *Journal of the Operational Research Society*, 62(1), 1–10 (2011).

17. Lee, S. M., Goal programming for decision analysis of multiple objectives, *Sloan Management Review*, 14, 11–24 (1973).

18. Lee, S. M. and Clayton, S.R., A goal programming model for academic resource allocation, *Management Science*, 18(8), B395–B408 (1972).

19. LINGO. Release 13.0. *LINDO Systems Inc.* (2011).

20. Martel, J.-M. and Aouni, B., Incorporating the Decision-Maker's preferences in the goal programming model, *Journal of the Operational Research Society*, 41, 1121–1132 (1990).

21. Martel, J.-M. and Aouni, B., Méthode multicritére de choise d'un emplacement: le cas d'un aéroport dans le Nouveau Québec, *Information System and Operations Research*, 30(2), 97–117 (1992).

22. Romero, C., *Handbook of critical issues in goal programming*, Pergamon Press, Oxford (1991).

23. del Sagrado, J., del Águila, I.M., and Orellana, F. J., Ant Colony Optimization for the Next Release Problem, *2nd International Symposium on Search Based Software Engineering*, pp.47–56 (2010).

24. Sawaragi, Y., Nakayama, H., and Tanino, T., *Theory of Multiobjective Optimization*, Academic Press, New York (1985).

25. Stancu-Minasian, I. M., *Stochastic programming with Multiple Objective Functions*, D. Reidel Publishing Company, Dordrecht (1984).

26. White, D. J., *Optimality and efficiency*, Wiley, Chichester (1982).

27. Zhang, Y., Harman, M., and Mansouri, S. A., The multiobjective next release problem, *Proceedings of the 9th Annual Conference on Genetic and Evolutionary Computation GECCO'07* (2007).

CHAPTER 17

DECISION THEORY UNDER RISK AND APPLICATIONS IN SOCIAL SCIENCES: II. GAME THEORY

E. V. PETRACOU[1] AND A. N. YANNACOPOULOS[2]

[1] Department of Geography, University of the Aegean, Greece
[2] Department of Statistics, Athens University of Economics and Business, Greece

Obviously, you are not a golfer—The Big Lebowski.

17.1 INTRODUCTION

In this chapter we continue our investigation of decision theory and its use in the social sciences, by looking at the problem of decision making when there is more than one agent involved and furthermore, the actions of one agent may interfere with the gain (or loss) of the other. This is a more realistic situation and is clearly needed for a better understanding of social phenomena.

This leads us to the theory of games. The theory of games was introduced in the 1940's by Von Neumann and Morgernstern as a model for economic behavior and has since become a dominant tool in the economic and social sciences. The mathematical content of the theory is very rich, and brilliant mathematicians such as, e.g., Von Neumann and Nash have provided the mathematical framework. Our approach is inspired by that of Aubin [1].

17.2 BEST REPLIES AND NASH EQUILIBRIA

The simplest possible situation we may envisage is that of two agents A (Athanasios) and E (Electra). Each of the players has a strategy set $\mathcal{S}_A = \{s_1, \cdots, s_n\}$ and $\mathcal{S}_E = \{s_1, \cdots, s_m\}$. Let us denote by $P_A(s_A, s_E)$ the payoff of agent A is he plays $s_A \in \mathcal{S}_A$ while E plays $s_E \in \mathcal{S}_E$ and by $P_E(s_A, s_E)$ the payoff of agent E is she plays $s_E \in \mathcal{S}_E$ while A plays $s_A \in \mathcal{S}_A$. Since P_A and P_E are considered as payoffs (gains) it is natural to give the next definition.

Definition 17 (Best replies) *Define*

$$P_A^\sharp(s_E) := \sup_{s_A \in \mathcal{S}_A} P_A(s_A, s_E),$$

the best payoff that agent A may obtain (over his strategy set) given that agent E plays strategy s_E. The strategies s_A^* such that

$$P_A(s_A^*, s_E) = P_A^\sharp(s_E)$$

are called the best replies of A to the strategy s_E of E and we denote that as $s_A^* \in BR(s_E)$.
In a similar manner we define

$$P_E^\sharp(s_A) := \sup_{s_E \in \mathcal{S}_E} P_E(s_A, s_E)$$

and the strategies s_E^* such that

$$P_E(s_A, s_E^*) = P_E^\sharp(s_A),$$

the best replies of E to the strategy s_A of A, denoted as $s_E^* \in BR(s_A)$.

Note that equivalently,

$$\text{for given } s_E \in \mathcal{S}_E,\ BR(s_E) = \arg\max_{s_A \in \mathcal{S}_A} P_A(s_A, s_E) \subset \mathcal{S}_A,$$
$$\text{for given } s_A \in \mathcal{S}_A,\ BR(s_A) = \arg\max_{s_E \in \mathcal{S}_E} P_E(s_A, s_E) \subset \mathcal{S}_E.$$

In general the sets of maximizers are not consisting of a single element.

Then we have the following fundamental definition.

Definition 18 (Nash equilibrium) *We call the strategy* (s_A^*, s_E^*) *a Nash equilibrium if*

$$s_A^* \in BR(s_E^*), \quad s_E^* \in BR(s_A^*),$$

i.e. a Nash equilibrium is a pair of strategies that has the property of being a best reply for both agents.

A Nash equilibrium has the property that none of the two agents wishes to leave it. Another way to define a Nash equilibrium is by

$$P_A^\sharp(s_E^*) = P_A(s_A^*, s_E^*),$$
$$P_E^\sharp(s_A^*) = P_E(s_A^*, s_E^*).$$

The above definitions may be rewritten in terms of loss functions rather that payoff functions, where in this case the suprema are exchanged by infima.

■ EXAMPLE 17.1 The Prisoner's Dilemma Game

Two social groups considered as agents A and E, may choose to oppose or accepting the government's decision on some issue. The strategy sets are discrete $\mathcal{S}_A = (s_{A,1}, s_{A,2})$, where $s_{A,1}$ corresponds to A opposing the government's decision and $s_{A,2}$ corresponds to A accepting the government's decision. Similarly with $\mathcal{S}_E$. For the sake of exposition let us assume that the government is trying to impose salary cuts to two different groups of employees (public sector A and private sector E) and the payoff is to be measured in terms of $-x$, where x is the monetary sum to be cut.

The payoff P_A is such that $P_A(s_{A,1}, s_{E,1}) = -a$, $P_A(s_{A,1}, s_{E,2}) = -c$, $P_A(s_{A,2}, s_{E,1}) = 0$, $P_A(s_{A,2}, s_{E,2}) = -b$, where $a < b < c$. As an illustrative example consider the case $a = 1, = 3, c = 4$. The payoff P_E is such that $P_E(s_{A,1}, s_{E,1}) = -a$, $P_E(s_{A,1}, s_{E,2}) = 0$, $P_E(s_{A,2}, s_{E,1}) = -c$, $P_E(s_{A,2}, s_{E,2}) = -b$.

For finite games we often display the payoff function in a single table

$A \setminus E$	$s_{E,1}$	$s_{E,2}$
$s_{A,1}$	$(-a, -a)$	$(-c, 0)$
$s_{A,2}$	$(0, -c)$	$(-b, -b)$

We now easily calculate

$$P_A^\sharp(s_{E,1}) = 0, \quad P_A^\sharp(s_{E,2}) = -b$$

(this is the maximum value of the first entry in the parentheses for the first column and the second column, respectively) and

$$P_E^\sharp(s_{A,1}) = 0, \ \ P_E^\sharp(s_{A,2}) = -b$$

(this is the maximum value of the second entry in the parentheses for the first row and the second row, respectively). Therefore the set of best replies are

$$BR(s_{E,1}) = s_{A,2}, \ \ BR(s_{E,2}) = s_{A,2}$$

(since if E plays $s_{E,1}$ then A by playing $s_{A,2}$ obtains the maximum payoff which is 0 and if E plays $s_{E,2}$ then A by playing $s_{A,2}$ obtains the maximum payoff which is $-b$ — we look at the first number in the parentheses) and

$$BR(s_{A,1}) = s_{E,2}, \ \ BR(s_{A,2}) = s_{E,2}$$

(since if A plays $s_{A,1}$ then E by playing $s_{E,2}$ obtains the maximum payoff which is 0, since if A plays $s_{A,2}$ then E by playing $s_{E,2}$ obtains the maximum payoff which is $-b$ — we look at the second number in the parentheses). The common element of the set of best replies is $(s_{A,2}, s_{E,2})$, which is the Nash equilibrium. This can be seen since $s_{A,2} \in BR(s_{E,2})$ and $s_{E,2} \in BR(s_{A,2})$. Observe that if both agents play this strategy they suffer a salary cut of b euros each. This is not the best solution since if they both played $(s_{A,1}, s_{E,1})$ they would both suffer a salary cut of $a < b$, so they would be better off. Observe that in this game both players would have mutually benefited if they had cooperated with each other, rather than played as individuals one against the other (for if they had communicated their intentions they would both choose strategy 1). This example furthermore shows that Nash equilibria maybe stable but not necessarily desirable, one should pretty much prefer to lock into $(s_{A,1}, s_{E,1})$, which Pareto dominates the other outcomes.

■ **EXAMPLE 17.2 The Game of Chicken**

Consider two agents A and E (two countries) which are in a hostile situation. The first strategy is to retreat whereas the second strategy is to attack. If both agents attack then they have losses of c, if both retreat they have losses of $a < c$. If one attacks and the other retreats the one that attacks loses nothing whereas the one that retreats has a loss of b. As before $a < b < c$.

This game has the matrix

$A \setminus E$	$s_{E,1}$	$s_{E,2}$
$s_{A,1}$	$(-a,-a)$	$(-b,0)$
$s_{A,2}$	$(0,-b)$	$(-c,-c)$

The set of best replies are

$$\begin{aligned} BR(s_{E,1}) &= s_{A,2}, \quad BR(s_{E,2}) = s_{A,1}, \\ BR(s_{A,1}) &= s_{E,2}, \quad BR(s_{A,2}) = s_{E,1}. \end{aligned}$$

It is seen that the pair $(s_{A,1}, s_{E,2})$ has the property that $s_{A,1} \in BR(s_{E,2})$ and $s_{E,2} \in BR(s_{A,1})$. Furthermore, the pair $(s_{A,2}, s_{E,1})$ also has the property that $s_{A,2} \in BR(s_{E,1})$ and $s_{E,1} \in BR(s_{A,2})$. Therefore, the Nash equilibria are now two $(s_{A,1}, s_{E,2})$ and $(s_{A,2}, s_{E,1})$, corresponding to one of the countries choosing to retreat while the other chooses to attack. These equilibria are Pareto ranked and there is *a priori* no way to pick one of the two, unless other mechanisms are included in the game.

■ EXAMPLE 17.3 The Battle of the Sexes

This is a game where the players have different preferences for what they wish to do (e.g., A prefers going to the ballet whereas E prefers going to see a football match) but at any rate they both prefer to be together. As Osborne and Rubinstein put it, this game "models a situation in which players wish to coordinate their behavior but have conflicting interests" [8].

This game has the payoff matrix

$A \setminus E$	$s_{E,1}$	$s_{E,2}$
$s_{A,1}$	$(0,-a)$	$(-b,-b)$
$s_{A,2}$	$(-b,-b)$	$(-a,0)$

The set of best replies are

$$\begin{aligned} BR(s_{E,1}) &= s_{A,1}, \quad BR(s_{E,2}) = s_{A,2}, \\ BR(s_{A,1}) &= s_{E,1}, \quad BR(s_{A,2}) = s_{E,2}. \end{aligned}$$

It is seen that the pair $(s_{A,1}, s_{E,1})$ has the property that $s_{A,1} \in BR(s_{E,1})$ and $s_{E,1} \in BR(s_{A,1})$. Furthermore, the pair $(s_{A,2}, s_{E,2})$ also has the property that $s_{A,2} \in BR(s_{E,2})$ and $s_{E,2} \in BR(s_{A,2})$. Therefore the Nash equilibria are two $(s_{A,1}, s_{E,1})$ and $(s_{A,2}, s_{E,2})$.

■ EXAMPLE 17.4 Coordination Game

This is a game in which again A and B wish to be together. This game has the payoff matrix

$A \setminus E$	$s_{E,1}$	$s_{E,2}$
$s_{A,1}$	$(-b,-b)$	$(0,-a)$
$s_{A,2}$	$(-a,0)$	$(-c,-c)$

The set of best replies are

$$BR(s_{E,1}) = s_{A,2}, \quad BR(s_{E,2}) = s_{A,1},$$
$$BR(s_{A,1}) = s_{E,2}, \quad BR(s_{A,2}) = s_{E,1}.$$

and the Nash equilibria are now two $(s_{A,1}, s_{E,2})$ and $(s_{A,2}, s_{E,1})$. The Nash equilibria are Pareto ranked, one of the two being inferior of the other, however, the two players may as well lock into the inferior one (there is no *a priori* reason of which to choose unless other mechanisms are included in the game). One may argue that this simple observation is already a challenge to our assumptions of rationality.

■ EXAMPLE 17.5 A Model for International Agreements

Assume that the two agents A and E represent countries that will either participate or not in an international agreement. The first strategy corresponds to participating, the second in not participating. If both do not participate they will face a loss of a units. The matrix of the game is as follows:

$A \setminus E$	$s_{E,1}$	$s_{E,2}$
$s_{A,1}$	$(0,0)$	$(-1,1)$
$s_{A,2}$	$(1,-1)$	$(-a,-a)$

The Nash equilibria depend on the value of a. If $a \in (0,1)$ then one can easily see that there is a unique Nash equilibrium $(s_{A,2}, s_{E,2})$ in which the two countries do not cooperate. Clearly, this is not the best possible of all decisions since if the countries were playing $(s_{A,1}, s_{E,1})$ instead this would be beneficial for both (Pareto optimality). However, unless the two countries do not act in a selfish way (or if they do not communicate concerning their intentions beforehand) they will not get to this equilibrium. If $a > 1$ then the situation changes and there are two Nash equilibria $(s_{A,1}, s_{E,2})$ and $(s_{A,2}, s_{E,1})$. There is no way beforehand to know which of the two will be chosen, unless other criteria or supplementary mechanism selections are included in the game.

In many cases of interest the strategy space may be continuous. The notions of best reply and Nash equilibria may be generalized in a natural way. We illustrate them with the following example:

■ EXAMPLE 17.6 A Model of Political Competition

Assume a local society whose spatial location is the interval $[0, 1]$. The members of the community must decide on the exact location $s \in [0, 1]$ where a site that is used by the community is to be built. All the members of the community participate in voting for the exact location to be decided, and we assume that each voter wishes for the facility to be built as close as possible to her living site. We further assume that all agents ideal points are distributed in $[0, 1]$ with a distribution density f, for example, the uniform distribution. Two politicians propose for the location of the facility as part of their electoral campaign. The winner of the game is the politician whose proposal goes through and the payoff is 1. The other candidate gets -1, so the game is a zero-sum game. The voters simply vote for the closest candidate and we assume that their actions may not affect the politicians proposals.

The strategy sets $\mathcal{S}_A = \mathcal{S}_E = [0, 1]$ are now uncountable sets containing an infinity of possible strategies. Suppose A chooses $s_A \in [0, 1]$ and E chooses $s_E \in [0, 1]$. The payoff of each politician clearly depends on the median $s_M := \frac{s_A + s_E}{2}$. The payoff of politician A is

$$P_A(s_A, s_E) = \begin{cases} 1 & \text{if} \quad s_A < s_E \text{ and } s_M > \frac{1}{2} \quad \text{or } s_E < s_A \text{ and } s_M < \frac{1}{2}, \\ 0 & \text{if} \quad s_A = s_E \text{ and } s_M = \frac{1}{2}, \\ -1 & \text{if} \quad s_A < s_E \text{ and } s_M < \frac{1}{2} \quad \text{or } s_E < s_A \text{ and } s_M > \frac{1}{2}, \end{cases}$$

whereas $P_E(s_A, s_E) = -P_A(s_A, s_E)$. This payoff models the fact that if, e.g., $s_A > s_E$ then all voters on the left of s_M will vote for candidate E. Since the voters are uniformly distributed the fraction of voters on the left of s_M is s_M, this corresponds to the probability for E being preferred. Then A gets the rest of the voters which are $1 - s_M$. Similarly for the other cases.

The best response correspondence for candidate A is

$$BR_A(s_E) = \begin{cases} (s_E, 1 - s_E) & \text{if} \quad s_E < \frac{1}{2}, \\ s_E & \text{if} \quad s_E = \frac{1}{2}, \\ (1 - s_E, s_E) & \text{if} \quad s_E > \frac{1}{2}. \end{cases}$$

Observe that the best reply is not a single-valued mapping, the image may be a whole interval and not a single number! This situation is quite common in optimization applications. The best response of candidate E is symmetric,

$$BR_E(s_A) = \begin{cases} (s_A, 1 - s_A) & \text{if} \quad s_A < \frac{1}{2}, \\ s_A & \text{if} \quad s_A = \frac{1}{2}, \\ (1 - s_A, s_A) & \text{if} \quad s_A > \frac{1}{2}. \end{cases}$$

The Nash equilibrium is the common point of these two correspondences, which is $s_A = s_E = \frac{1}{2}$. This is obvious since $\frac{1}{2} = BR_A(\frac{1}{2})$ and $\frac{1}{2} = BR_E(\frac{1}{2})$. It can further be shown that this Nash equilibrium is unique.

In this simple example it can be seen that the Nash equilibrium is for both candidates to adopt the same strategy and propose to build the facility in the middle. This of course reflects the symmetry of the distribution of the voters. An asymmetric distribution would lead to a different Nash equilibrium. This model has been proposed by Hotelling in 1927 (long before the introduction of game theory) as a model for location analysis and of course was treated in a different fashion than shown here. It was later taken up by Downs in 1957 as a model for electoral competition and has led to many variants and a lot of discussion (see, for example, Ref. [4]).

17.3 MIXED STRATEGIES AND MINIMAX

Consider two agents A (Athanasios) and E (Electra) playing a zero sum game. Each of the players has a strategy set $\mathcal{S}_A = \{s_{A,1}, \cdots, s_{A,n}\}$ and $\mathcal{S}_E = \{s_{E,1}, \cdots, s_{E,m}\}$ and may choose a mixed strategy which is represented by two probability vectors $p_A = \{p_{A,1}, \cdots, p_{A,n}\}$, $p_E = \{p_{E,1}, \cdots, p_{E,m}\}$ respectively. A mixed strategy p_A means that agent A will play strategy $s_{A,i}$ with probability $p_{A,i}$, $i = 1, \cdots, n$ and similarly for agent E. The payoff of player A if he plays (pure) strategy i when E plays (pure) strategy j is c_{ij}. Since it is a zero-sum game the payoff matrix has the property $c_{ij} = -c_{ji}$.

The expected payoff for player A if he plays the mixed strategy p_A while E plays the mixed strategy p_E is

$$P_A(p_A, p_E) := \sum_{i=1}^{n} \sum_{j=1}^{m} p_{A,i} p_{E,j} c_{ij},$$

whereas the expected payoff for player E, in the same situation, is $P_E(p_A, p_E) = -P_A(p_A, p_E)$. In contrast to what happens for pure strategies it can be shown that a Nash equilibrium always exists for mixed strategies. This result holds true for games which are not necessarily zero-sum games (i.e., when $c_{ij} \neq -c_{ji}$).

■ EXAMPLE 17.7 Matching Pennies

Consider the game with payoff matrix

$A \setminus E$	$s_{E,1}$	$s_{E,2}$
$s_{A,1}$	$(1,-1)$	$(-1,1)$
$s_{A,2}$	$(-1,1)$	$(1,-1)$

This game presents a situation where according to Osborne and Rubinstein "the interests of the players are diametrically opposed" or "strictly competitive," [8] and has no pure strategy Nash equilibrium.

If we consider mixed strategies then player A will pick strategy $s_{A,1}$ with probability $p_{A,1} = p$ and strategy $s_{A,2}$ with probability $p_{A,2} = 1-p$ whereas player E will pick strategy $s_{E,1}$ with probability $p_{E,1} = q$ and strategy $s_{E,2}$ with probability $p_{E,2} = 1 - q$. Then it can be seen that $p = q = \frac{1}{2}$ is a (mixed strategy) Nash equilibrium.

When working in terms of mixed strategies a game maybe rewritten as a linear programming problem.

■ EXAMPLE 17.8 Games and Linear Programming

Let us consider the following game theoretic situation:

Player A will choose the vector p_A so as to win at least λ on average, i.e., it must hold

$$\lambda \leq \sum_{i=1}^{n} c_{ij} p_{A,i}, \quad j = 1, \cdots, m.$$

His goal is to maximize this λ under the above constraints. This means that A will choose $(p_{A,1}, \cdots, p_{A,n}, \lambda)$ as the solution of the linear programming problem

$$\max_{(p_A,\lambda)} \lambda$$
subject to
$$\lambda - \sum_{i=1}^{n} c_{ij} p_{A,i} \leq 0, \quad j = 1, \cdots, m,$$
$$\sum_{i=1}^{n} p_{A,i} = 1,$$
$$p_{A,i} \geq 0, \quad i = 1, \cdots, n.$$

Player E will choose the mixed strategy $p_E = (p_{E,1}, \cdots, p_{E,m})$ so that she loses at most μ in the mean, that is $\mu \geq \sum_{j=1}^{m} c_{ij} p_{E,j}$ for $i = 1, \cdots, n$. Then she wishes to minimize this loss, so that she will choose p_E and μ as the solution of the linear programming problem

$$\min_{(p_E,\mu)} \mu$$
subject to
$$\mu - \sum_{j=1}^{m} c_{ij} p_{E,j} \geq 0, \quad i = 1, \cdots, n,$$
$$\sum_{j=1}^{m} p_{E,j} = 1,$$
$$p_{E,j} \geq 0, \quad j = 1, \cdots, m.$$

The students familiar with the theory of linear programming will soon notice that one problem is the dual of the other. Thus the optimal value of μ and the optimal value of λ are related. This observation is true in more general situations—see the minimax theorem.

Mixed strategies can be defined equally well for the case of continuous strategy spaces; in this case a mixed strategy is a probability distribution over the strategy space and the payoff is defined as the expectation over this distribution. For instance, if the strategy spaces S_A and S_E are continuous then a mixed strategy for A is a probability distribution G_A on S_A and a mixed strategy for E is a probability distribution G_E on S_E. Then the expected payoff is given by $\int_{S_A} \int_{S_E} f(x, y) dG_E(y) dG_A(x)$.

17.4 NASH EQUILIBRIA AND CONSERVATIVE STRATEGIES

Let us now consider more general strategy spaces, not necessarily discrete, so that the payoffs P_A and P_E of the two agents will be functions $f_A, f_E : S_A \times S_E \to \mathbb{R}$. Let $f_A(x, y)$ by the gain of A and $f_E(x, y)$ be the gain of E

when they play strategies $(x, y) \in \mathcal{S}_A \times \mathcal{S}_E$. Then

$$BR_A(y) = \arg\max_{x \in \mathcal{S}_A} f_A(x, y)$$

is the set of best replies of A to the strategy y of E and

$$BR_E(x) = \arg\max_{y \in \mathcal{S}_E} f_E(x, y)$$

is the set of best replies of E to the strategy x of A. A Nash equilibrium is a strategy (x^*, y^*) such that

$$x^* \in BR_A(y^*), \quad y^* \in BR_E(x^*).$$

Note that the set of best replies of one player to a strategy of another player may not consist of a single element (see, e.g., Example 17.6). Furthermore, note that a Nash equilibrium can be characterized as some sort of fixed point (for a multivalued map).

We now consider a slightly different approach to the problem of characterizing the best strategy to be followed by the two players. Suppose that each player chooses her or his moves so as to minimize the gain of the other. Then A knows that E will choose this strategy y that minimizes $f_A(x, y)$. So A has access to $\inf_{y \in \mathcal{S}_E} f_A(x, y)$. Then he must try to do his best to maximize that and thus he will choose a strategy $\bar{x}$ so as to solve the problem $\sup_{x \in \mathcal{S}_A} \inf_{y \in \mathcal{S}_E} f_A(x, y)$. Such a strategy $\bar{x}$ maximizes the "worst scenario" for the gains of A, i.e., the gains that he will have access to given that E is nasty, clever and lucky enough to mess A up as badly as she can. We will call this strategy for A, a conservative strategy and denote it by

$$\bar{x} \in \arg\max_{x \in \mathcal{S}_A} \left(\inf_{y \in \mathcal{S}_E} f_A(x, y) \right).$$

Similarly for E, assuming that A does his best to mess her up we may define a similar strategy $\bar{x}$ which is the conservative strategy for E as

$$\bar{y} \in \arg\max_{y \in \mathcal{S}_E} \left(\inf_{x \in \mathcal{S}_A} f_E(x, y) \right).$$

The above discussion motivates the following:

Definition 19 (Conservative strategies) *A strategy $(\bar{x}, \bar{y}) \in \mathcal{S}_A \times \mathcal{S}_E$ is called a conservative strategy if*

$$\bar{x} \in \arg\max_{x \in \mathcal{S}_A} \left(\inf_{y \in \mathcal{S}_E} f_A(x, y) \right),$$

$$\bar{y} \in \arg\max_{y \in \mathcal{S}_E} \left(\inf_{x \in \mathcal{S}_A} f_E(x, y) \right).$$

The obvious question that arises is: Under which circumstances are these two strategies related?

17.5 ZERO-SUM GAMES AND THE MINIMAX THEOREM

Consider the special class of games in which $f(x,y) := f_A(x,y) = -f_E(x,y)$, for all $(x,y) \in \mathcal{S}_A \times \mathcal{S}_E$. In such games the gain of A is the loss of E and vice versa. Such games are called zero-sum games.

Suppose that (x^*, y^*) is a Nash equilibrium for such a game. Recall that for any function f, $\sup(-f) = -\inf(f)$, $\inf(-f) = -\sup(f)$. Then,

$$BR_A(y) = \arg\max_{x \in \mathcal{S}_A} f_A(x,y) = \arg\max_{x \in \mathcal{S}_A} f(x,y)$$

and

$$BR_E(x) = \arg\max_{y \in \mathcal{S}_E} f_E(x,y) = \arg\min_{y \in \mathcal{S}_E} f(x,y),$$

i.e., the best replies of A to a strategy $y \in \mathcal{S}_E$ are these strategies that *maximize* $f(x,y)$ over x for a given y, whereas the best replies of E to a strategy $x \in \mathcal{S}_A$ are these strategies that *minimize* $f(x,y)$ over y for a given x. Therefore, in a zero-sum game player A maximizes $f(x,y)$ over x (for fixed y) whereas player E minimizes $f(x,y)$ over y (for fixed x). If we have a matrix game then one player maximizes over rows whereas the other player minimizes over columns. Then by definition the Nash equilibrium is a point (x^*, y^*) such that

$$f(x,y^*) \leq f(x^*,y^*) \leq f(x^*,y), \quad \forall\,(x,y) \in \mathcal{S}_A \times \mathcal{S}_E.$$

This is a saddle point for the function f. Therefore, the Nash equilibrium for a zero sum game is characterized as a saddle point for the function f.

■ EXAMPLE 17.9

Consider $\mathcal{S}_A \times \mathcal{S}_E = [-1,1] \times [-1,1]$ and $f(x,y) = -x^2 + y^2$. The Nash equilibrium is the strategy $(0,0)$.

Let us now consider the conservative strategies. From the definitions we obtain

$$\bar{x} \in \arg\max_{x \in \mathcal{S}_A} \left(\inf_{y \in \mathcal{S}_E} f_A(x,y) \right) = \arg\max_{x \in \mathcal{S}_A} \left(\inf_{y \in \mathcal{S}_E} f(x,y) \right),$$

and

$$\begin{aligned} \bar{y} \in \arg\max_{y \in \mathcal{S}_E} \left(\inf_{x \in \mathcal{S}_A} f_E(x,y) \right) &= \arg\max_{y \in \mathcal{S}_E} \left(\inf_{x \in \mathcal{S}_A} -f(x,y) \right) \\ &= \arg\min_{y \in \mathcal{S}_E} \left(\sup_{x \in \mathcal{S}_A} f(x,y) \right). \end{aligned}$$

The above considerations lead to the following:

Definition 20 *A strategy* $(\bar{x}, \bar{y}) \in \mathcal{S}_A \times \mathcal{S}_E$ *is a conservative strategy for a zero-sum game if*

$$\bar{x} \in \arg\max_{x \in \mathcal{S}_A} \left(\inf_{y \in \mathcal{S}_E} f(x,y) \right),$$

$$\bar{y} \in \arg\min_{y \in \mathcal{S}_E} \left(\sup_{x \in \mathcal{S}_A} f(x,y) \right).$$

The numbers

$$v_A := \sup_{x \in \mathcal{S}_A} \left(\inf_{y \in \mathcal{S}_E} f(x,y) \right), \text{ and } v_E := \inf_{y \in \mathcal{S}_E} \left(\sup_{x \in \mathcal{S}_A} f(x,y) \right)$$

are called the conservative values of the game for A and E, respectively.

In general $v_A \neq v_E$! Let us first obtain an obvious inequality. For every $y \in \mathcal{S}_E$, $f(x,y) \leq \sup_{x \in \mathcal{S}_A} f(x,y)$ and of course this inequality holds for the infimum over $y \in \mathcal{S}_E$ so that $f(x,y) \leq \inf_{y \in \mathcal{S}_E} \sup_{x \in \mathcal{S}_A} f(x,y) = v_E$. Furthermore, for every $x \in \mathcal{S}_A$, $\inf_{y \in \mathcal{S}_E} f(x,y) \leq f(x,y)$ and since this inequality holds for every x it holds also for the supremum of the quantity on the left-hand side over all x, $v_A := \sup_{x \in \mathcal{S}_A} (\inf_{y \in \mathcal{S}_E} f(x,y)) \leq f(x,y)$. Therefore,

$$v_A \leq f(x,y) \leq v_E, \quad \forall (x,y) \in \mathcal{S}_A \times \mathcal{S}_E.$$

The quantity $v_E - v_A$ is called the duality gap.

The important question that arises is when is this duality gap equal to 0? The second important question is whether there is any connection between Nash equilibria and conservative strategies.

Theorem 15 *A Nash equilibrium is also a conservative strategy for which* $v_A = v_E$ *and the converse also holds.*

Proof: Let (x^*, y^*) be a Nash equilibrium for the game. We will then show that $v_A = v_E$ and the (x^*, y^*) is also a cooperative equilibrium. Since $v_A \leq v_E$ it is enough to show that $v_A \geq v_E$ also holds. By the definition of a Nash equilibrium we have the saddle point property

$$f(x, y^*) \leq f(x^*, y^*) \leq f(x^*, y), \quad \forall (x,y) \in \mathcal{S}_A \times \mathcal{S}_E.$$

We start by the right-hand side of this inequality which since it holds for all y it holds also for the infimum over all y to yield $f(x^*, y^*) \leq \inf_{y \in \mathcal{S}_E} f(x^*, y)$ and when the right-hand side of this inequality is considered as a function of x its value at x^* is certainly less than the supremum of this quantity over all

$x \in \mathcal{S}_A$, therefore

$$f(x^*, y^*) \leq f(x^*, y) \leq \inf_{y \in \mathcal{S}_E} f(x^*, y) \leq \sup_{x \in \mathcal{S}_A} \left(\inf_{y \in \mathcal{S}_E} f(x, y) \right) = v_A. \quad (17.1)$$

We then take the left-hand side of the saddle inequality, which since it holds for all x it also holds for the supremum over all $x \in \mathcal{S}_A$ to yield $\sup_{x \in \mathcal{S}_A} f(x, y^*) \leq f(x^*, y^*)$ and when the left-hand side of the latter equality is considered as a function of y the value of this function at y^* (which is the number on the left-hand side) is certainly larger or equal to the infimum of this function over all $y \in \mathcal{S}_E$. The above reasoning gives

$$v_E = \inf_{y \in \mathcal{S}_E} \left(\sup_{x \in \mathcal{S}_A} f(x, y) \right) \leq \sup_{x \in \mathcal{S}_A} f(x, y^*) \leq f(x^*, y^*). \quad (17.2)$$

Combining (17.1) and (17.2) yields $v_E \leq v_A$ so that $v_E = v_A$.

By the saddle path property y^* minimizes $f(x^*, y)$ and x^* being a Nash equilibrium is chosen so as to maximize $f(x, y)$. Therefore y^* minimizes the maximum of $f(x, y)$ over x, i.e., $y^* \in \arg\min_{y \in \mathcal{S}_E} \left(\sup_{x \in \mathcal{S}_A} f(x, y)\right)$ and y^* is also a conservative equilibrium for E. By the saddle path property x^* maximizes $f(x, y^*)$ and since y^* is a Nash equilibrium it is chosen so as to minimize $f(x, y)$. Therefore x^* maximizes the minimum of $f(x, y)$ over y, i.e., $x^* \in \arg\max_{x \in \mathcal{S}_A} (\inf_{y \in \mathcal{S}_E} f(x, y))$ so that x^* is also a conservative equilibrium.

It remains to show the converse property. Suppose that $v_A = v_E$ and that $(\bar{x}, \bar{y})$ is a conservative equilibrium. We will show that $(\bar{x}, \bar{y})$ is also a Nash equilibrium. Let $v = v_A = v_E$ be the common value for the game. Since $\bar{x}$ maximizes the minimum of $f(x, y)$ over y we have that

$$\inf_{y \in \mathcal{S}_E} f(\bar{x}, y) = \sup_{x \in \mathcal{S}_A} \left(\inf_{y \in \mathcal{S}_E} f(x, y) \right) = v.$$

Similarly since $\bar{y}$ minimizes the maximum of $f(x, y)$ over x, we have that

$$\sup_{x \in \mathcal{S}_A} f(x, \bar{y}) = \inf_{y \in \mathcal{S}_E} \left(\sup_{x \in \mathcal{S}_A} f(x, y) \right) = v$$

so that $\sup_{x \in \mathcal{S}_A} f(x, \bar{y}) = \inf_{y \in \mathcal{S}_E} f(\bar{x}, y) = v$. We now have the obvious[17] inequalities

$$v = \inf_{y \in \mathcal{S}_E} f(\bar{x}, y) \leq f(\bar{x}, \bar{y}), \text{ and } v = \sup_{x \in \mathcal{S}_A} f(x, \bar{y}) \geq f(x, \bar{y}), \ \forall x \in \mathcal{S}_A,$$

[17]The inf over y is smaller or equal than the value of the function at any point y, and we pick $y = \bar{y}$.

so that

$$f(x,\bar{y}) \leq f(\bar{x},\bar{y}), \ \forall\, x \in \mathcal{S}_A.$$

Similarly, we have that[18]

$$v = \sup_{x\in\mathcal{S}_A} f(x,\bar{y}) \geq f(\bar{x},\bar{y}), \text{ and } v = \inf_{y\in\mathcal{S}_E} f(\bar{x},y) \leq f(\bar{x},y), \ \forall\, y \in \mathcal{S}_E,$$

so that

$$f(\bar{x},y) \geq f(\bar{x},\bar{y}), \ \forall\, y \in \mathcal{S}_E.$$

Therefore, the point $(\bar{x},\bar{y})$ has the property

$$f(x,\bar{y}) \leq f(\bar{x},\bar{y}) \leq f(\bar{x},y), \quad \forall\,(x,y) \in \mathcal{S}_A \times \mathcal{S}_E$$

and thus it is a saddle point, hence a Nash equilibrium. ∎

It remains to show that a Nash equilibrium exists. Since in the particular case of two players zero-sum games Nash equilibria can be characterized as saddle points of the payoff function, to do this we may use a general theorem for the existence of saddle points. This can be done using a class of important theorems called minimax theorems. The first such theorem was proved by John Von Neumann and bears his name.

At this point we present and provide a proof of the famous Von Neumann minimax theorem. These are numerous proofs of this theorem as well as numerous generalizations and extensions. Here we present one of the simplest versions of this theorem, and give a proof that was published by the mathematical economicist Nikaidō in 1954 [5]. For an alternative proof one may see, e.g., Ref. [2].

Theorem 16 (Von Neumann minimax) *Let $\mathcal{S}_A$ and $\mathcal{S}_E$ be compact convex*[19] *sets and let $f : \mathcal{S}_A \times \mathcal{S}_E \to \mathbb{R}$, continuous concave in x and convex in y. Then there exists a point $(\bar{x},\bar{y})$ such that*

$$f(\bar{x},\bar{y}) = \sup_{x\in\mathcal{S}_A} f(x,\bar{y}) = \inf_{y\in\mathcal{S}_E} f(\bar{x},y).$$

[18]The sup over x is greater or equal than the value of the function at any point x and we pick $x = \bar{x}$.

[19]Recall the following fundamental definitions: A subset M of a metric space is called closed if it contains all its accumulation points. A subset M is open if for every $x \in M$ there exists an open ball of radius $\epsilon > 0$ centered at x, $B_x(\epsilon)$ such that $B_x(\epsilon) \subset M$. A subset M is open if and only if its complement is closed and vice versa. A subset of a metric space is called compact if every open cover of this set has a finite subcover. Equivalently, a subset of a metric space is compact if every bounded sequence in this set has a convergent subsequence. A subset M of a metric space is called convex if it has the property that if $x, y \in M$ then $\lambda x + (1-\lambda)y \in M$ for all $\lambda \in [0,1]$. In finite-dimensional spaces compactness for a set is equivalent to this set being closed and bounded; however this is not true in infinite dimensions eventhough a compact set is always closed and bounded. A function $f : M \to \mathbb{R}$ is convex if $f(\lambda x + (1-\lambda)y) \leq \lambda f(x) + (1-\lambda)f(y)$ for all $x, y \in M$, $\lambda \in [0,1]$. A function is concave if $-f$ is convex.

Proof: For any $\lambda, \mu \in \mathbb{R}$ define the sets

$$A_\lambda := \{x \in \mathcal{S}_A \,:\, f(x,y) \geq \lambda, \ \forall\, y \in \mathcal{S}_E\} \subset \mathcal{S}_A,$$
$$E_\mu := \{y \in \mathcal{S}_E \,:\, f(x,y) \leq \mu, \ \forall\, x \in \mathcal{S}_A\} \subset \mathcal{S}_E.$$

By the properties of f these sets are closed convex sets. Next define

$$\lambda^* = \sup\{\lambda \,:\, A_\lambda \neq \emptyset\} \ \mu^* = \inf\{\mu \,:\, E_\mu \neq \emptyset\},$$

i.e., the smallest upper bound of the set of λ such that these exists an $x \in \mathcal{S}_A$ such that $f(x,y) \geq \lambda$ for all $y \in \mathcal{S}_E$ and the largest lower bound of the set of μ such that there exist a $y \in \mathcal{S}_E$ such that $f(x,y) \leq \mu$ for all $x \in \mathcal{S}_A$.

By the compactness of $\mathcal{S}_A$ and $\mathcal{S}_B$ it follows that[20]

$$A_{\lambda^*} \neq \emptyset, \ \ \lambda^* < \infty,$$
$$E_{\mu^*} \neq \emptyset, \ \ \mu^* > -\infty.$$

Since A_{λ^*} and E_{μ^*} are not empty let $x^* \in A_{\lambda^*}$ and $y^* \in E_{\mu^*}$. For the pair (x^*, y^*) it clearly holds that

$$\lambda^* \leq f(x^*, y^*) \leq \mu^*.$$

If $\lambda^* = \mu^*$ then by the definition of A_{λ^*} and E_{μ^*} it follows that

$$f(x^*, y) \geq \lambda^* = f(x^*, y^*) = \mu^* \geq f(x, y^*), \ \ \forall (x,y) \in \mathcal{S}_A \times \mathcal{S}_E,$$

therefore (x^*, y^*) is the saddle point we seek. It thus remain to show that $\lambda^* = \mu^*$.

To show that let $\epsilon > 0$. By the definitions of λ^*, μ^* if follows that $A_{\lambda^*+\epsilon} = \emptyset$ and $E_{\mu^*-\epsilon} = \emptyset$. By the definition of A_λ, we have that $A_{\lambda^*+\epsilon} = \emptyset$ implies that

$$\forall\, x \in \mathcal{S}_A, \exists\, y \in \mathcal{S}_E \,:\, f(x,y) < \lambda^* + \epsilon =: \lambda_1. \tag{17.3}$$

Similarly by the definition of E_μ, we have that $E_{\mu^*-\epsilon} = \emptyset$ implies that

$$\forall\, y \in \mathcal{S}_E, \exists\, x \in \mathcal{S}_E \,:\, f(x,y) > \mu^* - \epsilon =: \mu_1. \tag{17.4}$$

For a pair $(x,y) \in \mathcal{S}_A \times \mathcal{S}_E$ define now

$$\mathcal{S}_{A,y} := \{x \in \mathcal{S}_A \,:\, f(x,y) < \lambda_1\},$$
$$\mathcal{S}_{E,x} := \{y \in \mathcal{S}_E \,:\, f(x,y) > \mu_1\},$$

[20]By Weierstrass' theorem a continuous function on a compact set has a minimum and a maximum.

which by continuity of f are open sets. By (17.3) and (17.4) it follows that

$$\mathcal{S}_A \subset \bigcup_{y \in \mathcal{S}_E} \mathcal{S}_{A,y}, \quad \mathcal{S}_E \subset \bigcup_{x \in \mathcal{S}_A} \mathcal{S}_{E,x},$$

so that we may use the open sets $\mathcal{S}_{A,y}$, $\mathcal{S}_{E,x}$ for all possible x and y to construct open covers for S_A and S_E. But since S_A and S_E are compact sets, these open covers have finite subcovers. This means that there exist two finite sets $\{x_i\} \in \mathcal{S}_A$, $i = 1, \cdots, \ell_1$ and $\{y_j\} \in \mathcal{S}_E$, $j = 1, \cdots, \ell_2$ such that

$$\mathcal{S}_A \subset \bigcup_{j=1}^{\ell_2} \mathcal{S}_{A,y_j}, \quad \mathcal{S}_E \subset \bigcup_{i=1}^{\ell_1} \mathcal{S}_{E,x_i}$$

therefore

$$\min_j f(x, y_j) < \lambda_1, \ \forall x \in \mathcal{S}_A, \quad \max_i f(x_i, y) > \mu_1, \ \forall y \in \mathcal{S}_E. \tag{17.5}$$

Set

$$\phi_i(y) := \max(0, f(x_i, y) - \mu_1), \quad \psi_j(x) := \max(0, \lambda_1 - f(x, y_j)),$$

for $i = 1, \cdots, \ell_1$, $j = 1, \cdots, \ell_2$, and observe that by (17.5) it follows that

$$\sum_{i=1}^{\ell_1} \phi_i(y) > 0, \quad \sum_{j=1}^{\ell_2} \psi_j(x) > 0.$$

Define the map

$$\mathfrak{F} : (x, y) \mapsto \left(\frac{\sum_{i=1} \ell_1 \phi_i(y) x_i}{\sum_{i=1}^{\ell_1} \phi_i(y)}, \frac{\sum_{j=1} \ell_2 \psi_j(x) y_i}{\sum_{j=1} \ell_2 \psi_j(x)} \right),$$

which clearly is a continuous map $\mathfrak{F} : \mathcal{S}_A \times \mathcal{S}_E \to \mathcal{S}_A \times \mathcal{S}_E$. More importantly it is a continuous map of the set $co(\{x_i\}) \times co(\{y_j\})$ to itself where by co we denote the convex hull (or convex closure). Brouwer's fixed point theorem[21] guarantees the existence of a point $(x_*, y_*) \in co(\{x_i\}) \times co(\{y_j\})$ such that

$$x_* \sum_{i=1}^{\ell_1} \phi_i(y_*) = \sum_{i=1}^{\ell_1} \phi_i(y_*) x_i,$$

$$y_* \sum_{j=1}^{\ell_2} \psi_j(x_*) = \sum_{j=1}^{\ell_2} \psi_i(x_*) y_i.$$

[21]The Brouwer fixed point theorem states that any continuous function for a convex compact subset of a Euclidean space to itself has a fixed point.

For the i such that $\phi_i(y_*) > 0$ we have that $f(x_i, y_*) > \mu_1$. Taking convex combinations of these inequalities, recalling that x_* is in the convex hull of $\{x_i\}$ and using the concavity property of $f(x, y)$ with respect to the first variable we see that

$$f(x_*, y_*) > \mu_1.$$

Similarly, for the j such that $\psi_j(x_*) < 0$ we have that $f(x_*, y_j) < \lambda_1$. Following the same approach as above but now using the convexity of $f(x, y)$ with respect to the second variable we see that

$$f(x_*, y_*) < \lambda_1.$$

Combining these two inequalities $\mu_1 < \lambda_1$ therefore $\mu^* < \lambda^* + 2\epsilon$ for any $\epsilon > 0$ so that $\mu^* \leq \lambda^*$. That combined with the fact $\lambda^* \leq \mu^*$ guarantees $\lambda^* = \mu^*$ and the proof is complete. ■

One could argue that the concepts and techniques used in the proof are probably on the boundary of applied mathematics. At any rate we include it here and encourage the student to study it since applied mathematics may often have to resort to "pure" techniques and benefit considerably from them. As the eminent mathematician Henry Pollak once said, "There is no real distinction between pure mathematics and applied mathematics. There is only a difference between good mathematics and uninteresting mathematics." Furthermore, according to great geometer Nikolai Lobachevsky, "There is no branch of mathematics, however abstract, which may not someday be applied to the phenomena of the real world."

An important consequence of the minimax theorem is the following:

Theorem 17 *A Nash equilibrium always exists in the space of mixed strategies for finite games.*

Proof: The proof uses the von Neumann minimax theorem (see Theorem 16) and the convexity properties of the space of mixed strategies. ■

Remark 17.5.1 *Theorem 17 can be generalized, upon conditions, for games with more complicated strategy spaces.*

17.6 NASH EQUILIBRIA FOR MIXED STRATEGIES

Now let us consider a more general game where there are more than 2 players. Assume that $\mathcal{I} = \{1, 2, \cdots, I\}$ is the set of players. We adopt the following notation: By s_i we denote the strategy of player i and by s_{-i} the strategies of all the other players except i. Let $u_i(s_i, s_{-i})$ be the payoff of player i if she plays strategy s_i while the other players play strategy s_{-i}. The best reply of player i to the strategies of the other players is

$$BR_i(s_{-i}) = \arg\max_{s_i} u_i(s_i, s_{-i}).$$

Clearly, this is not necessarily a single-valued mapping, in general it is a correspondence.

Definition 21 *A strategy $s = (s_1, \cdots, s_i\}$ is a Nash equilibrium for this game if $s_i \in BR_i(s_{-i})$ for all $i \in \mathcal{I}$.*

In other words a Nash equlibrium is a best reply to itself.

The existence of Nash equilibrium may be shown using a general theorem due to Kakutani. This is a fixed point theorem for non-single-valued mappings (correspondences). Clearly, a correspondence may be considered as a set valued mapping. If S is a set we will use the notation 2^S for the power set, i.e., the set consisting of all the possible subsets of S.

Definition 22 *A correspondence has closed graph if for all sequences $\{x_n\}$ and $\{y_n\}$ such that $x_n \to x$ and $y_n \to y$ we have that $y_n \in f(x_n)$ implies $y \in f(x)$.*

The closed graph property is some type of continuity assumption for the correspondence.

Theorem 18 (Kakutani) *Let $S \subset \mathbb{R}^n$ be a nonempty, compact and convex set. Let $f : S \to 2^S$ be a correspondence (set valued map), which has closed, nonempty and convex graph. Then f has a fixed point, i.e., there exists an $x \in f$ (not necessarily unique) such that $x \in f(x)$.*

The existence of the Nash equilibrium for mixed strategies may then be obtained by an application of the Kakutani fixed point theorem.

Theorem 19 (Nash) *There always exist a Nash equilibrium for the mixed strategies.*

Proof: Assume, without loss of generality, that each player has n pure strategies. The mixed strategy space, for each player, is now the unit simplex Δ^{n-1} which clearly is nonempty, convex and compact. The total strategy space is $\Delta^{n-1} \times \cdots \times \Delta^{n-1}$ where we take the Cartesian product of the unit simplex with itself I times. Clearly this is also a nonempty, convex and compact set. The payoff function for each agent is the expected payoff which is a continuous function. By Weierstrass theorem, since this continuous function is defined on a compact set it achieves its maximum, therefore the best replies correspondences are nonempty. It can be seen that $BR_i(s_{-i})$ is convex. Indeed, let $s, \bar{s} \in BR_i(s_{-i})$. Then $\lambda\, s + (1-\lambda)\, \bar{s} \in BR_i(s_{-i})$ for all $\lambda \in [0, 1]$. Finally, we need to show that the best reply correspondence has the closed graph property. This is not a very easy task, it is guaranteed again by a general theorem called the Berge maximum theorem (see, e.g., Ref. [2]). According to this theorem since the payoff function is continuous and compact the best reply correspondence is upper hemicontinuous and this leads to the required closed graph property. Thus an application of the Kakutani fixed point theorem leads to the existence of a fixed point for the best reply correspondence $BR = (BR_1, \cdots, BR_I)$ which is the Nash equilibrium. ∎

17.7 COOPERATIVE GAMES

In many situations in social sciences and economics agents do not play for themselves but rather cooperate with other agents. This will only happen if it is to their benefit to do so. Assume that utility is transferable. Therefore, before addressing such situations we need to describe first the utility of the various possible coalitions between players, how these change as a function of the various possible strategies and then define a rule on how the common utility of a coalition is divided among the individuals. The above questions lead us to the mathematical modeling of cooperative games. For a complete introduction see, e.g., Ref. [6].

We first need to define the set of all players $\mathcal{I} = \{1, \cdots, I\}$. We then need to define the set of all possible coalitions. A coalition is a subset of $\mathcal{I}$, therefore the set containing all possible coalitions is the power set $2^{\mathcal{I}}$, the set of all possible subsets of $\mathcal{I}$. For a discrete set I we will denote by $card(I)$ its cardinality,[22] i.e., the number of elements in this set. Since $\mathcal{I}$ is a discrete set, the powerset $2^{\mathcal{I}}$ is also a discrete set of cardinality 2^I where $I = card(\mathcal{I})$ is the number of players. The set $\mathcal{I}$ is the largest possible coalition called the grand coalition and the empty set $\emptyset$ is called the empty coalition.

Definition 23 *A characteristic function is a function* $v : 2^{\mathcal{I}} \to \mathbb{R}$ *satisfying* $v(\emptyset) = 0$.

The characteristic function is a set valued function which assigns a real number to every coalition, which is called the worth of the coalition. This depending on the setup of the game will be considered either as loss or gain. The fact that the empty set is assigned the value 0 is just a convenient convention.

Definition 24 *The pair* $(\mathcal{I}, v)$ *is called a cooperative game in characteristic function form.*

The worth $v(C)$ of a coalition C has to be divided among the players. A division rule $a = (a_1, \cdots, a_I)$ is called payoff. A special case of payoff is an imputation.

Definition 25 *An imputation is a payoff* $a = (a_1, \cdots, a_I)$ *such that*

[22] Alternative notation for the cardinality of a set is either $|I|$ (which we avoid so as not to confuse the reader with absolute value) or $\sharp I$.

(i) $a_i \geq v(\{i\})$ *for all* $i = 1, \cdots, I$ *and*

(ii) $\sum_{i=1}^{I} a_i = v(\mathcal{I})$.

The first condition tells us that after joining the coalition each individual gets a share of the total earnings that is larger or equal to what she would get is she played on her own (this condition in called individual rationality), while the second is an efficiency or Pareto condition, stating that all the wealth obtained by the grand coalition is distributed to the individuals.

Rationality of a payoff may go beyond the individual level. In particular, an allocation is called coalitionally rational if $a(C) := \sum_{i \in C} a_i \geq v(C)$ for every coalition $C \subset 2^{\mathcal{I}}$, while it is called collectively rational if $a(\mathcal{I}) := \sum_{i \in C} a_i = v(\mathcal{I})$ (this coincides with the Pareto efficiency condition).

Definition 26 *An imputation* $a = (a_1, \cdots, a_I)$ *(effectively) dominates over an imputation* $b = (b_1, \cdots, b_I)$ *for the coalition* C*, denoted by* $a \succ_C b$*, if*

(i) $a_i > b_i$ *for all* $i \in C$ *and*

(ii) $\sum_{i \in C} a_i \leq v(C)$.

We will say that an imputation a *(effectively) dominates over an imputation* b *if* $a \succ_C b$ *for some coalition* C.

Clearly, if it was not for the second condition the above definition would not be possible.

The behavior of the agents, i.e., whether they will join a particular coalition or not depends on the form of the characteristic function. The following definitions are used.

Definition 27 *A game* $(\mathcal{I}, v)$ *is called*

(i) superadditive if $v(C_1 \cup C_2) \geq v(C_1) + v(C_2)$ *for every* $C_1, C_2 \subset 2^{\mathcal{I}}$ *such that* $C_1 \cap C_2 = \emptyset$,

(ii) convex if $v(C_1) + v(C_2) \leq v(C_1 \cup C_2) + v(C_1 \cap C_2)$ *for all* $C_1, C_2 \subset 2^{\mathcal{I}}$,

(iii) monotone if $C_1 \subseteq C_2 \subseteq 2^{\mathcal{I}}$ *implies* $v(C_1) \leq v(C_2)$ *(or the opposite inequality).*

A game $(\mathcal{I}, v)$ is called subadditive if $(\mathcal{I}, -v)$ is superadditive, and concave if $(\mathcal{I}, -v)$ is convex.

Definition 28 *A collection of coalitions* $\{C_k\}$ *is called balanced if there exist* $\lambda_k \in [0, 1]$ *such that for every* $i \in \mathcal{I}$, $\sum_{k: i \in C_k} \lambda_k = 1$ *(the numbers* $\{\lambda_k\}$ *are called balancing weights).*
A game $(\mathcal{I}, v)$ *is called balanced if for every balanced collection of coalitions* $\{C_k\}$ *with balancing weights* $\{\lambda_k\}$ *it holds that*

$$\sum_k \lambda_k v(C_k) \leq v(\mathcal{I}).$$

A convex game is superadditive. Superadditivity may be interpreted as that the coalition $C_1 \cup C_2$ is worth more in terms of the characteristic function than the groups C_1 and C_2 on their own. Therefore, if v represents gain, it is to their mutual benefits for the two coalitions C_1, C_2 to unite to a larger coalition $C_1 \cup C_2$. The notion of subadditivity has a similar meaning if the characteristic function represents loss.

Definition 29 *Let $\mathfrak{I}(\mathcal{I}, v)$ be the set of imputations for a game. The core of the game is defined as*

$$\mathfrak{C}(\mathcal{I}, v) := \{a \in \mathfrak{I}(\mathcal{I}, v) \; : \; \sum_{i \in C} a_i \geq v(C) \; \forall C \subset \mathcal{I}\}.$$

The core is defined in such a way that no coalition can improve upon the allocation a to its members.[23] Therefore, if we are in the core no member is willing to leave the core and join another coalition. An allocation is in the core if it is efficient and collectively (coalitionally) rational. It may also be seen that the core of a game $\mathfrak{C}(\mathcal{I}, v)$ is the set of all collectively rational payoffs. Alternatively, the core for a game is the set of imputations that are not dominated for any coalition.

The core may be empty as the next example illustrates.

■ EXAMPLE 17.10 Simple Majority Game

Assume the simple majority game with 3 players. A coalition is winning only if it contains at least two players. In terms of the characteristic function this gives $v(\{i\}) = 0$ for all $i \in \mathfrak{I} = \{1, 2, 3\}$ and $v(C) = 1$ if $card(C) \geq 2$. For this game $\mathfrak{C}(\mathcal{I}, v) = \emptyset$ as there is no imputation for which $\sum_{i \in C} a_i \geq v(C) \; \forall C \subset \mathcal{I}$ (unless $v(\mathcal{I}) \geq \frac{3}{2}$).

The non-emptyness of the core is guaranteed by the following theorem (see, e.g., Ref. [3] or [9].

Theorem 20 (Shapley-Bondareva) *The core of a cooperative game $(\mathcal{I}, v)$ is non empty if and only if it is balanced.*

Other types of solutions, more general, and thus easier to exist, are possible. The following notion is due to Von Neumann.

Definition 30 (Stable sets) *A stable set $\mathfrak{S}(\mathcal{I}, v)$ for a cooperative game $(\mathcal{I}, v)$ is the set of imputations, such that:*

(i) if $a, b \in \mathfrak{S}(\mathcal{I}, v)$ then neither $a \succ b$ nor $b \succ a$ (internal stability);

(ii) if $c \notin \mathfrak{S}(\mathcal{I}, v)$ then there exists $a \in \mathfrak{S}(\mathcal{I}, v)$ such that $a \succ c$ (external stability).

[23] If $\sum_{i \in C} a_i < v(C)$ for a coalition C then the members of C could improve their payoffs.

Remark 17.7.1 *The following alternative interpretation of the core and the stable set are useful: For any* $\mathfrak{X} \subseteq \mathfrak{I}(\mathcal{I}, v)$ *define*

$$D(\mathfrak{X}) := \{a \in \mathfrak{I}(\mathcal{I}, v) \ : \ a \succ b \ for\ some b \in \mathfrak{X}\}.$$

A stable set is a set $\mathfrak{S}(\mathcal{I}, v)$ *such that (i)* $\mathfrak{S}(\mathcal{I}, v) \cap D(\mathfrak{S}(\mathcal{I}, v)) = \emptyset$ *and (ii)* $\mathfrak{S}(\mathcal{I}, v) \cup D(\mathfrak{S}(\mathcal{I}, v)) = \mathfrak{I}(\mathcal{I}, v)$ *where these two conditions are the internal and external stability conditions, respectively. These two conditions can be expressed as one condition* $\mathfrak{S}(\mathcal{I}, v) = \mathfrak{I}(\mathcal{I}, v) \setminus D(\mathfrak{S}(\mathcal{I}, v))$*, which in fact is a fixed point condition for the set valued map* $f(\mathfrak{X}) = \mathfrak{I}(\mathcal{I}, v) \setminus D(\mathfrak{X})$*. On the other hand, the core for a game can be expressed as* $\mathfrak{C}(\mathcal{I}, v) = \mathfrak{I}(\mathcal{I}, v) \setminus D(\mathfrak{I}(\mathcal{I}, v))$.

The stable set may not be unique, but typically it is easier for a game to have a stable set than a core. It is clear that $\mathfrak{C}(\mathcal{I}, v) \subseteq \mathfrak{S}(\mathcal{I}, v) \subseteq \mathfrak{I}(\mathcal{I}, v)$. This means that the core is contained in every stable set.

■ EXAMPLE 17.11

The set $\mathfrak{S}(\mathcal{I}, v) = \{(\frac{1}{2}, \frac{1}{2}, 0), (\frac{1}{2}, 0, \frac{1}{2}), (0, \frac{1}{2}, \frac{1}{2})\}$ is a stable set for the simple majority game with three persons. However, this is not the only stable set. For instance, the set $\{(x_1, 1 - x_1 - x_3, x_3)\}$ for any $x_1, x_3 \in [0, \frac{1}{2})$ is also a stable set.

Definition 31 *The value of a cooperative game is an operator* $\phi = (\phi_1, \cdots, \phi_I) : \mathbb{S} \to \mathbb{R}^I$ *assigning a payoff to each player in the game for a particular characteristic function* v.

Definition 32 *The Shapley value for a game is a mapping* $\phi = (\phi_1, \cdots, \phi_I) : 2^{\mathcal{I}} \to \mathbb{R}^I$ *defined by*

$$\phi_i(v) = \sum_{C \in \mathcal{I} \setminus \{i\}} \frac{card(C)!(I - card(C) - 1)!}{I!} (v(C \cup \{i\}) - v(C)) .$$

Remark 17.7.2 *The factor* $\frac{card(C)!(I-card(C)-1)!}{I!}$ *has a probabilistic interpretation.* $card(C)!$ *is the total number of ways that the set* C *can be formed prior to player's* i *addition to it.* $(I - card(C) - 1)!$ *is the number of ways the remaining players can be added to* C *afterwards.* $I!$ *is the total number of all the orderings of the players.*

Remark 17.7.3 *An equivalent definition of the Shapley value is as*

$$\phi_i(v) = \sum_{C : i \in C} \frac{(card(C) - 1)!(I - card(C))!}{I!} (v(C) - v(C \setminus \{i\})) ,$$

where now the summation is over all the coalitions containing player i*. The prefactor* $\frac{(card(C)-1)!(I-card(C))!}{I!}$ *has a probabilistic interpretation. It is the* *probability that when i enters she will find the coalition $C \setminus \{i\}$ already there. The numerator is the number of ways in which the $card(C) - 1$ members of $C \setminus \{i\}$ come [i.e., the total number of ways that the coalition $C \setminus \{i\}$ first forms $(card(C) - 1)!$] and this is multiplied in the ways that player i and the remaining $I - card(C)$ players enter [there are $(I - card(C))!$ ways for that]. The denominator is the total number of permutations of I players ($I!$).*

■ EXAMPLE 17.12

Consider 3 players. Take player 1. All the coalitions not containing 1 are $\{2\}$, $\{3\}$, $\{2,3\}$. Assume that player 1 joins them with equal probability. The probability of when 1 joins she will find, e.g., $\{2\}$ already there is $\frac{(1-1)!(3-1)!}{3!} = \frac{1}{3}$. This is because there are 2 coalitions of 1 player (not containing player 1) and in a total of 6 and she will join each one with equal probability which is $\frac{1}{3}$. The probability of 1 when joining finding, e.g., $\{2,3\}$ already there is $\frac{(2-1)!(3-2)!}{3!} = \frac{1}{6}$. This is because there is only 1 coalition of 2 players not containing 1 and the probability of choosing that out of 6 possible coalitions is $\frac{1}{6}$.

The Shapley value is a concept of value for a cooperative game that satisfies certain axioms which are considered as reasonable modelling assumptions. These axioms are (i) symmetry [i.e., symmetric players – players i and j with the same marginal contribution $v(C \cup \{i\}) = v(C \cup \{j\})$ for all $C \subseteq \mathcal{I}$) are assigned the same Shapley value), (ii) dummy players (i.e. players i such that $v(C \cup \{i\}) = v(C)$ for all $C \subseteq \mathcal{I}$] are assigned zero value, (iii) additivity and (iv) efficiency, $\sum_{i \in \mathcal{I}} \phi_i(v) = v(\mathcal{I})$. Shapley in a seminal paper published in 1953 has proved that the Shapley value is the unique operator satisfying the above properties.

Theorem 21 *If $(\mathcal{I}, v)$ is a convex game then $\phi(v) \in \mathfrak{C}(\mathcal{I}, v)$.*

■ EXAMPLE 17.13 Measurement of Voting Power

Consider the simple majority game with characteristic function v as defined in Example 17.10. To calculate the Shapley value for, e.g., agent 1 we need to find all coalitions not including agent 1; these are $\{2\}$, $\{3\}$ and $\{2,3\}$. Then, the Shapley value is obtained as a sum over these 3 coalitions,

$$\begin{aligned}
\phi_1(v) &= \frac{1!(3-1-1)!}{3!}(v(\{1,2\}) - v(\{2\})) \\
&+ \frac{1!(3-1-1)!}{3!}(v(\{1,3\}) - v(\{3\})) \\
&+ \frac{2!(3-2-1)!}{3!}(v(\{1,2,3\}) - v(\{2,3\})) = \frac{1}{3}.
\end{aligned}$$

A similar calculation shows that $\phi_2(v) = \phi_3(v) = \frac{1}{3}$. That means that in this simple majority game every player has exactly the same power. This would not be true if for example one of the players had veto power.

■ EXAMPLE 17.14 Shapley Value for Dummy and Veto Players

If a player i is a dummy player, i.e., a player that does influence any coalition by joining it will get $\phi_i = 0$. If a player i is a veto player, i.e., a player that must be included in ay winning coalition then such a player will obtain a large Shapley value.

■ EXAMPLE 17.15 Cost Allocation Games

Assume that the full cost of a common facility (common good) is to be shared between I countries (players). The set of countries is denoted by $\mathcal{I} = \{1, \cdots, I\}$. The countries may cooperate and form coalitions in order to secure the common good. Let $v(C)$ be the benefit of the coalition of countries C. We may assume that this benefit can be expressed in terms of income and this total income will then be transferred to the members of the coalition as some sort of side payment. If B_i is the net benefit of country i from exploiting the common facility then the charge (participation cost) of i in the common project can be given as $C_i = B_i - \phi_i(v)$ where ϕ is the Shapley value. This can be supported by the interpretation of the Shapley value as the average (expected) marginal benefit of a country i by participating in a coalition assuming that coalitions are chosen in random and with equal probability. In other words, the Shapley value assigns to a country the expected contribution she is to have to a coalition.

The Shapley value has an interesting interpretation as an expected utility function as has been shown by Roth [7]. We need to introduce the following concepts first. A position in a game is a pair (i, v) where i is a player and v is a game. A player has preferences over mixtures of games (i.e., she may choose to play a game with a probability). In other words a player is choosing over lotteries in the space of games. The preference relation is assumed to satisfy 4 axioms:

(i) Let v be a game in which i is a dummy player. Then $(i, v) \sim (i, v_0)$ (where $\sim$ denotes indifference and v_0 is the null game assigning zero value to every coalition). Furthermore, $(i, v_i) \succ (i, v_0)$ where v_i is the game in which i is a dictator ($v(C) = 1$ if $i \in C$ and $v(C) = 0$ otherwise).

(ii) For any game v and permutation π, $(i, v) = (\pi(i), \pi(v))$ (symmetry).

(iii) $(i, (pw+(1-p)v)) \sim [p(i,w);(1-p)(i,v)]$ where the second term denotes the lottery where (i, w) occurs with probability p and (i, v) occurs with probability $1 - 0$.

(iv) $(i, v_R) \sim (i, (1/r)v_i)$ where v_R is defined by $v_R(C) = 1$ if $C \subset R$ and 0 otherwise[24].

We then have the following:

Theorem 22 *If u is an expected utility function over positions in games satisfying the above 4 axioms and normalized so that $u(i, v_i) = 1$ and $u(i, v_0) = 0$ then $u(i, v) = \phi_i(v)$ where ϕ is the Shapley value.*

17.8 CONCLUSION

In this short introduction we have tried to initiate the student to the fundamental notions of game theory and how it may be applied to understand and model phenomena in the social sciences.

Our presentation provided a glimpse of the major aspects in the field, trying to balance mathematics with social science, and of course has had to omit many interesting and exiting aspects of game theory. However, it is our hope that this short introduction will provide an incentive to some of the students reading this book to study this field into more detail.

Acknowledgments

The authors wish to thank Professor Nicholas Yannacopoulos for his constructive comments and enjoyable conversations that maximized their utility (in the Benthamian use of the concept). Furthermore, they acknowledge the useful comments of the three referees that led to a considerable improvement of the chapter. Last (but not least!) they also wish to thank Dr. Yang for putting this effort together and offering them the tribune from which to present this first introduction to this exciting subject.

EXERCISES

17.1 Work out the conservative strategies for the examples of the matrix games presented in this chapter and show that they coincide with the Nash equilibria.

17.2 Write down the characteristic function for a voting game with a veto player and work out the Shapley value for a simple example (3 or 4 voters).

[24] v_R can be interpreted as a unanimity game (recall that v_i is a game in which i is a dictator).

17.3 Let $\mathcal{I}$ be a set of agents each one with utility u_i. Suppose that utilities are transferable. Show that a possibility in defining a characteristic function for this game is by

$$v(C) = \min_{s_{\mathcal{I}\setminus C}} \max_{s_C} \sum_{i \in C} u_i(s_C, s_{\mathcal{I}\setminus C}).$$

REFERENCES

1. Aubin, J. P., *Optima and Equilibria: An Introduction to Nonlinear Analysis*, Springer, Berlin (1993).
2. Berge, C., *Topological Spaces: Including a Treatment of Multi-Valued Functions, Vector Spaces and Convexity*, Dover (2010).
3. Bondareva, O. N., Some applications of linear programming methods to the theory of cooperative games (In Russian). Problemy Kybernetiki, 10: 119-139 (1963).
4. McCarty, N. M. and Meirowitz, A., *Political Game Theory: An Introduction*, Cambridge University Press, Cambridge (2007).
5. Nikaidō, H., On von Neumann's minimax theorem, Pacific J. Math., 4:65–72 (1954).
6. Peleg, B. and Sudhölter, P., *Introduction to the Theory of Cooperative Games*, 2nd Ed., Springer, Heidelberg (2003).
7. Roth, A. E., The Shapley value as a von Neumann–Morgenstern utility, Econometrica, 45:657–664 (1977).
8. Osborne, M. J. and Rubinstein, A., *A Course in Game Theory*, The MIT Press (1994).
9. Shapley, L. S., On balanced sets and cores. Naval Research Logistics Quarterly, 14:453–460 (1967).

CHAPTER 18

CONTROL PROBLEMS ON DIFFERENTIAL EQUATIONS

CHUANG ZHENG

School of Mathematical Science, Beijing Normal University, China

18.1 INTRODUCTION

Control theory is an interdisciplinary branch of engineering and mathematics that deals with the behavior of dynamical systems. It is originated and has been strongly inspired by industrial applications such as airplanes, chemical plants, space vehicles and so on (see, for instance, [7], [19]). Control theory addresses a variety of different problems. The usual objective of control is to calculate solutions for the proper corrective action from the controller that results in system stability, that is, the system will hold the set point and not oscillate around it. One simple example is a car's cruise control, which is a device designed to maintain vehicle speed at a constant desired speed provided by the driver. However, not every system can be controlled. In addition to knowing the way to design controls, it is necessary to understand

Mathematical Modeling with Multidisciplinary Applications.
Edited by Xin-She Yang

the inner property of the system. Hence, one natural and significant concept is that of *controllability*. It can be formulated, roughly, as follows. Consider an evolution system [either described in terms of Ordinary or Partial Differential Equations (ODE/PDE)]. We are allowed to act on the trajectories of the system by means of a suitable control (the right-hand side of the system, the boundary conditions, etc.). Then, for a given time interval $t \in (0, T)$ and initial and final states, we have to find a control such that the solution matches both the initial state at time $t = 0$ and the final one at time $t = T$. To do that, we need to know whether there exists such a control. If the control does exist, the system is *controllable*. There is a large literature on these topics. The foundations of finite-dimensional control theory were established by R. E. Kalman [8,9] and since then, the theory has been greatly generalized, first to linear and nonlinear infinite dimensional systems and to stochastic systems, etc. (see Refs. [4,6,11–13,15,16] and the references therein). We refer, for instance, to the book by Lee and Marcus [10] for an introduction to those problems in the context of finite-dimensional systems. We also refer to the survey paper by Russell [15] and to the book of Lions [11] (also to his survey paper [12]) for an introduction to the controllability of PDEs, and to more recent works [14], [18], [17] and [20].

In 1988, Lions [12] introduced the so-called Hilbert Uniqueness Method (shorted as HUM). Roughly speaking, it is based on the principle that, whenever a system is controllable, the control can be built by minimizing a suitable quadratic functional defined on the class of solutions of the adjoint system. Suitable variants of this functional allow building controls of minimal L^2-norm, bang-bang controls, approximate controls, etc. The details will be shown in Section 18.3.3. The main difficulty when minimizing these functionals is to show that they are coercive. This turns out to be equivalent to the so-called observability property for the adjoint equation, which provides global estimates on the adjoint state everywhere in terms of partial measurements.

There are many journal papers and books studying control problems of differential equations. Due to limited space, in this chapter we address only some basic topics related to the controllability of ODEs/PDEs. Section 18.2 could be seen as an introduction to control problems in the context of finite dimensional systems (ODEs). We first establish an example of modeling an RLC series circuit with the electric current as the controller. Then we give a mathematical definition of controllability and show the Kalman condition, which is a criteria for determining whether the system is controllable or not. In Section 18.3, we show another example of modeling a vibration system and show the corresponding control problem in the context of infinite dimensional systems (PDEs). We reestablish the controllability conceptions and show the equivalence between the controllability of the controlled system and the observability of its adjoint system. Hence, the controllability problem can be transferred to a problem whether the solutions of its adjoint system satisfy some kind of inequalities, which is easy to deal with in the mathematical point of view. Although the definitions are similar, we emphasize that finite dimensional and infinite dimensional systems may have quite different properties from a control theoretical point of view. Interested readers can see the recent survey paper [19] and references therein for more details.

18.2 ORDINARY DIFFERENTIAL EQUATIONS

In the latter 18th century, the steam engine led to the Industrial Revolution and at its core was the centrifugal governor, which was designed by James Watt in 1788. The centrifugal governor is an automatic control device to maintain the normal operation of the steam engine: As the speed of the prime mover increases/decreases, it drives some kind of mechanical device such that the speed of the prime mover decreases/increases by reducing/improving the rate of the working-fluid entering the steam engine. Hence, the centrifugal governor finishes the control of maintaining the speed of the steam engine.

However, intuitive physical phenomena is not sufficient to study and analyze the control system and its dynamical characteristic. Mathematical modeling is necessary to describe the control system quantitatively. In general, instead of considering the mathematical modeling as a part of the control theory, people prefer to it as a part of other scientific subjects. Therefore, most of the materials we contribute in this section can be found elsewhere.

We emphasize that unlike the normal mathematical modeling process, the control device is designed by the designer. Hence, the corresponding mathematical description of the control device can be designed even with the same physical model. This is the key point the readers have to keep in mind when reading this part: read every material from the control point of view.

18.2.1 Model Formulation

We first show an example of a controller on RLC series circuit. The an RLC series circuit is an electrical circuit consisting of a resistor R, an inductor L, and a capacitor C, connected in series. In this circuit, the three components are all in series with the voltage source. The governing differential equation can be found by substituting into Kirchhoff's voltage law (KVL), the constitutive equation for each of the three elements. From KVL, we have

$$L\frac{di(t)}{dt} + Ri(t) + \frac{1}{C}\int_{-\infty}^{t} i(\tau)d\tau = e_i(t), \tag{18.1}$$

with

$$\frac{1}{C}\int_{-\infty}^{t} i(\tau)d\tau = e_C(t). \tag{18.2}$$

Now we consider this RLC circuit as a system with input function $e_i(t)$ and

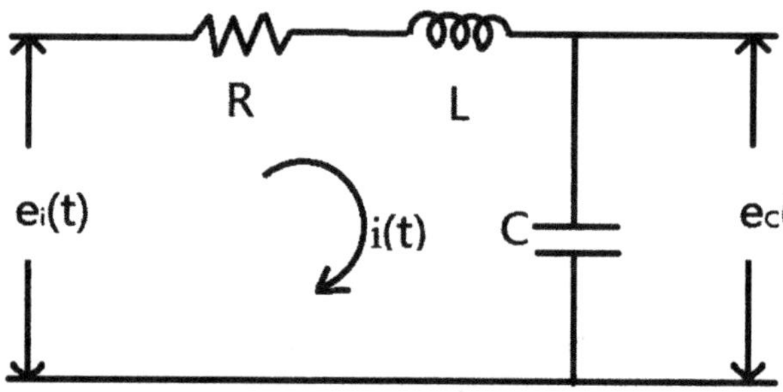

Figure 18.1 RLC series circuit with controller $e_i(t)$.

output $e_C(t)$. Setting

$$u(t) = e_i(t), \qquad y(t) = e_C(t), \qquad x_1(t) = \int_{-\infty}^{t} i(\tau)d\tau, \qquad x_2(t) = i(t),$$

system (18.1) and (18.2) corresponds to the following controlled system

$$\begin{cases} \dfrac{dx_1(t)}{dt} = x_2(t), \\ \\ \dfrac{dx_2(t)}{dt} = -\dfrac{1}{LC}x_1(t) - \dfrac{R}{L}x_2(t) + \dfrac{1}{L}u(t), \end{cases} \tag{18.3}$$

with the observer

$$y(t) = \frac{1}{C}x_1(t). \tag{18.4}$$

We can determine the charge and the current $x_1(t)$ and $x_2(t)$ during the time variation, once we give the input voltage $u(t)$, the charge and the current at t_0, i.e., $x_1(t_0)$ and $x_2(t_0)$. We say $x_1(t)$ and $x_2(t)$ are *states* of the system (18.3). Moreover, Eq. (18.3) is the *state equation* of the RLC system. (18.4) is the corresponding *observation equation*, which shows the relationship between the input function and the state.

We now rewrite system (18.3) and (18.4) by means of vectors. Denote by

$$\mathbf{x}(t) = \begin{pmatrix} x_1(t) \\ x_2(t) \end{pmatrix}.$$

System (18.3) and (18.4) can be written by

$$\frac{d\mathbf{x}(t)}{dt} = \begin{bmatrix} 0 & 1 \\ -\frac{1}{LC} & -\frac{R}{L} \end{bmatrix} \mathbf{x}(t) + \begin{bmatrix} 0 \\ \frac{1}{L} \end{bmatrix} u(t), \qquad (18.5)$$

$$y(t) = \begin{bmatrix} \frac{1}{C} & 0 \end{bmatrix} \mathbf{x},$$

respectively.

On the other hand, if we choose $x_1(t) = i(t), x_2(t) = e_C(t)$ as the states of the RLC series circuit, instead of $\int_{-\infty}^{t} i(\tau)d\tau$ and $i(t)$, the state equation of the system could be written by

$$\frac{d\mathbf{x}(t)}{dt} = \begin{bmatrix} -\frac{R}{L} & -\frac{1}{L} \\ \frac{1}{C} & 0 \end{bmatrix} \mathbf{x}(t) + \begin{bmatrix} \frac{1}{L} \\ 0 \end{bmatrix} u(t), \qquad (18.6)$$

with the observation equation

$$y(t) = [0 \quad 1]\,\mathbf{x}.$$

The example shows that the state equations can be different even though they describe the same dynamical system, due to the choices of the state variables. In other words, even though the state equation is not unique, the relationship between the control and the output is eternal for the same controlled system, no matter what kind of state variable is chosen.

In general, suppose that the controlled system has m inputs $u_1, u_2, \cdots, u_m$, l outputs $y_1, y_2, \cdots, y_l$ and n variables $x_1, x_2, \cdots, x_n$. Denote by

$$\mathbf{u} = \begin{pmatrix} u_1 \\ u_2 \\ \vdots \\ u_m \end{pmatrix}, \qquad \mathbf{y} = \begin{pmatrix} y_1 \\ y_2 \\ \vdots \\ y_l \end{pmatrix}, \qquad \mathbf{x} = \begin{pmatrix} x_1 \\ x_2 \\ \vdots \\ x_n \end{pmatrix},$$

the system can be formulated as follows:

$$\frac{d\mathbf{x}(t)}{dt} = \mathbf{f}(\mathbf{x}(t), t, \mathbf{u}(t)),$$

where each component of $\mathbf{f}$ could be a nonlinear function. The observation equation has the form

$$\mathbf{y}(t) = \mathbf{g}(\mathbf{x}(t), t, \mathbf{u}(t)).$$

In control theory, nowadays the most well developed controlled system has the form

$$\begin{cases} \dfrac{d\mathbf{x}(t)}{dt} = A(t)\mathbf{x}(t) + B(t)\mathbf{u}(t), \\ \mathbf{y}(t) = C(t)\mathbf{x}(t) + D(t)\mathbf{u}(t), \end{cases} \tag{18.7}$$

where $A(t), B(t), C(t)$ and $D(t)$ are $n \times n, n \times m, l \times n$ and $l \times m$ matrices, respectively. System (18.7) is the so-called *linear system*.

From the above discussion we find out that if even most of the specific control systems are nonlinear, they can be described by linear systems (or linear constant-coefficient systems). In this chapter, we will introduce some basic tools of ordinary differential equations to develop the controllability property of the corresponding control systems.

18.2.2 Controllability

As we mentioned before, the state equation of the system can be represented as first-order linear ordinary differential equations. In the simplified case when A and B are invariant matrices, the nonhomogeneous equation is given by

$$\begin{cases} \dfrac{d\mathbf{x}(t)}{dt} = A\mathbf{x}(t) + B\mathbf{u}(t), t \in (t_0, t_1), \\ \mathbf{x}(t_0) = \mathbf{x}_0. \end{cases} \tag{18.8}$$

In (18.8), A is a real $n \times n$ matrix, B is a real $n \times m$ matrix and $\mathbf{x}_0$ a vector in $\mathbb{R}^n$. The function $\mathbf{x} : [t_0, \infty) \longrightarrow \mathbb{R}^n$ represents the *state* and $\mathbf{u} : [t_0, \infty) \longrightarrow \mathbb{R}^m$ represents the *control*. Both are vector functions of n and m components, respectively, depending exclusively on time t. Obviously, in practice $m \leq n$. The most desirable goal is, of course, controlling the system by means of a minimum number of m controls.

To start with, we first give the definition of e^{At}, which is defined by the Series Expansion method:

$$e^{At} = I + At + \frac{A^2t^2}{2!} + \frac{A^3t^3}{3!} + \cdots + \frac{A^nt^n}{n!} + \cdots. \tag{18.9}$$

Here we omit the proof of the convergence of the right-hand side of (18.9) and assume that A is a constant $n \times n$ matrix.

Most of the properties of e^{At} are the same as e^{at} with a constant a. We will not enter in all of the details since they can be found in any classical textbook of ordinary differential equations (for instance, Ref. [1]).

The algorithm of computing e^{At} is given as follows:

1. Obviously, it is necessary to find the series $A^2, A^3, \cdots$. We apply the method to transform A to a Jordan normal form:

Let P be a normal matrix. From the definition of e^{At}, the following equality holds (see Exercise 1):

$$P^{-1}e^{At}P = e^{P^{-1}APt}.$$

Applying the above property, we compute e^{At} as follows.

Step 1 Choose an appropriate P such that A turns to be a Jordan normal form, i.e.,

$$P^{-1}AP = J = \begin{pmatrix} J_1 & & & \mathbf{0} \\ & J_2 & & \\ & & \ddots & \\ \mathbf{0} & & & J_q \end{pmatrix},$$

with

$$J_i = \begin{pmatrix} \lambda_i & 1 & 0 & \cdots & 0 \\ 0 & \lambda_i & 1 & \cdots & 0 \\ & & \cdots\cdots & & \\ 0 & 0 & 0 & \cdots & 1 \\ 0 & 0 & 0 & \cdots & \lambda_i \end{pmatrix}, \quad i = 1, 2, \cdots, q.$$

Step 2 The matrix Jt has the form

$$Jt = \begin{pmatrix} Q_1 & & & \mathbf{0} \\ & Q_2 & & \\ & & \ddots & \\ \mathbf{0} & & & Q_q \end{pmatrix},$$

with

$$Q_i = \begin{pmatrix} e^{\lambda_i t} & te^{\lambda_i t} & \cdots & \frac{t^{l_i-1}}{(l_i-1)!}e^{\lambda_i t} \\ 0 & e^{\lambda_i t} & \cdots & \frac{t^{l_i-2}}{(l_i-2)!}e^{\lambda_i t} \\ & \cdots\cdots & & \\ 0 & 0 & \cdots & e^{\lambda_i t} \end{pmatrix}, \quad i = 1, 2, \cdots, q.$$

Step 3 Implement the matrix e^{At} with the formula $e^{At} = Pe^{Jt}P^{-1}$.

Characterized by the variation of constants formula, the solution of system (18.8) has the form

$$\mathbf{x}(t) = e^{A(t-t_0)}\mathbf{x}_0 + \int_{t_0}^{t} e^{(t-\tau)A}B\mathbf{u}(\tau)d\tau, \tag{18.10}$$

where e^{At} represents a matrix of an exponential function. The exact form of the solution and the way to compute it can be found in the exercises of this chapter.

In this section, we are interested in the exact controllability of system (18.8), which is stated as follows: Let $\mathbf{x}_0$ and $\mathbf{x}_1$ be two given points in $\mathbb{R}^n$ and $t_1 > t_0$, so is there a function $\mathbf{u}(x,t)$, such that the solution of system (18.8) satisfies $\mathbf{x}(t_0) = \mathbf{x}_0$ and $\mathbf{x}(t_1) = \mathbf{x}_1$? If such a control $\mathbf{u}$ exists, we say that the system is *exactly controllable* from $\mathbf{x}_0$ to $\mathbf{x}_1$ at time t_1 by the control $\mathbf{u}$.

In controlled system (18.8), A represents the inherent structure of the physical phenomena and B is designed by a human being. As we mentioned before, in applications it is desirable to make the number of controls m to be as small as possible. We hope it can be done by setting an appropriate control mechanism B. Later, we will show an example in which n components of the state can be controlled with one control only. However, in order to achieve this goal, B needs to be chosen in a strategic way depending on the matrix A.

Let us look at two examples. The first one is the previous model we established for the RLC circuit. Both components will be controlled by means of a scalar control. In the second one controllability does not hold since one of the components of the system is insensitive to the control.

■ EXAMPLE 18.1

In system (18.5), we set the parameters $R = 0$ and $L = C = 1$, and then it can be written as

$$\frac{d\mathbf{x}(t)}{dt} = A\mathbf{x}(t) + B\mathbf{u}(t), \tag{18.11}$$

where the matrices are now, respectively,

$$A = \begin{pmatrix} 0 & 1 \\ -1 & 0 \end{pmatrix}, \qquad B = \begin{pmatrix} 0 \\ 1 \end{pmatrix},$$

or, equivalently,

$$\begin{cases} x_1' &= x_2, \\ x_2' &= -x_1 + u. \end{cases} \tag{18.12}$$

Here we only have one control u for both components x_1 and x_2. In other words, we hope to control the charge and the current during the time variation with the input voltage. On one hand, the input voltage u does not appear directly in the first equation of (18.12), which seems to be bad for controlling x_1. On the other hand, both components are present in the second equation with the control, which is desirable. Therefore we cannot conclude immediately whether system (18.12) is controllable or not. Fortunately, the answer is positive. Indeed, given some arbitrary initial and final data, (x_1^0, x_2^0) and (x_1^1, x_2^1), respectively, it is easy to choose a cubic polynomial function $z = z(t)$ such that

$$\begin{cases} z(t_0) = x_1^0, & z(t_1) = x_1^1, \\ z'(t_0) = x_2^0, & z'(t_1) = x_2^1. \end{cases} \tag{18.13}$$

We can then define $u = z'' + z$ as being the control.

■ **EXAMPLE 18.2**

In system (18.11), we now directly set

$$A = \begin{pmatrix} a & 0 \\ 0 & b \end{pmatrix}, \quad B = \begin{pmatrix} c \\ 0 \end{pmatrix}$$

where a, b, c are constants. Obviously the solution has the form

$$\begin{cases} x_1 &= e^{a(t-t_0)}x_1^0 + c\int_{t_0}^t e^{a(t-s)}u(s)ds, \\ x_2 &= x_2^0 e^{b(t-s)}, \end{cases} \tag{18.14}$$

where $x(t_0) = (x_1^0, x_2^0)$ are the initial data at $t = t_0$.

The system is not controllable since the control u does not act on the second component x_2 and it is completely determined by the initial data x_2^0. Hence the system is not controllable.

18.2.3 Kalman's Rank Condition

In this section, we will show the classical Kalman's rank condition, which completely solves the controllability problem of finite dimensional linear systems. The theorem shows a simple fact: the system is exactly controllable if and only if some kind of matrix deduced by the system (P here) is of full rank. In other words, the Kalman's rank condition gives us a criteria to determine whether the system is controllable or not. The proof is theoretical and technical, and readers who are not interested in this could skip this part and go to the theorem directly. Anyway, to understand the proof, readers only need linear algebra and a little patience.

Theorem 23 (Kalman's rank condition) *System (18.8) is exactly controllable in some time t_1 if and only if the rank of the matrix*

$$P = [B, AB, \cdots, A^{n-1}B]$$

is n. Consequently, if system (18.8) is exactly controllable in some time t_1, it is exactly controllable in any time.

To prove the above theorem, we need the following lemma:

Lemma 18.2.1 *Suppose $K(t), \mathbf{u}(t)$ are $n \times r$ matrix and r-dimensional vector, respectively. Each element of them is a continuous function on the interval $[t_1, t_2]$. Then for any n-dimensional vector $\mathbf{a}$, the following two assertions are equivalent:*

- *There exists a $\mathbf{u}(t)$ such that $\int_{t_0}^{t_1} K(t)\mathbf{u}(t)dt = \mathbf{a}$;*
- *$\mathbf{x} = 0$ is the only n-dimensional vector such that $\mathbf{x}^T K(t) = 0, \forall\ t \in [t_1, t_2]$.*

Proof: We first state that $\mathbf{x}^T K(t) = 0, \forall\ t \in [t_1, t_2]$ is equivalent to that $\mathbf{x}^T \int_{t_0}^{t_1} K(t)\mathbf{u}(t)dt = 0$ holds for all $\mathbf{u}(t)$.

In fact, if $\mathbf{x}^T K(t) = 0$, obviously

$$\mathbf{x}^T \int_{t_0}^{t_1} K(t)\mathbf{u}(t)dt = \int_{t_0}^{t_1} \mathbf{x}^T K(t)\mathbf{u}(t)dt = 0.$$

Conversely, if $\mathbf{x}^T \int_{t_0}^{t_1} K(t)\mathbf{u}(t)dt = 0$ holds for all $\mathbf{u}(t)$, we take $\mathbf{u}(t) = K^T(t)\mathbf{x}$, then we have $\int_{t_0}^{t_1} ||K(t)||^2 dt = 0$. Based on the property of the integration, we directly deduce $\mathbf{x}^T K(t) = 0, \forall\, t \in [t_1, t_2]$.

Now we prove the lemma with the above statement. Suppose for any n-dimensional vector $\mathbf{a}$, there exists a $\mathbf{u}(t)$ such that $\int_{t_0}^{t_1} K(t)\mathbf{u}(t)dt = \mathbf{a}$. Especially, by choosing $\mathbf{a}$ as

$$\mathbf{a}_1 = \begin{pmatrix} 1 \\ 0 \\ \vdots \\ 0 \end{pmatrix}, \quad \mathbf{a}_2 = \begin{pmatrix} 0 \\ 1 \\ \vdots \\ 0 \end{pmatrix}, \quad \cdots, \quad \mathbf{a}_1 = \begin{pmatrix} 0 \\ 0 \\ \vdots \\ 1 \end{pmatrix},$$

we obtain the corresponding $\mathbf{u}$ as $\mathbf{u}_1(t), \mathbf{u}_2(t), \cdots, \mathbf{u}_n(t)$, such that

$$\int_{t_0}^{t_1} K(t)\mathbf{u}_i(t)dt = \mathbf{a}_i, \qquad i = 1, 2, \cdots, n.$$

If $\mathbf{x}^T K(t) = 0, \forall\, t \in [t_1, t_2]$, then $\mathbf{x}^T \int_{t_0}^{t_1} K(t)\mathbf{u}(t)dt = 0$ holds for all $\mathbf{u}(t)$. By taking $\mathbf{u}$ as $\mathbf{u}_1(t), \mathbf{u}_2(t), \cdots, \mathbf{u}_n(t)$ into account we obtain

$$\mathbf{x}^T \int_{t_0}^{t_1} K(t)\mathbf{u}_i(t)dt = \mathbf{x}^T \mathbf{a}_i = x_i = 0, \qquad i = 1, 2, \cdots, n,$$

where x_i is the ith component of the vector $\mathbf{x}$. Hence, $\mathbf{x} = 0$.

Now we suppose $\mathbf{x} = 0$ is the only solution of $\mathbf{x}^T K(t) = 0, \forall\, t \in [t_1, t_2]$. We want to prove: for any n-dimensional vector $\mathbf{a}$, there exists a $\mathbf{u}(t)$ such that

$$\int_{t_0}^{t_1} K(t)\mathbf{u}(t)dt = \mathbf{a}. \tag{18.15}$$

Let

$$S = \{\int_{t_0}^{t_1} K(t)\mathbf{u}(t)dt : \ u(t) \text{ is continuous on } [t_1, t_2]\}.$$

Obviously S is a linear space and the maximum of its dimension is n.

It is sufficient to prove that S is a n-dimensional space. In fact, if $\mathbf{dim}S = n$, then $\mathbf{a}$ belongs to S and there exists a $\mathbf{u}(t)$ such that (18.15) holds.

We assume that the dimension of S is less than n. There must exist a $\mathbf{x} \neq 0$ such that

$$\mathbf{x}^T \int_{t_0}^{t_1} K(t)\mathbf{u}(t)dt = 0$$

holds for all $\mathbf{u}(t)$. Consequently, $\mathbf{x}^T K(t) = 0, \forall\, t \in [t_1, t_2]$. However, it is a contradiction that $\mathbf{x} = 0$ is the only solution.

Proof of Theorem 23. The solution of (18.8) is given by

$$\mathbf{x}(t) = e^{A(t-t_0)}\mathbf{x}_0 + \int_{t_0}^{t} e^{(t-\tau)A} B\mathbf{u}(\tau)d\tau.$$

System (18.8) is exactly controllable if and only if there exists a $\mathbf{u}(\cdot)$ such that

$$e^{A(t_1-t_0)}\mathbf{x}_0 + \int_{t_0}^{t} e^{(t_1-\tau)A} B\mathbf{u}(\tau)d\tau = x_1,$$

or, equivalently,

$$-x_0 + e^{-A(t_1-t_0)}x_1 = \int_{t_0}^{t_1} e^{-(t-t_0)A} B\mathbf{u}(t)dt.$$

From Lemma 18.2.1, system (18.8) is exactly controllable if and only if $\mathbf{x} = 0$ is the only solution of $\mathbf{x}^T e^{-At} B = 0$, $\forall t \in [0, t_1 - t_0]$.

Now we show that $\mathbf{x}^T e^{-At} B = 0$ is equivalent to $\mathbf{x}^T A^m B = 0$ for all $m = 0, 1, \cdots, n-1$. Indeed, let $t = 0$ in $\mathbf{x}^T e^{-At} B = 0$ we have $\mathbf{x}^T B = 0$.

Since $f(t) = \mathbf{x}^T e^{-At} B = 0$ on the interval $[0, t_1 - t_0]$. Consequently $f'(t) = 0$ at $t = 0$ which leads to $\mathbf{x}^T AB = 0$. Step by step we have $\mathbf{x}^T A^m B = 0$ for all $m = 0, 1, \cdots, n-1$. Conversely, if $\mathbf{x}^T A^m B = 0$ holds, by the Cayley-Hamilton Theorem it is true for any $m \in \mathbb{R}$. It means

$$\mathbf{x}^T e^{-At} B = \sum_{m=0}^{\infty} (-1)^m \frac{t^m}{m!} \mathbf{x}^T A^m B \equiv 0.$$

Finally, since $\mathbf{x} = 0$ is the only solution of $\mathbf{x}^T A^m B = 0$, it is equivalent to the rank of the matrix

$$[B, AB, \cdots, A^{n-1}B]$$

is n.

18.3 PARTIAL DIFFERENTIAL EQUATIONS

18.3.1 Model Formulation

In this section we will formulate a physical control model for a vibrating string. More precisely, consider a string staying on its position of equilibrium. Some small vibrations lead the displacements of some articles. The internal tension will cause the displacements of the neighborhoods of these articles and finally will lead to a wave movement of the whole string. We hope the string can move to any exact prespecified position by adding an external force on the string.

We first make some reasonable assumptions: The string is thin and can be seen as a line with a constant linear mass. The string is soft and stretchy and the tension satisfies Hooke's law (strain is directly proportional to stress) while bending. The movement of the string proceeds on a flat surface and the displacement of each particle is perpendicular to the equilibrium position. Moreover, all displacements are small.

Set the x-axis as the direction of the equilibrium position, and the y-axis as the direction of the displacement of the articles (see Figure 18.3.1). The unknown function $y = y(t, x)$ is the displacement of the particle at position x in time t. The external force density $u(t, x)$ describes the force acting on each unit length of x. ρ is the density of the string and is a constant.

We apply the *Micro-element analysis*. Choose any small particle $[x, x+dx]$ where dx represents an infinitesimal variable. Obviously it has the quality ρdx and obeys Newton's second law

$$\mathbf{F} = m\mathbf{a}.$$

The force of the particle has the tension on the left endpoint $-\mathbf{T}(t, x)$, the tension on the right endpoint $\mathbf{T}(t, x+dx)$ and the external force on the particle perpendicular to x-axis $\mathbf{G}(t, x; dx)$. Let $\mathbf{T}(t, x)$ be differentiable with respect

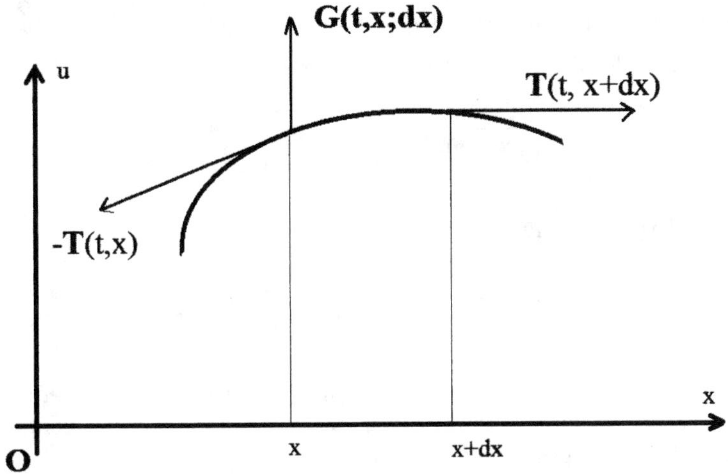

Figure 18.2 Vibrating string.

to x. The Newton's second law can be specifically represented as

$$\begin{aligned}\rho dx \frac{\partial^2 y}{\partial t^2}\mathbf{y}^\circ &= -\mathbf{T}(t,x) + \mathbf{T}(t,x+dx) + \mathbf{G}(t,x;dx) \\ &= \frac{\partial \mathbf{T}}{\partial x}dx + g(t,x)dx\mathbf{u}^\circ,\end{aligned}$$

where the higher-order infinitesimal is omitted in the second equality. Denote by (T_1, T_2) where T_1 and T_2 are components on the $\mathbf{x}^\circ$ and $\mathbf{y}^\circ$ directions. The previous quality comes to be

$$\frac{\partial T_1}{\partial x} = 0, \tag{18.16}$$

$$\rho\frac{\partial^2 y}{\partial t^2} = \frac{\partial T_2}{\partial x} + u(t,x). \tag{18.17}$$

Moreover, since the tension $\mathbf{T}$ acts on the tangential direction of the string, we have the third equation

$$T_2 = T_1\frac{\partial y}{\partial x}. \tag{18.18}$$

Taking (18.18) into (18.16) and (18.17), it holds

$$\rho\frac{\partial^2 y}{\partial t^2} = T_1(t)\frac{\partial^2 y}{\partial x^2} + u(t,x). \tag{18.19}$$

Under the assumption that the displacements are small, we have $\left|\frac{\partial y}{\partial x}\right| \ll 1$, and the magnitude of the force equals

$$T = \sqrt{T_1^2 + T_2^2} = T_1\sqrt{1 + \left(\frac{\partial y}{\partial x}\right)^2} \approx T_1.$$

The length of the particle equals

$$ds = \sqrt{dx^2 + dy^2} = dx\sqrt{1 + \left(\frac{\partial y}{\partial x}\right)^2} \approx dx.$$

Hence, during the dynamical process, the length of the particle remains constant. Due to Hooke's law, the magnitude of the force $T \approx T_1$ remains constant when time changes. Consequently (18.19) comes to be

$$\frac{\partial^2 y}{\partial t^2} = a^2\frac{\partial^2 y}{\partial x^2} + \frac{1}{\rho}u(t,x), \qquad a = \sqrt{\frac{T}{\rho}}. \tag{18.20}$$

System (18.20) is the *string's vibrating equation.* The coefficient a describes the velocity of the wave depending on the string itself. The control $u(t,x)/\rho$ is the force acting on the string.

We now first establish the well-posedness of (18.20), i.e., the existence and uniqueness of the solution, under some suitable initial data and boundary conditions.

Initial data For a physical process starting at time $t = 0$ and evolving over time interval $t \in [0, T]$, the status at $t = 0$ will affect the movement when time evolves. In our case, two statuses of the string will be taken into account:

$$\text{Initial displacement } y(0,x) = y_0(x); \ \text{ initial velocity } \frac{\partial y}{\partial t}(t,x)\,|_{t=0} = y_1(x).$$

Mathematically speaking, the initial conditions are the values of $y(t,x)$ and its partial derivatives with respect to t at time $t = t_0$. In general, if m is the highest order of $y(t,x)$ with respect to t, the initial data should be given as $y, \frac{\partial y}{\partial t}, \cdots, \frac{\partial^{m-1} y}{\partial t^{m-1}}$ at $t = t_0$.

Boundary conditions In the string vibration problem, let the string locate in the interval $[a, b]$. There are several kinds of boundary conditions such as the Dirichlet boundary condition, Neumann boundary condition, mixed boundary condition, etc. Here we will consider the simplest Dirichlet case in which the movement of the boundary points satisfies

$$y(t,x)|_{x=a} = y(t,x)|_{x=b} = 0.$$

Equation (18.20) is deduced from the vibration process and is called as the simplest one-dimensional *wave equation.* Its importance lies in the fact that not only does it arise in fields like acoustics, electromagnetics, and fluid dynamics but also because it is the basic hyperbolic partial differential equation for the description of waves—as they occur in physics—such as sound waves, light waves and water waves. In the sequel, we will introduce the well-posedness of (18.20) under suitable initial data and boundary conditions. Moreover, we will establish the controllability conception and describe some properties. As we shall see in the following section, the main properties of the hyperbolic equations such as time-reversibility and the lack of regularizing effects have some very important consequences in control problems too.

18.3.2 Controllability

Let ω be a nonempty subset of the interval $\Omega = (0,1)$. 1_ω is the characteristic function of ω. We consider the nonhomogeneous controlled wave equation:

$$\begin{cases} \dfrac{\partial^2 y}{\partial t^2} - \dfrac{\partial^2 y}{\partial x^2} = u 1_\omega, & (t,x) \in (0,T) \times \Omega, \\ y(t,0) = y(t,1) = 0, & t \in (0,T), \\ y(0,x) = y_0, \ \dfrac{\partial y}{\partial t}(0,x) = y_1, & x \in \Omega. \end{cases} \tag{18.21}$$

In (18.21) $y = y(t,x)$ is the state and $u = u(t,x)$ the control. Since u is multiplied by 1_ω the action of the control is restricted on the ω. Our aim is to change the dynamics of the system in the whole domain by means of the control u.

Since system (18.21) is a system with a partial differential equation, some notations and basic properties are necessary. We put them here, and more details can be found in [5].

Denote by $U = (0,T) \times \Omega$. We have

Definitions. Let $u : U \to \mathbb{R}$, $(t,x) \in U$.

[i] $\dfrac{\partial u}{\partial t}(t,x) = \lim\limits_{h \to 0} \dfrac{u(t+h,x) - u(t,x)}{h}$, provided this limit exists.

[ii] $\dfrac{\partial u}{\partial x}(t,x) = \lim\limits_{h \to 0} \dfrac{u(t,x+h) - u(t,x)}{h}$, provided this limit exists.

[iii] $C^k(\Omega) = \{u : \Omega \to \mathbb{R} \mid u \text{ is } k\text{-times continuously differentiable}\}$.

[iv] $\mathcal{D}(\Omega) = C^\infty(\Omega) = \{u : \Omega \to \mathbb{R} \mid u \text{ is infinitely differentiable}\} = \cap_{k=0}^{\infty} C^k(\Omega)$.

[v] $L^2(U) = \{u : U \to \mathbb{R}$ is Lebesgue measurable and $\|u\|_{L^2(U)} < \infty\}$, where

$$\|u\|_{L^2(U)} = \left(\int_U |u|^2 dxdt\right)^{1/2}.$$

[v] $H_0^1(\Omega) = \{u : \Omega \to \mathbb{R}$ is Lebesgue measurable and $u \in L^2(\Omega), u_x \in L^2(\Omega), u|_{x=0,1} = 0\}$, where

$$\|u\|_{H_0^1(\Omega)} = \left(\int_\Omega |u|^2 dx + \int_\Omega |u_x|^2 dx\right)^{1/2}.$$

The following theorem is a consequence of the classical results of existence and uniqueness of solutions of nonhomogeneous evolution equations. All the details may be found, for instance, in [3].

Theorem 24 *For any $u \in L^2((0,T) \times \omega)$ and $(y_0, y_1) \in H_0^1(\Omega) \times L^2(\Omega)$ equation (18.21) has a unique finite energy solution:*

$$y \in C([0,T]; H_0^1(\Omega)) \cap C^1([0,T]; L^2(\Omega)).$$

For any initial data $(y_0, y_1) \in H_0^1(\Omega) \times L^2(\Omega)$, the set of reachable states

$$R(T; (y_0, y_1)) = \{(y(T), \frac{\partial y}{\partial t}(T)) : \ y \text{ solution of } (18.21)$$

with $u \in L^2((0,T) \times \omega)\}$.

Remark that, for any $(y_0, y_1) \in H_0^1(\Omega) \times L^2(\Omega)$, $R(T; (y_0, y_1))$ is an affine subspace of $H_0^1(\Omega) \times L^2(\Omega)$ when the control u varies all over $L^2((0,T) \times \Omega)$. However the action of the control is localized in ω. Thus the controls may be viewed to belong to $L^2((0,T) \times \omega)$.

The problem of *controllability* consists in describing the set of reachable states. There are different notions of controllability that need to be distinguished:

(A) *Approximate controllability.* System (18.21) is said to be approximately controllable in time T if the set of reachable states is dense in $H_0^1(\Omega) \times L^2(\Omega)$ for every $(y_0, y_1) \in H_0^1(\Omega) \times L^2(\Omega)$.

(B) *Exact controllability.* System (18.21) is said to be exactly controllable in time T if the set of reachable states equal to $H_0^1(\Omega) \times L^2(\Omega)$ for every $(y_0, y_1) \in H_0^1(\Omega) \times L^2(\Omega)$.

(C) *Null controllability.* System (18.21) is said to be null controllable in time T if the set of $(0,0) \in R(T; (y_0, y_1))$ for every $(y_0, y_1) \in H_0^1(\Omega) \times L^2(\Omega)$.

Some remarks are in order.

- The definition of the exact controllability is equal to the one in Section 18.2.2. In fact, the above three definitions are all equivalent for the finite-dimensional system (18.8).

- A very important property of the wave equation is the so-called *finite speed of the propagation.* Roughly speaking, the influence of the vibration from the left point $x = 0$ will require some time duration passing to the right side $x = 1$(for instance, sound propagation from one point to another). Due to the finite speed of propagation of solutions of the wave equation, for any of these properties to hold, the control time T has to be sufficiently large, except for the trivial case in which the control subdomain ω coincides with the whole interval Ω.

- The null and exact controllability are equivalent notions due to another crucial property, *time reversibility.* More precisely, if we expect to transfer the state from **A** (y_0, y_1) at $t = 0$ to the state **B** (y_0^T, y_1^T) at $t = T$, we have the following methodology: we first solve system (18.21) backwards in time with the initial data given by $(y(T), y_t(T)) = (y_0^T, y_1^T)$ without control u. The corresponding solution at time $t = 0$ is denoted by **C** and is provided by the time reversible of the system. Moreover, since system (18.21) is linear, it obvious that transferring **A** to **B** is equivalent to transferring **A**$-$**C** to $(0, 0)$.

- Null controllability is a commonly used notion since the state $(0, 0)$ is an equilibrium for system (18.21). Once the system reaches the equilibrium at time $t = T$ by a control u, we can stop controlling and the system naturally stays in the equilibrium for all $t > T$. It also means, once the control exists, there are infinite ways to construct more controls.

18.3.3 Adjoint System and Observability

In this section we transform the exact controllability property of system (18.21) to another equivalent condition which is the *exact observable* of its dual system.

We denote by $H^{-1}(\Omega)$ the dual space to $H_0^1(\Omega)$. In other words, f belongs to $H^{-1}(\Omega)$ provided f is a bounded linear functional on $H_0^1(\Omega)$. We write $\langle \, , \rangle$ to denote the pairing between $H^{-1}(\Omega)$ and $H_0^1(\Omega)$. We define the norm

$$\|f\|_{H^{-1}(\Omega)} = \sup\{\langle f, u\rangle \mid u \in H_0^1(\Omega), \|u\|_{H_0^1(\Omega)\leq 1}\}.$$

Assume $f \in H^{-1}(\Omega)$. Then there exist functions f^0, f^1 in $L^2(\Omega)$ such that

$$\langle f, v\rangle = \int_\Omega (f^0 v + f^1 v_x)dx \qquad (v \in H_0^1(\Omega)). \tag{18.22}$$

Furthermore,

$$\|f\|_{H^{-1}(\Omega)} = \inf\left\{ \left(\int_\Omega (|f^0|^2 + |f^1|^2) dx \right)^{1/2} \mid f \text{ satisfies } (18.22) \right\}$$

for $f^0, f^1 \in L^2(\Omega)$. The existence of the functions can be found on p. 284 of Ref. [5] as the characterization of H^{-1} .

For $(\varphi_0^T, \varphi_1^T) \in L^2(\Omega) \times H^{-1}(\Omega)$, consider the following backward homogenous equation:

$$\begin{cases} \dfrac{\partial \varphi^2}{\partial t^2} - \dfrac{\partial^2 \varphi}{\partial x^2} = 0, & (t,x) \in (0,T) \times \Omega, \\ \varphi(t,0) = \varphi(t,1) = 0, & t \in (0,T), \\ \varphi(T,x) = \varphi_0^T, \ \dfrac{\partial \varphi}{\partial t}(T,x) = \varphi_1^T, & x \in \Omega. \end{cases} \tag{18.23}$$

Let $(\varphi, \frac{\partial \varphi}{\partial t}) \in C([0,T]; L^2(\Omega) \times H^{-1}(\Omega))$ be the solution of (18.23). We have the following lemma.

Lemma 18.3.1 *The following assertions are equivalent:*

- *System (18.21) can be driven from $(y_0, y_1) \in H_0^1(\Omega) \times L^2(\Omega)$ at $t = 0$ to the rest $(0,0)$ at $t = T$.*

- *For all $(\varphi_0^T, \varphi_1^T) \in L^2(\Omega) \times H^{-1}(\Omega)$, there exists $u \in L^2((0,T) \times \omega)$ such that*

$$\int_0^T \int_\omega \varphi u dx dt = \left\langle \frac{\partial \varphi}{\partial t}(0), y_0 \right\rangle_{-1,1} - \int_\Omega \varphi(0) y_1 dx, \tag{18.24}$$

where $\varphi(t)$ is the corresponding solution of (18.23) with initial data $(\varphi_0^T, \varphi_1^T)$.

Proof: We will first prove the assertions are equivalent under the assumption that all functions involved are smooth. Then by a density argument we finish the proof.

We suppose that $(y_0, y_1), (\varphi_0^T, \varphi_1^T) \in \mathcal{D}(\Omega) \times \mathcal{D}(\Omega)$, $u \in \mathcal{D}((0,T) \times \omega)$ and let y and φ be the solutions of (18.21) and (18.23), respectively. Multiply the equation of y by φ, integrate on $(0,T)$, and we obtain

$$\int_0^T \int_\Omega \varphi \left(\frac{\partial^2 y}{\partial t^2} - \frac{\partial^2 y}{\partial x^2} \right) = \int_0^T \int_\omega \varphi u dx dt. \tag{18.25}$$

Integrating by parts the left hand-side of (18.25) equals

$$\text{LHS of (18.25)} = \int_\Omega \left(\varphi\frac{\partial y}{\partial t} - \frac{\partial \varphi}{\partial t}y\right)dx\,|_0^T + \int_0^T\int_\Omega y\left(\frac{\partial^2\varphi}{\partial t^2} - \frac{\partial^2\varphi}{\partial x^2}\right)dxdt$$
$$= \int_\Omega \left(\varphi(T)\frac{\partial y}{\partial t}(T) - \frac{\partial \varphi}{\partial t}(T)y(T)\right)dx - \int_\Omega \left(\varphi(0)\frac{\partial y}{\partial t}(0) - \frac{\partial \varphi}{\partial t}(0)y(0)\right)dx.$$

Hence,

$$\int_0^T\int_\omega \varphi u dxdt = \int_\Omega \left(\varphi_0^T\frac{\partial y}{\partial t}(T) - \varphi_1^T y(T)\right)dx - \int_\Omega \left(\varphi(0)y_1 - \frac{\partial \varphi}{\partial t}(0)y_0\right)dx. \tag{18.26}$$

From a density argument we deduce, by passing to the limit in (18.26), it holds

$$\begin{aligned}\int_0^T\int_\omega \varphi u dxdt = & \int_\Omega \varphi_0^T\frac{\partial y}{\partial t}(T)dx - \langle\varphi_1^T, y(T)\rangle_{-1,1} \\ & + \int_\Omega \varphi(0)y_1 dx - \left\langle\frac{\partial\varphi}{\partial t}(0), y_0\right\rangle_{-1,1}\end{aligned} \tag{18.27}$$

for any $(y_0, y_1) \in H_0^1(\Omega) \times L^2(\Omega)$ and $(\varphi_0^T, \varphi_1^T) \in L^2(\Omega) \times H^{-1}(\Omega)$.

It follows directly from (18.27) that the two assertions are equivalent and the proof is complete.

The second assertion in Lemma 18.3.1 may be seen as an optimality condition for the critical points of the functional $\mathcal{J} : L^2(\Omega) \times H^{-1}(\Omega) \to \mathbb{R}$,

$$\mathcal{J}(\varphi_0^T, \varphi_1^T) = \frac{1}{2}\int_0^T\int_\omega |\varphi|^2 dxdt + \left\langle\frac{\partial\varphi}{\partial t}(0), y_0\right\rangle_{-1,1} - \int_\Omega \varphi(0)y_1 dx, \tag{18.28}$$

where φ is the solution of (18.23) with initial data $(\varphi_0^T, \varphi_1^T) \in L^2(\Omega)\times H^{-1}(\Omega)$. In fact, if we denote by $(\hat\varphi_0^T, \hat\varphi_1^T)$ the minimizer of $\mathcal{J}$ and $\hat\varphi$ the corresponding solution of (18.23) with initial data $(\hat\varphi_0^T, \hat\varphi_1^T)$, then it holds

$$\begin{aligned}0 &= \lim_{h\to 0}\frac{1}{h}(\mathcal{J}((\hat\varphi_0^T, \hat\varphi_1^T) + h(\varphi_0^T, \varphi_1^T)) - \mathcal{J}(\hat\varphi_0^T, \hat\varphi_1^T)) \\ &= \int_0^T\int_\omega \varphi\hat\varphi dxdt - \left\langle\frac{\partial\varphi}{\partial t}(0), y_0\right\rangle_{-1,1} + \int_\Omega \varphi(0)y_1 dx,\end{aligned}$$

for any $(\varphi_0^T, \varphi_1^T) \in L^2(\Omega) \times H^{-1}(\Omega)$ where φ is the corresponding solution of (18.23).

We now show that the sufficient condition ensuring the existence of a minimizer for $\mathcal{J}$ is the observability property of system (18.23), which is defined as follows:

Definition 18.1 *System (18.23) is* **observable in time** T *if there exists a positive constant* $C > 0$ *such that*

$$C \left\|(\varphi_0^T, \varphi_1^T)\right\|^2_{L^2(\Omega)\times H^{-1}(\Omega)} \leq \int_0^T \int_\omega |\varphi|^2 dxdt, \tag{18.29}$$

holds for any $(\varphi_0^T, \varphi_1^T) \in L^2(\Omega) \times H^{-1}(\Omega)$ *where* φ *is the corresponding solution of (18.23) with initial data* $(\varphi_0^T, \varphi_1^T)$.

Inequality (18.29) is called an *observability inequality.* It shows that the whole energy of the system (18.23) can be observed by the partial measurement on the subdomain ω. The following theorem holds:

Theorem 25 *Let* $(y_0, y_1) \in H_0^1(\Omega) \times L^2(\Omega)$ *and suppose that (18.23) is observable in time* T. *Then the functional* $\mathcal{J}$ *defined by (18.28) has a unique minimizer* $(\hat{\varphi}_0^T, \hat{\varphi}_1^T) \in L^2(\Omega) \times H^{-1}(\Omega)$.

To prove Theorem 25, we need the following fundamental result in the Calculus of Variations:

Theorem 26 *Let* H *be a reflexive Banach space,* K *a closed convex subset of* H *and* $\psi : K \to \mathbb{R}$ *a convex, lower semicontinuous and coercive function. Then* ψ *attains its minimum in* K, *i.e. there exists* $x_0 \in K$ *such that*

$$\psi(x_0) = \min_{x\in K} \psi(x).$$

The proof of Theorem 26 can be found in Ref. [2].

Proof of Theorem 25: It is easy to confirm that $\mathcal{J}$ is continuous and convex (see Exercise 18.3). To provide the existence of the minimizer of functional $\mathcal{J}$, it is sufficient to verify that $\mathcal{J}$ coercive. More precisely, we need to prove that

$$\lim_{\left\|(\varphi_0^T, \varphi_1^T)\right\|_{L^2(\Omega)\times H^{-1}(\Omega)}\to\infty} \mathcal{J}(\varphi_0^T, \varphi_1^T) = \infty.$$

Recalling the definition of $\mathcal{J}$ in (18.28) and the observability property (18.29), we have

$$\begin{aligned}
\mathcal{J}(\varphi_0^T, \varphi_1^T) \;&\geq\; \frac{1}{2}\int_0^T \int_\omega |\varphi|^2 dxdt \\
&\quad -\frac{1}{2}\left\|(y_0, y_1)\right\|_{H_0^1(\Omega)\times L^2(\Omega)} \left\|(\varphi_0^T, \varphi_1^T)\right\|_{L^2(\Omega)\times H^{-1}(\Omega)} \\
&\geq\; \frac{C}{2}\left\|(\varphi_0^T, \varphi_1^T)\right\|_{L^2(\Omega)\times H^{-1}(\Omega)} \\
&\quad -\frac{1}{2}\left\|(y_0, y_1)\right\|_{H_0^1(\Omega)\times L^2(\Omega)} \left\|(\varphi_0^T, \varphi_1^T)\right\|_{L^2(\Omega)\times H^{-1}(\Omega)}.
\end{aligned}$$

Hence, it follows from Theorem 26 that $\mathcal{J}$ has a minimizer $(\hat{\varphi}_0^T, \hat{\varphi}_1^T) \in L^2(\Omega) \times H^{-1}(\Omega)$. The uniqueness can be shown as a consequence of the strictly convex of $\mathcal{J}$, which we also leave to an exercise (see Exercise 18.3).

Combining Lemma 18.3.1 and Theorem 25, we guarantee that, under the hypothesis system (18.23) being observable at time T, system (18.3) is exactly controllable. Moreover, a control could be obtained as the solution of the homogenous system (18.23) with the initial data minimizing the functional $\mathcal{J}$. Hence, the controllability problem is reduced to a minimization problem that may be solved by the Direct Method of the Calculus of Variations.

Furthermore, the following corollary shows that the control obtained by this method is of minimal $L^2((0,T) \times \omega)$-norm.

Corollary 1 *Let $u = \hat{\varphi} 1_\omega$ be the control given by minimizing the functional $\mathcal{J}$. If $v \in L^2((0,T) \times \omega)$ is another control driving the solution of (18.3) to zero at $t = T$ with initial data (y_0, y_1), then*

$$\|u\|_{L^2((0,T)\times\omega)} \leq \|v\|_{L^2((0,T)\times\omega)} . \tag{18.30}$$

Proof: We leave it as an exercise.

EXERCISES

18.1 Deduce the formula (18.10) and compute the solution of (18.8) with this formula.

18.2 In fact, there are infinite ways to construct function z satisfying (18.13). Try to find one function z with a cubic polynomial function with initial data at $t_0 = 0$, i.e., $(x(0), x'(0)) = (1,0)$ and final data at $t_1 = 1$, i.e., $(x(1), x'(1)) = (0,0)$. Then write down the corresponding control for system (18.12).

18.3 Explain the functional $\mathcal{J}$ as (18.28) is continuous and strictly convex.

18.4 Prove Corollary (1).

REFERENCES

1. Arnold, V. I., *Ordinary Differential Equations.* Springer-Verlag, Berlin, (2006). Translated from the Russian by Roger Cooke, Second printing of the 1992 edition.
2. Brezis, H., *Functional Analysis, Sobolev Spaces and Partial Differential Equations.* Springer, New York (2011).
3. Cazenave, T. and Haraux, A., *An Introduction to Semilinear Evolution Equations*, Vol. 13 of *Oxford Lecture Series in Mathematics and its Applications.* The Clarendon Press Oxford University Press, New York (1998).
4. Delfour, M. C. and Mitter, S. K., *Controllability and Observability for Infinite-Dimensional Systems.* SIAM J. Control, 10:329–333 (1972).

5. Evans, L. C., *Partial Differential Equations*, Vol. 19 of *Graduate Studies in Mathematics*. American Mathematical Society, Providence, RI (1998).

6. Fattorini, H. O., *Boundary Control Systems.* SIAM J. Control, 6:349–385 (1968).

7. Fernández-Cara, E. and Zuazua, E., *On the history and perspectives of control theory.* Matapli, (74):47–73 (2004).

8. Kalman, R. E., *On the general theory of control systems.* Proceedings of the First IFAC Congress on Automatic Control, 1:481–492 (1961).

9. Kalman, R. E., Ho, Y. C., and Narendra, K. S., *Controllability of linear dynamical systems.* Contributions to Differential Equations, 1:189–213 (1963).

10. Lee, E. B. and Markus, L., *Foundations of Optimal Control Theory.* Robert E. Krieger Publishing Co. Inc., Melbourne, FL, second edition (1986).

11. Lions, J.-L., *Contrôlabilité Exacte, Perturbations et Stabilisation de Systèmes Distribués. Tome 1*, Vol. 8 of *Recherches en Mathématiques Appliquées [Research in Applied Mathematics]*. Masson, Paris, (1988).

12. Lions, J.-L., *Exact controllability, stabilization and perturbations for distributed systems.* SIAM Rev., 30(1):1–68 (1988).

13. Markus, L., *Controllability of nonlinear processes.* J. Soc. Indust. Appl. Math. Ser. A Control, 3:78–90 (1965).

14. Micu, S. and Zuazua, E., *An introduction to the controllability of partial differential equations.* In *"Quelques Questions de Théorie du Contròle"*, Sari, T., ed., pages 69–157. Collection Travaux en Cours Hermann (2004).

15. Russell, D. L., *Controllability and stabilizability theory for linear partial differential equations: recent progress and open questions.* SIAM Rev., 20(4):639–739 (1978).

16. Sussmann, H. J., *A general theorem on local controllability.* SIAM J. Control Optim., 25(1):158–194 (1987).

17. Zuazua, E., *Controllability of partial differential equations and its semi-discrete approximations.* Discrete Contin. Dyn. Syst., 8(2):469–513 (2002).

18. Zuazua, E., *Some problems and results on the controllability of partial differential equations.* In European Congress of Mathematics, Vol. II (Budapest, 1996), volume 169 of Progr. Math., pp. 276–311. Birkhäuser, Basel (1998).

19. Zuazua, E., *Propagation, observation, and control of waves approximated by finite difference methods.* SIAM Rev., 47(2):197–243 (electronic) (2005).

20. Zuazua E., *Controllability and observability of partial differential equations: Some results and open problems.* In *Evolutionary Differential Equations*, Vol. 8, pages 527–621. Elsevier Science (2006).

CHAPTER 19

MARKOV-JUMP STOCHASTIC MODELS FOR TROPICAL CONVECTION

BOUALEM KHOUIDER

Mathematics and Statistics, University of Victoria, Canada

19.1 INTRODUCTION

Atmospheric convection is the process through which warm and moist air parcels rise from the surface, condense liquid water and form cumulus clouds. This results in precipitation and heavy storms. The process of condensation is accompanied by the release of latent heat, which is associated with the phase change of water from vapor to liquid and/or ice. In the tropics, moist convection constitutes a major source of energy for both local and large-scale circulations. Precipitation patterns in the tropics are organized into cloud clusters and superclusters on a wide range of scales; they range from the convective cell (the cumulus cloud) of 1 to 10 km, to planetary scale waves with oscillation periods of 40 to 60 days. Due to the complex interactions between the local processes of convection and the large-scale waves, climate models fail

Edited by Xin-She Yang

to properly capture tropical circulation patterns and their effect on the global circulation. In a climate model, the governing equations are discretized on a coarse mesh of roughly 100 km to 200 km and the effects of processes that are not resolved on such grids are represented by a parameterization also called a subgrid model. According to the last report of the United Nations' Intergovernmental Panel on Climate Change (IPCC), the interactions of clouds and the climate system is one of the major challenges in climate research.

The phenomenon of convection is in essence due to the very simple physical mechanism of buoyancy, which was discovered thousands of years ago by Archimedes. Buoyancy is the force that pushes light fluid to rise and heavy fluid to sink. When light fluid lies over heavy fluid as in normal atmospheric and oceanic conditions the situation is stable and if a fluid parcel is displaced mechanically in the vertical it will quickly sink or rise back toward its initial position and would normally undergo an oscillatory motion around its initial position. When the sun heats the surface (sea or land), the air parcels near the ground become quickly warm and moist, due to the evaporation of sea or land water. Because the warm and moist air near the surface is lighter than the dry and cold air above it, a turbulent motion begins and quickly mixes the air layer near the surface; because of the strong tropospheric stratification (the large discrepancy between the air density at the surface and at the top of the troposphere), the day is not long enough for this mixing process to penetrate very high before sun set. The cold surface-stable conditions are quickly restored at night as the ground loses its heat to space as long-wave radiation and the cycle continues. This process is called dry convection because it does not involve the phase change of water. It mainly serves to form what is known as the planetary boundary or mixed layer where the air density and *potential temperature*[25] are relatively constant in the vertical. It leaves the upper troposphere unperturbed—happy and cool. When some "lucky" parcels are able to make it beyond the mixed layer they expand and cool down because of the pressure drop and potentially become saturated with water vapor. At this point the rising parcel starts to condense liquid water and releases latent heat, which in turn warms the parcel and partially compensates for the cooling by expansion. When this diabatic heating is large enough, the parcel becomes positively buoyant and will eventually rise high enough and entrain further convection and form a cloud.

The level at which a rising parcel starts to condense water is called the lifted condensation level (LCL) while the level at which a parcel becomes positively buoyant because of condensational heating is called the level of free convection (LFC). The path of a hypothetical air parcel rising from the surface is shown by the thick solid line in Figure 19.1. The thick dashed line is

[25] The potential temperature is the temperature that an unsaturated air parcel would have if it is displaced adiabatically, i.e, without exchange of heat with the environment, to a reference pressure level, usually near the surface.

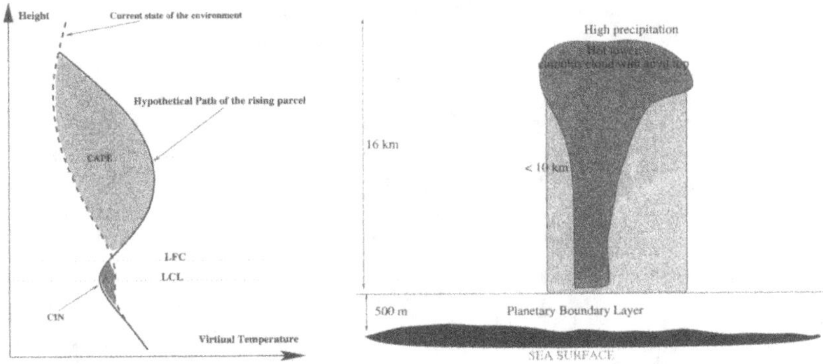

Figure 19.1 Left: Path followed by a hypothetical parcel of air rising from the surface through the atmospheric column (solid line). The dashed line represents the environmental virtual temperature. The green area represents the convective available potential energy (CAPE) while the red area is the negative energy (CIN) that the rising parcel needs to overcome in order to reach its level of free convection and become freely buoyant. The dotted lines show the LCL and LFC levels. Right: A cartoon of a hot tower cumulus cloud formed by air parcels rising from the mixed boundary layer.

the environmental *virtual temperature*,[26] which yields a good approximation for the buoyancy that takes into account the water loading of the moist parcel. At any given point on its path, the parcel becomes positively buoyant, i.e., forced upward, if it finds itself virtually warmer than the environmental air around it and negatively buoyant if it is virtually colder. Given that potential energy associated with the convecting parcel is the vertical integral of the buoyancy force (which is proportional to the virtual temperature difference between the rising parcel and the environment), typically, the rising parcel needs to overcome a certain amount of negative energy before it reaches its LFC. The negative energy is known as convective inhibition (CIN) and is represented by the red area in Figure 19.1 while the positive (green) area above it is called convective available potential energy (CAPE). Observations showed that various mechanisms can contribute to provide the energy necessary for the rising parcel to overcome the CIN barrier. These include both local effects such as turbulent fluctuations in the boundary layer moisture and temperature, gust fronts, cold pools, and density currents and large scale effects such as organized convergence and propagating waves. Due to its complexity, the effect of CIN is still very poorly understood and as such it is not very well represented in climate models. The deterministic parameterization of CIN is at best unrealistic!

[26]The virtual temperature is the temperature that a dry air parcel would have if it has the same pressure and density as the moist parcel.

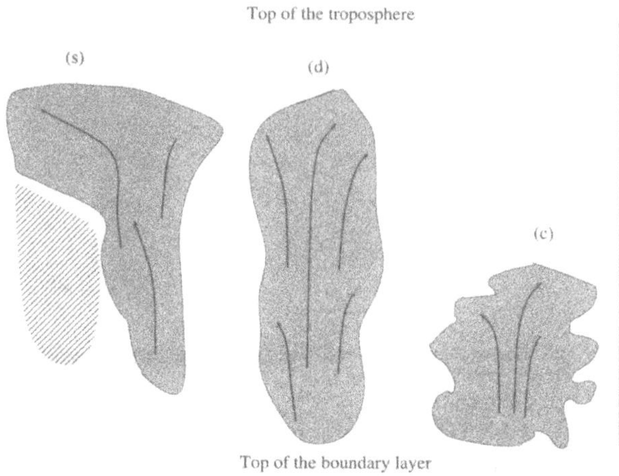

Figure 19.2 A cartoon of the three cloud types showing congestus (c), deep convective (d), and a decaying deep convective tower with a lagging large stratiform anvil (s), with stratiform rain falling into a dry region below it where it eventually evaporates and cools the environment (hatched area). The arrows indicate convective motion within the cloud.

Another conundrum of atmospheric convection, especially in the tropics, is its organization over a wide range of scales. Recent satellite and radar observations showed that organized convection involves three cloud types that interact with each other and help define a self-similar morphology for convectively coupled waves of various sizes that are often embedded in each other like Russian dolls. It is suggested in some research papers that this can be possible only if the cloud-cloud interactions occur in some stochastic manner so that when they are averaged locally or globally they would preserve their self-similar structure. This common structure consists of cumulus congestus clouds that prevail in front of the wave followed by deep convective clouds that penetrate to the upper troposphere (near the tropopause) which in turn are lagged by icy stratiform anvils that prevail in the upper troposphere. Arguably a consensual explanation for the main physical mechanisms for this behavior is still an active research area but a more or less accepted explanation (which is somewhat biased toward the author's own research) is as follows. 1) Because of the entrainment of surrounding dry air, the rising parcels lose their buoyancy typically below the freezing level and form congestus clouds. 2) As they form and dissipate in the middle of the troposphere, congestus clouds deposit and converge moisture in the horizontal and thus serve to precondition the environment for future parcels to penetrate higher and form deep convective clouds. 3) Stratiform clouds are believed to be a continuation of the deep penetrative clouds which start to dissipate from below and leave their icy anvil tops behind that continue to produce ice, thus heat. A cartoon of this three cloud-type paradigm is given in Figure 19.2.

In this chapter, we present two stochastic models to represent, respectively, the random fluctuations of CIN and the random interactions of the three cloud types of organized convection using Markov jump processes. We start by presenting in section 19.2 a very basic introduction to random variables and the notion of Monte Carlo integration to familiarize the reader with the use of random numbers in computer simulations. In Section 19.3, we provide a crash-course introduction to the notion of Markov chains and birth-death processes. The stochastic model for CIN based on a simple birth-death process derived from an Ising-type model is then presented in Section 19.4 while the stochastic multicloud model is discussed in Section 19.5. A list of exercises is given at the end of the chapter to help improve the understanding of the theory in Sections 19.2 and 19.3.

19.2 INTRODUCTION TO RANDOM NUMBERS: THEORY AND SIMULATIONS

19.2.1 Random Variables

A random variable $X = X(\omega)$ is by definition a function on a probability space Ω that takes values in a discrete set or a continuous interval of real numbers, according to a given probability distribution, P, defined of Ω. A random variable with values in a discrete set of real numbers is called *a discrete random variable* and a random variable that takes value in an interval is called *a continuous random variable.*

Examples of discrete random variables:

a) Perhaps the simplest example of a discrete random variable is that of tossing a fair coin, which results in heads or tails with a 50/50% chance. We can thus associate a random variable X that takes the value $X = 0$ if the outcome is heads and $X = 1$ if it is tails with the probability distribution

$$P(\{X = 0\}) = \frac{1}{2}, \ P(\{X = 1\}) = \frac{1}{2}.$$

b) A similar example is that of throwing a die. The associated random variable takes its values in the discrete set $\{1, 2, 3, 4, 5, 6\}$ according to which face of the die appears. It's probability distribution is given by

$$P(\{X = 1\}) = p_1, P(\{X = 2\}) = p_2, \cdots, P(\{X = 6\}) = p_6,$$

where $0 \leq p_1, p_2, \cdots, p_6 \leq 1$ satisfy $p_1 + p_2 + \cdots + p_6 = 1$. If the die is unloaded, then $p_1 = p_2 = \cdots = p_6 = 1/6$.

c) If instead we consider throwing two dice at a time. Then the sum X of the two faces of the two dice is a random variable that takes values in

$\{2, 3, \cdots, 12\}$. If we assume that the two dice are unbiased, then

$$P\{X=2\} = P\{die1=1\}P\{die2=1\} = \frac{1}{36},$$

$$P\{X=3\} = P\{die1=1\}P\{die2=2\} + P\{die1=2\}P\{die2=1\} = \frac{2}{36},$$

$$\cdots,$$

$$P\{X=12\} = P\{die1=6\}P\{die2=6\} = \frac{1}{36}.$$

Examples of continuous random variables:
Concrete examples of continuous random variables are ubiquitous in nature and human life. Often a continuous random variable represents an idealized approximation of a discrete random variable taking values in a large discrete set. Common examples include that of the price of an asset in the stock market or the exact dimensions of a manufactured object—there is always some deviations from the aimed dimensions due to imperfections in the devices used for making the product, etc.

The probability distribution of a continuous random variable is given in terms of the probability that X lies in a given interval $[a, b]$: $P(\{a \leq X \leq b\})$.
a) Uniform random variable on $[0, 1]$*:*
The uniform random variable is perhaps the simplest continuous random variable. It takes values in a specific interval $[\alpha, \beta]$. The distribution of the uniform random variable on $[0, 1]$ is as follows. If $[a, b] \subset [0, 1]$, then $P(\{a \leq X \leq b\}) = b - a$, the length of the interval $[a, b]$.
If $(a, b) \cap (0, 1) = \varnothing$, then $P(\{a \leq X \leq b\}) = 0$.

Note that accordingly, we have in general $P(\{a \leq X \leq b\}) = |(a, b) \cap (0, 1)|$ and $P(\{0 \leq X \leq 1\}) = 1$.

Very often, the probability distribution for a continuous random variable is given by an integral

$$P(\{a \leq X \leq b\}) = \int_a^b f(x)dx,$$

where $f(x)$ is a non-negative real valued function with the important property

$$\int_{-\infty}^{+\infty} f(x)dx = 1.$$

$f(x)$ is called the *probability density function* (pdf for short). For the example of a uniform random variable on $[0, 1]$ we have

$$f(x) = \begin{cases} 1 & \text{if } 0 \leq x \leq 1 \\ 0, & \text{otherwise.} \end{cases}$$

The function

$$F(x) = \int_{-\infty}^{x} f(x)dx,$$

which is the probability that $-\infty \leq X \leq x$ is known as the *cumulative distribution function* (CDF). For simplicity in exposition, the uniform distribution on an interval $[\alpha, \beta]$ is denoted below by $\mathcal{U}(\alpha, \beta)$.

b) Gaussian or normal distributed random variable:

A Gaussian random variable takes its values in $(-\infty, +\infty)$ according to the Gaussian probability density function

$$f(x) = \frac{1}{\sqrt{2\pi}\sigma} \exp\left(-\frac{(x-\mu)^2}{2\sigma^2}\right),$$

where $\mu, \sigma > 0$ are two real parameters known as the mean and standard deviation of the Gaussian distribution. Note that

$$P(\{a \leq X \leq b\}) = \frac{1}{\sqrt{2\pi}\sigma} \int_a^b \exp\left(-\frac{(x-\mu)^2}{2\sigma^2}\right) dx$$

is not known in closed form but it is easily evaluated using quadrature technique-based algorithms that are already implemented in many of the available softwares. For example in Matlab one can use the function *normcdf* which serves to evaluate the cumulative distribution function

$$F(x) = \frac{1}{\sqrt{2\pi}\sigma} \int_{-\infty}^{x} \exp\left(-\frac{(x-\mu)^2}{2\sigma^2}\right) dx.$$

c) Exponentially distributed random variable:

The pdf of an exponential random variable is given by

$$f(x) = \lambda e^{-\lambda x} \text{ for } x \geq 0,$$

where $\lambda > 0$ is a real parameter. The CDF takes the form

$$F(x) = 1 - e^{-\lambda x}, \quad x \geq 0.$$

d) The Gamma distribution:

Similarly to the exponential random variable a Gamma random variable takes its values on $(0, +\infty)$. The pdf of a Gamma distribution with two parameters α, λ is given by

$$f(x) = \frac{1}{\Gamma(\alpha)} x^{\alpha-1} \lambda^{\alpha} e^{-\lambda x} \text{ for } x \geq 0.$$

Here $\Gamma(\alpha) = \int_0^{\infty} x^{\alpha-1} e^{-x} dx$ is a normalization constant with $\Gamma(n) = (n-1)!$, the function Γ is some sort of a generalization of the factorial operation to real numbers. Just like the normal distribution, the CDF of the Gamma distribution cannot be computed by hand. An approximate value however can be obtained numerically through the in*complete Gamma function:* $\frac{1}{\Gamma(\alpha)}\int_0^x x^{\alpha-1}e^{-x}\,dx$ which can be found in most of the available numerical libraries. In Matlab, for example, it is called *gammainc.*

The exponential and Gamma distributions are commonly used to model the time of occurrence of rare events and highly *skewed* random variables. They are part of the family of *fat tail distributions.*

19.2.2 Mean, Variance, and Expectation

Let X denote a continuous random variable with a probability distribution $P(x) \equiv P\{X \leq x\}$. The *mean or expectation* of X is given by

$$E[X] = \int_{-\infty}^{+\infty)} x dP(x),$$

where the integral is in the sense of Stieltjes integration and $dP(x)$ can be understood as the infinitesimal probability that $x \leq X \leq x + dx$. If X has a a PDF, say $f(x)$, then

$$E[X] = \int_{-\infty}^{+\infty} x f(x) dx.$$

Sometimes the mean is denoted by an overbar or angle brackets

$$E[X] \equiv \bar{X} \equiv \langle X \rangle.$$

For any given real valued function g we can define the expectation of $g(X)$ as

$$E[g(X)] = \int_{-\infty}^{+\infty} g(x) dP(x) = \int_{-\infty}^{+\infty} g(x) f(x) dx.$$

The *variance* of X is given by $Var[X] = E[(X - E[X])^2] = E[X^2] - E[X]^2$ and the standard deviation is given by $\sigma = \sqrt{Var[X]}$. Using simple integration rules, we can easily show that for

- $\mathcal{U}(0, 1)$, we have

$$E[X] = \int_0^1 x dx = \frac{1}{2} \text{ and } Var[X] = \int_0^1 x^2 dx - \frac{1}{4} = \frac{1}{12}.$$

- $\mathcal{N}(\mu, \sigma)$, Gaussian distribution with parameters μ, σ, we have

$$E[X] = \frac{1}{\sqrt{2\pi}\sigma} \int_{-\infty}^{+\infty} x \exp\left(-\frac{(x-\mu)^2}{2\sigma^2}\right) dx = \mu$$

and $Var[X] = \frac{1}{\sqrt{2\pi}\sigma} \int_{-\infty}^{+\infty} (x-\mu)^2 \exp\left(-\frac{(x-\mu)^2}{2\sigma^2}\right) dx - \mu^2 = \sigma^2.$

- the exponential distribution with a parameter $\lambda > 0$, we have

$$E[X] = \lambda \int_0^{+\infty} xe^{-\lambda x} dx = \frac{1}{\lambda} \text{ and } Var[X] = \frac{1}{\lambda^2}.$$

19.2.3 Conditional Probability

Let X, Y be two random variables with their respective probability density functions f_X and f_Y. The probability $P(\{a \leq X \leq b, c \leq Y \leq d\})$ that both $a \leq X \leq b$ and $c \leq Y \leq d$ happen at the same time, is known as the *joint probability distribution.* The joint cumulative distribution function is the two variable function

$$F(x, y) = P(\{X \leq x, Y \leq y\}).$$

The joint probability density function of X, Y is given by

$$f(x, y) = \frac{\partial^2 F(x, y)}{\partial x \partial y},$$

so that

$$F(x, y) = \int_{-\infty}^{y} \int_{-\infty}^{x} f(t, s) dt ds.$$

We have the following compatibility conditions between the *marginals* and the joint distribution.

$$F_X(x) \equiv \int_{-\infty}^{x} f_X(t) dt = \int_{-\infty}^{+\infty} \int_{-\infty}^{x} f(t, s) dt ds \tag{19.1}$$

$$\text{and } F_Y(y) \equiv \int_{-\infty}^{y} f_Y(s) ds = \int_{-\infty}^{+\infty} \int_{-\infty}^{y} f(t, s) ds dt.$$

The two random variables X, Y are said to be *independent* if

$$P(\{a \leq X \leq b, c \leq Y \leq d\}) = P(\{a \leq X \leq b\}) \times P(\{c \leq Y \leq d\}).$$

If X, Y are independent then their joint probability density function satisfies

$$f(x, y) = f_X(x) f_Y(y).$$

The probability that $a \leq X \leq b$ given that $c \leq Y \leq d$ denoted by $P(\{a \leq X \leq b/c \leq Y \leq d\}$ is known as the conditional probability of the random variable X given Y. We have

$$P(\{a \leq X \leq b, c \leq Y \leq d\}) = P(\{a \leq X \leq b/c \leq Y \leq d\}) \times P(\{c \leq Y \leq d\}).$$

If X, Y are independent, then the conditional probability satisfies

$$P(\{a \leq X \leq b/c \leq Y \leq d\}) = P(\{a \leq X \leq b).$$

The covariance of two random variables X, Y is given by $Cov(X,Y) \equiv E[(X - E[X])(Y - E[Y])] = E[XY] - E[X]E[Y]$. It is easy to verify that $Cov(X,X) = Var[X]$ and if X, Y are independent then $Cov(X,Y) = 0$. Note that $Cov(X,Y) = 0$ doesn't necessarily mean that X, Y are independent. If $Cov(X,Y) > 0$, then X, Y are said to be positively correlated and if $Cov(X,Y) < 0$, they are said to be negatively correlated. Moreover, we have $E[aX+bY] = aE[X]+bE[Y]$, $Var[X+Y] = Var[X]+Var[Y]+2Cov(X,Y)$, and $Var[cX] = c^2 Var[X]$.

19.2.4 Law of Large Numbers

Let $X_1, X_2, \cdots, X_n, n \geq 2$ be a sequence of independent and identically distributed (i.i.d) random variables, each having a mean μ and a standard deviation σ: $E[X_j] = \mu, Var[X_j] = \sigma^2, \forall j$. We define the sample mean as the average random variable:

$$\bar{X}_n = \frac{1}{n}(X_1 + X_2 + \cdots X_n) = \frac{1}{n}\sum_{j=1}^{n} X_j.$$

Then the expected sample mean equals the population mean:

$$E[\bar{X}_n] = \frac{1}{n}\sum_{j=1}^{n} E[X_j] = \frac{1}{n}n\mu = \mu.$$

The sample mean is said to be unbiased. Moreover, we have $Cov(X_j, X_k) = 0, j \neq k$ (because the r.v.'s are independent), implies

$$Var[\bar{X}_n] = \frac{1}{n^2}\sum_{j=1}^{n} Var[X_j] = \frac{\sigma^2}{n}.$$

As a result the sample mean converges to the population mean μ in the probability or weak sense, i.e.,

$$\lim_{n \longrightarrow +\infty} P\left\{\left|\frac{1}{n}\sum_{j=1}^{n} X_j - \mu\right| > \epsilon\right\} = 0, \text{ for all } \epsilon > 0 \text{ fixed.}$$

This is known as *the weak law of large numbers.* It results directly from Chebyshev's inequality:

$$\forall \epsilon > 0, P(\{|\bar{X}_n - \mu| \geq \epsilon\}) \leq \frac{Var[\bar{X}_n]}{\epsilon^2} = \frac{\sigma^2}{n\epsilon^2}.$$

We have also the *strong law of large numbers* which states

$$P\left\{\lim_{n \longrightarrow +\infty} \frac{1}{n} \sum_{j=1}^{n} X_j = \mu\right\} = 1$$

but the proof of the latter result is much more involved. It can be found in any standard textbook on probability theory.

19.2.5 Monte Carlo Integration

Consider the integral $I = \int_0^1 g(x)dx$. This integral can be thought of as the expectation $E[g(U)]$ where U is a random variable uniformly distributed on $[0, 1]$ $(U \sim \mathcal{U}(0, 1))$. According to the law of large numbers above the expected value $I = E[g(U)]$ can be estimated by a sample mean. Let $U_1, U_2, \cdots, U_n$ be a sequence of random numbers *generated* according to the uniform distribution $\mathcal{U}(0, 1)$. Then,

$$I \approx I_n = \frac{1}{n} \sum_{j=1}^{n} g(U_j).$$

The strong law of large numbers guarantees that with probability one,

$$\lim_{n \longrightarrow \infty} I_n = I.$$

This is the basis for Monte-Carlo integration. We note that the sampling of genuinely random numbers is not possible on a digital computer. Instead, most programming languages and computing environments such as Matlab have one or more built in functions that can generate sequences of *pseudo-random* numbers. The function *rand()* of Matlab for example can be used to generate sequences of pseudo-random numbers that are $\mathcal{U}(0, 1)$ while *randn* samples the standard normal distribution $\mathcal{N}(0, 1)$ (Gaussian distribution of mean zero and variance one). Pseudo-random number generators are based on sequences of *floating-point* numbers that are drawn from some recurrent formula. Given that the number of *floating-point* numbers that are represented on any given computer is finite, all the pseudo-random numbers are periodic but their periods are usually very long.

■ **EXAMPLE 19.1**

Consider

$$I = \int_0^1 e^x dx = e - 1 \approx 1.7183.$$

To use Monte Carlo integration, we view this integral as the expectation

$$E[e^U] \approx \frac{1}{n}\sum_{j=1}^{n} e^{U_j}, \text{ with } \mathcal{U}(0,1)$$

and use the function *rand()* of Matlab to generate a sequence of $\mathcal{U}(0,1)$ pseudo-random numbers. Note that in Matlab *rand(N,M)* returns an $N \times M$ matrix of random numbers, all uniformly distributed in $[0,1]$. We consider the two cases with $n = 10$ and $n = 1000$ and produce two replicas in each case in order to highlight the random character of the computation.

```
>>rand('state',0) %this (re)initializes the function rand
>>I = mean(exp(rand(1,10))
ans=
      1.8318
>>I = mean(exp(rand(1,10))
ans=
      2.0358
>>I = mean(exp(rand(1,1000000))
ans=
      1.7189
>>I = mean(exp(rand(1,1000000))
ans=
      1.7178
```

Increasing the number of samples clearly improves the estimated integral.

EXAMPLE 19.2

Here we show how the Monte Carlo method can be used to compute an improper integral. We consider the integral

$$I = \int_{-\infty}^{+\infty} \cos(x) e^{-x^2/2} dx.$$

We have

$$I = \int_{-\infty}^{+\infty} \sqrt{2\pi}\cos(x)\frac{1}{\sqrt{2\pi}} e^{-x^2/2} dx.$$

Recognizing the normal probability density function $f(x) = \frac{1}{\sqrt{2\pi}} e^{-x^2/2}$, we view the given integral as the expectation $E[\cos(X)]$, where X is an

$\mathcal{N}(0,1)$ random variable. Therefore

$$I \approx \frac{\sqrt{2\pi}}{n}\sum_{j=1}^{n}\cos(X_j),$$

where the X_j's are random numbers sampled (drawn) from the standard normal distribution. This can be easily accomplished in Matlab by using the *randn()* function. The Matlab lines below show how this is implemented. To produce a benchmark value, we first use the deterministic *quad* function (i.e., the composite Simpson rule) on a finite interval $[-A, A]$ with $A = 10$ and $A = 100$.

```
>> quad('cos(x).*exp(-x.^2/2)',-10,+10)
ans =
    1.5203
>> quad('cos(x).*exp(-x.^2/2)',-100,+100)
ans =
    1.5203
>> sqrt(2*pi)*mean(cos(randn(1,100)))
ans =
    1.5682
>> sqrt(2*pi)*mean(cos(randn(1,100)))
ans =
    1.4070
>> sqrt(2*pi)*mean(cos(randn(1,1000000)))
ans =
    1.5212
>> sqrt(2*pi)*mean(cos(randn(1,1000000)))
ans =
    1.5225
```

The two examples above illustrated how to use (pseudo-) random numbers generated by the functions *rand* and *randn* to perform Monte Carlo integration based on the uniform and normal distributions, respectively. In practice, we may need to generate random numbers from an arbitrary probability distribution such as the exponential, the gamma, or the beta distribution. Many methods have been developed and made available in the literature to deal with such general cases as well as very particular ones. In the sequel, we assume that a uniform random number generator such as *rand* is given and present two standard techniques that can be used to generate pseudo-random numbers from an arbitrary distribution.

19.2.6 Inverse Transform Method

The inverse transform method, for generating pseudo-random numbers, from an arbitrary distribution, is based on the following basic statement.

Given a random variable X and its probability density f_X and cumulative distribution function, $F_X(x)$, then the random variable

$$U = F_X(X) = \int_{-\infty}^{X} f_X(x)dx$$

is uniformly distributed on $(0,1)$. Conversely, if $U \sim \mathcal{U}(0,1)$, then we have

$$P(\{X \leq x\}) = P(\{F^{-1}(U) \leq x\}) = P(\{U \leq F(x)\}) = F(x).$$

This is represented schematically in Figure 19.3. If we are able to invert F easily, then the inverse transform method for f_X can be implemented in two easy steps.

1. Draw a uniform pseudo-random variate $U \sim \mathcal{U}(0,1)$.

2. Set $X = F^{-1}(U)$.

■ EXAMPLE 19.3

We consider the standard exponential distribution $X \sim \exp(\lambda)$. We have $F(x) = 1 - e^{-\lambda x}$ and for a given uniform variate, the inverse transform method yields

$$X = F^{-1}(U) = -\frac{1}{\lambda}\ln(1-U).$$

Given that the probability for drawing U and $1-U$ from the uniform distribution is the same, in practice exponential variates are usually generated by drawing uniform variates $U \in (0,1)$, then returning $X = -\ln(U)/\lambda$. We now use the algorithm above to approximate the integral $I = \int_0^{+\infty} 2xe^{-2x}dx = \frac{1}{2}$, which is the expectation of the exponential random variable $X \sim \exp(2)$.

```
>> u = rand(1,1000000);
>> x = - log(u)/2;
>> mean(x)
ans =
    0.4993
>> u = rand(1,1000000);
>>x = - log(u)/2;
>>mean(x)
ans =
    0.5001
>> u = rand(1,1000000);
>>x = - log(u)/2;
>>mean(x)
ans =
    0.5004
```

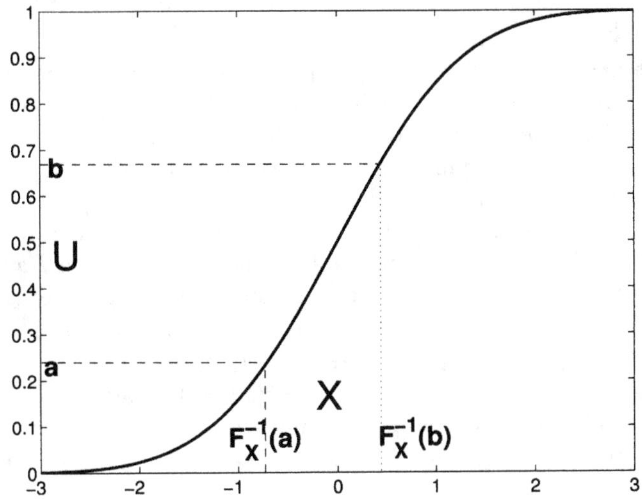

Figure 19.3 Schematic of the inverse method: $P(\{a \leq U \leq b\}) = P(\{F_X^{-1}(a) \leq X \leq F_X^{-1}(b)\})$.

19.2.7 Acceptance-Rejection Method

As mentioned above, the inverse transform method is only feasible when the CDF $F(x)$ can be easily inverted. Although one can always resort to numerical root-finding techniques such as Newton's method to invert $F(x)$, it is not always a good idea because this can be very costly especially when we need to generate a large number of variates—to perform a Monte Carlo integration, for example. A better approach for the case when the inverse of $F(x)$ is not known in closed form or is expensive to evaluate is the *acceptance-rejection method* discussed here.

To simulate a random variable X obeying a given probability distribution with a pdf $f(x)$, the acceptance-rejection method starts by finding a function $t(x)$ such that $f(x) \leq t(x), \forall x \in \mathbb{R}$ whose CDF is easy to invert. Once such a function $t(x)$ is found, the acceptance-rejection method consists of the three main steps listed below. Let $g(y) = t(y)/K$ where $K = \int_{-\infty}^{+\infty} t(y)dy$.

1. Use the inverse transform method to generate a pseudo-random number Y corresponding to $g(y)$.

2. Draw a uniform variate U from $\mathcal{U}(0, 1)$, independent of Y.

3. If $U \le f(Y)/t(Y)$, then return $X = Y$ (accept), otherwise go to step 1 (reject).

Recall that for a given (fixed) Y, the probability for a uniformly distributed random number U to satisfy $U \le f(Y)/t(Y)$ is $P(\{U \le f(Y)/t(Y)\}) = f(Y)/t(Y)$. Therefore, the more this ratio is close to one the better are the chances for the random number Y to be accepted and the above procedure to be terminated. The points Y where this ratio is close to 1 are very likely to be accepted while those with a small $f(Y)/t(Y)$ are very unlikely to be accepted. To gain efficiency, it is thus important to choose a function $t(x)$ which is as close as possible to $f(x)$. Also, it can be shown that the average number of iterations (acceptance and rejection trails) to terminate the procedure with an accepted value X is given by $K = \int_{-\infty}^{+\infty} t(x)dx$.

If the support of $f(x)$ is bounded, i.e., $f(x) = 0$ outside a bounded interval $[\alpha, \beta]$, then a natural choice for $g(x)$ is simply the uniform distribution on $[\alpha, \beta]$ and choose $t(x) = \text{ constant } = \max_{[\alpha,\beta]} f(x)$ for $\alpha \le x \le \beta$ and $t(x) = 0$ otherwise.

EXAMPLE 19.4

As an example we consider the *beta-distribution* on $[0, 1]$.

$$f(x) = \frac{x^{\alpha_1 - 1}(1-x)^{\alpha_2 - 1}}{B(\alpha_1, \alpha_2)}, \qquad x \in [0,1],$$

where

$$B(\alpha_1, \alpha_2) = \int_0^1 x^{\alpha_1 - 1}(1-x)^{\alpha_2 - 1} dx.$$

For $\alpha_1 = \alpha_2 = 3$, we have $f(x) = 30(x^2 - 2x^3 + x^4), \quad x \in [0, 1]$. Note that its CDF is a fifth-order polynomial that is not easy to invert and thus the inverse transform method would be hard to apply here.

Let $t(x) = \max_{[0,1]} f(x) = 30/16$. Using the uniform distribution as the reference density $g(x)$, we have the following algorithm:

1. Draw two independent uniform random variates U_1, U_2 from $\mathcal{U}(0, 1)$.
2. If $U_2 \le 16(U_1^2 - 2U_1^3 + U_1^4)$ accept $X = U_1$; otherwise, reject and go back to step 1.

19.3 MARKOV CHAINS AND BIRTH-DEATH PROCESSES

In a general manner a stochastic process is a collection of random variables, $X_t, t \in S$, where S is a discrete set or an interval of $\mathbb{R}$. The X_t's can be either correlated or uncorrelated with each other. Markov chains lie somehow in between the fully correlated and fully uncorrelated extremes. In this section, we will review Markov chains in general and in particular the case of the birth and death process, which will be applied in Sections 1.3 and 1.4 below to model some features of tropical convection. In fact, birth and death processes are widely used in biology and computer science and many other disciplines.

19.3.1 Discrete-Time Markov Chains

Consider a discrete stochastic process, $X_{t_0}, X_{t_1}, X_{t_2}, \cdots$ where $t_0 < t_1 < t_2 < \cdots$ is an increasing sequence of discrete times and the X_{t_i}'s are discrete random variables defined on a common *state space*, $x_0, x_1, x_2, \cdots$. For simplicity we will omit the subscript t and denote the discrete process simply as $X_0, X_1, X_2, \cdots$.

The stochastic process $X_n, n \geq 0$ is called a Markov chain if in addition it satisfies the Markov property

$$P\{X_{n+1} = x_j / X_n = x_i, X_{n-1} = x_k, \cdots, X_0 = x_l\} = P\{X_{n+1} = x_j / X_n = x_i\}. \tag{19.2}$$

The Markov chain is said to be stationary or homogeneous if $P\{X_{n+1} = x_j / X_n = x_i\}$ is independent of n. The conditional probabilities $P_{ij} = P\{X_{n+1} = x_j / X_n = x_i\}$ are called the transition probabilities and the matrix $P = [P_{ij}]_{i,j \geq 0}$ is called the transition probability matrix.

Two important properties of the matrix P are 1) all of its entries lie between 0 and 1 and 2) all of its rows sum to one, namely, $0 \leq P_{ij} \leq 1$ and $\sum_{j=0}^{\infty} P_{ij} = 1$. These two properties result directly from the fact that for each fixed i, $P_{i,j}, j = 0, 1, \cdots$ is a probability distribution on the state space $x_0, x_1, x_2, \cdots$. A matrix that satisfies these two properties is called a stochastic matrix.

The Chapman-Kolmogorov equations

Consider the n-step transition probabilities $P_{ij}^{(n)} = P\{X_{n+k} = x_j / x_k = i\} = P\{X_n = x_j / X_0 = x_i\}$. According to the definition of conditional probabilities, we have for all $i, j \geq 0$

$$\begin{aligned} P_{ij}^{(n+m)} &\equiv P\{X_{n+m} = x_j / X_0 = x_i\} \\ &= \sum_{k=0}^{\infty} P\{X_{n+m} = x_j / X_m = x_k\} P\{X_m = x_k / X_0 = x_i\} = \sum_{k=0}^{\infty} P_{ik}^{(n)} P_{kj}^{(m)}. \end{aligned}$$

Thus

$$P_{ij}^{(n+m)} = \sum_{k=0}^{\infty} P_{ik}^{(n)} P_{kj}^{(m)}, i, j = 0, 1, 2, \cdots.$$

These identities are known as the Chapman-Kolmogorov equations. If we let $P^{(n)}$ denote the n-step transition probability matrix, then it is easy to see, by the Chapman-Kolmogorov equations, that $P^{(n+1)} = P^{(n)} \times P$ and $P^{(n)} = P^n = P \times P \times \cdots \times P$ (n times).

Limiting and stationary distribution

A Markov chain X_n is said to have a limiting distribution if the limit $\lim_{n\longrightarrow\infty} P_{ij}^n$ exists for all i, j, and is independent of i. The limit $\pi_j = \lim_{n\longrightarrow\infty} P_{ij}^n$ when it exists is called the limiting distribution of X_n. As a consequence of the Chapmann-Kolmogorov equations, the limiting distribution satisfies

$$\pi_j = \sum_{i=0}^{\infty} \pi_i P_{ij} \text{ or } \pi = \pi P, \tag{19.3}$$

i.e., π is a left eigenvector of the matrix P associated with the eigenvalue $\lambda = 1$. Any row vector $\pi_j, j = 0, 1, 2\cdots$ of real numbers which satisfies (19.3) and the two properties of a probability distribution: $\sum_{i=0}^{\infty} \pi_j = 1,\ 0 \leq \pi_j \leq 1$ is called a stationary or invariant distribution of the Markov chain. The limiting distribution when it exists is an invariant distribution but the converse is not always true. Sufficient conditions for the existence of the limiting distribution and for the uniqueness of the stationary distribution are known but they are beyond the scope of this brief introduction. They involve the notion of ergodicity which intuitively amounts to saying that all the states of the chain are visited equally infinitely many times when the chain is run for an infinitely long time. The interested reader is referred to the books by S. M. Ross [12] and G. F. Lawler [2].

Time reversible chains and detailed balance

Consider a stationary ergodic (i.e., that has a limiting and unique stationary distribution π_j) Markov chain with transition probabilities P_{ij}. Assume that the chain is run for a very long time and is in its equilibrium state, i.e., it satisfies $P\{X_n = x_j\} = \pi_j, j = 0, 1, 2, \cdots$. Consider the backward process $X_n, X_{n-1}, X_{n-2}, \cdots$ when n goes to infinity. By some manipulations we can show that the time reversed process is also a Markov chain with the transition probabilities

$$Q_{ij} \equiv P\{X_m = x_j / X_{m+1} = x_i\} = \frac{\pi_j}{\pi_i} P_{ji}.$$

A Markov chain is said to be time reversible if $Q_{ij} = P_{ij}$ for all $i, j \geq 0$. In other words, X_n is time reversible if

$$\pi_i P_{ij} = \pi_j P_{ji}, i, j = 0, 1, 2, \cdots. \tag{19.4}$$

The equations in (19.4) are known as the detailed balance equations. They basically state that, in the long run, the rate of transition from state x_i to x_j equals the rate of transition from x_j to x_i.

■ EXAMPLE 19.5

A random walk on a finite set. Imagine a person who takes random steps between the positions $0, 1, 2, \cdots, M$. If at time n the person finds herself in a state $i, 0 \leq i \leq M$, she will take a step randomly to the left or to the right according to the following transition probabilities:

$$P_{i,i+1} = \alpha_i, \;\; P_{i,i-1} = 1 - \alpha_i, i = 1, 2, \cdots, M-1, P_{0,1} = 1, P_{M,M-1} = 1.$$

Here $0 < \alpha_i < 1$ for $i = 1, \cdots, M-1$ to guarantee that the induced Markov chain is ergodic and time reversible. When the random walker hits 0 or M, with probability one, she will return to 1 or to $M-1$, respectively, at the next step. This is a bounded random walk with reflecting boundaries. Bounded and unbounded random walks are widely used in the theory and applications of Markov chains and probability theory. While the $P_{i,j}$'s define the transition probability matrix of the discrete time Markov chain, we can easily write down the detailed balance equations:

$$\pi_0 = (1 - \alpha_1)\pi_1, \;\; \pi_i \alpha_i = \pi_{i+1}(1 - \alpha_{i+1}), \; i = 1, 2, \cdots M-2,$$

$$\pi_{M-1}\alpha_{M-1} = \pi_M.$$

Setting $\alpha_0 = 1$ and $\alpha_M = 0$, we can easily see that the solution to these equations is given by

$$\pi_0 = \left[1 + \sum_{j=1}^{M} \prod_{l=1}^{j} \frac{\alpha_{l-1}}{1 - \alpha_l}\right]^{-1} \quad \text{and } \pi_j = \pi_0 \prod_{l=1}^{j} \frac{\alpha_{l-1}}{1 - \alpha_l}.$$

For the particular case $\alpha_j = 0.5, j = 1, 2, \cdots, M-1$, we have $\pi_0 = \pi_M = 1/2M$ and $\pi_j = 1/M$, for $1 \leq j \leq M-1$.

19.3.2 The Poisson Process

The Poisson process, N_t, is a counting process, i.e., the process of counting the number of events that occur in sequence by time t, such that:

i) $N_0 = 0$.

ii) N_t has independent and stationary increments. In other words, given $t_1 < t_2 \leq t_3 < t_4$ the number of events that occur between times t_1 and t_2, $N_{t_2} - N_{t_1}$, and the number of events that occur between t_3 and t_4, $N_{t_4} - N_{t_3}$, are independent random variables and the distribution of $N_{t+s} - N_t$, for $s, t > 0$, depends only on the length of the interval, s, and not on t.

iii) N_t is Poisson distributed with mean λt where λ is a positive parameter called the rate of the process:

$$P\{N_{t+s} - N_t = k\} = P\{N_s - N_0 = k\} = P\{N_s = k\} = e^{-\lambda s}\frac{(\lambda s)^k}{k!}.$$

Using Taylor expansion and the moment generating function of the Poisson distribution, we can show that a counting process that satisfies i) and ii) above is a Poisson process with rate $\lambda > 0$ if and only if it also satisfies, for all small $h > 0$,

$$P\{N_{t+h} - N_t = 1\} = \lambda h + o(h) \text{ and } P\{N_{t+h} - N_t \geq 2\} = o(h),$$

where $o(h)$ is an arbitrary function of h that satisfies $\lim_{h \longrightarrow 0} o(h)/h = 0$.

Inter-arrival times
Consider a Poisson process $N_t, t \geq 0$ with a rate $\lambda > 0$. Let T_1 be the time at which the first event of N_t occurs, T_2 the time spent between the first and the second events, T_3 the times between the second and third events, etc. Then $T_1, T_2, \cdots$ is a sequence of independent exponentially distributed random variables with a common rate $\lambda > 0$. This follows directly from the independence of increments property and the fact that $P\{T_1 > t\} = P\{N_t = 0\} = e^{-\lambda t}$. The waiting time $S_n = T_1 + T_2 + \cdots + T_n$ until the nth event occurs is Gamma distributed with parameters n and λ.

This intimate relationship between the Poisson process and the exponential random variables is due to the so-called *memory less* characterization-property of the exponential distribution, namely, T is an exponentially distributed random variable if and only if

$$P\{T > t + s\} = P\{X > t\}P\{X > s\}.$$

■ EXAMPLE 19.6

Assume certain types of batteries are being used in a certain electronic device, one at a time. A battery in operation is replaced by a new one upon its failure. Assume that the lifetime of a battery in operation is an exponential random variable with rate $\lambda > 0$. Then the number of batteries being used by time t, N_t, is a Poisson process with rate $\lambda > 0$, namely, we have $P\{N_t = k\} = e^{-\lambda t}(\lambda t)^k/k!$.

Assume that in a certain day, customers enter a store according to a Poisson process with rate λ, i.e., the number N_t of customers that enter the store by time t of the day is Poisson distributed with mean λt. Then the time spent between the arrivals of two successive customers is an exponential random variable with mean $1/\lambda$, i.e., the average time spent between two successive arrivals is $1/\lambda$ (λ has units of one over time). Assume $1/\lambda = 1$ hour. Then, if we have waited 30 minutes after the nth customer arrived and nobody showed up, then the expected time for the $(n+1)$th customer is still 1 hour, because of the memoryless property of the exponential distribution.

19.3.3 Continuous-Time Markov Chains

A stochastic process X_t where t is in $(0, +\infty)$ defined on a discrete state space $x_0, x_1, x_2, \cdots$ is called a Markov chain if it satisfies the Markov property (in continuous form)

$$P\{X_{t+s} = x_j / X_s = x_i, X_u = x_u, 0 \leq u \leq s\} = P\{X_{t+s} = x_j / X_s = x_i\}.$$

If $P\{X_{t+s} = x_j / X_s = x_i\}$ is independent of s then X_t is said to be a stationary or homogeneous Markov chain. Similarly to the discrete case, the conditional probabilities $P_{i,j}(t) = P\{X_{t+s} = x_j / X_s = x_i\}$ are called the transition probabilities and the time-dependent matrix $P(t) = [P_{ij}(t)]$ is called the transition probability matrix.

The continuous version of the Chapman-Kolmogorov equations, which also follows from the elementary definition of conditional probabilities, reads

$$P_{ij}(t+s) = \sum_{k=0}^{\infty} P_{ik}(s) P_{kj}(t),$$

which results from the conditional probability formula $P\{X_{t+s} = x_j / X_0 = x_i\} = \sum_{k=0}^{\infty} P\{X_{t+s} = x_j / X_t = x_k\} P\{X_t = x_k / X_0 = x_i\}$. Notice the analogy with the discrete case.

Waiting time and transition rates

An important property of continuous-time Markov chains is one that links them directly to the Poisson process. Assume that at time t the Markov chain X_t is in state x_i, i.e., $X_t = x_i$. For a stationary stochastic process, the Markov property is equivalent to the fact that the time T_i the process stays in state x_i before it makes a transition (or jump) to another state $x_j \neq x_i$, i.e., $T_i = \inf\{s > 0 \text{ such that } X_{t+s} \neq x_i \text{ given that } X_t = x_i\}$, is an exponential random variable. Moreover, the times T_{ij}, it takes the chain to make a transition from x_i to $x_j, j \neq i$ are independent exponential random variables. By construction we have $T_i = \min\{T_{ij}, j = 0, 1, 2 \cdots, j \neq i\}$ and if $q_{ij}, i \neq j$ are the rates of the T_{ij}'s and v_i denote the rates of the *waiting times* $T_i, i = 0, 1, 2, \cdots$. Then according to the properties of exponential random variables, $v_i = \sum_{j \neq i} q_{ij}$. The matrix R whose diagonal $R_{ii} = -v_i$ and non-diagonal entries are the q_{ij}'s is called the *infinitesimal generator* of the chain. As we will see below, the matrix R completely determines the Markov chain. The entries q_{ij} are often called the transition rates of the chain. As an immediate consequence of this, we have, for sufficiently small $h > 0$,

$P_{ii}(h) = P\{T_i < h\} = 1 - v_i h + o(h)$ and $P_{ij}(h) = P\{T_{ij} > h\} = q_{ij}h + o(h)$.

Kolmogorov forward and backward equations
By the Chapman-Kolmogorov equations, we have for $h, t > 0$

$$P_{ij}(t+h) - P_{ij}(t) = \sum_{k=0}^{\infty} P_{ik}(h)P_{kj}(t) - P_{ij}(t)$$

$$= \sum_{k\neq i} P_{ik}(h)P_{kj}(t) - [1 - P_{ii}(h)]P_{ij}(t).$$

Dividing both sides by h and letting $h \longrightarrow 0$ and using the identities above yields the Kolmogorov backward equations

$$\frac{d}{dt}P_{ij}(t) = \sum_{k\neq i} q_{ik}P_{kj}(t) - v_i P_{ij}(t).$$

If instead we write $P_{ij}(t+h) = \sum_{k=0}^{\infty} P_{ik}(t)P_{kj}(h)$, then similar manipulations yield the Kolmogorov forward equations

$$\frac{d}{dt}P_{ij}(t) = \sum_{k\neq j} q_{kj}P_{ik}(t) - v_j P_{ij}(t).$$

In matrix notation, the backward equations become $P' = RP$ while the forward equations read $P' = PR$. The backward and/or forward equations provide a system of linear ordinary differential equations for the transition probabilities. It is closed with the initial conditions $P_{ij}(0) = \delta_{ij} = 1$ if $i = j$ and 0 otherwise. Following the standard theory of differential equations, its solution is given by $P(t) = \exp(tR)$. However evaluating the matrix exponential can be problematic in practice, especially, when the matrix R is very large, i.e., when the Markov chain has a large number of states. A lot of research has been done in order to come up with practical solvers for this large ODE system. The interested reader is referred to work of Ref. [13].
Limiting distribution and detailed balance
Similarly to the discrete case, the limiting distribution of X_t is given by $P_j = \lim_{t\longrightarrow\infty} P_{ij}(t)$ when this limit exists and is independent of i. It satisfies the steady state forward equations

$$\sum_{k\neq j} q_{kj}P_k = v_j P_j,\ 0 \leq P_j \leq 1,\ \sum_{j=0}^{\infty} P_j = 1.$$

A solution $P_j, j \geq 0$ to these equations is called a stationary or an equilibrium distribution. Intuitively, these equations express the fact that, in the long run, the rate of transition away from state j, $v_j P_j$, is balanced by the rate of transitions to state j, $\sum_{k \neq j} q_{kj} P_k$.

Let $\mathbb{P}_{ij}$ be the probability that the Markov chain makes a transition from x_i to $x_j \neq x_i$ the first time it leaves state i, i.e., $\mathbb{P}_{ij} = P\{T_{ij} = \min(T_{ik}, k \neq i)\}$. According to the properties of exponential random variables, we have $\mathbb{P}_{ij} = q_{ij}/v_i$, $i \neq j$ and $\mathbb{P}_{ii} = 0$. $\mathbb{P}$ is a stochastic matrix and the associated discrete Markov chain is called the *embedded discrete* chain of the original continuous-time Markov chain X_t.

Let $\pi_j, j = 0, 1, \cdots$, be the limiting distribution of the embedded chain. Then we have the following relationships between π_j and P_j:

$$\pi_j = \frac{v_j P_j}{\sum_k v_k P_k}, \quad j = 0, 1, \cdots .$$

We note that these equalities express simply the fact that $(\pi_j) \propto (v_j P_j)$ with the term in the denominator being the normalization constant. Therefore if we know π_j for the embedded discrete chain we can find P_j and vis versa.

The continuous process X_t is said to be time reversible if the embedded discrete chain is time reversible, i.e., if $\pi_i \mathbb{P}_{ij} = \pi_j \mathbb{P}_{ji}$ for all i, j, i.e., $v_i P_i \mathbb{P}_{ij} = v_j P_j \mathbb{P}_{ji}$ for all i, j. But $\mathbb{P}_{ij} = q_{ij}/v_i$, therefore we have

$$P_i q_{ij} = P_j q_{ji}, \quad i, j = 0, 1, 2, \cdots , \tag{19.5}$$

which are the detailed balance equations for the continuous chain X_t. Intuitively, these equations express the fact that the rate of transitions from state i to state j and from j to i are balanced in the long run.

■ EXAMPLE 19.7

Consider a certain machine that operates for an exponentially distributed random time before it breaks down at a rate $\mu > 0$. Upon its failure the machine is sent to the repair shop where it takes an exponential time with rate $\lambda > 0$ before it is repaired and put back in service. Let X_t be the random variable that takes the value 1 if the machine is in service at time t and 0 otherwise. Then X_t is a continuous-time Markov chain with the state space reduced to 0 and 1, a two-state Markov chain. The infinitesimal generator is given by $R = \begin{bmatrix} -\lambda & \lambda \\ \mu & -\mu \end{bmatrix}$ and the backward equations are given by

$$\begin{aligned} P'_{00} &= \lambda[P_{10} - P_{00}], \quad & P'_{10} &= \mu[P_{00} - P_{10}], \\ P'_{01} &= \lambda[P_{11} - P_{01}], \quad & P'_{11} &= \mu[P_{01} - P_{11}]. \end{aligned}$$

We first note that this system splits into two two-by-two subsystems, one for P_{00} and P_{10} and one for P_{01} and P_{11}. Combining the two first equations yields $\mu P'_{00} + \lambda P'_{10} = 0$. i.e., $P_{00}(t) + P_{10}(t) = c$ is constant and substitution in the second equation yields a single (linear) equation for P_{00} which can be easily solved by the method of integrating factors. With the initial conditions $P_{ij}(0) = \delta_{ij}$, we find

$$P_{00}(t) = \frac{u}{\lambda+\mu} + \frac{\lambda}{\lambda+\mu}e^{-(\lambda+\mu)t} \text{ and } P_{10} = \frac{\mu}{\lambda+\mu} - \frac{\mu}{\lambda+\mu}e^{-(\lambda+\mu)t}$$

and using the identities $P_{i0} + P_{i1} = 1, \quad i = 0, 1$ yields the other two probabilities without having to solve the remaining equations:

$$P_{01}(t) = \frac{\lambda}{\lambda+\mu} - \frac{\lambda}{\lambda+\mu}e^{-(\lambda+\mu)t} \text{ and } P_{11} = \frac{\lambda}{\lambda+\mu} + \frac{\mu}{\lambda+\mu}e^{-(\lambda+\mu)t}.$$

Letting $t \longrightarrow \infty$ leads to the limiting distribution of the chain

$$P_0 = \frac{\mu}{\lambda+\mu}, P_1 = \frac{\lambda}{\lambda+\mu}.$$

Notice that detailed balanced is evidently satisfied: $\lambda P_0 = \mu P_1$.

■ EXAMPLE 19.8 The Birth and Death Process

Assume that customers arrive in a shop according to a Poisson process with rate λ. The customers are served by m tellers. Upon arrival, a customer proceeds to the first available teller or waits in a queue until a teller is freed. Assume that the service time of each teller is an exponential random variable with rate $\mu >$. Then, the number of customers, X_t, in the store at any given time, $t \geq 0$, is a continuous-time Markov chain with transition rates

$$\begin{aligned} q_{n,n+1} &= \lambda, \text{ for } n \geq 0, q_{n,n-1} = n\mu \text{ for } 1 \leq n \leq m, \\ q_{n,n-1} &= m\mu \text{ for } n \geq m+1. \end{aligned}$$

The rates of the waiting times between two transitions (i.e., before the next arrival or departure occurs) are given by $v_0 = \lambda, v_n = \lambda + n\mu$, for $1 \leq n \leq m$, and $v_n = \lambda + m\mu$ if $n \geq m+1$. In queuing theory this is known as the M/M/m model. The infinitesimal generator of the chain is a tridiagonal matrix. Similar models are ubiquitous in practice. They are called birth and death processes. $X_t = n$ is the regarded as the number of members in a population at time t. The rate at which arrivals occur, $q_{n,n+1} = \lambda_n, n \geq 0$, is called the birth rate and the rate at which departures occur, $q_{n,n-1} = \mu_n, n \geq 1$, is called the death rate. Notice that in the general case both arrival and departure rates are assumed dependent on n and that we implicitly assumed $\mu_0 = 0$. If in addition

$\lambda_k = 0$ for a certain integer $k \geq 1$ then the process becomes bounded to the states $0, 1, 2, \cdots, k$. If $X_t = k$, then new arrivals are not accepted.

The steady state forward equations for the birth-death process are

$$\lambda_0 P_0 = \mu_1 P_1,$$
$$(\lambda_n + \mu_n) P_n = \mu_{n+1} P_{n+1} + \lambda_{n-1} P_{n-1}, n \neq 1.$$

Substitution of the first equation into the second, the second into the third, and so on, yields

$$\lambda_n P_n = \mu_{n+1} P_{n+1},\ n \geq 0,$$

which are nothing but the detailed balance equations (19.5) for the birth-death process. With the constraint $\sum_{j=0}^{\infty} P_j = 1$, the solution to this equations is

$$P_n = \frac{\lambda_{n-1}\lambda_{n-2}\cdots\lambda_0}{\mu_n \mu_{n-1} \mu_1} P_0, \quad P_0 = \left[1 + \sum_{j=0}^{\infty} \frac{\lambda_{n-1}\lambda_{n-2}\cdots\lambda_0}{\mu_n \mu_{n-1} \mu_1}\right]^{-1}.$$

Therefore, a necessary condition for the birth-death process to admit a limiting distribution is $\sum_{j=0}^{\infty} \frac{\lambda_{n-1}\lambda_{n-2}\cdots\lambda_0}{\mu_n \mu_{n-1} \mu_1} < \infty$. This is guaranteed for the M/M/m process if the ratio $\lambda/m\mu < 1$, i.e., the rate at which customers arrive is smaller than the rate at which they are being served. In the case of the finite state birth-death process this condition is always satisfied because the summation terminates at $n = k$; $\lambda_k = 0$.

19.4 A BIRTH-DEATH PROCESS FOR CONVECTIVE INHIBITION

19.4.1 The Microscopic Stochastic Model for CIN: Ising Model

Here a give a brief description of the microscopic stochastic model for CIN, which is used as a basis for the birth-death process for CIN. The interested reader is referred to Ref. [9] for more details. It is based on the Ising model for magnetization of statistical mechanics. As stated above, CIN is an energy barrier for spontaneous deep penetrative convection in the tropic and it is known to have important fluctuations in the horizontal on the order of 1 km to 10 km. Hence, we consider sites that are uniformly distributed on a lattice (which can be though of as spanning one horizontal gridbox of the climate model) on which we define an order parameter σ_I on a finite lattice $\Lambda \subset \{0, 1\}^{\mathbb{Z}}$.

$$\sigma_I(x) = 1 \text{ at a site if convection is inhibited (a CIN site),} \tag{19.6}$$
$$\sigma_I(x) = 0 \text{ at a site if there is potential for deep convection (a PAC site).}$$

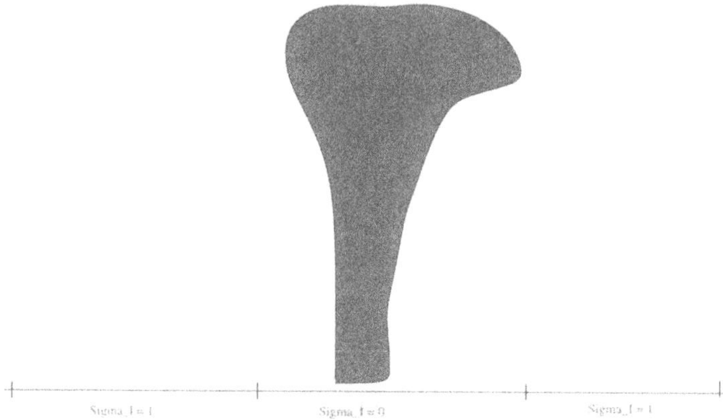

Figure 19.4 A Cartoon of a deep penetrative hot-tower cloud represented at a PAC site. The order parameter takes values 0 or 1 on a given site according to whether it is a CIN site or there is potential for deep convection.

A cartoonist picture of this representation is shown in Figure 19.4. On the coarse grid of mesh size Δx of a climate model, the value of CIN at a coarse mesh point, $j\Delta x$, is given by the average

$$\bar{\sigma}_I(j\Delta x) = \frac{1}{\Delta x}\int_{(j-1/2)\Delta x}^{(j+1/2)\Delta x} \sigma_I(x)dx. \tag{19.7}$$

Here we assume a simple 1D domain for simplicity. The model's cumulus parameterization then "decides" according to this average CIN value on whether to allow deep convection to occur at that grid point or not. As explained in the introduction section, factors to overcome CIN are very complex and can be both local or external. Instead of trying to follow the details of such interactions, the microscopic CIN sites are set to interact with each other and with the external large-scale values of the deterministic flow variables according to the following probabilistic rules.

A) If a CIN site is surrounded by mostly CIN sites, then it has higher probability to remain a CIN site.

B) If a PAC site is surrounded by mostly CIN sites, then it has higher probability to switch to a CIN site.

C) The external large-scale values, $\vec{u}_j$, supply an external potential $h(\vec{u}_j)$ that can modify the dynamics according whether external conditions favor CIN or PAC.

Following the standard theory of the Ising model of statistical physics, the microscopic energy for CIN is given by the Hamiltonian

$$H_h(\sigma_I) = -\frac{1}{2}\sum_x \sum_{y \neq x} J(\gamma(x-y))\sigma_I(x)\sigma_I(y) - h\sum_x \sigma_I(x). \tag{19.8}$$

Here $H_h(\sigma_I)$ is the microscopic energy associated with a given configuration σ_I with $J \geq 0$ is the symmetric interaction potential and γ defines the range of microscopic interactions

$$J(\gamma r) = \frac{1}{L+1} U\Big(\frac{N}{L+1} r\Big)$$

with $U(r) = U(-r), r \in \mathbb{R}$, $U(r) = 0, |r| \geq 1$, for example,

$$U(r) = \begin{cases} U_0 & \text{if } r < 1, \\ 0 & \text{otherwise.} \end{cases} \tag{19.9}$$

$\gamma = \frac{N}{L+1}$, where, in the 1D setting, $2L$ is the number of interacting neighboring sites and N is the total number of sites. Note that, in the absence of the external factor, $h = 0$, the minimum energy level is achieved when $\sigma_I(x) = 1\ \forall x$, i.e., an all-CIN configuration, while the all-PAC configuration, $\sigma_I(x) = 0$, has the highest level—zero energy.

If we regard the mixed boundary layer as a heat bath for CIN, then according to the theory of statistical mechanics, the equilibrium distribution of the configurations σ_I for the Hamiltonian dynamics (that oscillate about the minimum energy level) is given by the Gibbs measure

$$G(\sigma_I) = \frac{1}{Z_\Lambda} e^{-\beta H_h(\sigma_I)}, \tag{19.10}$$

where β is a positive parameter that depends on the "temperature" of the microscopic system and Z_Λ is a normalization constant (which can be very large and hard to compute!).

The external potential h modifies the CIN configuration according to whether the large scales are favorable to CIN or not. The Hamiltonian in (19.8) is a monotonic function of h, this is very helpfull when it comes to chosing the proper dependence on the large-scale variables. In practice, h, can be represented by either the large-scale subsidence of upper troposphere air, that cools and dries the boundary layer and thus increases the energy for CIN, or the large-scale fluctuations in the boundary layer temperature and moisture, that tends to destroy CIN energy or a certain combination of the two. Here, we omit the discussion about the actual coupling of the stochastic CIN to an actual climate model. The interested reader is referred to the research papers [6,9,10].

Consistent with the Gibbs distribution, a simple dynamical model that systematically obeys the rules in A), B), and C), is given next.

A configuration randomly flips at a site x,

$$\sigma_I^x(y) = \begin{cases} 1-\sigma_I(x) & \text{if } y = x, \\ \sigma_I(y) & \text{if } y \neq x, \end{cases} \tag{19.11}$$

according a Markov jump process where the rate $c(\sigma_I, x)$ is given by the so-called Arrhenius adsorption and desorption rates

$$c(\sigma,x) = \begin{cases} \frac{1}{\tau}e^{-\beta V(x)}, & \sigma_I = 1, \\ \frac{1}{\tau}, & \sigma_I = 0, \end{cases} \tag{19.12}$$

for which $G(\sigma)$ in (19.10) is the invariant measure. Here $V(x) \equiv H[\sigma_I^{new}(x = 0)] - H[\sigma_I^{old}(x = 1)] = \frac{1}{2}\sum_{z\neq x} J(\gamma(x-z))\sigma_I(z) + h$ is the energy difference between the new configuration and the old configuration of σ_I, when one single CIN site is destroyed, i.e., the energy that the rising parcel needs to provide in order for it to potentially penetrate deep in the troposphere. We note that a CIN site surrounded by CIN sites has a high energy to overcome in order to become a PAC site (rule A)) while a PAC site naturally decays into a CIN site and would remain so for a long time if it is surrounded by CIN sites (rule B)). Finally, the inclusion of the external potential h account for the effects for the effect of the large-scale dynamics, i.e., rule C).

Here τ is a parameter that represents the time scale of CIN which is typically on the order of a few minutes to a few hours. This in effect defines a two-state Markov chain at each site x of the lattice whose transition rates, $q_{10} = e^{-\beta V(x)}/\tau$ and $q_{01} = 1/\tau$, depend on the state of the neighboring sites through the energy potential J. The construction of such Markov chains takes its roots from a more systematic procedure called Markov Chain Monte Carlo (or MCMC for short) that provides an efficient way to sample (draw random numbers) from a given distribution (here the Gibbs measure), especially, even when it is only known up to some normalization constant. It amounts to constructing an ergodic Markov chain whose equilibrium distribution is the probability distribution we wish to sample. It is easy to verify in our case that, at each site, our two-state Markov chain is in detailed balance with respect to the Gibbs measure:

$$q_{10}G[\sigma_I(x = 1)] = q_{01}G[\sigma_I(x = 0)].$$

In practice, the climate model gridbox is on the order of 100 to 200 km, which would require up to 40×40 microscopic CIN sites (in the full three-dimensional setting) that are 2 to 5 km apart. For each grid box, we thus need to sample 1600 Gibbs measure at each time step of the climate model. This would induce astronomical computations in addition to the already high-order complexity of the large-scale model that solves for the flow components. Below, we present a systematic coarse graining strategy that permits the derivation of a single birth-death process (that counts the number of active CIN sites) on each climate model grid box.

19.4.2 The Coarse-Grained Mesoscopic Stochastic Model: Birth-Death Process

We now use a systematic coarse-graining strategy to reduce the complexity of the microscopic CIN model. This method is first developed in [3] and used for the CIN model in [6]. The coarse-grained procedure starts by the definition of a coarse-grained stochastic process that tracks the total number of CIN sites in a given mesoscopic region, the climate model grid box. We introduce a coarse lattice Λ_c. Let m, q be two integers. The fine and coarse lattices are given by

$$\Lambda \equiv \frac{1}{mq}\mathbb{Z} \cap [0,1] \text{ and } \Lambda_c \equiv \frac{1}{m}\mathbb{Z} \cap [0,1].$$

Each cell, $D_k, k = 1, ..., m$, on the coarse lattice is divided onto q microscopic cells:

$$D_k \equiv \frac{1}{q}\{1, 2, ..., q\}, \forall k = 1, ..., m.$$

We introduce the coarse-grained sequence of random variables (stochastic process)

$$\eta_t(k) = \sum_{y \in D_k} \sigma_{I,t}(y). \tag{19.13}$$

Then, the sequence $\eta = \{\eta_t(k)\}_{k,t}$ in (19.13) is a birth-death Markov process defined on the configuration space

$$\mathcal{H}_{m,q} = \{0, 1, ..., q\}^{\Lambda_c},$$

such that $\eta_t(k)$ increases/decreases by one according to the transition probabilities

$$\begin{aligned}
&\text{Prob}\{\eta_{t+\Delta t}(k) = n+1 | \eta_t(k) = n\} = C_a(k,n)\Delta t + o(\Delta t),\\
&\text{Prob}\{\eta_{t+\Delta t}(k) = n-1 | \eta_t(k) = n\} = C_d(k,n)\Delta t + o(\Delta t),\\
&\text{Prob}\{\eta_{t+\Delta t}(k) = n | \eta_t(k) = n\} = 1 - (C_a(k,n) + C_d(k,n))\Delta t + o(\Delta t),
\end{aligned} \tag{19.14}$$

where the coarse-grained absorption and desorption rates are given, respectively, by

$$\begin{aligned}
C_a(k,n) &= \frac{1}{\tau_I}[q - \eta(k)]\\
C_d(k,n) &= \frac{1}{\tau_I}\eta(k)e^{-\beta\bar{V}(k)}
\end{aligned} \tag{19.15}$$

where

$$\bar{V}(k) = \sum_{\substack{l \in \Lambda_c \\ l \neq k}} \bar{J}(k,l)\eta(k) + \bar{J}(0,0)\big(\eta(k) - 1\big) + h, \tag{19.16}$$

with $\bar{J}$ is the coarse-grained interaction potential.

It is shown in Ref. [3] that $\bar{J}$ satisfies

$$J(x-y) = \bar{J}(k,l) + \frac{1}{L+1}O(\frac{q}{L+1}), x \in D_l, y \in D_k, k \neq l,$$

where

$$\begin{aligned}
\bar{J}(k,l) &= m^2 \int\int_{D_l \times D_k} J(\gamma(r-s))drds \\
&= \frac{1}{q^2} \sum_{x \in D_k} \sum_{y \in D_l} J(\gamma(x-y)) \\
&= \frac{1}{q^2} \frac{1}{L+1} \sum_{x \in D_k} \sum_{y \in D_l} U(\frac{N}{L+1}|x-y|), \quad k \neq l,
\end{aligned} \tag{19.17}$$

and

$$\bar{J}(0,0) = \frac{1}{q(q-1)} \frac{1}{L+1} \sum_{x \in D_l} \sum_{\substack{y \in D_l \\ y \neq x}} U(\frac{N}{L+1}|x-y|), \tag{19.18}$$

and the coarse-grained Hamiltonian is given by

$$\bar{H}(\eta) = -\frac{1}{2} \sum_{l \in \Lambda_c} \sum_{\substack{k \in \Lambda_c \\ k \neq l}} \bar{J}(k,l)\eta(k)\eta(l) - \frac{1}{2}\bar{J}(0,0) \sum_{l \in \Lambda_c} \eta(l)\big(\eta(l) - 1\big) - h \sum_{l \in \Lambda_c} \eta(l). \tag{19.19}$$

Notice the presence of the term representing the interactions between mesoscopic (coarse-grained) cells in the definition of the coarse-grained Hamiltonian.

The canonical Gibbs measure for the coarse-grained process is given by

$$G_{m,q,\beta}(\eta) = \frac{1}{Z_{m,q,\beta}} e^{-\beta \bar{H}(\eta)} P_{m,q}(d\eta)$$

where $P_{m,q}(d\eta)$ is the prior distribution. It is easily verified that this distribution satisfies detailed balance with respect to the coarse grained adsorption desorption rates:

$$\begin{aligned}
C_a(k,\eta)G_{m,q,\beta}(\eta) &= C_d(k,\eta+\delta_k)G_{m,q,\beta}(\eta+\delta_k) \\
C_d(k,\eta)G_{m,q,\beta}(\eta) &= C_a(k,\eta+\delta_k)G_{m,q,\beta}(\eta+\delta_k).
\end{aligned} \tag{19.20}$$

If the interactions between coarse-grained sites are ignored, which amounts to reducing the number of coarse cells to one that occupies the whole climate model grid box, the expressions for the coarse-grained potential, $\bar{V}$, and the coarse-grained Hamiltonian, $\bar{H}$, in (19.16) and (19.19), respectively, at an isolated coarse site k with a spin $\eta(k)$ simplify to

$$\begin{aligned}\bar{V}_h(\eta(k)) \equiv \bar{H}_h(\eta(k)) - \bar{H}_h(\eta(k)+1) &= \bar{J}(0,0)(\eta(k)-1)+h,\\ \bar{H}_h(\eta(k)) &= -\frac{1}{2}\bar{J}(0,0)\eta(k)\big(\eta(k)-1\big) - h\eta(k).\end{aligned} \quad (19.21)$$

According to the definition of the internal potential U in (19.9) and the definition of $\bar{J}(0.0)$ in (19.18) we have in the the case where only nearest neighbor interactions ($L=1$) are allowed between microscopic sites

$$U(N|x-y|/L+1) \neq 0 \Longleftrightarrow x=y \text{ or } x=y\pm\frac{1}{N}$$

(with $1/N$ being the actual mesh size on the microscopic lattice), hence

$$\bar{J}(0,0) = \frac{2U_0}{2(q-1)} = \frac{U_0}{q-1}.$$

Transition Probability Matrix

According to the theory of Markov chains in Section 19.3 the transition probabilities, $P_t(i,j), 0 \le i,j \le q$, to go from a state $\eta_0(k)=i$ at time $t=0$ to a state $\eta_t(k)=j$ at time $t>0$ satisfies the forward equations

$$\begin{aligned}P'_{i,j}(t) = C_d(j+1,k)P_{i,j+1}(t) &+ C_a(j-1,k)P_{j,j-1}(t)\\ &-\big(C_a(j,k)+C_d(j,k)\big)P_{i,j}(t),\ j=0,\cdots,q\end{aligned} \quad (19.22)$$

The solution of this linear ODE is easily computed through the standard exponential formula; the transition matrix is given by

$$[p_t(j,j')] = e^{tA}, \quad (19.23)$$

where A is the tridiagonal–infinitesimal generator matrix; its upper and lower diagonals are formed by desorption and adsorption rates, $C_d(j+1,k), C_d(j-1,k)$, respectively, while the main diagonal is the negatives of the waiting time rates, $-(C_d(j,k)+C_d(j,k))$.

However, in practice, there is no need to compute this exponential matrix when using this model for the purpose of climate simulations. The birth-death Markov process can be easily simulated without having to solve directly for the transition probabilities. Next, we present two algorithms to accomplish this. An approximate algorithm that uses the acceptance-rejection technique based on a fixed but small time step and an exact algorithm that uses the inverse method to advance with random time steps. The latter is known as

Gillespie's exact algorithm after the physicist and mathematical chemist who first used it to simulate chemical reactions.

19.4.3 Acceptance-Rejection Algorithm for the Birth-Death Markov Process

Given a time interval $[0, \Delta T]$, which can be thought of as one time step of the global climate model simulation, and given the external potential h at time $t = 0$, the acceptance-rejection algorithm starts by dividing the time interval into K equal time steps of size $\delta t = \Delta T/K$. We assume that δt is small enough so that the probability for having more than one transition, i.e., one birth or one death, is negligible. Approximately, at any given time step $t_n = n\delta t, n = 0, 1, \cdots, K-1$, we can have either one birth, one death, or none. Let T_1, T_2 be the times until the next birth and next death, respectively. These are independent exponential random variables with rates $\lambda = C_a(\eta_n)$ and $\mu = C_d(\eta_n)$, respectively, where η_n is the state of the birth-death process at time t_n. The approximate algorithm consists on figuring out first whether a transition (a birth or a death) actually occurs in the time interval $[t_n, t_n + \delta t]$. This is accomplished by sampling the random variable $S = \min(T_1, T_2)$, which is an exponential random variable with rate $\lambda + \mu$. If a transition occurs it is then further classified as a birth or a death according to the probabilities $P\{T_1 < T_2\} = \lambda/(\lambda + \mu)$ and $P\{T_2 < T_1\} = \mu/(\lambda + \mu)$. This algorithm is based on the acceptance-rejection method of Section 19.2.

Acceptance-Rejection algorithm

1) Given the state η_n of the process at time t_n, compute the birth and death rates $\lambda = C_a(\eta_n)$ and $\mu = C_d(\eta_n)$.

3) Draw a uniform random number, r_1, on the interval $[0, 1)$.

4) First test: if $r_1 \leq 1 - (\mu + \lambda)\delta t$ then (no transition occurs in time δt).

   Set $\eta(t_n + \delta t) = \eta_n$ and $t_n = t_n + \delta t$.

   If $t_n < \Delta T$ then goto 1)

   else continue (exactly one transition occurs)

5) Draw a second random number, r_2, uniformly distributed on the interval $[0, 1)$.

6) Second test: if $r_2 \leq \lambda/(\mu + \lambda)$ then (a birth occurs) set $\eta_{t_n+\delta t} = \eta_n + 1$

   else (a death occurs) set $\eta_{t_n+\delta t} = \eta_n - 1$

7) Set $t_n = t_n + \delta t$. If $t_n < \Delta T$ then goto 1.

In the transition classification step we have divided the interval $[0, 1]$ into two subintervals of sizes $\lambda/(\lambda+\mu)$ and $\mu/(\lambda+\mu)$, respectively. The probability for a birth to occur is equal to the probability that the uniform random number r_2 is in $[0, \lambda/(\lambda+\mu)]$ and that of a death is equivalent to r_2 being in $(\lambda/(\lambda+\mu), 1]$.

19.4.4 Gillespie's Exact Algorithm

Instead of dividing the time interval into fixed small subintervals, we directly sample the exponential distribution $S = \min(T_1, T_2)$ to compute the (random) time, s, at which the first transition occurs, by using the inverse-method presented in the previous section. If $s \leq \Delta T$ (where again ΔT is the large-scale time step of the climate model), then we accept the transition and further classify it as a birth or a death as in the previous algorithm. This is repeated until the cumulative time reaches or exceeds ΔT.

Gillespie's Exact-Inverse-Method Algorithm

1) Given the state η_t of the process at time $t, 0 \leq t \leq \Delta T$.

2) Draw a uniform random number r_1 from $[0, 1]$ and set $s = -\frac{1}{\lambda+\mu}\ln(r_1)$.

3) If $s + t > \Delta T$, then set $t = \Delta T$ and terminate the algorithm.

   Otherwise (the transition is accepted) we draw a second uniform random number r_2 in $[0, 1]$.

4) If $r_2 < \lambda/(\lambda + \mu)$, set $\eta_{t+s} = \eta_t + 1$.

   otherwise set $\eta_{t+s} = \eta_t - 1$.

5) Set $t = t + s$. If $t < \Delta T$ goto 1.

Notice that this algorithm does not assume that at most one transition occurs at each time step, instead, it finds the exact time when the first transition occurs. This is the reason why it is called the exact algorithm. When the time step δt is chosen carefully so that the probability for more than one transition is very small, the two algorithms provide practically the same results. However, in addition to not having to worry whether δt is sufficiently small, the exact algorithm has the advantage of not using unnecessary steps where no transitions occur—which is typical for the acceptance-rejection method, and it is thus more efficient.

19.4.5 Numerical Tests

Here we implement and test the Monte Carlo algorithm above for the birth and death process with the adsorption and desorption rates in (19.15) in the absence of the external field: $h \equiv 0$. We assume a single-uncoupled mesoscopic cell of size q.

In Figures 19.5, 19.6 and 19.7 we plot the time evolution of η_t/q for both a single realization (top of the figure) and the average over 100 realizations (bottom panel) for the values $\beta J_0 = -4, 0, 2$ and $q = 5$, $q = 10$ and $q = 40$, respectively.

Independently on q, we observe that when βU_0 is positive (attractive potential) the process tends to equilibrate around the maximum level q (CIN site), when $\beta U_0 = 0$ (no local interactions) it equilibrates around the middle point $q/2$, and when $\beta U_0 < 0$ (repulsive potential) it tends to the low CIN level 0, i.e., a PAC state. This behavior can be explained as follows:

- When $\beta U_0 > 0$ the factor $e^{-\beta J_0}$ is a small positive number and so the birth (adsorption) rate dominates the (death) desorption rate expect at the highest level q when it is zero. Thus, the process oscillates somewhere close to the state $\eta_t = q$.

- When $\beta U_0 = 0$, the two rates are comparable; the desorption rate is higher for $j > q/2$ whereas the adsorption rate is higher when $j < q/2$, and they are equal at $q/2$. Hence, the process oscillates around the middle $q/2$ where the births and deaths are balanced.

- When $\beta U_0 > 0$, the adsorption rate dominates except near the state $j = 0$, where the death rate is effectively zero. Hence, the process will oscillate in the vicinity of the low CIN level (PAC state).

19.5 A BIRTH-DEATH PROCESS FOR CLOUD-CLOUD INTERACTIONS

This section presents a multidimensional birth-death stochastic process to capture the random interactions between the three cloud types that characterize organized tropical convection. This model has first appeared in [4] and it is somewhat in the refining stage in order to be implemented in an actual climate model. At this point, it is mainly used by the research group who created it and their collaborators in the context of idealized climate models used for pure research work.

We aim to represent the unresolved variability of organized tropical convection in a typical large-scale climate simulation with a mesh size of 100 to 200 km. We consider a horizontal grid box for the tropical troposphere, above the planetary boundary layer, of rectangular shape, divided onto a lattice of $n \times n$ lattice points or sites. The parameter n is a positive integer on the order 100 or less so that the lattice sites are 1 to 5 kilometers apart, the typical scale for an individual cloud. We assume that each lattice site is either occupied by a certain cloud type (congestus, deep, or stratiform) or it is a clear sky site. A given site will switch from a given configuration to an other according

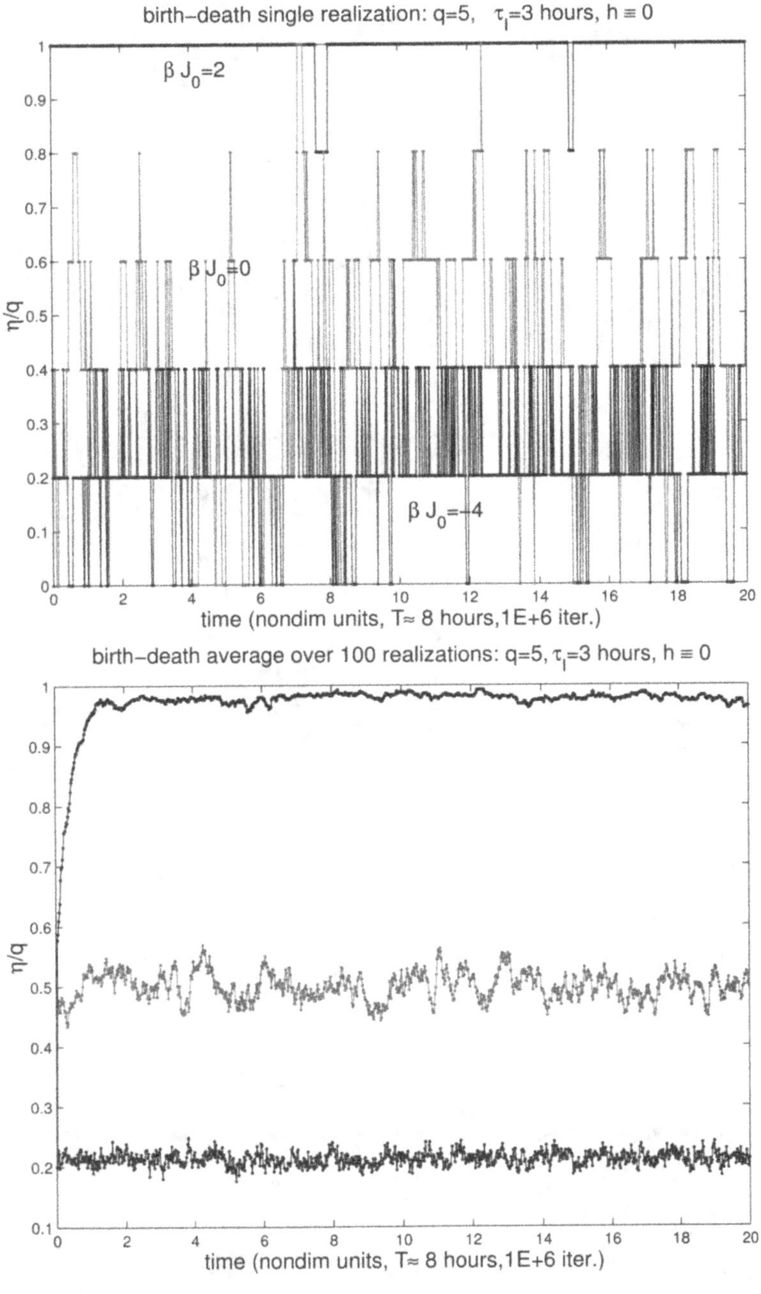

Figure 19.5 Evolution in time of the random process η_t/q. Top: single realization, bottom: average over 100 realizations, $\tau_I = 3$ hours, $q = 5$.

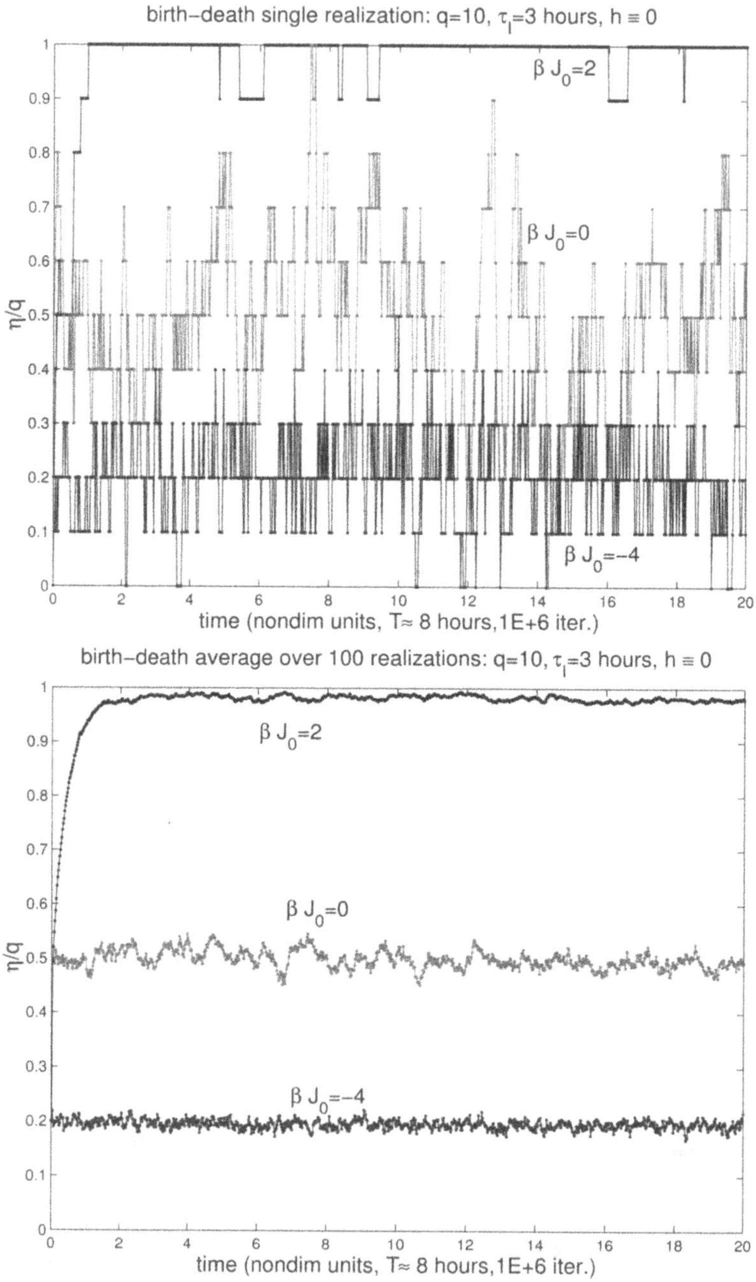

Figure 19.6 Same as in Figure 19.5 but for $q = 10$.

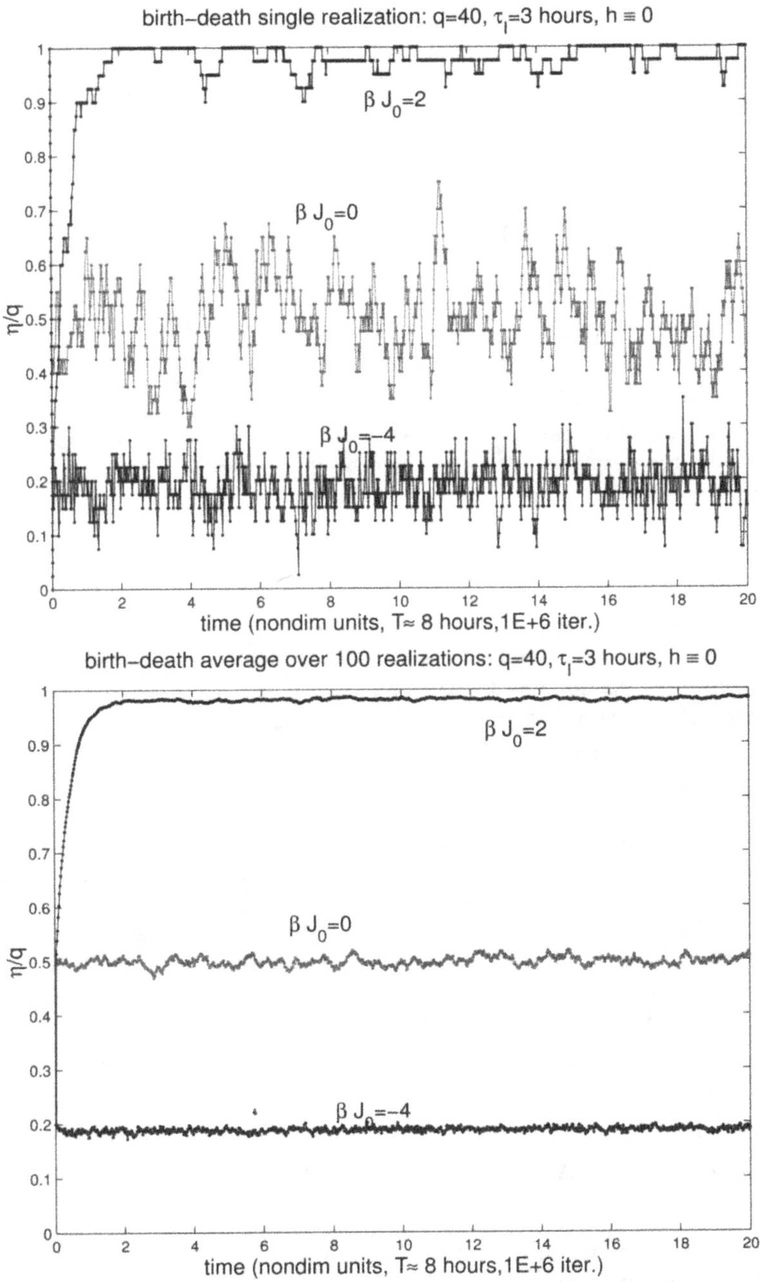

Figure 19.7 Same as in Figure 19.5 but for $q = 40$.

0	1	0	2
2	3	2	0
0	0	1	0
1	1	0	3

Figure 19.8 Lattice cloud model. A given lattice site is either clear sky (0) or occupied by a congestus cloud (1), a deep convective cloud (2), or a stratiform anvil cloud (3).

to some probability rules, which depend on the large-scale resolved variables. We thus construct a stochastic process at each lattice site taking the discrete values from 0 to 3 according to whether it is a clear sky site or it is occupied by a certain cloud type, as shown in Figure 19.8.

Let X_t^i denote the state of site i of the lattice, $i = 1, \cdots, n \times n$, at time t:

$$X_t^i = \begin{cases} 0 & \text{if site } i \text{ is clear sky,} \\ 1 & \text{if site } i \text{ is occupied by a congestus cloud,} \\ 2 & \text{if site } i \text{ is occupied by a deep convective cloud,} \\ 3 & \text{if site } i \text{ is occupied by a stratiform anvil.} \end{cases} \tag{19.24}$$

X_t^i is a Markov chain with the transition probabilities,

$$\begin{aligned} &P_{lk}^i \equiv \text{Prob}\left\{X_{t+\Delta t}^i = k/X_t^i = l\right\} = R_{lk}^i \Delta t + o(\Delta t), \\ &\text{for } l, k = 0, 1, 2, 3, \text{ and } l \neq k \end{aligned} \tag{19.25}$$

and

$$P_{ll}^i \equiv \text{Prob}\left\{X_{t+\Delta t}^i = l/X_t^i = l\right\} = 1 - \sum_{k=0,k\neq l}^{3} P_{lk}^i, \tag{19.26}$$

where $\Delta t > 0$ is a small time increment and the R_{lk}^i's are prescribed transition rates. For simplicity, we ignore the direct-local interactions between sites and assume that the rates R_{lk}^i depend solely on the large-scale resolved variables according to the following intuitive rules of switching back and forth between cloud type to cloud type and from cloudy and noncloudy sites.

1. A clear site turns into a congestus site with high probability if CAPE is positive and the middle troposphere is dry.

2. A congestus or clear sky site turns into a deep convective site with high probability if CAPE is positive and the middle troposphere is moist.

3. A deep convective site turns into a stratiform site with high probability with a prescribed conversion rate, which may or may not depend on the state of the environment.

4. A cloudy site turns back to a clear sky with a certain probability according to a prescribed decay time scale for each cloud type.

5. It is very unlikely, during the short period of time Δt, for a clear sky or a congestus site to turn into a stratiform site, for a deep convective or stratiform site to turn into a congestus site, or for a stratiform site to turn into a deep convective site.

Notice that the assumption that the transition rates depend only on the large scale variables, which amounts to ignoring interactions between the lattice sites all together, implies that the stochastic processes associated with the different sites are independent and statistically identical. Therefore, unless otherwise stated, in the rest of the chapter, we drop the superscript i and consider only the generic process X_t with the transition probabilities P_{lk} and transition rates R_{lk}.

It follows immediately from Assumption 5 that

$$R_{03} = R_{13} = R_{21} = R_{31} = R_{32} = 0. \tag{19.27}$$

For fixed large-scale conditions, the stochastic process is a stationary Markov chain with the infinitesimal generator

$$R = \begin{bmatrix} -R_{01} - R_{02} & R_{01} & R_{02} & 0 \\ R_{10} & -R_{10} - R_{12} & R_{12} & 0 \\ R_{20} & 0 & R_{20} - R_{23} & R_{23} \\ R_{30} & 0 & 0 & -R_{30} \end{bmatrix}. \tag{19.28}$$

Among all the physical quantities used to describe the state of the atmosphere, in a given large scale numerical model, two are considered to be important for both triggering and maintaining tropical convection, i.e., for the formation and decay of the three cloud types (congestus, deep, and stratiform). These quantities are the convective available potential energy (CAPE) and the relative moisture content, i.e., moistness or rather dryness of the middle of the troposphere. In practice both CAPE and the atmospheric dryness are well-defined functions of the large-scale moist thermodynamic variables. Here, we assume that both CAPE and dryness are two external parameters, denoted here by the letters C and D, respectively, varying roughly between 0 and 2.

Let

$$\Gamma(x) \equiv \begin{cases} 1 - e^{-x} & \text{if } x > 0, \\ 0 & \text{otherwise.} \end{cases}$$

Then according to the assumptions 1,2,3,4 given above, we define

$$\begin{aligned} R_{01} &= \frac{1}{\tau_{01}}\Gamma(C)\Gamma(D), & R_{02} &= \frac{1}{\tau_{02}}\Gamma(C)(1-\Gamma(D)), \\ R_{10} &= \frac{1}{\tau_{10}}\Gamma(D), & R_{12} &= \frac{1}{\tau_{12}}\Gamma(C)(1-\Gamma(D)), \\ R_{20} &= \frac{1}{\tau_{20}}(1-\Gamma(C)), R_{23} = 1/\tau_{23}, & R_{30} &= 1/\tau_{30}. \end{aligned} \tag{19.29}$$

Note for instance that R_{01} is zero when $C \leq 0$ or $D \leq 0$ and approaches τ_{01}^{-1} when C and D are sufficiently large and positive, consistent with Assumption 1 above. Here the τ_{lk}'s are prescribed time scales of formation or decay of the corresponding cloud type or of conversion of cloud type l to cloud type k. There is no obvious way to chose their values. Based on physical intuition gained from observations, numerical simulations, and theory of tropical convection, the rule of thumb is that the cloud lifetime is on the order of hours, that the rate of cloud formation is much faster than that of their decay, and that stratiform clouds should decay much more slowly than either congestus or deep. Here we consider the two extreme cases depicted in table 19.1, to highlight some interesting features of the stochastic multicloud model parameterization.

In (19.29), we assumed for simplicity that the stratiform generation and decay rates, R_{23} and R_{30}, are both independent of the large scale parameters C, D. However, there is no physical reason why this should be the case and obviously, the results would be sensitive to such dependence. To illustrate this point we also consider, in addition to (19.29), an example where R_{23} increases slowly with CAPE, using the time scales associated with Case 2 of Table 19.1,

$$R_{23} = \frac{1}{\tau_{23}}\Gamma(\sqrt{C}). \tag{19.30}$$

19.5.1 The Stationary Distribution, Cloud Area Fractions, and the Equilibrium Statistics of the Lattice Model

The equilibrium distribution, $\mathcal{P}_e$, of the multistate Markov chain X_t introduced above, is given by the left eigenvalue of the infinitesimal generator. We have

$$\mathcal{P}_e = \frac{1}{Z}\left(1, \frac{R_{01}}{R_{10}+R_{12}}, \frac{1}{R_{20}+R_{23}}\left(R_{02}+\frac{R_{12}R_{01}}{R_{10}+R_{12}}\right), \frac{R_{23}}{R_{30}}\frac{1}{R_{20}+R_{23}}\left(R_{02}+\frac{R_{12}R_{01}}{R_{10}+R_{12}}\right),\right) \tag{19.31}$$

where Z is a normalization constant, so that the entries of $\mathcal{P}_e$ sum to one.

Table 19.1 Example of prescribed values of the time scale of formation or decay of each cloud type or of conversion of one cloud type to another.

Time	Description	Case 1	Case 2
τ_{01}	formation of congestus	1 hour	3 hours
τ_{10}	decay of congestus	5 hours	2 hours
τ_{12}	conversion of congestus to deep	1 hour	2 hours
τ_{02}	formation of deep	2 hours	5 hours
τ_{23}	conversion of deep to stratiform	3 hours	0.5 hour
τ_{20}	decay of deep	5 hours	5 hours
τ_{30}	decay of stratiform	5 hours	24 hours

Next, we define the area fractions $\sigma_c, \sigma_d, \sigma_s$ occupied by clouds of type congestus, deep, or stratiform at any given time t, as the number of lattice sites for which $X_t = 1, X_t = 2, X_t = 3$, respectively, divided by the total number of sites $N = n \times n$:

$$\sigma_c = \frac{1}{N}\sum_{i=1}^{N} \mathbb{1}_{\{X_t^i=1\}}, \quad \sigma_d = \frac{1}{N}\sum_{i=1}^{N} \mathbb{1}_{\{X_t^i=2\}}, \quad \sigma_s = \frac{1}{N}\sum_{i=1}^{N} \mathbb{1}_{\{X_t^i=3\}}, \tag{19.32}$$

where

$$\mathbb{1}_{\{X_t^i=k\}} = \begin{cases} 1 & \text{if } X_t^i = k, \\ 0 & \text{otherwise.} \end{cases}$$

The clear sky area fraction is given by

$$\sigma_{cs} \equiv 1 - \sigma_c - \sigma_d - \sigma_s.$$

In effect, the area fraction vector $(\sigma_{cs}, \sigma_c, \sigma_d, \sigma_s)$ is given by the probability distribution of the generic stochastic process X_t at time t. Therefore, the equilibrium distribution $\mathcal{P}_e$ in (19.31) yields the long-time statistical equilibrium for the filling fractions $\sigma_c, \sigma_d, \sigma_s$.

We now use Monte Carlo to simulate the sequence of Markov chain's $X_t^i, i = 1, 2, \cdots N$, associated with each one of the lattice sites. We use the acceptance-rejection algorithm for a birth-death process discussed in the previous section where the transition times of the different sites are assumed to be independent exponential random variables. A maximum of two random numbers are thus generated at each iteration and for each lattice site, conditional on the states 0,1,2,3. Recall that state 0 can change to state 1 or state 2, state 1 can go to either 0 or 2, and 2 can go to either 0 or 3, while state 3 can go only to 0. The first random number determines whether we make a change or not and the second random number determines if we got up or down, accordingly in the hierarchy of states. Only one random number is generated for state 3, since only one change (3 to 0) is permitted.

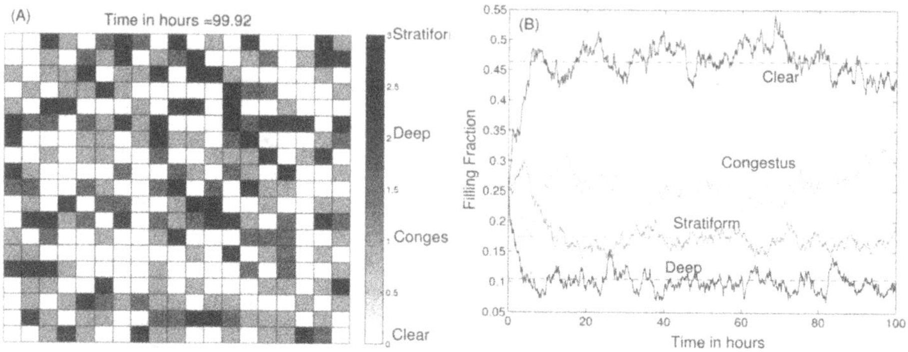

Figure 19.9 An example of Monte Carlo simulation of stochastic multicloud model with $n = 20, C = 0.25, D = 0.75$, and the cloud time scales are as in Table 19.1, Case 1. (A) A snapshot picture of one typical lattice configuration and (B) time series of the total coverages associated with each cloud type with the equilibrium values overlaid (dashed lines).

As a test case, we let $C = 0.25$ and $D = 0.75$: a relatively moist middle troposphere with a moderate but positive CAPE value. Starting with a random initial lattice configuration, we integrate the stochastic lattice model for about 100 hours, with $n = 20$ and the typical time scales displayed in Table 19.1, Case 1. A snapshot (single realization at a fixed time) of the lattice state is shown in Figure 19.9(a) while the associated time series of the area fractions for each cloud type are shown in Figure 19.9(b), with the corresponding equilibrium values, from (19.31), are overlaid. Starting initially with a random lattice configuration, the cloud coverage fractions relax quickly to their corresponding equilibrium values and fluctuate around them with a significant variability of about 5% to 25% of the total area.

19.5.2 Coarse-Grained Birth-Death Stochastic Model and the Mean-Field Equations

Clearly, for a large number of sites of up to 100×100, the full Monte Carlo simulation of evolving the 100×100 Markov chains all at once is impractical. However, since in practice we do not need to know the microscopic configuration of the lattice but only large-scale macroscopic features such as the cloud fractions are needed. Here, we derive a multidimensional stochastic birth-death process for the three cloud species using a coarse-graining methodology similar to the one used for the CIN model, though much simpler because local interactions are ignored.

Let $N = n \times n$ be the total number of lattice sites. Let N_c^t be the number of congestus sites, N_d^t the number of deep convective sites, and N_s^t the number of stratiform sites, inside the lattice, at any given time $t \geq 0$. The number of clear sky sites is $N^t_{cs} = N - N^t_c - N^t_d - N^t_s$, by conservation of the total number of sites. Next, we compute the (transition) probabilities for the numbers (random variables) N^t_c, N^t_d, N^t_s to go up or down by one during the small interval of time $(t, t+\delta t]$.

We have

$$\text{Prob}\{N_c^{t+\delta t} \geq k+1/N^t_c = k\} = \sum_{i=1}^{N} \text{Prob}\{X^i_t = 0\}\ P^i_{01} + o(\Delta t),$$

i.e., the probability that the number of congestus sites goes up by at least one is the sum of all the probabilities that a given clear sky site will turn into a congestus site. Given that all the sites are identical, i.e., the transition probability P^i_{lk} is independent of i. Moreover, as stated above, we have [see (19.32)]

$$\text{Prob}\{X^i_t = 1\} = \frac{N^t_c}{N} = \sigma^t_c, \quad \text{Prob}\{X^i_t = 2\} = \frac{N^t_d}{N} = \sigma^t_d, \tag{19.33}$$
$$\text{Prob}\{X^i_t = 3\} = \frac{N^t_s}{N} = \sigma^t_s, \quad \text{Prob}\{X^i_t = 0\} = \frac{N^t_{cs}}{N} = \sigma^t_{cs}, \quad \forall i = 1, 2, \cdots, N.$$

Using the fact that the sites are independent, and the fact that the min of m exponential random variables of equal rates λ is an exponential random variable with rate $m\lambda$, we arrive at

$$\text{Prob}\{N_c^{t+\Delta t} = k+1/N^t_c = k\} = N_{cs}P_{01} + o(\Delta t) = N_{cs}R_{01}\Delta t + o(\Delta t). \tag{19.34}$$

Similarly, we have

$$\begin{aligned}\text{Prob}\{N_c^{t+\Delta t} = k-1/N^t_c = k\} &= \sum_{i=1}^{N} \text{Prob}\{X^i_t = 1\}\ (P^i_{10} + P^i_{12}) + o(\Delta t)\\ &= N_c(R_{10} + R_{12})\Delta t + o(\Delta t),\\ \text{Prob}\{N_d^{t+\Delta t} = k+1/N^t_d = k\} &= \sum_{i=1}^{N} \text{Prob}\{X^i_t = 0\}\ P^i_{02} +\\ \text{Prob}\{X^i_t = 1\}P^i_{12} + o(\Delta t) &= (N_{cs}R_{02} + N_cR_{12})\Delta t + o(\Delta t),\end{aligned} \tag{19.35}$$

$$\begin{aligned}
\text{Prob}\{N_d^{t+\Delta t} = k-1/N_d^t = k\} &= \sum_{i=1}^{N} \text{Prob}\{X_t^i = 2\}\ (P_{20}^i + P_{23}^i) + o(\Delta t) \\
&= N_d(R_{20} + R_{23})\Delta t + o(\Delta t), \\
\text{Prob}\{N_s^{t+\Delta t} = k+1/N_s^t = k\} &= \sum_{i=1}^{N} \text{Prob}\{X_t^i = 2\}\ P_{23}^i + o(\Delta t) \\
&= N_d R_{23}\Delta t + o(\Delta t), \\
\text{Prob}\{N_s^{t+\Delta t} = k-1/N_s^t = k\} &= \sum_{i=1}^{N} \text{Prob}\{X_t^i = 3\}\ P_{30}^i + o(\Delta t) \\
&= N_s R_{30}\Delta t + o(\Delta t). \qquad (19.36)
\end{aligned}$$

The stochastic process N_x^t, $x = cs, c, d, s$ form a coupled system of birth-death Markov processes whose transition probabilities are given by (19.34) to (19.36), which can be easily evolved in time using Gillespie's exact algorithm for which the cloud coverages are recovered according to (19.33), consistent with (19.32). In practice, we can also view this coupled birth-death system as a multistate/multivariable Markov chain undergoing one of the following seven transitions at a time: one congestus is formed from a clear sky, one deep is formed from a clear sky, one congestus is converted to deep, one deep is converted to stratiform, or one cloudy site of type 1,2, or 3 turns to a clear sky site. The associated transition probabilities are given by the the original rates in (19.29) multiplied by the total number of sites that are subject to the given transition in a way which is consistent with the formulas (19.34) to (19.36). For example the rate of transition from clear sky to congestus is $N_{cs}R_{01}$ and the rate of conversion of congestus to deep is $N_c R_{12}$, etc.

The vector (N_c^t, N_d^t, N_s^t) is in effect *a multidimensional birth-death process with immigration* for the three cloud populations. The birth rates are given by the spontaneous formation of congestus and deep clouds from clear sky sites, $N_{cs}R_{01}$ and $N_{cs}R_{0,2}$, and the death rates are given by the natural decay rates of the three cloud types, $N_c R_{10}, N_d R_{2,0}, N_s R_{30}$. The rates of conversion from congestus to deep and from deep to stratiform, $N_c R_{1,2}, N_d R_{2,3}$, represent the rates of immigration from one population to another. We note that by construction of the multicloud model the birth rate of stratiform clouds and the immigration from deep to congestus, from congestus to stratiform, from stratiform to deep, and from stratiform to congestus are all set to zero. Accordingly, we can easily write down the multidimensional forward equations of this 3D birth-death process.

Let $\epsilon_1 = (1,0,0), \epsilon_2 = (0,1,0), \epsilon_3 = (0,0,1)$ denote the three canonical unit vectors of $\mathbb{R}^3$ and let $\mathbf{Z}_t = (N_c^t, N_d^t, N_s^t)$ denote our three-dimensional birth-death process. Let $\mathbf{i} = (i,j,k)$ be a generic element of the state space $\{0,1,\cdots,N\}^3$ and $\mathbf{P}_{\mathbf{i},\mathbf{j}}$ denote the transition probability matrix of $\mathbf{Z}_t$. Then the backward equations are given by

$$\begin{aligned}\frac{d}{dt}\mathbf{P}_{\mathbf{i},\mathbf{j}} =& N_{cs}R_{01}\mathbf{P}_{\mathbf{i}+\epsilon_1,\mathbf{j}} + N_{cs}R_{02}\mathbf{P}_{\mathbf{i}+\epsilon_2,\mathbf{j}} + N_cR_{12}\mathbf{P}_{\mathbf{i}+\epsilon_2-\epsilon_1,\mathbf{j}} \\ &+ N_dR_{23}\mathbf{P}_{\mathbf{i}+\epsilon_3-\epsilon_2,\mathbf{j}} + N_cR_{10}\mathbf{P}_{\mathbf{i}-\epsilon_1,\mathbf{j}} + N_dR_{20}\mathbf{P}_{\mathbf{i}-\epsilon_2,\mathbf{j}} + N_sR_{30}\mathbf{P}_{\mathbf{i}-\epsilon_3,\mathbf{j}} \\ &- \left[N_{cs}(R_{01}+R_{02}) + N_c(R_{10}+R_{12}) + N_d(R_{23}+R_{20}) + N_sR_{30}\right]\mathbf{P}_{\mathbf{i},\mathbf{j}} \\ &\mathbf{i},\mathbf{j} \in \{0,1,\cdots N\}^3. \qquad (19.37)\end{aligned}$$

We note that (19.37) is a very large system of differential equations of dimension N^3. With a typical $N = 40 \times 40$ sites this forms a $\approx (4 \times 10^9) \times (4 \times 10^9)$ dimensional system. Though the solution is known is closed form as the exponential of the infinitesimal matrix, its actual computation is very difficult and even impossible using conventional methods. Sophisticated high performance computing technique are needed; although the infinitesimal generator is a very sparse matrix (only 8 entries are non-zero on each row), its exponential is a full matrix.

Nonetheless, for the purpose of climate simulations, we do not actually need to compute the transition matrix $\mathbf{P}_{\mathbf{i},\mathbf{j}}$ but we need only to evolve the three-dimensional process $\mathbf{Z}_t$ using a Monte Carlo algorithm involving only the seven non-zero transition rates at each time step. The Gillespie's exact algorithm version for simulating the 3D birth-death process, over one climate model time step $[0, \Delta T]$, can be formulated as follows.

Multidimensional Gillespie's exact algorithm:

0) Let $\lambda = N_{cs}(R_{01} + R_{02}) + N_c(R_{10} + R_{12}) + N_d(R_{23} + R_{20}) + N_sR_{30}$.

1) Let $Z_t = (N_c, N_d, N_s)$ be the state of the system at time t, $0 \le t < \Delta T$.

2) Generate a random number r_1 uniformly from $(0,1)$. Set $s = -\frac{1}{\lambda}\ln(r_1)$.

3) If $t + s > \Delta T$, then no transition occurs. Set $t = \Delta T$. Stop.

4) If $t + s \le \Delta T$. Divide the interval into seven subintervals $I_1, I_2, \cdots, I_7$ of sizes

$$\frac{N_{cs}R_{01}}{\lambda}, \frac{N_{cs}R_{02}}{\lambda}, \frac{N_cR_{10}}{\lambda}, \frac{N_cR_{12}}{\lambda}, \frac{N_dR_{23}}{\lambda}, \frac{N_dR_{20}}{\lambda}, \frac{N_sR_{30}}{\lambda},$$

respectively.

5) Generate a random number r_2. Select the subinterval I_k such that $r_2 \in I_k$. Then make the corresponding transition as follows.

   - If $r_2 \in I_1$, then $N_c = N_c + 1$.
   - If $r_2 \in I_2$, then $N_d = N_d + 1$.
   - If $r_2 \in I_3$, then $N_c = N_c - 1$.

- If $r_2 \in I_4$, then $N_c = N_c - 1,\ N_d = N_d + 1$.
- If $r_2 \in I_5$, then $N_d = N_d - 1,\ N_s = N_s + 1$.
- If $r_2 \in I_6$, then $N_d = N_d - 1$.
- If $r_2 \in I_7$, then $N_s = N_s - 1$.

6) Set $t = t + s$. If $t < \Delta T$ got to 1.

As one would expect, the dynamics of the area fractions obtained by evolving the full microscopic lattice model, described in the previous section, through the detailed description of each one of the stochastic processes, X_t^i are statistically equivalent to those obtained by evolving the coarse-grained birth-death processes just described. However, it is important to note that the computations are orders of magnitude cheaper in the latter case: Compare generating two random numbers and testing them against seven transition rates versus simulating each one of the $n \times n$ sites.

19.5.3 The Deterministic Mean-Field Equations and Numerical Simulations

As for all evolving physical quantities, one can easily use standard calculus to derive deterministic differential equations for the cloud coverages $\sigma_c, \sigma_d, \sigma_s$. According to the discussion above, we have the following three-by-three system of ODEs. In the jargon of stochastic modeling they are called mean-field equations and can be obtained rigorously as the continuous limit when the number of lattice sites goes to infinity:

$$\begin{aligned}\dot{\sigma}_c &= (1 - \sigma_c - \sigma_d - \sigma_s)R_{01} - \sigma_c(R_{10} + R_{12}),\\ \dot{\sigma}_d &= (1 - \sigma_c - \sigma_d - \sigma_s)R_{02} + \sigma_c R_{12} - \sigma_d(R_{20} + R_{23}),\\ \dot{\sigma}_s &= \sigma_d R_{23} - \sigma_s R_{30}.\end{aligned} \tag{19.38}$$

Note that the growth and decay rates of the mean-field variables in (19.38) are given respectively by the birth, death and immigration rates in (19.37) that are simply normalized by the total number of sites N. This is a non-homogeneous linear system of ODEs with a unique equilibrium solution given by the stationary distribution in (19.31).

The stability properties of the mean-field equations can be used to learn something about the behavior of the stochastic system. In Figure 19.10, we plot the contours of the real and imaginary parts of the eigenvalues of the matrix (of the ode system) in (19.38) as functions of the parameters C and D, using the time scales from Table 19.1, Case 1. Recall from the standard theory of differential equations that an equilibrium point is said to be an asymptotically stable node if all the associated eigenvalues are real negative. Small perturbations of the equilibrium decay and the equilibrium is recovered when $t \longrightarrow \infty$. If at least one eigenvalue is positive it is an unstable node and small perturbations grow without bound. If the matrix admit complex eigenvalues, then the equilibrium is called an asymptotically stable spiral if the real parts of all the eigenvalues are negative. It is an unstable spiral if one complex eigenvalue has a positive real part. In the case of an asymptotically stable spiral, small perturbations of the equilibrium spiral in toward the equilibrium position while in an unstable spiral they spiral out away from the equilibrium position; in the first case the solution exhibits decaying oscillations while in the second it is characterized by oscillations that grow in amplitude. An equilibrium with pure imaginary eigenvalues is called a center. In this case the solution is characterized by constant amplitude oscillations.

As we see from Figure 19.10, the equilibrium of the mean-field equations (19.38) goes from an asymptotically stable node to a stable spiral as C is increased from 0 to 2. In other words this system bifurcates from an exponentially damped regime to an oscillatory damped regime; for large values of C, we have one real negative eigenvalue and a pair of complex conjugate eigenvalues whose real part is negative while for small values of C, and only slightly depending on the values of D, we have three negative real eigenvalues. The imaginary part increases significantly with increasing values of C, especially for slightly moist conditions corresponding to D between 0.2 and 0.3. The damping strength is also sensitive to changes in C and D.

An important nondimensional number for the stochastic multicloud model is given by the ratio of the frequency to the damping rate, for the complex conjugate pair, plotted in Figure 19.11(E). In Figure 19.11, we display two time series of the area coverages obtained by evolving the stochastic model with $D = 0.4$ and the two different values of $C = 0.1$ and $C = 1.5$. According to Figure 19.11(E), the case $C = 0.1$ has a frequency to damping ratio near zero (below 0.1) while in the second case this ratio is above 0.6. As it is anticipated, the two time series are qualitatively different with the one corresponding to $C = 1.5$ having sharper peaks while the one corresponding to $C = 0.1$ has much longer excursions. This suggests that a large frequency to damping ratio, in a complex conjugate pair for the mean-field equations, would yield sharp and rapid oscillations, for the associated stochastic system, while a small ratio would yield smoother oscillations with much longer excursions from equilibrium.

19.6 FURTHER READING

To learn more about moist thermodynamics and moist convection in general, we refer to the excellent book of K. Emanuel [1]. More on organized convection and convectively coupled waves can be found in the review papers by G. Kiladis *et al.* [7] and C. Zhang [15]. For the theory of Markov chains and birth death processes, we refer to the two pedagogical books of S. M. Ross [12] and G. F. Lawler [2]. A good text on statistical mechanics and the Ising model used in Section 19.4 is found in Ref. [14]. To learn more about the

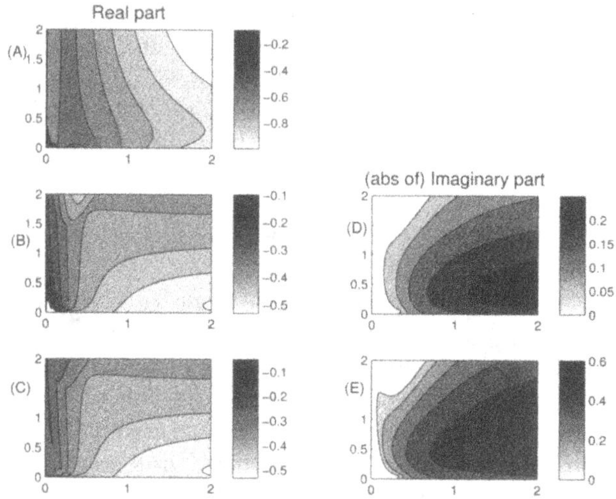

Figure 19.10 Equilibrium eigenvalues of the mean-field equations. Panels (A), (B), and (C) represent the contours of the real parts of the three eigenvalues, respectively, as CAPE C (horizontal axis) and dryness D (vertical axis) are varied from 0 to 2, Panel (D) shows the imaginary part of the complex conjugate pair, and Panel (E) displays the ratio of the frequency over the damping rate.

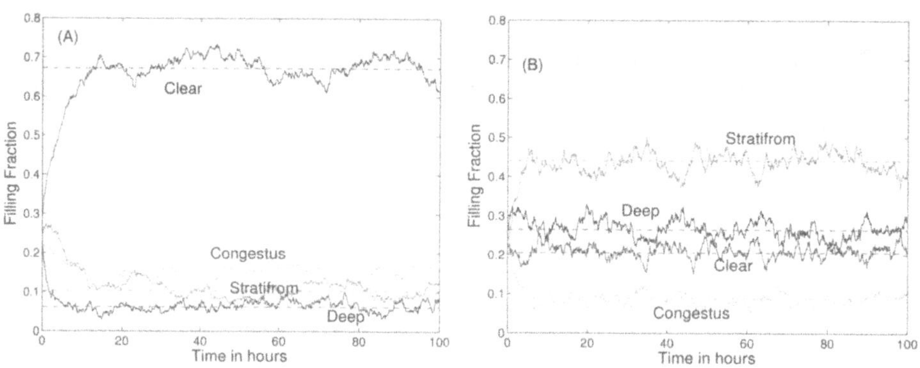

Figure 19.11 Stochastic oscillations for both (a) when the frequency to damping ratio is small and (b) when it is large for the parameter values $D = 0.4$ and $C = 0.1$ and $C = 1.5$, respectively, and the τ_{lk}'s are as in Table 19.1.

practice and theory of Markov Chain Monte Carlo, we suggest the book by C. P. Robert and G. Casella [11] and the more practically oriented text of W. Stewart [13]. Finally, for more general (deterministic) mathematical models of the atmosphere and ocean we refer to the book by J. A. Majda [8] and a review on recent developments in mathematical modeling for climate science in the tropics is found in Ref. [5].

EXERCISES

19.1 Verify statistically that the average number of iterations to generate one random number using the acceptance rejection method for the beta-distribution in Example 19.4 above is $30/16 \approx 1.875$. Use Monte Carlo integration based on the acceptance-rejection method to estimate the expectation $E[X] = \int_0^1 x f(x) dx$.

19.2 In this exercise we test three different approaches to generate pseudo-normal variates, with mean zero and variance one. One is based on the central limit theorem, one is the polar method discussed in the notes, and the third is the function *randn* of Matlab.

1. **Method based on the central limit theorem.**
   Recall that according to the central limit theorem, if $x_k, k = 1, 2, \cdots, n$ are n random numbers from a distribution with mean μ and variance σ^2 then as n increases

$$y = \frac{\sum_{k=1}^{n} x_k - n\mu}{\sigma\sqrt{n}}$$

   approaches a normal distribution with mean zero and variance one. If the x_k's are uniformly distributed on $[0, 1]$ then $\mu = 1/2$ and $\sigma^2 = 1/12$. If in addition, we choose $n = 12$ then we obtain a simple formula for y:

$$y = \sum_{k=1}^{12} x_k - 6$$

   is approximately normally distributed with mean zero and variance one. Write a short Matlab code to generate normal variates according to this formula by generating 12 uniform variates using the function *rand* of Matlab, sum them together and substrate 6. E.g., `>> y = sum(rand(1,12))- 6;` will generate one pseudo-random number, which is approximately $\mathcal{N}(0, 1)$. Use this as a building block to write a Matlab code to generate sequences of normally distributed random numbers of arbitrary size.

2. **The polar method.**
   Let x_1, x_2 be two independent uniformly distributed (pseudo-) random

numbers on $[0, 1)$. Then,

$$y_1 = \sin(2\pi x_1)\sqrt{-2\log(x_2)} \text{ and } y_2 = \cos(2\pi x_1)\sqrt{-2\log(x_2)}$$

are normally distributed with mean zero and variance one. Write a Matlab code to generate a sequence of normally distributed random numbers of an arbitrary size, according to this algorithm.

3. Generate 1000 normally distributed $\mathcal{N}(0, 1)$ according to each one of the algorithms above and according to the Matlab function *randn* and save the three sequences as three different vectors, which you may call *Ncentral, Npolar, Nrandn*, respectively. Then use the Matlab function *hist* to bin each one of three random vectors in bins of size 1. Normalize the bin numbers by the total samples (1000). Use the *bar* command of Matlab to plot the histogram and plot the normal density $f(x) = e^{-x^2/2}/\sqrt{2\pi}$ on top on each one of the histogram.

   Follow the following simple Matlab instructions for the function *randn* of Matlab, as a guideline example.

```
>> N =1000;
>>Nrandn = randn(N,1);
>> x=-3:1:3;
>> nh = hist(Nrandn,x); %counts number of random numbers
                        %in each subinterval centered
>> nhnormalized =  nh/N; %
>> figure
>> bar(x, nhnormalized)
>> hold on
>>ezplot('exp(-x^2/2)/sqrt(2*pi)',[-4,4]);
```

Check the validity of each one of the methods above by comparing the numerical values of unit-binned histograms (nhnomalized in the Matlab code above) to the exact normal distribution:

$$\bar{h}_i = \int_{i-.5}^{i+.5} \frac{e^{-x^2/2}}{\sqrt{2\pi}} dx, \quad i = \cdots, -2, -1, 0, 1, 2, \cdots$$

$$= \cdots, 0.5977, 6.0598, 24.1730, 38.2925, 24.1730, 6.0598, 0.5977, \cdots \%.$$

19.3 Use the Chapman-Kolmogorov equation for $P^{(n+1)}$ to show that the limiting distribution of a Markov chain satisfies $\pi_j = \sum_{i=0}^{\infty} \pi_i P_{ij}$ for all $j \geq 0$.

19.4 A taxi driver conducts his business in three different towns 1, 2, and 3. On any given day, when he is in town 1, the probability that the next passenger he picks up is going to a place in town 1 is 0.3, the probability that the passenger is going to town 2 is 0.2 and the probability that he is going to town 3 is 0.5. When he is in town 2, the next passenger he picks up is going to a place in town 1 with probability 0.1, to town 2 with probability 0.8 and to town 3 with probability 0.1. When he is in town 3, these probabilities are 0.4 to go to towns 1 and 2 and 0.2 to go to a place in town 3.

1. Argue why the underlying process of tracking the location of the taxi driver after dropping a passenger is a Markov chain. Write down the associated transition probability matrix.

2. Given that the taxi driver is currently in town 1 and is waiting to pick up his first customer for the day, what is the probability that he picks his third customer of the day in town 2?

3. In the long run, which of towns 1,2, 3 does the taxi driver visit the most? Justify your answer.

19.5 Let T_1, T_2 be two independent exponentially distributed random variables with rates $\lambda > 0, \mu > 0$, respectively.

(a) Show that $S = \min(T_1, T_2)$ is an exponential random variable with rate $\lambda + \mu$.

(b) Show that $P\{T_1 < T_2\} = \frac{\lambda}{\lambda+\mu}$.

(c) Show that if $\mu = \lambda$ then $T_1 + T_2$ is a Gamma random variable with parameters $\alpha = 2$ and λ and in general if $T_1, T_2, \cdots, T_n$ are n independent and exponentially distributed random variables, with the same rate λ, then the sum $S_n = T_1 + T_2 + \cdots T_n$ is Gamma distributed with parameters n and λ; $f_{S_n}(x) = x^{n-1}\lambda^n e^{-\lambda x}/(n-1)!$.

19.6 Show that the product of two stochastic matrices is a stochastic matrix.

19.7 A matrix P is said to be doubly stochastic if both all of its rows and all of its columns sum to 1. Show that the limiting distribution of Markov chain on a finite state space $\{x_0, x_1, \cdots, x_M\}$ with a doubly stochastic probability transition matrix is uniform, i.e., $\pi_j = 1/M + 1$ for $j = 0, 1, 2, \cdots, M$.

19.8 Let X be a non-negative random variable. Show that X is exponentially distributed if and only if it satisfies the memoryless property

$$P\{X > s + t\} = P\{X > s\}P\{X > t\}, \quad \text{for all } s, t > 0.$$

19.9 Write down the forward and backward equations for a bounded birth death process with birth rates λ_n and death rates μ_n where $\mu_0 = 0$ and $\lambda_k = 0$ for $k \geq 1$. Give the infinitesimal generator matrix.

19.10 Find the transition probabilities for a birth only process, i.e., a birth only process for which $\lambda_n > 0$ and $\mu_n = 0$ for all n. Start with the case $\lambda_n = \lambda$, i.e., the birth rate is independent of n.

19.11 Let X_t be a continuous-time Markov chain with state space 1,2,$\cdots$ and associated waiting rates $v_1, v_2, \cdots$ and transition rates $q_{ij}, i \neq j$. Consider the first passage time T_k into state k, given by

$$T_k = \min\{t \geq 0, X_t = k\}.$$

Let m_{ik} be the expected first passage time from state i to state k: $m_{ik} = E[T_k/X_0 = i]$.

1. Show that $v_i\, m_{ik} = 1 + \sum_{j \neq k} q_{ij}\, m_{jk}.$

2. Find m_{14} if X_t is a four state Markov chain with rates

$$q_{1,2} = 2, q_{1,3} = 2, q_{1,4} = 1, q_{2,1} = 3, q_{2,3} = 3, q_{2,4} = 0,$$

$$q_{3,1} = 0, q_{3,2} = 2, q_{3,4} = 2, q_{4,1} = 1, q_{4,2} = 0, q_{4,3} = 3.$$

19.12 Let X_t be a continuous-time Markov chain with state space $\{1, 2, 3\}$ and rates $q_{1,2} = 1, q_{2,1} = 4, q_{2,3} = 1, q_{3,2} = 4, q_{1,3} = q_{3,1} = 0$. Find the probability transition matrix $P(t)$ of X_t and the limiting distribution if it exists.

REFERENCES

1. Emanuel, K., *Atmospheric Convection*, Oxford University Press, Oxford (1994).
2. Lawler, G. F., *Introduction to Stochastic Processes*, 2nd Edition, Chapman and Hall/CRC (2006).
3. Katsoulakis M. A., Majda A. J., and Vlachos D. G., "Coarse-grained stochastic processes and Monte Carlo simulations in lattice systems", *Journal of Computational Physics*, Vol **186**(1), 250 – 278 (2003).
4. Khouider, B., Biello, J., and Majda, A., " A stochastic multicloud model for tropical convection" in *Commun. Math. Sci.*, Vol. **8** (1), 187–216 (2010).
5. Khouider, B., Majda, A. J., and Stechmann, S., "Climate Science, Waves, and PDEs for the Tropics" to appear in *Nonlinearity*, (2012)
6. Khouider, B., Majda, A. J., and Katsoulakis, M. A., "Coarse-grained stochastic models for tropical convection and climate", *Proc. Nat. Acad. Sci.*, Vol. **100**(21), pages 11941–11946 (2003).
7. Kiladis, J. N., Wheeler, M. C., Haertel, P. T., Straub, K. H., and Roundy, P. E., "Convectively coupled equatorial waves", *Rev. Geophys.*, Vol **47** RG2003, doi:10.1029/2008RG000266 (2009).

8. Majda, A. J., *Introduction to PDEs and Waves for the Atmosphere and Ocean*, American Mathematical Society (2003).

9. Majda, A. J. and Khouider, B., "Stochastic and mesoscopic models for tropical convection", *Proc. Nat. Acad. Sci. USA*, Vol. **99**, pages 1123–1128 (2002).

10. Majda, A. J., Franzke, C., and Khouider, B., "An applied mathematics perspective on stochastic modelling for climate", *Philos. Trans. Roy. Soc.*, Vol. **366A**, pages 2427–2453 (2008).

11. Robert, C. P. and Casella, G., *Monte Carlo Statistical Methods*, Springer-Verlag, New York (1999).

12. Ross, S. M., *Introduction to Probability Models*, 10th Edition, Academic Press (2010).

13. Stewart, W. J., *An Introduction to the Numerical Solution of Markov Chains*, Princeton University Press, New Jersey (1994).

14. Thompson, C., *Mathematical Statistical Mechanics*, Princeton Univ. Press, Princeton (1972).

15. Zhang, C., "Madden-Julian Oscillation", *Reviews of Geophysics*, Vol. **43**, RG2003 (2005).

PROBLEM SOLUTIONS

SOLUTIONS FOR CHAPTER 1

1.1 Let y be the axis perpendicular to the path of its movement towards the east. From Newton's second law, we know that

$$\ddot{y} = a = 2\omega v,$$

where v is the vertical velocity, and ω is the angular velocity of the Earth.

Here we have assume that the horizontal velocity due to the Coriolis effect is small, and no air is present. From basic physics, we know that $v = gt$ where g is the acceleration due to gravity. We have $h = \frac{1}{2}gt^2$, or $t = \sqrt{2h/g}$.

Now the governing equation becomes

$$\ddot{y} = 2\omega gt.$$

Mathematical Modeling with Multidisciplinary Applications.
Edited by Xin-She Yang

By integrating $\ddot{y}$ once, we have

$$\dot{y} = \omega g t^2.$$

Integrating it again, we obtain

$$y = \frac{\omega}{3} g t^3 = \frac{\omega}{3} \sqrt{\frac{8h^3}{g}}.$$

Since $\omega = 2\pi/(24 \times 3600)$ radian/s, $g = 9.8$ m/s^2, we have $y \approx 7.7$ mm for $h = 50$ m.

1.2 The solution is $y(x) = A/x + Bx^2$ where A and B are arbitrary constants.

1.3 Using a trial solution $u \sim \exp(rt)$, we have

$$(r^2 + 2\eta\omega_0 r + \omega_0^2)e^{rt} = 0,$$

or

$$r^2 + 2\eta\omega_0 r + \omega_0^2.$$

Its solutions are

$$r = [-\eta \pm \sqrt{\eta^2 - 1}]\omega_0.$$

Clearly, $\Delta = \eta^2 - 1 = 0$ defines a critical case. If $\eta = 1$, it is called critical damping, while $\eta > 1$ corresponds to overdamping and $\eta < 1$ corresponds to underdamping. When $\eta = 0$, the system becomes undamped simple harmonic motion, and ω_0 is its natural frequency. When $\eta > 0$, the amplitude of the system will decrease with time.

1.4 In the polar coordinate system, the Laplace equation becomes

$$\frac{1}{r}\frac{\partial}{\partial r}\left(r\frac{\partial u}{\partial r}\right) + \frac{1}{r^2}\frac{\partial^2 u}{\partial \theta^2} = 0.$$

When converting from Cartesian coordinates, differentiation rules should be used.

1.5 Verify this by substitution. For more information, please refer to more advanced literature.

1.6 Verify the solutions by direct substitutions. For the later case, we have $\beta = a^2\lambda^2 + \omega^2$.

1.7 By direct substitution $u(x, t) = \exp(-vt/2)w(x, t)$, we have

$$\frac{\partial^2 w}{\partial t^2} = a^2\frac{\partial^2 w}{\partial x^2} + \left(b + \frac{v^2}{4}\right)w,$$

Polyanin A. D. and Zaitsev V., *Handbook of Nonlinear Partial Differential Equations*, Chapman & Hall/CRC, Boca Raton, (2004).

which is indeed the Klein-Gordon equation.

1.8 Using the transformation $u(x,t) = \frac{2}{\phi}\frac{\partial \phi}{\partial x}$ and differentiation rules, the Burgers equation becomes $\frac{\partial \phi}{\partial t} = \frac{\partial^2 \phi}{\partial x^2}$.

SOLUTIONS FOR CHAPTER 2

2.1 From the kinetic energy of a moving object $E_k = \frac{1}{2}mv^2$ and the potential energy due to the Earth's gravity $E_p = -\frac{GM_E m}{r}$, the total energy must be zero when an object is just able to escape the Earth, so we have

$$E_k + E_p = \frac{1}{2}mv^2 - \frac{GM_E m}{r} = 0,$$

whose solution is $v = \sqrt{\frac{2GM_E}{r}}$. As $g = GM_E/r^2 = 9.8$ m/s^2, we have $v = \sqrt{2gr} = 11.17$ km/s, where we have used $r = 6370$ km.

2.2 Raindrops vary in size from about 0.1 mm to 5.5 mm. Now the fluid is the air with density $\rho_a = 1.2$ kg/m^3, and viscosity $\mu_a = 1.8 \times 10^{-5}$ Pa s. For a very small cloud drop or raindrop $d = 0.15$ mm $= 1.5 \times 10^{-4}$ m with a density of $\rho_w = 1000$ kg/m^3, we can estimate its terminal velocity as

$$v = \frac{(\rho_w - \rho_a)gd^2}{18\mu_a} = \frac{(1000 - 1.2) \times 9.8 \times (1.5 \times 10^{-4})^2}{18 \times 1.8 \times 10^{-5}} \approx 0.68 \text{ m/s},$$

which is about the same value as observed by experiment. In this case, the Reynolds number is approximately $Re = \frac{\rho_a v d}{\mu_a} \approx 6.8$, which is bigger than 1. There will be some difference between estimated values and the real velocity.

However, for larger raindrops, their falling velocities are high, and Stokes' law is no longer valid.

2.3 For the waves in deep waters when the wavelength λ is much smaller than h (or $\lambda \ll h$), we have $\frac{2\pi h}{\lambda}\rangle\infty$, and $\exp[-\frac{2\pi h}{\lambda}]\rangle 0$. Now we get

$$\tanh\left(\frac{2\pi h}{\lambda}\right) = \frac{e^{\frac{2\pi h}{\lambda}} - e^{-\frac{2\pi h}{\lambda}}}{e^{\frac{2\pi h}{\lambda}} + e^{-\frac{2\pi h}{\lambda}}} \approx \frac{e^{\frac{2\pi h}{\lambda}} - 0}{e^{\frac{2\pi h}{\lambda}} + 0} \approx 1,$$

which leads to

$$v = \sqrt{\frac{g\lambda}{2\pi}\tanh\left(\frac{2\pi h}{\lambda}\right)} \approx \sqrt{\frac{g\lambda}{2\pi}}.$$

A tsunami is a giant water wave whose wavelength and speed are constantly changing as it travels towards the shore. In deep ocean waters, the wave height is typically less than half a meter, but its wavelength is in the range of 25 km to 50 km. Let us now estimate its speed in deep water using the typical values of $\lambda = 25$ km to 50 km (or 2.5×10^4 to 5.0×10^4 meters), $g = 9.8$ m/s^2. We have the phase speed

$$v = \sqrt{\frac{g\lambda}{2\pi}} = \sqrt{\frac{9.8 \times 2.5 \times 10^4}{2\pi}} \approx 197 \text{ m/s} \approx 444 \text{ mph},$$

for $\lambda = 25$ km. For longer wavelength $\lambda = 50$ km, its speed is about 630 mph. This means that the first arrival of tsunami waves is always of long wavelength. As they travel shorewards, their wavelengths become typically in the range of 1.5 km to 5 km, but their wave heights can reach up to 30 metres. Their speed can reduce to 230 down to 25 mph.

2.4 We assume that the air pressure is hydrostatic. That is, the increase of the pressure dp is balanced by the increment of the weight $-\rho g dz$ for a thin layer with a unit area. Here z is the altitude above the Earth's surface and the $-$ sign indicates the fact that the z increases and pressure decreases in atmosphere. Therefore, we have $dp = -\rho g dz$ or

$$\frac{dp}{dz} = -\rho g.$$

From $p = \rho RT/M$, we have $\rho = pM/RT$, and we now have

$$\frac{dp}{dz} = -\frac{pMg}{RT},$$

or

$$\frac{dp}{p} = -\frac{Mg}{RT} dz.$$

Integrating from $z = 0$ to $z = h$, we have

$$\int_{p_0}^{p} dp = \ln p - \ln p_0 = -\int_0^h \frac{Mg}{RT} dz = -\frac{Mg}{RT} h,$$

where p_0 is the pressure on the Earth's surface at $z = 0$. This means that

$$\ln \frac{p}{p_0} = -\frac{Mg}{RT} h.$$

Taking the logarithms, we have

$$p = p_0 e^{-\gamma h}, \qquad \gamma = \frac{Mg}{RT}.$$

We can see that the air pressure decreases exponentially as the height h increases. We can define a characteristic height

$$L = \frac{1}{\beta} = \frac{RT}{Mg},$$

so that

$$p = p_0 e^{-h/L}.$$

For the typical values of $M = 0.0289$ kg/mole, $g = 9.8$ m/s^2, and $T = 293$ K (or 20°C), we have

$$L = \frac{8.31 \times 393}{0.0289 \times 9.8} \approx 8597 \text{ m} = 8.597 \text{ km},$$

which corresponds to $\gamma \approx 0.00116$ m^{-1}. This means that the air pressure at $z = L$ will becomes $1/e \approx 36.8\%$ of the pressure on the Earth's surface.

SOLUTIONS FOR CHAPTER 3

3.1 In the Runge-Kutta method, the steps can be reasonably large, compared with other methods. Fore example, one can use $h = \Delta x = 0.5$, then the Runge-Kutta steps will still give very accurate results.

3.2 From $dw/dt = aw - bw^2$, we have

$$w_{n+1} - w_n = (aw_b - bw_n^2)\Delta t,$$

which can be written as

$$w_{n+1} = \lambda w_n - \beta w_n^2, \quad \lambda = 1 + a\Delta t, \beta = b\Delta t.$$

Rescaling w_n as $u_n = \frac{\beta}{\lambda} w_n$, we have

$$u_{n+1} = \lambda u_n (1 - u_n),$$

which is a well-known chaotic mapping. Try to vary λ from 1 to 4 and see how u_n behavior. You will see that when $\lambda \approx 4$, you will see chaos and the final u_n is very sensible to the small change in u_1. In essence, you will observe the butterfly effect. Write a simple program and see what you can observe.

3.3 Use an explicit scheme to solve this set of equations. Try to vary the time steps so as to produce smooth trajectories.

3.4 The Crank-Nicolson scheme for the heat conduction gives

$$-ru_{j+1}^{n+1} + (1 + 2r)u_j^{n+1} - ru_{j-1}^{n+1} = ru_{j+1}^n + (1 - 2r)u_j^n + ru_{j-1}^n,$$

where $r = \lambda \Delta t / 2(\Delta x)^2$. This forms a tridiagonal matrix system. Its von Newmann stability condition is

$$A = \frac{1 - 4r \sin^2(k\Delta x/2)}{1 + 4r \sin^2(k\Delta x/2)}.$$

Since $|A| < 1$ for all values of k, this method is unconditionally stable.

3.5 This equation is a nonlinear reaction-diffusion equation, it can generate stable patterns such as ribbons, rings and stripes under the right conditions. Write a simple program to show to demonstrate this. This pattern formation is an important characteristics of nonlinear reaction-diffusion systems.

SOLUTIONS FOR CHAPTER 4

4.1 What happen if you chose to start with numbers which are related to each other, such as $f(x) = p^x$ for $x = 1, 2, 3, 4$? Try out

$$2 \quad 4 \quad 8 \quad 16$$

$$0 \quad 2 \quad 4 \quad 8$$

4.2 Obviously, there is no fixed answer for this open problem.

4.3 A reasonable value for the CO is 3.70 liters/minute.

4.4 No fixed answer.

4.5 What we see here is that we have to reshape the results from Excel and realize that $1.125 = 9/8$ and that $0.125 = 1/8$. This insight can be seen as an important part of instrumental genesis. Of course, the expression $(9x^2 - 1)/8$ is purer and maybe more beautiful in the eyes of most viewers.

4.6 Obviously, there is no fixed answer, but an exponential decreasing model is to be expected.

SOLUTIONS FOR CHAPTER 5

5.1 In conductors we obtain from Ohm's Law (5.7),

$$\mathbf{E} = \frac{1}{\sigma}\mathbf{J}.$$

Using (5.4) this equation yields

$$\mathbf{E} = \frac{1}{\sigma}\mathbf{curl}\,\mathbf{H}.$$

www.mathworks.com/matlabcentral/fileexchange/29726-pattern-formation-and-kpp-equation
See Lingefjärd, T. (2002). Mathematical modeling for preservice teachers: A problem from anesthesiology. *The International Journal of Computers for Mathematical Learning* 7(2), pp. 117–143.

Now we replace this expression for $\mathbf{E}$ in Faraday's Law (5.3). We deduce

$$i\omega\mathbf{B} + \mathbf{curl}\left(\frac{1}{\sigma}\mathbf{curl}\,\mathbf{H}\right) = 0.$$

Finally, Eq.(5.8) is obtained by using the constitutive law (5.6).

5.2 For a general vector field $\mathbf{F} = F_r\mathbf{e}_r + F_\theta\mathbf{e}_\theta + F_z\mathbf{e}_z$, by developing (5.9) we get

$$\mathbf{curlF} = \left(\frac{1}{r}\frac{\partial F_z}{\partial\theta} - \frac{\partial F_\theta}{\partial z}\right)\mathbf{e}_r + \left(\frac{\partial F_r}{\partial z} - \frac{\partial F_z}{\partial r}\right)\mathbf{e}_\theta + \left(\frac{1}{r}\frac{\partial}{\partial r}(rF_\theta) - \frac{1}{r}\frac{\partial F_r}{\partial\theta}\right)\mathbf{e}_z.$$

Then, for a field of the form (5.2) this formula yields

$$\mathbf{curlA} = -\frac{\partial A_\theta}{\partial z}\mathbf{e}_r + \frac{1}{r}\frac{\partial}{\partial r}(rA_\theta)\,\mathbf{e}_z.$$

Again we can obtain

$$\mathbf{curlcurlA} = -\left(\frac{\partial^2 A_\theta}{\partial z^2} + \frac{\partial}{\partial r}\left(\frac{1}{r}\frac{\partial(rA_\theta)}{\partial r}\right)\right)\mathbf{e}_\theta.$$

5.3 From (5.10) and the previous exercise,

$$\mathbf{curlH} = -\frac{\partial H_\theta}{\partial z}\mathbf{e}_r + \frac{1}{r}\frac{\partial}{\partial r}\Big(rH_\theta\Big)\mathbf{e}_z.$$

Multiplying this equality by $1/\sigma$ and then applying the **curl** operator we get

$$\mathbf{curl}\left(\frac{1}{\sigma}\mathbf{curlH}\right) = -\left(\frac{\partial}{\partial z}\left(\frac{1}{\sigma}\frac{\partial H_\theta}{\partial z}\right) - \frac{\partial}{\partial r}\left(\frac{1}{\sigma r}\frac{\partial(rH_\theta)}{\partial r}\right)\right)\mathbf{e}_\theta.$$

Finally, by replacing this expression in (5.8) we easily deduce (5.11).

5.4 Let $T = \frac{2\pi}{\omega}$. Let us recall that for any complex number a the real part of a is given by

$$\mathrm{Re}(a) = \frac{1}{2}(a + \bar{a}),$$

where $\bar{a}$ denotes the complex conjugate of a.

Then,

$$\frac{1}{T}\int_0^T \mathcal{A}(\mathbf{x},t)\cdot\mathcal{C}(\mathbf{x},t)\,dt = \frac{1}{T}\int_0^T \mathrm{Re}(e^{i\omega t}\mathbf{A}(\mathbf{x}))\cdot\mathrm{Re}(e^{i\omega t}\mathbf{C}(\mathbf{x}))\,dt$$

$$= \frac{1}{4}\frac{1}{T}\int_0^T (e^{i\omega t}\mathbf{A}(\mathbf{x}) + e^{-i\omega t}\overline{\mathbf{A}(\mathbf{x})})\cdot(e^{i\omega t}\mathbf{C}(\mathbf{x}) + e^{-i\omega t}\overline{\mathbf{C}(\mathbf{x})})\,dt$$

$$= \frac{1}{4}\frac{1}{T}\left(\mathbf{A}(\mathbf{x})\cdot\mathbf{C}(\mathbf{x})\int_0^T e^{2i\omega t}\,dt + \mathbf{A}(\mathbf{x})\cdot\overline{\mathbf{C}(\mathbf{x})}\int_0^T dt + \overline{\mathbf{A}(\mathbf{x})}\cdot\mathbf{C}(\mathbf{x})\int_0^T dt\right.$$

$$\left.+\overline{\mathbf{A}(\mathbf{x})}\cdot\overline{\mathbf{C}(\mathbf{x})}\int_0^T e^{-2i\omega t}\,dt\right) = \frac{1}{4}\left(\mathbf{A}(\mathbf{x})\cdot\overline{\mathbf{C}(\mathbf{x})} + \overline{\mathbf{A}}(\mathbf{x})\cdot\mathbf{C}(\mathbf{x})\right)$$

$$= \frac{1}{4}\left(\mathbf{A}(\mathbf{x})\cdot\overline{\mathbf{C}(\mathbf{x})} + \overline{\mathbf{A}(\mathbf{x})\cdot\overline{\mathbf{C}(\mathbf{x})}}\right) = \frac{1}{2}\mathrm{Re}\left(\mathbf{A}(\mathbf{x})\cdot\overline{\mathbf{C}(\mathbf{x})}\right).$$

5.5 According to the stirred tank model, the volume of a pit lake is calculated as

$$\frac{dV(t)}{dt} = \sum_{j=1}^{N^s} q_j(t) - q_0(t),$$

where

$$q_0(t) = \begin{cases} 0, & \text{if the lake is flooding} \quad (V(t) < V_f), \\ \sum_{j=1}^{N^s} q_j(t), & \text{if the lake is full} \quad (V(t) = V_f). \end{cases}$$

Here V_f refers to the final volume of the lake.

Focusing on the period $V(t) < V_f$, the lake volume does not remain constant, therefore the mass $m_i(t) = y_i(t)V(t)$ [see Eq. (5.31)]. Taking the time derivative of $m_i(t)$, we get

$$\frac{dm_i(t)}{dt} = V(t)\frac{dy_i(t)}{dt} + y_i(t)\frac{dV(t)}{dt}.$$

By writing $\frac{dy_i(t)}{dt}$ as a function of $\frac{dm_i(t)}{dt}$ and after replacing $\frac{dV(t)}{dt}$ by its expression in the above equations, we have

$$\frac{dy_i(t)}{dt} = \frac{1}{V(t)}\frac{dm_i(t)}{dt} - \frac{1}{V(t)}y_i(t)\sum_{j=1}^{N^s} q_j(t).$$

During the lake flooding ($V(t) < V_f \Rightarrow q_0 = 0$) $\frac{dm_i(t)}{dt}$ is written as

$$\frac{dm_i(t)}{dt} = V(t)r_i^c(t,\mathbf{y}(t)) + S^a r_i^a(t,\mathbf{y}(t)) + \sum_{j=1}^{M^r} S_j^b r_{ij}^b(t,\mathbf{y}(t)) + \sum_{j=1}^{N^s} a_{ij}(t)q_j(t).$$

By replacing (19) into (19), we obtain the equation we are looking for

$$\frac{dy_i(t)}{dt} = r_i^c(t, \mathbf{y}(t)) + \frac{1}{V(t)}\Big[S^a r_i^a(t, \mathbf{y}(t)) + \sum_{j=1}^{M^r} S_j^b r_{ij}^b(t, \mathbf{y}(t)) + \sum_{j=1}^{N^s}(a_{ij}(t) - y_i(t))q_j(t)\Big].$$

As it can be seen, it is exactly the same as (5.34).

5.6 In order to solve this exercise, we will go back to the situation in which the chemical species of interest E_i, $i = 1, \ldots, N$ are involved in a set of J couples of reversible reactions. The generic notation of Eq.(5.35) is used for this reactions, with $l = 1, \ldots, 2J$, being characterized by the fact that the stoichiometric coefficients satisfy Eq. (5.41).

In this situation, the evolution of the concentration of the ith chemical species is given by

$$\frac{dy_i}{dt} = \sum_{j=1}^{J}(\lambda_i^{2j-1} - \nu_i^{2j-1})\delta_j^*(t),$$

where

$$\delta_j^* = \left(k_{2j-1}\prod_{i=1}^{N} y_i^{\nu_i^{2j-1}} - k_{2j}\prod_{i=1}^{N} y_i^{\lambda_i^{2j-1}}\right).$$

Notice that we are following a kinetic approach to describe equilibrium conditions. These conditions are attained when $\delta_j^* = 0$, $j = 1, \ldots, J$, therefore the equilibrium constant associated with the jth equilibrium reaction is given by

$$K_j^e = \prod_{i=1}^{N} y_i^{\lambda_i^{2j-1} - \nu_i^{2j-1}}.$$

On the other hand, the extension of Eq. (5.52) to J couples of reversible reactions is written as

$$n_i(t) = n_{i,init} + \sum_{j=1}^{J}(\lambda_i^{2j-1} - \nu_i^{2j-1})\xi_j(t).$$

Hence, as $y_i = n_i\rho$,

$$y_i(t) = \rho\left(n_{i,init} + \sum_{j=1}^{J}(\lambda_i^{2j-1} - \nu_i^{2j-1})\xi_j(t)\right).$$

By replacing (19) into (19) we get

$$K_j^e = \rho^{\sum_{i=1}^{N}(\lambda_i^{2j-1}-\nu_i^{2j-1})} \prod_{i=1}^{N} \left[n_{i,init} + \sum_{j=1}^{J} (\lambda_i^{2j-1} - \nu_i^{2j-1}) \xi_j \right]^{\lambda_i^{2j-1}-\nu_i^{2j-1}},$$

where K_j^e and ξ_j are the equilibrium constant and reaction extent of the jth chemical reaction. Equation (19) constitutes a nonlinear system of algebraic equations that must be solved for ξ_j, $j = 1, \ldots, J$ in order to obtain the concentration at equilibrium.

SOLUTIONS FOR CHAPTER 6

6.1 Taking the derivative of $F_L(\nu) = e^\nu/(1+e^\nu)$ and employing the product and chain rules from calculus, we get

$$\begin{aligned} f_L(\nu) &= \frac{dF_L(\nu)}{d\nu} \\ &= \frac{e^\nu}{1+e^\nu} - \frac{e^\nu e^\nu}{(1+e^\nu)^2} \\ &= \left(\frac{e^\nu}{1+e^\nu}\right)\left(1 - \frac{e^\nu}{1+e^\nu}\right) \\ &= F(\nu)\left[1 - F(\nu)\right], \end{aligned}$$

as required.

6.2 Working with expression (6.1) and letting $\beta = \beta_1 - \beta_0$, we have

$$\begin{aligned} \Pr(y_i = 1|\beta) &= P(U_{i1} > U_{i0}) \\ &= \Pr(\varepsilon_{i0} < \varepsilon_{i1} + x_i'\beta) \\ &= \int_{-\infty}^{\infty} F_{EV}(\varepsilon_{i1} + x_i'\beta)\, f_{EV}(\varepsilon_{i1})\, d\varepsilon_{i1} \\ &= \int_{-\infty}^{\infty} \exp\left(-e^{-(\varepsilon_{i1}+x_i'\beta)}\right) e^{-\varepsilon_{i1}} \exp\left(-e^{-\varepsilon_{i1}}\right) d\varepsilon_{i1} \\ &= \int_{-\infty}^{\infty} \exp\left(-e^{-\varepsilon_{i1}} - e^{-(\varepsilon_{i1}+x_i'\beta)}\right) e^{-\varepsilon_{i1}}\, d\varepsilon_{i1} \\ &= \int_{-\infty}^{\infty} \exp\left(-e^{-\varepsilon_{i1}}\left(1 + e^{-x_i'\beta}\right)\right) e^{-\varepsilon_{i1}}\, d\varepsilon_{i1}. \end{aligned}$$

Letting $t = e^{-\varepsilon_{i1}}$, we have that $dt = -e^{-\varepsilon_{i1}} d\varepsilon_{i1}$. As $\varepsilon_{i1} \to \infty$, $t \to 0$ and as $\varepsilon_{i1} \to -\infty$, $t \to \infty$. Therefore, we can rewrite the integral as

$$\begin{aligned}
\Pr(y_i = 1|\beta) &= \int_{\infty}^{0} -\exp\left(-t\left(1+e^{-x_i'\beta}\right)\right) dt \\
&= \int_{0}^{\infty} \exp\left(-t\left(1+e^{-x_i'\beta}\right)\right) dt \\
&= -\frac{1}{1+e^{-x_i'\beta}} \exp\left(-t\left(1+e^{-x_i'\beta}\right)\right)\Big|_{t=0}^{\infty} \\
&= -\frac{1}{1+e^{-x_i'\beta}}(0-1) \\
&= \frac{1}{1+e^{-x_i'\beta}},
\end{aligned}$$

as required. A more general version of the proof for the case of multinomial outcomes is available in Ref. [25].

6.3 The full conditional distribution $\pi(\beta|y, z)$ is proportional to $f(z|\beta)\pi(\beta)$ and its kernel can be written as

$$\pi(\beta|y,z) \propto \exp\left[-\frac{1}{2}\left\{(z - X\beta)'(z - X\beta) + (\beta - \beta_0)' B_0^{-1} (\beta - \beta_0)\right\}\right]$$
$$\propto \exp\left[-\frac{1}{2}\left\{-z'X\beta - \beta'X'z + \beta'X'X\beta + \beta'B_0^{-1}\beta - \beta'B_0^{-1}\beta_0 - \beta_0'B_0^{-1}\beta\right\}\right],$$

where we have omitted terms that do not involve β. Collecting terms and using the definitions of $\hat{B}$ and $\hat{\beta}$, we have that $\pi(\beta|y, z)$ is proportional to

$$\exp\left[-\frac{1}{2}\left\{\beta'\left(X'X + B_0^{-1}\right)\beta - \beta'\left(X'z + B_0^{-1}\beta_0\right) - \left(z'X + \beta_0'B_0^{-1}\right)\beta\right\}\right]$$
$$= \exp\left[-\frac{1}{2}\left\{\beta'\hat{B}^{-1}\beta - \beta'\hat{B}^{-1}\hat{\beta} - \hat{\beta}'\hat{B}^{-1}\beta\right\}\right].$$

Adding and subtracting $\hat{\beta}'\hat{B}^{-1}\hat{\beta}$ inside the curly braces, we can complete the square and write

$$\begin{aligned}
\pi(\beta|y,z) &\propto \exp\left[-\frac{1}{2}\left\{\left(\beta-\hat{\beta}\right)'\hat{B}^{-1}\left(\beta-\hat{\beta}\right) - \hat{\beta}'\hat{B}^{-1}\hat{\beta}\right\}\right] \\
&\propto \exp\left[-\frac{1}{2}\left\{\left(\beta-\hat{\beta}\right)'\hat{B}^{-1}\left(\beta-\hat{\beta}\right)\right\}\right],
\end{aligned}$$

where the last line follows by recognizing that $\hat{\beta}'\hat{B}^{-1}\hat{\beta}$ does not involve β and can therefore be absorbed in the constant of proportionality. The result is the kernel of the Gaussian density and hence we have shown that

$$\beta | y, z \sim N(\hat{\beta}, \hat{B}),$$

as required.

SOLUTIONS FOR CHAPTER 7

7.1 The definition of continuity of T requires that for any $\epsilon > 0$ there is a $\delta > 0$ such that $d(Tx, Ty) < \epsilon$ whenever $d(x, y) < \delta$. For a contraction map T, we know that $d(Tx, Ty) \leq cd(x, y)$, where $c \in [0, 1)$. So, for any given $\epsilon > 0$ we choose $\delta = \frac{\epsilon}{c}$ if $c \neq 0$ and δ =anything we wish if $c = 0$. Then we have

$$d(Tx, Ty) \leq cd(x, y) < c\delta = \epsilon \text{ if } c \neq 0,$$

and

$$d(Tx, Ty) \leq cd(x, y) = 0 < \epsilon \text{ if } c = 0,$$

proving that T is continuous.

7.2 We must prove that $d_\infty(x, y) = \|x - y\|_\infty = \max_{t \in I} |x(t) - y(t)|$ the three properties of a metric hold.

1. $d_\infty(x, y) \geq 0$ for all $x, y \in C(I)$, and $d_\infty(x, y) = 0$ if and only if $x = y$.

   Because of the absolute value, $d_\infty(x, y) \geq 0$ for all $x, y \in C(I)$. In addition, if $\max_{t \in I} |x(t) - y(t)| = 0$, then $|x(t) - y(t)| = 0$, so $x(t) = y(t)$. And, of course, $d_\infty(x, x) = 0$.

2. $d_\infty(x, y) = d_\infty(y, x)$ for all $x, y \in C(I)$;

   We see that

   $$d_\infty(x, y) = \|x - y\|_\infty = \max_{t \in I} |x(t) - y(t)| = \max_{t \in I} |y(t) - x(t)| = d_\infty(y, x).$$

3. $d_\infty(x, z) \leq d_\infty(x, y) + d_\infty(y, z)$ for all $x, y, z \in C(I)$.

   We have

   $$\begin{aligned} d_\infty(x, z) &= \max_{t \in I} |x(t) - y(t)| \\ &\leq \max_{t \in I} (|x(t) - y(t)| + |y(t) - z(t)|) \\ &\qquad \text{(triangle inequality for absolute value)} \\ &\leq \max_{t \in I} |x(t) - y(t)| + \max_{t \in I} |y(t) - z(t)| \\ &= d_\infty(x, y) + d_\infty(y, z). \end{aligned}$$

7.3 If $x(t)$ is a solution to the the IVP (7.1) and (7.2), then by the steps that developed (7.3) in the main text, we know that (7.3) is satisfied. On the other hand, suppose that (7.3) holds:

$$x(t) = x_0 + \int_{t_0}^{t} f(x(s), s)\, ds.$$

Differentiating with respect to t, we find that

$$x'(t) = \frac{d}{dt}\left(x_0 + \int_{t_0}^{t} f(x(s), s)\, ds\right) = f(x(t), t),$$

using the Fundamental Theorem of Calculus. Furthermore, setting $t = t_0$ in the integral equation, we find

$$x(t_0) = x_0 + \int_{t_0}^{t_0} f(x(s), s)\, ds = x_0,$$

the definite integral above is zero.

7.4 The answers are as follows:

(a) We can choose a and b in the definition of D as we wish. Since $x(0) = 1$ as opposed to $x(0) = 0$, we shift the inequality for x in the definition of D. Suppose we let $D = \{(x,t) || x - 1| < 1,\ |t| < \frac{1}{3}\}$. Then we see that

$$\max_{(x,t)\in D} |f(x)| = \max_{(x,t)\in D} |x| = 2 < 3 = \frac{b}{a},$$

so condition 1 in our set up is satisifed. Furthermore, we see that

$$|f(x) - f(y)| = |x - y|,$$

so the Lipschitz condition 2 is satisfied with $K = 1$ and $c = Ka = \frac{1}{3} < 1$.

(b) We calculate that $x'(t) = \frac{d}{dt}e^t = e^t = x(t)$, so the ODE is satisfied. Since $x(0) = e^0 = 1$, the initial condition is also satisfied. We conclude the $x(t) = e^t$ is a solution of the IVP.

(c) We define the Picard operator

$$(Tx)(t) = x(0) + \int_0^t f(x(s))\, ds = 1 + \int_0^t x(s)\, ds$$

and construct the sequence of iterates $x_{n+1} = Tx_n$ as requested. Beginning with $x_0 = 1$, we calculate that

$$\begin{aligned}
x_1 &= Tx_0 = 1 + \int_0^t x_0\,ds \\
&= 1 + \int_0^t 1\,ds = 1 + t, \\
x_2 &= Tx_1 = 1 + \int_0^t x_1(s)\,ds \\
&= 1 + \int_0^t (1+s)\,ds = 1 + \left(s + \frac{s^2}{2}\right)_0^t \\
&= 1 + t + \frac{t^2}{2}, \\
x_3 &= Tx_2 = 1 + \int_0^t x_2(s)\,ds \\
&= 1 + \int_0^t \left(1 + s + \frac{s^2}{2}\right) ds = 1 + \left(s + \frac{s^2}{2} + \frac{s^3}{6}\right)_0^t \\
&= 1 + t + \frac{t^2}{2} + \frac{t^3}{6}, \\
x_4 &= Tx_3 = 1 + \int_0^t x_3(s)\,ds \\
&= 1 + \int_0^t \left(1 + s + \frac{s^2}{2} + \frac{s^3}{6}\right) ds = 1 + \left(s + \frac{s^2}{2} + \frac{s^3}{6} + \frac{s^4}{24}\right)_0^t \\
&= 1 + t + \frac{t^2}{2} + \frac{t^3}{6} + \frac{t^4}{24}.
\end{aligned}$$

The pattern makes it fairly clear that the sequence of iterates approaches the Taylor of $e^t = \sum_{n=0}^{\infty} \frac{t^n}{n!}$.

7.5 The answers are as follows:

(a) We find that $x'(t) = \frac{d}{dt}\frac{1}{1-t} = \frac{1}{(1-t)^2} = x^2$, so the ODE is satisfied. Since $x(0) = \frac{1}{1-0} = 1$, the initial condition is also satisfied. This means that the $x(t) = \frac{1}{1-t}$ is a solution of the IVP.

(b) We define the Picard operator

$$(Tx)(t) = x(0) + \int_0^t f(x(s))\,ds = 1 + \int_0^t x^2(s)\,ds$$

and construct the sequence of iterates $x_{n+1} = Tx_n$ for $n-1, 2, 3, 4$. Beginning with $x_0 = 1$, we calculate that

$$\begin{aligned}
x_1 &= Tx_0 = 1 + \int_0^t (x_0)^2\, ds \\
&= 1 + \int_0^t 1\, ds = 1 + t, \\
x_2 &= Tx_1 = 1 + \int_0^t (x_1(s))^2\, ds \\
&= 1 + \int_0^t (1+s)^2\, ds = 1 + \int_0^t \left(1 + 2s + s^2\right) ds \\
&= 1 + \left(s + s^2 + \frac{s^3}{3}\right)_0^t = 1 + t + t^2 + \frac{t^3}{3}, \\
x_3 &= Tx_2 = 1 + \int_0^t (x_2(s))^2\, ds \\
&= 1 + \int_0^t \left(1 + s + s^2 + \frac{s^3}{3}\right)^2 ds \\
&= 1 + \int_0^t \left(1 + 2s + 3s^2 + \frac{8}{3}s^3 + \frac{5}{3}s^4 + \frac{2}{3}s^5 + \frac{1}{9}s^6\right) ds \\
&= 1 + \left(s + s^2 + s^3 + \frac{2}{3}s^4 + \frac{1}{3}s^5 + \frac{1}{9}s^6 + \frac{1}{63}s^7\right)_0^t \\
&= 1 + t + t^2 + t^3 + \frac{2}{3}t^4 + \frac{1}{3}t^5 + \frac{1}{9}t^6 + \frac{1}{63}t^7, \\
x_4 &= Tx_3 = 1 + \int_0^t (x_3(s))^2\, ds \\
&= 1 + \int_0^t \left(1 + s + s^2 + s^3 + \frac{2}{3}s^4 + \frac{1}{3}s^5 + \frac{1}{9}s^6 + \frac{1}{63}s^7\right)^2 ds \\
&= 1 + \int_0^t \left(1 + 2s + 3s^2 + 4s^3 + \frac{13}{3}s^4 + 4s^5 + \frac{29}{9}s^6\right. \\
&\quad + \frac{142}{63}s^7 + \frac{86}{63}s^8 + \frac{44}{63}s^9 \\
&\quad \left. + \frac{55}{189}s^{10} + \frac{2}{21}s^{11} + \frac{13}{567}s^{12} + \frac{2}{567}s^{13} + \frac{1}{3969}s^{14}\right) ds \\
&= 1 + \left(s + s^2 + s^3 + s^4 + \frac{13}{15}s^5 + \frac{2}{3}s^6 + \frac{29}{63}s^7 + + \frac{71}{252}s^8 + \frac{86}{567}s^9\right. \\
&\quad \left. + \frac{22}{315}s^{10} + \frac{5}{189}s^11 + \frac{1}{126}s^{12} + \frac{1}{567}s^{13} + \frac{1}{3969}s^{14} + \frac{1}{59535}s^{15}\right)_0^t \\
&= 1 + t + t^2 + t^3 + t^4 + \frac{13}{15}t^5 + \frac{2}{3}t^6 + \frac{29}{63}t^7 + + \frac{71}{252}t^8 + \frac{86}{567}t^9 \\
&\quad + \frac{22}{315}t^{10} + \frac{5}{189}t^11 + \frac{1}{126}t^{12} + \frac{1}{567}t^{13} + \frac{1}{3969}t^{14} + \frac{1}{59535}t^{15}.
\end{aligned}$$

It is harder to see that the sequence approaches the Taylor series of the solution in part (a), namely

$$x(t) = \frac{1}{1-t} = \sum_{n=0}^{\infty} t^n.$$

But the early terms are certainly correct, and the number of correct terms increases as we progress through the sequence.

7.6 Starting at (7.7),

$$d_2^2(Tx, Ty) \leq \int_I \left[\int_0^t K\,|x(s) - y(s)|\,ds\right]^2 dt,$$

we now suppose that $t < 0$. The Cauchy-Schwarz inequality gives

$$\int_t^0 g(s)h(s)\,ds \leq \left[\int_t^0 (g(s))^2\,ds\right]^{\frac{1}{2}} \left[\int_t^0 (h(s))^2\,ds\right]^{\frac{1}{2}}.$$

We apply the inequality with $g(s) = 1$ and $h(s) = |x(s) - y(s)|$. Then

$$\begin{aligned}\int_t^0 1 \cdot |x(s) - y(s)|\,ds &\leq \left[\int_t^0 1\,ds\right]^{\frac{1}{2}} \left[\int_t^0 |x(s) - y(s)|^2\,ds\right]^{\frac{1}{2}} \\ &\leq (-t)^{\frac{1}{2}} \left[\int_t^0 |x(s) - y(s)|^2\,ds\right]^{\frac{1}{2}}.\end{aligned}$$

Plugging (19) into (7.7) gives

$$\begin{aligned}d_2^2(Tx, Ty) &\leq K^2 \int_{-a}^0 \left[(-t)^{\frac{1}{2}} \left[\int_t^0 |x(s) - y(s)|^2\,ds\right]^{\frac{1}{2}}\right]^2 dt \\ &= K^2 \int_{-a}^0 \int_t^0 (-t)|x(s) - y(s)|^2\,dsdt \\ &= -K^2 \int_{-a}^0 \int_a^s t|x(s) - y(s)|^2\,dtds \\ &= -K^2 \int_a^s t\,dt \int_{-a}^0 |x(s) - y(s)|^2\,ds \\ &= -K^2 \left(\frac{s^2}{2} - \frac{a^2}{2}\right) \int_{-a}^0 |x(s) - y(s)|^2\,ds \\ &= K^2 \left(\frac{a^2}{2} - \frac{s^2}{2}\right) \int_{-a}^0 |x(s) - y(s)|^2\,ds\end{aligned}$$

$$\leq \frac{K^2a^2}{2}\int_{-a}^{0}|x(s)-y(s)|^2\,ds,$$

and square rooting gives the result.

7.7 For $x, y \in \bar{C}(I)$, we calculate that

$$\begin{aligned} d_1(Tx,Ty) &= \int_I \left| \int_0^t f(x(s),s)\,ds - \int_0^t f(y(s),s)\,ds \right| dt \\ &= \int_I \left| \int_0^t (f(x(s),s) - f(y(s),s))\,ds \right| dt \\ &\leq \int_I \left| \int_0^t |f(x(s),s) - f(y(s),s)|\,ds \right| dt \\ &\leq \int_I \left| \int_0^t K\,|x(s)-y(s)|\,ds \right| dt. \end{aligned}$$

We consider $t > 0$ first. Then

$$\begin{aligned} d_1(Tx,Ty) &\leq \int_I \int_0^t K\,|x(s)-y(s)|\,dsdt \\ &\leq K\int_I t\,|x(s)-y(s)|\,dsdt \\ &\leq Kad_1(x,y). \end{aligned}$$

In the other hand, if $t < 0$ then

$$\begin{aligned} d_1(Tx,Ty) &\leq \int_I \int_t^0 K\,|x(s)-y(s)|\,dsdt \\ &\leq K\int_I (-t)\,|x(s)-y(s)|\,dsdt \\ &\leq Kad_1(x,y). \end{aligned}$$

We conclude that the Picard operator T is contractive in the d_1 metric with contractivity factor $c = Ka < 1$.

SOLUTIONS FOR CHAPTER 8

8.1 Denoting the two timing observations as $\mathbf{y}[1]$ and $\mathbf{y}[2]$, we have (from the calculations presented in the Example)

$$\mathbf{x}\,|\,(\mathbf{y}[1] = y[1]) \stackrel{.}{\sim} \mathrm{N}(m[1], P[1]),$$

with $m[1] = 149.91$ and $P[1] = 147.97$. The linearization of the observation equation about $\hat{x} = m[1]$ has

$$H = \frac{1}{\sqrt{2gm[1]}} = 0.01844.$$

The observation $[5.45, 5.60] = 5.525 \pm 0.075$ is interpreted as a realization $y[2] = 5.525$ of a observation $\mathbf{y}[2]$ having a variance of $R[2] = (\frac{0.075}{1.96})^2 = 1.464 \cdot 10^{-3}$. Finally, applying the updating formulas (8.14), we obtain the new posterior $\mathbf{x} | (\mathbf{y}[1{:}2] = y[1{:}2]) \mathrel{\dot{\sim}} \mathrm{N}(m[2], P[2])$ with

$$\begin{aligned}
K[2] &= P[1]H[2]^T(R[2] + H[2]P[1]H[1]^T)^{-1} \\
&= \frac{147.97 \cdot 0.01844}{1.464 \cdot 10^{-3} + 147.97 \cdot (0.01844)^2} = 52.7, \\
P[2] &= P[1] - K[2]H[2]P[1] = 147.97 \cdot (1 - 52.7 \cdot 0.01844) = 4.185, \\
m[2] &= m[1] + K[2](y[2] - h(m[1])) \\
&= 149.91 + 52.7 \cdot (5.525 - \sqrt{\frac{2 \cdot 149.91}{9.81}}) = 151.2.
\end{aligned}$$

Frodo is now 95% certain that the chasm depth (in meters) is a number in the interval $151.2 \pm 1.96\sqrt{4.185} = [147,\ 155]$.

8.2 In general, the distance from a line $x_2 = mx_1 + b$ and a point (x_1, x_2) is $|mx_1 + b - x_2|/\sqrt{m^2+1}$.

The equation of the line for wall 1 is $x_2 = 0$, and the distance is $|0 \cdot x_1 + 0 - x_2|/\sqrt{0^2+1} = |-x_2|$. For a point inside the room, $x_2 > 0$ and the distance is $y_1 = x_2$.

The equation of the line for wall 2 is $x_2 = \sqrt{3}(12 - x_1)$, and the distance is $|-\sqrt{3}x_1 + 12\sqrt{3} - x_2|/\sqrt{3+1} = |6\sqrt{3} - \frac{\sqrt{3}}{2}x_1 - \frac{1}{2}x_2|$. For a point inside the room, the distance is $y_2 = 6\sqrt{3} - \frac{\sqrt{3}}{2}x_1 - \frac{1}{2}x_2$.

The equation of the line for wall 3 is $x_2 = \sqrt{3}x_1$, and the distance is $|\sqrt{3}x_1 - x_2|/\sqrt{3+1} = |\frac{\sqrt{3}}{2}x_1 - \frac{1}{2}x_2|$. For a point inside the room, the distance is $y_3 = \frac{\sqrt{3}}{2}x_1 - \frac{1}{2}x_2$.

Thus $\mathbf{y} \,|\, (\mathbf{x} = x) \sim \mathrm{N}(Hx + b, R)$ with

$$H = \begin{bmatrix} 0 & 1 \\ -\frac{\sqrt{3}}{2} & -\frac{1}{2} \\ \frac{\sqrt{3}}{2} & -\frac{1}{2} \end{bmatrix}, \quad b = \begin{bmatrix} 0 \\ 6\sqrt{3} \\ 0 \end{bmatrix}, \quad R = \sigma^2 I,$$

where σ is the standard deviation of a distance observation.

Using a flat prior, the position estimate is the posterior $\mathbf{x}|(\mathbf{y}=y) \sim \mathrm{N}(q, Q)$ with

$$Q = (H^T R{-}1 H)^{-1} = \sigma^2 \begin{bmatrix} 2/3 & 0 \\ 0 & 2/3 \end{bmatrix},$$

$$q = QH^T R^{-1}(y - b) = \begin{bmatrix} 7.1547 \\ 1.4641 \end{bmatrix}.$$

According to the Chebyshev inequality, the q-centered disk containing at least 95% of the probability has radius

$$\sqrt{\frac{\mathrm{trace}(P)}{0.05}} = \sqrt{\frac{(4/3)(0.5)^2\sigma^2}{0.05}} = 2.58\sigma.$$

The 95% ellipse is the set $\{x : (x-q)^T P^{-1}(x-q) < 5.99\}$. Here, $P = \frac{2\sigma^2}{3} I$, so the ellipse is a q-centred circle with radius $\sqrt{5.99 \cdot \frac{2}{3} \cdot 0.5^2 \sigma^2} = \sigma$.

8.3 The observation function and its 2×2 Jacobian matrix are

$$h(x) = \begin{bmatrix} \|s[1] - x\| \\ \|s[2] - x\| \end{bmatrix}, \quad H = \begin{bmatrix} (x - s[1])^T / \|x - s[1]\| \\ (x - s[2])^T / \|x - s[2]\| \end{bmatrix}$$

The cost function is

$$\phi(x) = \frac{1}{2}(y - h(x)) R^{-1} (y - h(x)) + \frac{1}{2} x^T P^{-1} x,$$

with $R = 10^2 I$ and $P = 30^2 I$. A contour plot of cost function values is shown below.

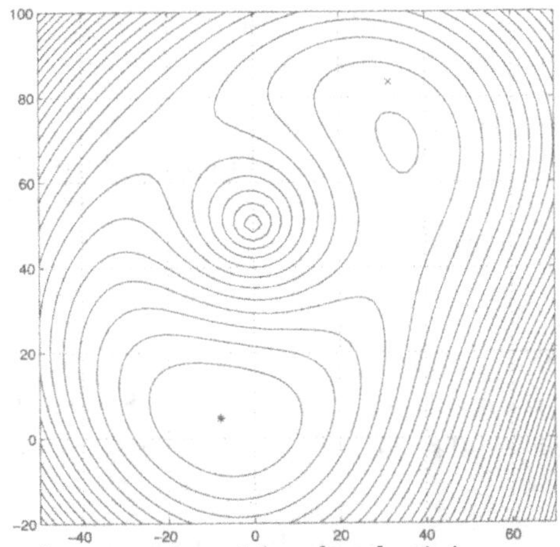

From the plot it can be seen that ϕ has local minima near the two locations that satisfy the range observations exactly. If we only use the range observations, there would be ambiguity in the position solution, but because of the prior, the cost function minimum near the origin is the unique global minimum.

The Gauss-Newton method's steps can be computed using this Matlab/Octave script.

```
m=[0;0]; P=30^2*eye(2);
y=[108;46]; R=10^2*eye(2);

h=@(x) [ norm(x-[100;0]); norm(x-[0;50])];
H=@(x) diag(h(x))\([x-[100;0],x-[0;50]]');

x=m; % initial estimate is the prior mean
nt=3; % number of GN steps
for it=1:nt
    HH=H(x);
    S=HH*P*HH'+R;
    K=P*HH'/S;
    d=m-x-K*HH*(m-x)-K*(h(x)-y)
    x=x+d
 end
```

The first three iterands of the Gauss-Newton method, starting from the prior mean, are $\begin{bmatrix} -7.2 \\ 3.6 \end{bmatrix}$, $\begin{bmatrix} -7.0700 \\ 4.1084 \end{bmatrix}$, $\begin{bmatrix} -7.0696 \\ 4.1121 \end{bmatrix}$.

SOLUTIONS FOR CHAPTER 9

9.1 Apply Itō's lemma to the function $f(t,x) = exp(\lambda t + \sigma x)$ for a proper choice of λ and then set $x = W(t)$. Alternatively, apply Itō's lemma to the function $g(x) = \ln(x)$ and then set $x = S(t)$.

9.2 First year calculus after you notice that $(x-K)^+$ only contributes to the integral if $x > K$. But x is $S(T)$ and this imposed restrictions on $W(T)$. This is a random variable with known distribution (the normal distribution) and this leads to the stated results.

9.3 Elementary.

9.4 Since $dS = \mu S + \sigma S dW$ taking expectations we see that $d\mathbb{E}[S(t)] = \mu\mathbb{E}[S(t)]dt$ from which it follows that $\mathbb{E}[S(t)] = S(0)\exp(\mu t)$. For the second moment, apply Itō's lemma on $f(x) = x^2$ calculated at $x = S(t)$ and then take expectations.

9.5 The solution of the static optimization problem is

$$\frac{1}{c} = e^{-\delta t} V_x,$$

$$w = \frac{-V_x}{x \cdot V_{xx}} \cdot \frac{\mu - r}{\sigma^2}.$$

Look for V in the form

$$V(t, x) = e^{-\delta t} h(t) \ln(x),$$

and upon substitution into the HJB

$$\dot{h}(t) + \left[\frac{x}{\ln(x)} \frac{(\mu - r)^2}{\sigma^2} \left(1 - \frac{1}{2}x\right) - \delta \right] h(t) - \frac{\ln(h(t))}{\ln(x)} + 1 = 0,$$

with final condition $h(T) = 0$, which can be solved explicitly.

SOLUTIONS FOR CHAPTER 10

10.1 Let E=number of European cabinets produced each day, and C=number of Chinese cabinets produced each day. Then we have

$$\begin{aligned}
\textit{Maximize}\text{ revenue} &= \$30E + \$26C \\
\text{subject to} & \\
2.80E + 2.60C &\leq 330 \text{ (carpentry department)}, \\
1.40E + 1.20C &\leq 220 \text{ (painting department)}, \\
0.70E + 0.70C &\leq 130 \text{ (finishing department)}, \\
E &\geq 58 \text{ (contract requirement)}, \\
C &\geq 58 \text{ (contract requirement)}, \\
E, C &\geq 0 \text{ (non-negativity)}.
\end{aligned}$$

Using the Solver add-in in Excel we obtain that the optimal solution is to produce 64 units of European cabinets and 58 units of Chinese cabinets each day, which can lead to a total revenue of \$3,428.

10.2 Let M_1=number of B1 to make. M_2 and M_3 are defined similarly. Let B_1=number of B_2 to buy. B_2 and B_3 are defined similarly. Then we have

$$\begin{aligned}
\text{Minimize cost} &= \$16.5M_1 + \$20.4B_1 + \$19M_2 + \$20.85B_2 \\
&\quad + \$22.5M_3 + \$24.76B_3, \\
\text{subject to} & \\
2.40M_1 + 3.30M_2 + 3.90M_3 &\leq 16,800 \text{ (Fabrication)}, \\
0.26M_1 + 0.32M_2 + 0.48M_3 &\leq 1,800 \text{ (Inspection)},
\end{aligned}$$

$$\begin{aligned} M_1 &\geq 0.65(M_1 + B_1) \text{ (Min make B1)}, \\ M_2 &\geq 0.65(M_2 + B_2) \text{ (Min make B2)}, \\ M_3 &\geq 0.65(M_3 + B_3) \text{ (Min make B3)}, \end{aligned}$$

$$\begin{aligned} M_1 + B_1 &= 2,100 \text{ (B1 Demand)}, \\ M_2 + B_2 &= 3,820 \text{ (B2 Demand)}, \end{aligned}$$

$$\begin{aligned} M_3 + B_3 &= 1,820 \text{ (B3 Demand)}, \\ M_1, M_2, M_3, B_1, B_2, B_3 &\geq 0 \text{ (non-negative integers)}. \end{aligned}$$

With the Excel solver, we obtain that the optimal solution is to make 1,663 units of B1, 2,483 units of B2, and 1,183 units of B3; buy 437 units of B1, 1,337 units of B2, and 637 units of B3, and the total cost is $153,797.

10.3 Let A=number of tonnes of Cargo A loaded. B, C, D and E can be defined similarly. Then we obtain that

$$\begin{aligned} \text{Maximize value of shipment} &= \$1,400A + \$1,780B + \$1,280C, \\ &\quad +\$920D + \$1,380E \\ \text{subject to} & \\ A + B + C + D + E &\leq 2,450 \text{ (Weight limit)}, \\ 28A + 56B + 32C + 48D + 38E &\leq 110,000 \text{ (Volume limit)}, \end{aligned}$$

$$\begin{aligned} 0.22 * 980 \leq A &\leq 980 \text{ (Min and Max of Cargo A)}, \\ 0.22 * 880 \leq B &\leq 880 \text{ (Min and Max of Cargo B)}, \\ 0.22 * 1,980 \leq C &\leq 1,980 \text{ (Min and Max of Cargo C)}, \end{aligned}$$

$$\begin{aligned} 0.22 * 2,300 \leq D &\leq 2,300 \text{ (Min and Max of Cargo D)}, \\ 0.22 * 3,680 \leq E &\leq 3,680 \text{ (Min and Max of Cargo E)}, \\ A, B, C, D, E &\geq 0 \text{ (non-negativity)}. \end{aligned}$$

Using the spreadsheet, we find that the optimal solution is to load 215.6 tonnes of A, 483.2 t of B, 435.6 t of C, 506 t of D, and 809.6 t of E, which can achieve to the total value of $3,302,272.

10.4 Let X_i=number of workers starting at period i (with $i = 1, 2, 3, 4, 5, 6$). Thus the model can be described as follows.

$$\begin{aligned}
\mathit{Minimize} \text{ staff size} &= X_1 + X_2 + X_3 + X_4 + X_5 + X_6, \\
\text{subject to} & \\
X_1 + X_6 &\geq 4 \text{ (3AM-7AM needs)}, \\
X_1 + X_2 &\geq 12 \text{ (7AM-11AM needs)}, \\
X_2 + X_3 &\geq 17 \text{ (11AM-3PM needs)}, \\
X_3 + X_4 &\geq 10 \text{ (3PM-7PMm needs)}, \\
X_4 + X_5 &\geq 14 \text{ (7PM-11PM needs)}, \\
X_5 + X_6 &\geq 5 \text{ (11PM-3AM needs)}, \\
X_1, X_2, X_3, X_4, X_5, X_6 &\geq 0 \text{ (non-negativity)}.
\end{aligned}$$

Using Excel and the Solver, we obtain that the optimal solution is to employ 35 workers, 4 of whom start at 3 a.m., 16 start at 7 a.m., 1 start at 11 a.m., 9 start at 3 p.m., and 5 start at 7 p.m.

SOLUTIONS FOR CHAPTER 11

11.1 Let $v(t) = u(t)\exp(\alpha t)$ and differentiate with respect to t to get

$$\begin{aligned}
v'(t) &= \alpha v(t) + u'(t)\mathrm{e}^{\alpha t} \\
&= \alpha v(t) - \alpha v(t) + g(t)\mathrm{e}^{\alpha t} \\
&= g(t)\exp(\alpha t).
\end{aligned}$$

Integrating both sides yields

$$v(t) = v(0) + \int_0^t g(s)\mathrm{e}^{\alpha s}\, ds.$$

This implies that

$$u(t) = u_0\mathrm{e}^{-\alpha t} + \int_0^t g(s)\mathrm{e}^{-\alpha(t-s)}\, ds.$$

11.2 First, recall that the stochastic integral is a Gaussian random variable with mean zero and variance equal to

$$\mathrm{Var}\left(\int_0^t \mathrm{e}^{\alpha s}\, dB(s)\right) = \int_0^t \mathrm{e}^{2\alpha s}\, ds = \frac{1}{2\alpha}\left(\mathrm{e}^{2\alpha t} - 1\right).$$

But then the stochastic integral has the same probability distribution as the random variable

$$\frac{1}{\sqrt{2\alpha}}\left(\mathrm{e}^{2\alpha t}-1\right)^{1/2}\times Y$$

where Y is a standard normal random variable, that is, Y has mean zero and variance equal to one and probability distribution

$$P(Y\leq y)=\frac{1}{2\pi}\int_{-\infty}^{y}\mathrm{e}^{-z^2/2}\,dz\,.$$

We compute the expectation to be

$$\mathbf{E}\left[\exp\left(\mathrm{i}x\int_0^t \mathrm{e}^{\alpha s}\,dB(s)\right)\right]=\mathbf{E}\left[\exp\left(\mathrm{i}\frac{1}{\sqrt{2\alpha}}\left(\mathrm{e}^{2\alpha t}-1\right)^{1/2}Y\right)\right]$$

$$=\frac{1}{\sqrt{2\pi}}\int_{-\infty}^{\infty}\exp\left(\frac{\mathrm{i}x}{\sqrt{2\alpha}}\left(\mathrm{e}^{2\alpha t}-1\right)^{1/2}y-\frac{1}{2}y^2\right)\,dy=\exp\left(-\frac{x^2}{4\alpha}\left(\mathrm{e}^{2\alpha t}-1\right)\right).$$

Hence, we find the cumulant function of the stochastic integral to be

$$\psi(x)=-\frac{x^2}{4\alpha}\left(\mathrm{e}^{2\alpha t}-1\right)\,.$$

11.3 Proceed like in Example 11.2, but now with the finite difference approximations for $u^{(3)}(t), u''(t)$ and $u'(t)$.

11.4 This is straightforward using the definition of an eigenvalue from linear algebra.

SOLUTIONS FOR CHAPTER 12

12.1 You win the sum 2^n at the $n+1$ play with probability $(\frac{1}{2})^{n+1}$. The expected winning is a nonconverging sum. The expected logarithmic utility on the other hand is the sum $\sum_{i=1}\infty(\frac{1}{2})^{n+1}\ln(2^n)$ which is a convergent sum.

12.2 Assume without loss of generality that $u_0=0$. The first option provides utility $u_0(1)$. The second option provides utility $\delta u_0(p)$. If the agent prefer the second it must be that $\delta u_0(p)>u_0(1)$ so that $p=u_0^{-1}\left(\frac{u_0(1)}{\delta}\right)$. It is easy to show that since $\delta<1$ and u_0 is strictly increasing that $p>1$.

SOLUTIONS FOR CHAPTER 15

15.1 The resulting closed-loop transfer function $G_c(s)$ becomes

$$G_c(s)=\frac{Y(s)}{W(s)}=\frac{12.46s+64.47}{39.69s^{1.25}+12.46s+65.068}.$$

The analytical solution (impulse response) of the control system (19) is

$$\begin{aligned} y(t) &= \frac{12.46}{39.69}\sum_{k=0}^{\infty}\frac{(-1)^k}{k!}\left(\frac{12.46}{39.69}\right)^k \times \mathcal{E}_k\Big(t,-\frac{65.068}{39.69};1.25,0.25-k\Big) \\ &+ \frac{64.47}{39.69}\sum_{k=0}^{\infty}\frac{(-1)^k}{k!}\left(\frac{65.068}{39.69}\right)^k \times \mathcal{E}_k\Big(t,-\frac{12.46}{39.69};1.25-1,1.25+k\Big), \end{aligned}$$

with zero initial conditions.

15.2 The characteristic equation of this system is

$$39.69s^{1.25}+12.46s+65.068=0 \;\Rightarrow\; 39.69s^{\frac{5}{4}}+12.46s^{\frac{4}{4}}+65.068=0.$$

Using the notation $w=s^{\frac{1}{m}}$, where LCM is $m=4$, we obtain a polynomial of complex variable w in the form

$$39.69w^5+12.46w^4+65.068=0.$$

Solving the polynomial (19) we get the following roots and their arguments:

$$w_1=-1.17474, |\arg(w_1)|=\pi,$$

$$w_{2,3}=-0.40540\pm 1.0426j, |\arg(w_{2,3})|=1.9416,$$

$$w_{4,5}=0.83580\pm 0.64536j, |\arg(w_{4,5})|=0.6575.$$

This first Riemann sheet is defined as a sector in w-plane within interval $-\pi/4<\arg(w)<\pi/4$. Complex conjugate roots $w_{4,5}$ lie in this interval and satisfied the stability condition given as $|\arg(w)|>\frac{\pi}{8}$, therefore the system is stable. The region where $|\arg(w)|>\frac{\pi}{4}$ is not physical.

15.3 Characteristic equation of the system (15.91) with the parameters (15.94), orders $q_1=q_2=0.98=98/100$, $q_3=0.99=99/100$, $q_4=0.97=97/100$, with $m=100$, for Jacobian (15.90) and slope a is

$$\lambda^{392}-\lambda^{294}+0.1\lambda^{293}-12\lambda^{196}+12.9\lambda^{195}-27.2\lambda^{97}=0,$$

and for Jacobian (15.90) and slope b it has form

$$\lambda^{392}+4\lambda^{294}+0.1\lambda^{293}-7\lambda^{196}+13.4\lambda^{195}+38.3\lambda^{97}=0.$$

Both above characteristic equations are polynomials of very high order and it is difficult to find the roots of such polynomials analytically. Because of this reason we can use a Matlab routine `roots()`. To remain system chaotic, there should be at least one root λ in unstable region, it means that $|\arg(\lambda)|<\pi/(2m)=\pi/200$. This condition is satisfied for roots $\lambda_1=0$ and $\lambda_2\approx 1.0120565137$ of slope a and $\lambda_1=0$ and $\lambda_{2,3}\approx 1.0107809162\pm 0.0153011315j$

of slope b. Such an equilibrium point is an unstable focus-node. These results confirm the results obtained via simulations and presented in Section 15.5.2.

15.4 General numerical solution of the fractional-order Chen's system (15.72), obtained by method (15.14), has the following form:

$$\begin{aligned} x(t_k) &= (a(y(t_{k-1}) - x(t_{k-1})))\, h^{q_1} - \sum_{j=v}^{k} c_j^{(q_1)} x(t_{k-j}), \\ y(t_k) &= (dx(t_k) - x(t_k)z(t_{k-1}) + cy(t_{k-1}))\, h^{q_2} - \sum_{j=v}^{k} c_j^{(q_2)} y(t_{k-j}), \\ z(t_k) &= (x(t_k)y(t_k) - bz(t_{k-1}))\, h^{q_3} - \sum_{j=v}^{k} c_j^{(q_3)} z(t_{k-j}), \end{aligned}$$

where $d = (c - a)$, T_{sim} is the simulation time, $k = 1, 2, 3 \ldots, N$, for $N = [T_{sim}/h]$, and $(x(0),\ y(0),\ z(0))$ is the start point (initial conditions). The binomial coefficients $c_j^{(q_i)}$, $\forall i$, are calculated according to relation (15.13). The parameters of the Chen system are: $a = 35$, $b = 3$, $c = 28$, $d = -7$, and orders: $q_1 = 0.8$, $q_2 = 1.0$, $q_3 = 0.9$.

SOLUTIONS FOR CHAPTER 16

16.1 LINGO provides the following solutions: $x_1 = 175$, $x_2 = 0$, and $x_3 = 108.3333$. All deviations are equal to zero, which means that the achievement levels match exactly the goals.

16.2 This model differs from the previous one because the second goal has been modified. In this case LINGO provides the following solutions: $x_1 = 750.0000$, $x_2 = 0$, and $x_3 = 0$. Furthermore, $d_1^+ = 250$, which means that the achievement level of the first objective does not match its goal.

16.3 LINGO provides the following solutions: $x_1 = 740$, $x_2 = 0$ and $x_3 = 0$. The values of the deviations are, respectively, $\delta_1^- = 0$, $\delta_1^+ = 240.0000$, $\delta_2^- = 20$, and $\delta_2^+ = 0$.

16.4 **a)** The random variable Y assigns the following values with associated probabilities:

$$Y = \begin{cases} 1 & p_1 = 0.1, \\ 4 & p_2 = 0.2, \\ 9 & p_3 = 0.7. \end{cases}$$

The expected value and the variance are equal to 7.2 and 14.304, respectively.

b) The deterministic equivalent problem is the following:

$$\min 0.1\delta_1^+ + 0.1\delta_1^- + 0.4\delta_2^+ + 0.4\delta_2^-,$$

subject to

$$\begin{cases} 7.2x_1 + x_2 + 3x_3 + \delta_1^- - \delta_1^+ = 500, \\ 2x_1 - 4x_2 - 7.2x_3 + \delta_2^- - \delta_2^+ = 25. \end{cases}$$

LINGO provides the following solutions: $x_1 = 63.53734$, $x_2 = 0$, and $x_3 = 14.17704$. All deviations are equal to zero.

SOLUTIONS FOR CHAPTER 17

17.1 Elementary.

17.2 Let player 1 be the veto player without loss of generality. Then $v(C) = 0$ if $1 \notin C$ and $v(C) = 1$ if $1 \in C$.

17.3 This is the maximum sum of utilities the members of the coalition may guarantee against the worst possible attack on their profit by those outside the coalition. Transferable utility means we are allowed to add utilities so this suggestion makes sense. In terms of mixed strategies this is the Von Neumann–Morgernstern suggestion.

SOLUTIONS FOR CHAPTER 18

18.1 The solution of $\mathbf{x}'(t) = A\mathbf{x}(t)$ can be expressed as

$$\mathbf{x}(t) = e^{At}C, \qquad C = (C_1, \cdots, C_n)^T,\ C_i \in \mathbb{R},\ i = 1, \cdots, n. \tag{19.39}$$

We try to find the solution of (18.8) as the form of (19.39) with

$$C(t) = (C_1(t), \cdots, C_n(t))^T.$$

Substituting $\mathbf{x}(t)$ into (18.8). We have

$$e^{At}C'(t) = B\mathbf{u}(t).$$

Hence,

$$C(t) = \int_{t_0}^{t_1} e^{-As}B\mathbf{u}(s)ds + C(t_0).$$

Moreover, taking the initial data $\mathbf{x}(t_0) = \mathbf{x}_0$ into account, the solution of (18.8) could be the form we expected in the formula (18.10).

18.2 Suppose $z(t) = a_1x^3 + a_2x^2 + a_3x + a_4$. Since $z(0) = 1, z'(0) = 0, z(1) = 0$ and $z'(1) = 0$, we have the following linear equations:

$$\begin{cases} a_4 &= 1, \\ a_3 &= 0, \\ a_1 + a_2 + a_3 + a_4 &= 0, \\ 3a_1 + a_2 + a_3 &= 0. \end{cases}$$

Hence,

$$z(t) = \frac{1}{2}x^3 - \frac{3}{2}x^2 + 1.$$

Consequently,

$$u(t) = z''(t) + z = \frac{1}{2}x^3 - \frac{3}{2}x^2 + 3x - 2$$

is one of the controls driving $(1, 0)$ at $t = 0$ to the state $(0, 0)$ at $t = 1$.

18.3 $\mathcal{J}$ is continuous if

$$\lim_{h \to 0} (\mathcal{J}((\varphi_0, \varphi_1) + h(\psi_0, \psi_1)) - \mathcal{J}(\varphi_0, \varphi_1)) = 0$$

for any $(\varphi_0, \varphi_1), (\psi_0, \psi_1) \in L^2(\Omega) \times H^{-1}(\Omega)$. Indeed, it is true since

$$\begin{aligned} &\mathcal{J}((\varphi_0, \varphi_1) + h(\psi_0, \psi_1)) - \mathcal{J}(\varphi_0, \varphi_1) \\ = \quad & \frac{1}{2}\int_0^T \int_\omega ((\varphi + h\psi)^2 - \varphi^2) dx dt - h\langle \frac{\partial \psi}{\partial t}(0), y_0 \rangle_{-1,1} - h \int_\Omega \psi(0) y_1 dx \\ = \quad & h \left(\int_0^T \int_\omega (\varphi\psi - \frac{h}{2}\psi^2) dx dt - \langle \frac{\partial \psi}{\partial t}(0), y_0 \rangle_{-1,1} - \int_\Omega \psi(0) y_1 dx \right). \end{aligned}$$

Now we show that $\mathcal{J}$ is strictly convex. Set $t \in (0, 1)$. We compute

$$\begin{aligned} &\mathcal{J}(t(\varphi_0, \varphi_1) + (1 - t)(\psi_0, \psi_1)) \\ &= t\mathcal{J}(\varphi_0, \varphi_1) + (1 - t)\mathcal{J}(\psi_0, \psi_1) - \frac{t(1 - t)}{2} \int_0^T \int_\omega |\varphi - \psi|^2 dx dt. \end{aligned}$$

However, the observability inequality (18.29) tells us that

$$C \left\| (\varphi_0 - \psi_0, \varphi_1 - \psi_1) \right\|^2_{L^2(\Omega) \times H^{-1}(\Omega)} \leq \int_0^T \int_\omega |\varphi - \psi|^2 dx dt.$$

Since $(\varphi_0, \varphi_1) \neq (\psi_0, \psi_1)$, the left-hand side of (19) is strictly positive. Substituting (19) into (19), we arrive at

$$\mathcal{J}(t(\varphi_0, \varphi_1) + (1 - t)(\psi_0, \psi_1)) > t\mathcal{J}(\varphi_0, \varphi_1) + (1 - t)\mathcal{J}(\psi_0, \psi_1)$$

and $\mathcal{J}$ is strictly convex.

18.4 Let $(\hat{\varphi}_0, \hat{\varphi}_1)$ be the minimizer of $\mathcal{J}$. Recalling (18.24) and let the control function be $u = \hat{\varphi} 1_\omega$. By taking $(\hat{\varphi}_0, \hat{\varphi}_1)$ as the test function of (18.24), it holds

$$\|u\|^2_{L^2((0,T)\times\omega)} = \int_0^T \int_\omega |\hat{\varphi}|^2 dx dt = \langle \frac{\partial \hat{\varphi}}{\partial t}(0), y_0\rangle_{-1,1} - \int_\Omega \hat{\varphi}(0) y_1 dx.$$

Meanwhile, for the control v and test function $(\hat{\varphi}_0, \hat{\varphi}_1)$, (18.24) gives

$$\langle \frac{\partial \hat{\varphi}}{\partial t}(0), y_0\rangle_{-1,1} - \int_\Omega \hat{\varphi}(0) y_1 dx = \int_0^T \int_\omega v \hat{\varphi} dx dt.$$

Substituting the above equation into the previous equation, we compute

$$\begin{aligned} \|u\|^2_{L^2((0,T)\times\omega)} = \int_0^T \int_\omega v\hat{\varphi} dx dt &\leq \|v\|_{L^2((0,T)\times\omega)} \|\hat{\varphi}\|_{L^2((0,T)\times\omega)} \\ &= \|v\|_{L^2((0,T)\times\omega)} \|u\|_{L^2((0,T)\times\omega)} \end{aligned}$$

and Corollary 1 holds.

SOLUTIONS FOR CHAPTER 19

19.1
```
%%%accept_reject.m
%%%%%%%%%%%%%%%%%%%%%%%%%
function [Ex,ANT]=accept_reject(N)
%%Ex=expectation, ASR=avergae success rate
rand('state',0)
X=zeros(N,1);
NT=X;
for I=1:N
    nr =1;
    nt=0;
    while(nr==1)
    nt=nt+1;
    U1=rand(1);U2=rand(1);
    if(U2<=16*(U1^2-2*U1^3+U1^4))
        X(I)=U1;
        nr=0;
    end, end
    NT(I)=nt;
end
ANT=mean(NT);
Ex=mean(X);
%%%%%%%%%%%%%%%%%%%%%%
>> [E,N]=accept_reject(100)
E = 0.4998 N =  1.7100
>> [E,N]=accept_reject(1000)
E =    0.5022  N =    1.9250
```

```
>> [E,N]=accept_reject(10000)
E =     0.5007 N =      1.8865
>> [E,N]=accept_reject(100000)
E =   0.4999   N =    1.8746
```

19.5 Consider two independent exponential random variables, T_1, T_2, with rates, μ, λ. We can think of two alarm clocks and consider the times T_1, T_2 at which each one of them goes off.

(a) First, let us compute the probability that at least one of the two clocks goes off during the time interval $[0, t]$:

Consider the random variable:

$$T = \min\{T_1, T_2\}$$

$$\begin{aligned}
\text{Prob}\{T \geq t\} &= \text{Prob}\{T_1 \geq t, T_2 \geq t\} \\
&= \text{Prob}\{T_1 \geq t\}\text{Prob}\{T_2 \geq t\} \\
&= \int_0^{+\infty} \mu e^{-\mu\tau} d\tau \int_0^{+\infty} \lambda e^{-\lambda\tau} d\tau \\
&= e^{-\mu t} e^{-\lambda t} = e^{-(\mu+\lambda)t}
\end{aligned}$$

which is the probability that no one of the alarm clocks goes during the time interval $[0, t]$, hence the probability that at least one of the alarm clocks goes off during that time interval is

$$\text{Prob}\{T < t\} = 1 - e^{-(\mu+\lambda)t}$$

which is the CDF of an exponential random variable with rate $\lambda + \mu$.

(b) The probability that clock T_1 goes off first:

$$\begin{aligned}
\text{Prob}\{T_1 < T_2\} &= \int_0^{\infty} \text{Prob}\{T_1 < t\} d\text{Prob}\{T_2 = t\} \\
&= \int_0^{\infty} (1 - e^{-\lambda t})\mu e^{-\mu t} dt \\
&= 1 - \int_0^{\infty} \mu e^{-(\lambda+\mu)t} dt \\
&= 1 - \frac{\mu}{\lambda + \mu} = \frac{\lambda}{\lambda + \mu}.
\end{aligned}$$

INDEX

Mathematical Modeling with Multidisciplinary Applications.
Edited by Xin-She Yang

www.ingramcontent.com/pod-product-compliance
Lightning Source LLC
LaVergne TN
LVHW011856200926
848484LV00002B/459
* 9 7 8 1 1 1 8 2 9 4 4 1 3 *